Cathodic Protection
[Second Edition]

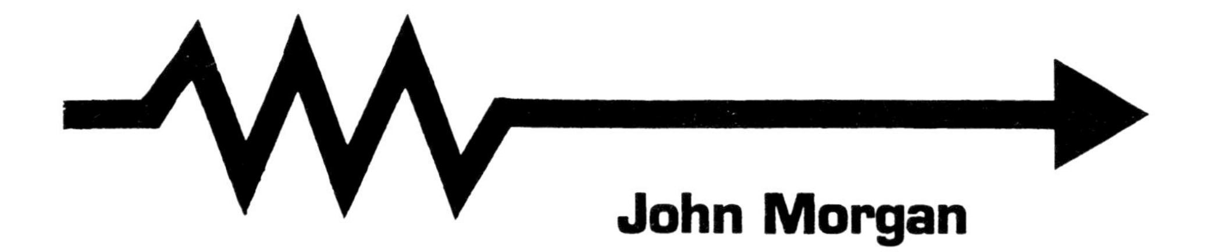

John Morgan

National Association of Corrosion Engineers

Published by

National Association of Corrosion Engineers
1440 South Creek Drive
Houston, Texas 77084

Library of Congress Catalog Card Number: 87-061750
ISBN 0-915567-28-8

2nd Printing, January 1993

Preface

Corrosion has been known to man since the earliest metallurgical times and has been a constant drain on his enterprise. It has only comparatively recently received a scientific treatment which can date its foundation less than a century and a half ago and its major developments to the last few decades.

The necessary presence of a corroding fluid has meant that many preventative methods rely upon isolating the metal from the harmful environments either by the use of a foreign coating or by the formation of a resistant corrosion or chemical film upon the metal. The electrochemical theory of corrosion suggests, that in bulk electrolytes at least, an electrical method of corrosion prevention could be used and this method is called cathodic protection.

The object of this book is to explain the techniques employed with this method rather than to establish the electrochemistry of the metal/electrolyte interface. The subject has been broadly divided into two parts, the first of which treats cathodic protection as a technique, while the second considers the methods of achieving protection on a variety of structures. These sections are not exhaustive nor are they intended as a do-it-yourself manual; rather it is hoped to bring to the reader a clearer understanding of how cathodic protection may be used and to give him some critical faculty with which to compare the results of its application and the engineering methods employed.

The criterion of protection has been assumed to be one of achiving a particular potential, or possibly a potential change, between the structure under protection and its local electrolyte. While there may be little theoretical justification for this criterion the ease with which it can be used makes it preferable even when employed only as a secondary criterion. This change in potential is caused by a flow of direct current from the electrolyte on to the metal and the engineer's job is so to arrange this that protection is effective everywhere. As such perhaps he will need more than anything else a training in light electrical engineering.

Metallic corrosion is, almost without exception, solely an economic loss and any method or combination of methods of its prevention must prove to be of economic advantage. To this end cathodic protection will often find its most acceptable application as a complement to some form of coating. While it is not the answer to every corrosion problem its use can cause great savings but more particularly removing the problem of corrosion from the designer will allow new freedoms, developments and economics in the engineering.

I could not have undertaken to write this book without a great deal of assistance. Much of the information in the book is taken from the original work of my many friends in the cathodic protection industry. Some of that which is my own research work has been stimulated by discussions with these same people, who have always been most helpful, though they are by no means to be held responsible for my heresies and errors. I am grateful to the firms who have supplied the photographs and these sources are acknowledged in the captions. Mr. J. T. Crennell, Dr. V. S. Griffiths and Mr. H. M. Powell have read various parts of the manuscript and I am particularly indebted to Mr. Crennell who has not only read all the proofs but has also written the foreword.

J. H. Morgan
Imber Grove, Esher
July, 1959

Preface to 2nd Edition

On reading the preface for the first edition I am only too aware of the last two paragraphs in which I mentioned my indebtedness to the people I have met in cathodic protection. My acquaintanceship has widened and this has enlarged the debt. Since writing, Mr. Powell has died, Mr. Crennell has left the Admiralty, and Professor Griffiths is now Vice Chancellor at Surrey University. Mr. Crennell has kindly given me permission to reproduce the foreword from the first edition which as appropriate now as it was then. Many of the illustrations and photographs in the book show work which I did when I was Chief Executive at Morgan Berkeley & Company Ltd. and the successors to this company, Corrintec U.K. Ltd., have kindly given me permission to reproduce them. The number of publications in cathodic protection has grown and the excellent abstracts published by NACE would be the most fruitful source of reference for further reading. Some historic papers and some of my own are referred to in the appropriate chapters.

J. H. Morgan

ERRATA

Cathodic Protection, 2nd Edition (reprint)

John Morgan—Author

Page	Location	Incorrect	Correct
p. 20	Para. 4, line 3	$E_a = E_a$	$E_a = \epsilon_a$
	Para. 4, line 5	E_c	ϵ_c
	Equation (1.3a)	E_a and E_c	ϵ_a and ϵ_c
p. 21	Equation (1.4a), (1.5a)	E_a and E_c	ϵ_a and ϵ_c
	Para. 1, line 4	E_a	ϵ_a
p. 39	Para. 4, line 3	1.80 V	1.05 V
p. 42	Equation (2.1), (2.2)	$i0$	i_0
p. 53	Para. 2, line 4	$\frac{\tau-15}{255}$	$\frac{\tau-15}{55}$
p. 74	Para. 1, line 2	$<$	$=$
p. 81	Figure 47	τ	ℓ
p. 88	Para. 1, line 11	$\frac{1}{1}$	$\frac{1}{\ell}$
p. 96	Equation (3.24)	$\frac{\rho}{n\ell}$	$\frac{\rho}{\pi\ell}$
p. 98	Figure 58	P_1 and P_2 (all)	ρ_1 and ρ_2 (all)
p. 122	Figure 72a and b		Figure a is Figure b, and vice versa
p. 209	Figure 118	E_A V, assuming Em=0.4 V	E_A, V (assuming Em=0.4 V)
p. 305	Lines 3-5	"... and the controller placed in the center castle of the ship, the cable entering the hull in the engine compartment. The majority of the anodes were placed outside the engine room, Fig 180."	"... and the controller and power unit located in the centre castle of the ship, Fig 180. The majority of the anodes were attached outside the engineroom the cable entering the hull in the engine compartment."

Contents

Foreword

Although the first practical application of cathodic protection is of a truly venerable age—long before the first iron-hulled ship—its general recognition and its wide industrial application have been only in the last quarter of a century.

The beautiful simplicity of the principle on which it operates is in contrast to the complexity of the technique, which must be adapted to fit the varying needs of each type of subject.

Whereas all other anticorrosive measures are dealloying rear-guard actions, aimed at saving as much as possible of the main body from destruction for as long as possible, cathodic protection offers the possibility of perfection, of complete freedom from corrosion; and this not merely as a theoretical ideal, but as a practical target.

The justification of any protective measures must be that they save more than they cost. It is sometimes difficult to balance the cost of installing and running a system of cathodic protection against the losses that would have been suffered due to corrosion in its absence; but in many cases the effects of even local corrosion can be so costly, in loss of operating time of plant or ship, that the cost of cathodic protection is a very modest premium to pay for immunity.

To ensure this immunity is the task of the cathodic protection engineer or specialist. Many of his practical problems to-day arise from the need to graft cathodic protection on to structures designed with no thought of its use. The growing practice of planning for cathodic protection at the design stage not only makes protection easier and more certain, but may save its cost by reducing the margin of safety that was left for losses by corrosion, in pipe-thickness, plated piles, underwater ties and the like.

This first publication of an English text-book on cathodic protection will contribute to the informed and effective use of a newly developed and most powerful weapon against the ravages of corrosion.

J. T. Crennell
R.N.S.S., Admiralty

Electrochemistry, Corrosion and Cathodic Protection

CHAPTER 1

Historical

Corrosion

Corrosion is the degradation of a metal by its chemical combination with a non-metal such as oxygen, sulfur, etc. Generally, this means a return of the metal to the form in which it originally existed as an ore with complete loss of its metallic properties. Most naturally occurring ores are oxides, sulfides or carbonates, and energy must be expended in converting these ores to metal. Corrosion, the reverse process, requires no such supply of energy so that the formation of the sulfide, oxide or carbonate occurs readily, or even spontaneously under certain conditions.

Those metals that are most easily obtained from their ores and require least energy in smelting are generally least prone to corrosion, and those that are won with most difficulty, tend to revert more readily to their natural state. Since the earliest metallurgy was concerned with the metals that could be obtained easily either because they occurred naturally or only required simple smelting, corrosion was very much less a problem than in more recent times. One exception was meteoric iron which was used in the early bronze age, being found in its metallic state in the center of small meteors. This source of iron was quickly exhausted, the available metal was consumed, probably suffering considerable corrosion, and bronze again became the most important metal.

Gold, which is found in its metallic state, does not corrode and silver, which is readily obtainable, enjoys almost equal immunity. Copper and bronze are much more readily won from their ores than iron and so not only did they precede iron chronologically, but they have a much higher corrosion resistance. This does not mean that these metals do not corrode, nor that corrosion was unknown to the ancients. The Romans were aware of it and Pliny (circa 100 B.C.) mentions methods of preventing the corrosion of bronze and iron, the former being protected by oil or tar and the latter by pitch, gypsum and white lead; red lead was also used by the Romans. About this time copper vessels, particularly those used for the preparation of food, were being tinned and poisoning was reported from the use of lead.

Metallic iron was used extensively in Roman ships and other structural equipment, as well as in tools and arms. Considerable corrosion must have occurred and iron was then being reported as inferior to that produced in the time of Alexander the Great. This is a familiar complaint today, though there is considerable evidence to support a decline in the corrosion resistance of recent iron. Some ancient works, particularly the Delhi Pillar, erected a millenium and a half ago, are in an excellent state of preservation. When considering the corrosion resistance of any archeological specimens, it will be only those which have, possibly by chance, extremely good corrosion resistance that will survive as relics and these cannot fairly be taken as a true cross section.

The introduction of modern smelting techniques using coke instead of charcoal and the contemporary use of coal and oil as the major fuels have led to sulfur contamination of both the iron and the atmosphere. These sulfur additions have now been established as some of the major causes of ferrous corrosion. The Delhi Pillar and several other Eastern iron relics enjoy a dry unpolluted atmosphere and have acquired a highly protective initial corrosion film. Small samples of these relics exposed to the industrial atmosphere of this country have corroded as rapidly as modern iron. Several iron bridges erected in South Wales during the early iron era showed remarkable corrosion resistance in what became a highly polluted atmosphere; their initial corrosion must have occurred before coal was extensively used and a protective, probably pure oxide, film was formed which gave them superior resistance to recently erected steel work. Thus, whether the metal or the atmosphere has deteriorated is questionable but the problem of corrosion has grown with the increased use of metals.

By the mid-eighteenth century the corrosion problem must have become appreciable though it was not until the early nineteenth century that any scientific approach was made towards a solution.

Electrochemistry

Wollaston (circa 1815) regarded corrosion by acids to be an electrochemical process, and a few years later, in 1819, a French writer suggested that rusting was also an electrochemical phenomenon. In 1824, Davy showed that when two dissimilar metals were electrically connected and immersed in water, the corrosion of one was accelerated while the other received a degree of protection. From this work he suggested that the copper bottoms of ships could be protected by attaching iron or zinc plates to them, the earliest example of practical cathodic protection.

In 1681 similar accelerated corrosion was noted and the Navy Board decided locally to remove the lead sheathing from ships' hulls to prevent the rapid corrosion of the rudder irons and bolt heads, Charles II and Samuel Pepys being the instigating experts. In 1830 de la Rive published a paper showing that impure zinc was corroded rapidly by the great number of bimetallic junctions that it contained, the corrosion cells being formed between the zinc and the impurities. This work was followed by the investigations of Faraday into the correlation of electrical and chemical phenomena. Much of Faraday's work could be described as corrosion experiments, and from these he was able to derive his laws of electrochemical action which give the relationship between the current flowing and the associated rate of corrosion.

The metals were arranged in decreasing order of activity by de la Rive who also showed that this order was dependent upon the electrolyte. The generally accepted theory assumed that an electrochemical reaction demanded the presence of two metals or a metal and a metal oxide. Sturgeon (circa 1830) considered that a single metal could have a surface that was 'unequally electrical and consequently electropolar' and Faraday set about to prove this by his experiments involving a single metal. In these he was able to produce potential differences by variations in the electrolyte concentration and temperature.

In 1837, the British Association for the Advancement of Science commissioned Robert Mallet to investigate the effects of 'sea water at various temperatures and of foul river water whether fresh or salt' on cast and wrought iron. During his tests he exposed a great number of specimens to these types of waters all over the British Isles and he observed the differential concentration cell effect on the corrosion of extended iron structures where sea and river water became stratified. At about this time considerable interest was being aroused by Davy's work on the protection of iron by zinc anodes and the development of hot dip galvanizing that followed. Mallet showed that zinc so used became covered with a thick layer of zinc oxide and calciferous crystals 'which retards or prevents its further corrosion and thus permits the iron to corrode.' The variation in the corrosion rate of alloyed zinc reported by de la Rive led Mallet to experiment with zinc alloy anodes. He found that metals cathodic to zinc decreased its efficiency while those which were anodic, notably sodium, tended to increase this and that an addition of mercury was an advantage. A workable anode could be made from zinc when alloyed with mercury and sodium and this produced superior galvanizing.

Towards the end of the century electrochemical corrosion received little attention and the view that two metals were required to produce this type of corrosion became accepted, the corrosion resistance of metals in aqueous solutions being associated with their purity. The corrosion of two metals in contact was investigated in detail by Heyn and Bauer about 1910. The nobler metal (the cathode) was known to corrode at a slower rate while the more base metal (the anode) corroded more rapidly. The investigators were able to establish that this corrosion increased with the relative separation of the metals in the electrochemical potential table. Other factors played an important part including the relative areas of the metals and the rate of arrival of oxygen at the cathode.

In a paper published in 1924, Evans described several mechanisms which led to the establishment of corrosion currents on a single metal surface and called this 'The Newer Electrochemical View on the Corrosion of Metals.' Some of these principles had been noted earlier, including observations of the differential temperature cell by Walcker (1825), the differential stress cell by Davy (1826), the differential concentration cell by Becquerel (1827) and the differential aeration cell by Marianini (1830). Evans and his colleagues, Hoar, Thornhill and Agar, continued their work at Cambridge and produced direct quantitative evidence of these electrochemical corrosion mechanisms. In 1938 Hoar published a discussion on the basic electrochemical theory of cathodic protection and, independently, a similar theory was suggested by Brown and Mears. Anodic protection of metals that passivate was proposed by Edeleanu in Cambridge in 1955.

Cathodic Protection

Having (in 1823) commissioned Sir Humphrey Davy to investigate the corrosion of the copper sheathing of the hulls of wooden naval ships, the Admiralty were the first users of cathodic protection. Davy experimented with anodes of tin, iron and zinc to protect the copper. The last two metals were used and in a later paper (1824) he favored the use of cast iron because it lasted longer and remained electrically more active than zinc. Zinc remained in use, however, and no doubt gave considerable protection to the copper sheathing. When wooden hulls were superseded by iron and steel, zinc anodes or protectors were still fitted. Though there was every reason to believe that zinc would successfully protect steel, its continued use seems to have rested more on tradition. The zincs were placed close to the stern gear and 'yellow' metal parts, such as circulating pipe inlets, as these areas proved to be the most susceptible to corrosion. The practice became universal in shipping circles and protectors were even placed in boilers, though it is doubtful whether any complete protection resulted. Zincs were reported as being in sound order, that is uncorroded, and this was often regarded as good practice.

Edison tried to achieve cathodic protection of a ship at sea from trailing impressed current anodes but the materials and techniques available to him in the eighteen nineties proved to be inadequate. Most early users of impressed current in sea water were concerned with attempts to effect antifouling or to prevent the scaling that would occur in boilers which were replenished with sea water. The polarity of this current was often considered unimportant and anticipation of the present cathodic protection trends can hardly be claimed.

Since the beginning of the present century liquid and gaseous fuels have been pumped through underground pipelines made of steel or iron. The extensive networks of oil pipelines that were installed in America in the nineteen twenties presented a vast corrosion problem. To an oil company a single leak from a pipeline can cause numerous losses and may include: loss of commodity, property damage including fire, expensive repairs, service interruptions, contamination of water supplies and loss of livestock, all of which leads to a deterioration of public relations.

By the late twenties leaks were few and could have been tolerated had not the leak frequency curve begun to rise alarmingly. In the early thirties all the major pipeline owners were applying anti-corrosive measures to the external protection of their pipes, including various coatings and cathodic protection. The earliest schemes were applied to the worst sections where the pipes had been laid in corrosive soils, and great success was achieved. The cathodic protection was derived from zinc anodes or from impressed current supplied either by d c wind generators or by transformers and copper oxide rectifiers from a c power supplies.

In 1936 the Mid-Continent Cathodic Protection Association was formed to discuss and exchange information on cathodic protection. This association later became the foundation of the National Association of Corrosion Engineers.

The other area where oil pipelines were used extensively was the Middle East; the first cathodic installation protected a group of sea water loading pipelines at Bahrain in 1939.

There are a great number of patents on methods of preventing the corrosion of buried metals, particularly pipes and cable sheaths. Seventy or so years ago a major cause of corrosion of buried metal pipes was the electrolysis effect, or interference, caused by stray currents from the electric traction systems. The first patents describe the connection of the pipes to the negative pole of the station generator; this method was universally adopted and is still used. The introduction of a further d c generator between the negative return of the electric traction and the structure was claimed to give superior results. In 1911 a German, Herman Geppert, obtained letters patent on 'a method of protecting articles from earth currents' and substantially described cathodic protection. Since then patents have applied to more specific devices such as reverse current switches, anodes, boosters, etc.

From these early beginnings cathodic protection has developed rapidly and its use has become widespread. New materials such as sacrificial alloys of magnesium and aluminum and superior impressed current anodes together with developments in electrical and electronic engineering have allowed great advances in the techniques. Cathodic protection is now established as an essential engineering service with a sound and comprehensive scientific background.

Corrosion

Electrochemical Theory

Modern theory describes the atom as having a massive positively charged central nucleus surrounded by a cloud of negatively charged electrons. The cloud of electrons is divided into a series of shells; the inner ones are filled first and require two, eight, eighteen, or thirty-two electrons to fill them. When the electrons just fill the shells completely with no electrons left over, then the element is very stable.

Usually, however, the atom either has a few electrons left over after filling the last completed shell, or too few to complete the outer shell. The atom tends to obtain a cloud of completed shells by gaining or losing electrons; this loss or gain tendency gives the element its chemical properties. A metal sheds electrons and a non-metal accepts electrons. Thus metals and non-metals will combine to form salts by the metal atom donating electrons and the non-metal receiving them, the molecule so formed having two or more such atoms, each modified to have an electron cloud of completed shells. These modified atoms are called ions.

The common salt molecule consists of an atom of sodium which has given an electron to its associated chlorine atom; this produces a positively charged metal ion and a negatively charged chlorine ion. If the salt is dissolved in water these ions separate and the solution contains a mixture of sodium ions and chlorine ions. The difference, therefore, between the metallic sodium and the sodium ion is one electron, which is little more than a small electric charge, and the difference between a chlorine atom and a chlorine ion is similarly a small electric charge.

The corrosion of a piece of metal may be summarized as the change from the metal to the metal ion or the loss of one or more electrons from the metallic atom. Electrically this can be written.

M	$\rightarrow$	M^{n+}	$+$	ne^-
metal		Positively charged metal ion		negatively charged electron

In the case of iron there are two electrons lost from each atom in forming ferrous ions.

$$\underset{\text{iron}}{Fe} \rightarrow \underset{\text{Ferrous ion}}{Fe^{++}} + \underset{\text{electrons}}{2e^{-}}$$

If a piece of iron is placed in water, the metallic iron goes into solution as ferrous ions and the metal assumes a negative charge from the excess electrons that remain in it. The passage of the metal atom into the solution as an ion is thus the equivalent of a flow of electric current from the metal into the solution.

metal		surface		solution
$2e^{-}$	←	$[Fe^{++} + 2e^{-}]$	→	Fe^{++}
electrons		current flow		ions

Faraday was the first to investigate this effect and was able to show that the rate of dissolution of a metal electrode was directly proportional to the amount of current that flowed and that the amount of metal that was dissolved was proportional to the total charge passed. Further, that metals could be plated out by a reversal of the current and that the same proportionality between the charge (ampere hours) and the weight of metal deposited existed. The metal electrode through which current flows from the metal into the solution or electrolyte is called the anode, and the metal electrode where the current flows from the electrolyte into the metal is called the cathode. Thus in the electrolyte current flows from the anode to the cathode while in the electrical circuit current flows, or is made to flow, from the cathode to the anode. Two electrodes, anode and cathode, and the electrolyte form a cell.

Figure 1 shows a simple cell in which current is caused to flow from the d c source to the anode, then through the electrolyte to the cathode and from there back to the negative pole of the source.

As the metal to metal-ion reaction is accompanied by a flow of electric charge, so similarly is the reaction of the non-metal to the non-metal ion. In the latter case, a non-metal ion is negatively charged so that the current flows in the direction of the reaction:- non-metal ion to non-metal; for example the change of the chloride ion into the chlorine atom is accompanied by a conventional flow of electricity from the anode into the electrolyte.

In the simple cell shown in Figure 1, the anodic reaction would be

$$\text{metal} \rightarrow \text{metal ion} + \text{electron}$$

and/or

$$\text{non-metal ion} \rightarrow \text{non-metal} + \text{electron}$$

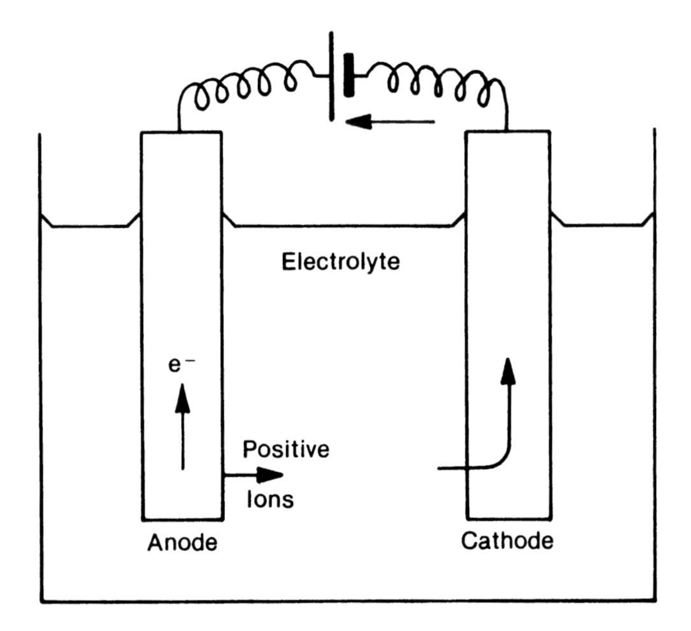

FIGURE 1 - Simple cell with anode, cathode and electrolyte.

The reaction that will take place, or the one that will take place more readily, will depend upon the various conditions that exist at the anode and in the electrolyte. Commonly the anode metal will corrode by its dissolution into positive metal ions.

The cathodic reaction in the cell will similarly be:-

metal ion + electron → metal

and/or

non-metal + electron → non-metal ion

In concentrated solutions of metal ions the metal itself will be formed and this is the basis of electro-plating. Where the hydrogen ion concentration is high, that is in acidic solutions, the 'metal' formed will be hydrogen which will be liberated at the cathode. In solutions that have considerable dissolved oxygen or other oxidizing agents, these will be converted into ions, oxygen forming hydroxyl ions $(OH)^-$.

In a cell which has iron electrodes and a dilute solution of common salt as the electrolyte, the reactions would be: (i) at the anode the iron would corrode

Fe	$\rightarrow$	Fe^{++}	+	$2e^-$
iron		Ferrous ion		Electrons

(ii) at the cathode either (a) hydrogen would be evolved

$2e^-$	+	$2H^+$	$\rightarrow$	$2H$	$\rightarrow$	H_2
electrons		hydrogen ions		hydrogen atoms		hydrogen molecule

or (b) the available oxygen would be converted into ions

H_2O	+	$2e^-$	+	$1/2\ O_2$	$\rightarrow$	$2(OH)^-$
water molecule		electrons		oxygen atom		hydroxyl ions

Potential Series

If two different metals, say iron and copper, are immersed in a salt solution containing ions of both metals, then an electrical potential difference will exist between them and if they are connected electrically, current will flow between them.

Each isolated piece of metal in the electrolyte will corrode on its surface and this means that the metal will go into solution as positively charged ions and leave the remainder of the metal negatively charged by virtue of the excess of electrons in it. The negatively charged metal will attract the positively charged ions and so reduce the tendency for further corrosion of the metal. An electrical balance will be reached when the metal has sufficient negative charge to attract as many positive ions back to its surface as are naturally formed by the metal dissolving. These positive ions need not be of the corroding metal.

The two metals immersed in the salt solution will come to equilibrium at different electrical potentials. This difference can be measured by normal potentiometric means. If successive pairs of metals are immersed in solutions of their salts their potential differences can be measured. A particular potential value can be ascribed to each, relative to one metal, and any pair will display a potential difference which will be the algebraic difference between their potentials relative to the reference metal. From these values a table can be made where metals superior in position will be negative in potential to those below, the amount of potential difference increasing with spacing: such a table was drawn up by de la Rive in 1830. To standardize this classification it is possible to immerse each metal in a solution of its own salt and to connect the two solutions through a salt bridge as in Figure 2. Further, the salt solutions are arranged to be at a specific, termed normal,

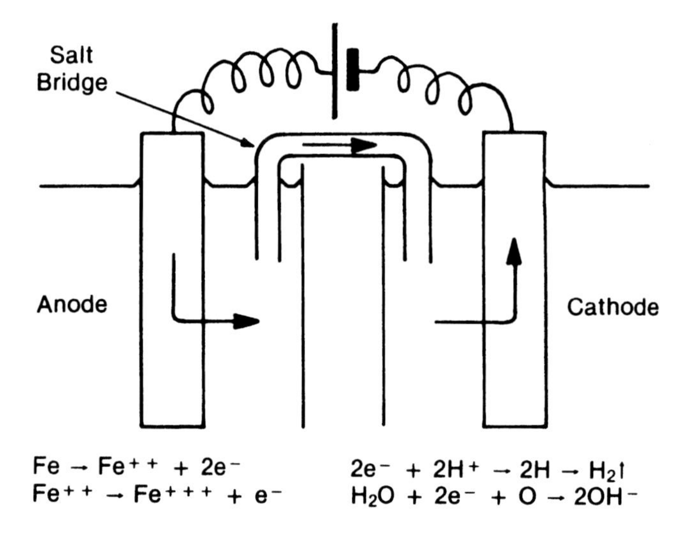

FIGURE 2 - Cell with anode and cathode metals in solutions of their own salts.

concentration or, rather, at normal activity which defines the ionic concentration.

The metal electrode in its salt solution is termed a half cell and the two pieces of metal first considered are both half cells joined by the bulk of the electrolyte to make the complete cell. The half cells formed by metals in solutions of their own salts at normal activity are called standard half cells. Half cells may be made by placing any metal in any salt solution or electrolyte. A convenient scientific standard is the standard hydrogen half cell or electrode. This half cell consists of a hydrogen film on a metal conductor immersed in an electrolyte which has a normal activity of hydrogen ions.

The potentials of the standard half cells relative to the hydrogen electrode can now be determined. This table of potentials is called the electrochemical series and is listed in Table I.

Table II shows the relative positions of the common metals in sea water.

The relative positions of the metals, as well as their electrode potentials relative to the standard half cell, differ when the metals are immersed in a standard solution of their salt and in sea water. This might have been predicted from the physical origin of the potential as the electrolyte will affect the equilibrium at the metal surface.

The potential of a particular half cell can be changed by varying the electrolyte concentration, its temperature, the dissolved gases (particularly oxygen), or by changing the physical state of the metal. This means that in-

TABLE I — Electrochemical Force Series

Electrode Reaction Atom to Ion	Potential, in Volts Standard Electrode at 25 °C
$K \rightarrow K^{+} + e^{-}$	-2.92
$Ca \rightarrow Ca^{++} + 2e^{-}$	-2.87
$Na \rightarrow Na^{+} + e^{-}$	-2.71
$Mg \rightarrow Mg^{++} + 2e^{-}$	-2.34
$Be \rightarrow Be^{++} + 2e^{-}$	-1.70
$Al \rightarrow Al^{+++} + 3e^{-}$	-1.67
$Mn \rightarrow Mn^{++} + 2e^{-}$	-1.05
$Zn \rightarrow Zn^{++} + 2e^{-}$	-0.76
$Cr \rightarrow Cr^{+++} + 3e^{-}$	-0.71
$Ga \rightarrow Ga^{+++} + 3e^{-}$	-0.52
$Fe \rightarrow Fe^{++} + 2e^{-}$	-0.44
$Cd \rightarrow Cd^{++} + 2e^{-}$	-0.40
$In \rightarrow In^{+++} + 3e^{-}$	-0.34
$Tl \rightarrow Tl^{+} + e^{-}$	-0.34
$Co \rightarrow Co^{++} + 2e^{-}$	-0.28
$Ni \rightarrow Ni^{++} + 2e^{-}$	-0.25
$Sn \rightarrow Sn^{++} + 2e^{-}$	-0.14
$Pb \rightarrow Pb^{++} + 2e^{-}$	-0.13
$H_2 \rightarrow 2H^{+} + 2e^{-}$	0.00
$Cu \rightarrow Cu^{++} + 2e^{-}$	0.34
$Cu \rightarrow Cu^{+} + e^{-}$	0.52
$2Hg \rightarrow Hg_2^{++} + 2e^{+}$	0.80
$Ag \rightarrow Ag^{+} + e^{-}$	0.80
$Pd \rightarrow Pd^{++} + 2e^{-}$	0.83
$Hg \rightarrow Hg^{++} + 2e^{-}$	0.85
$Pt \rightarrow Pt^{++} + 2e^{-}$	ca 1.2
$Au \rightarrow Au^{+++} + 3e^{-}$	1.42
$Au \rightarrow Au^{+} + e^{-}$	1.68

stead of creating a potential difference by using two pieces of dissimilar metals in a salt solution, a difference of potential can be achieved by using two pieces of the same metal and varying one of the other characteristics that affect the half cell potential.

If the two pieces of metal which display a difference in potential—be

TABLE II — Relative Positions of the Common Metals in Seawater

Magnesium	Lead
Magnesium alloys	Tin
Zinc	Muntz metal
Galvanized iron	Manganese bronze
	Naval brass
Aluminum 52SH	Nickel (active)
Aluminum 4S	78% Ni, 13.5% Cr, 6% Fe (active)
Aluminum 3S	Yellow brass
Aluminum 2S	Admiralty brass
Aluminum 53S-T	Aluminum bronze
Aclad	Red brass
	Copper
Cadmium	Silicon bronze
	5% Zn, Ni, Bal. Cu
Aluminum A17S-T	70% Cu, 30% Ni
Aluminum 17S-T	88% Cu, 2% Zn, 10% Sn
Aluminum 24S-T	88% Cu, 3% Zn, 6.5% Sn, 1.5% Pb
	—
Mild steel	Nickel (passive)
Wrought iron	78% NI, 13.5% Cr, 6% Fe (passive)
Cast iron	
Ni-Resist	70% Ni, 30% Cu (Monel)
13% chromium stainless steel (active)	18-8 stainless steel (passive)
	18-8 3% Mo stainless steel (passive)
50-50 lead tin solder	
18-8 stainless steel (active)	Silver
18-8, 3% Mo stainless steel (active)	Gold

they the same metal or not—are electrically connected, then current will flow through this connection and through the cell. At the metal/electrolyte surface of the negative electrode the anodic reaction, probably the dissolution of the metal into metal ions, will occur and an excess of electrons will be created in that piece of metal. At the positive electrode, electrons will either combine with positive ions or reduce oxygen to negative ions, in either case creating a deficiency of electrons. Electrons will flow from the anode to the cathode through the electrical connection. Conventional current will flow from the cathode to the anode through the metallic connection and will be transported ionically through the electrolyte from anode to cathode. The probable metal reaction at the anode will be the formation of metal ions and the electrode will corrode. This is the mechanism of electrochemical corrosion and it occurs whenever cells such as those described above are formed.

Corrosion Cell

The simple cell described is formed when any piece of metal is immersed in an electrolyte as small variations in potential will occur over the surface of the metal caused by differences either in the metal or in the electrolyte. These variations may be caused by a change of the metal properties, due perhaps to a partially plated surface, to an appendage such as a screw of different metal, to a difference in the physical working of the metal surface such as one area having been shotblasted and another ground, to variations in the heat treatment, or to residual stresses in the metal. Variations in the concentration of the electrolyte either of one particular salt or ion or of a dissolved reactive gas, will cause similar potential variations which set up anodic and cathodic areas. Potential variations may also be established by temperature differentials or other variations in the cell.

Some or all of these conditions will exist whenever moisture, even as the smallest drops, is in contact with a metal. No one variation will be exclusive but often one or two differentials will be predominant and dictate the form and type of corrosion. The surface of a metal may be divided simply into large anodic and cathodic areas or the whole surface may consist of a multitude of small cells, the anodes and cathodes being as small as the metal grain crystals.

Any single cell can be represented by an equivalent electrical circuit as in Figure 3. E_a and E_c are the potentials associated with the anode and cathode respectively. These are represented as full cells as it is only as such that they can be measured. The metal has a resistance R_m, R_a is the resistance associated with the anode and its close electrolyte and R_c the cathode resistance. By Ohm's law the current flowing in the cell will be

$$\frac{E_a - E_c}{R_m + R_a + R_c}$$

and this will cause the corrosion of the anode, the rate of which will be proportional to the current.

The potential around the circuit can be considered relative to the potential that is displayed at the anode and at the cathode, that is $(E_a - iR_a)$ and $(E_c - iR_c)$: these are the practical potential measurements that would be observed in a cell. If this potential relative to a standard half cell is

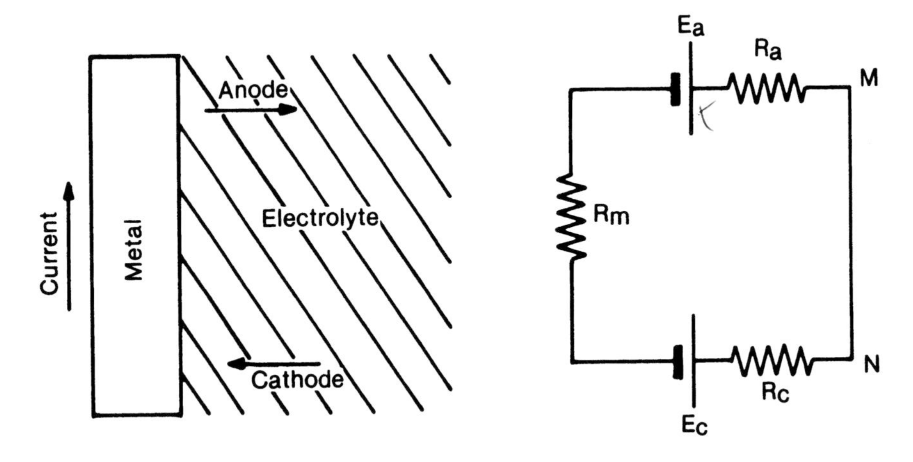

FIGURE 3 - Equivalent electrical circuit of simple cell.

plotted against the cell current, then the result will be as shown in Figure 4.

Where R_a is equal to R_c the potential plot will be as Figure 5a and the corrosion is said to be under mixed control; that is the influence of the anode resistance is the same as the influence of the cathode resistance. Anodic control operates where the anode resistance is much greater than the cathode resistance and the plot of potentials in this case is illustrated in Figure 5b. Cathodic control, where the cathode resistance is greater, has a plot of potential against current as in Figure 5c.

The anode or cathode resistance may be influenced by several factors. Generally the resistance is the inverse of the electrode size varying as its linear dimensions. It may be influenced by an oxide or other insulating film such as a paint or by the resistivity of the anolyte, that is the electrolyte near the anode, or by the resistivity of the catholyte. While the rate of corrosion—that is the actual rate of loss of metal—depends upon the current that flows in the cell, various coatings or the anode size will dictate the rate of loss of metal per unit area.

Where the loss per unit area is great, failure of thin metal sheet may occur by rapid penetration (pitting corrosion) while the loss of metal and the total charge (amp/hours) which has passed, may be comparatively small. This means that the mode of anode control may be more important than the actual control it exerts. If a paint coating on an anode decreases the cell current to 10% of its original value giving a high degree of anodic control the loss of metal though diminished tenfold may occur over 1% of the original anode area so that the penetration rate is increased ten times. Had the same decrease in cell current been achieved by a change in the

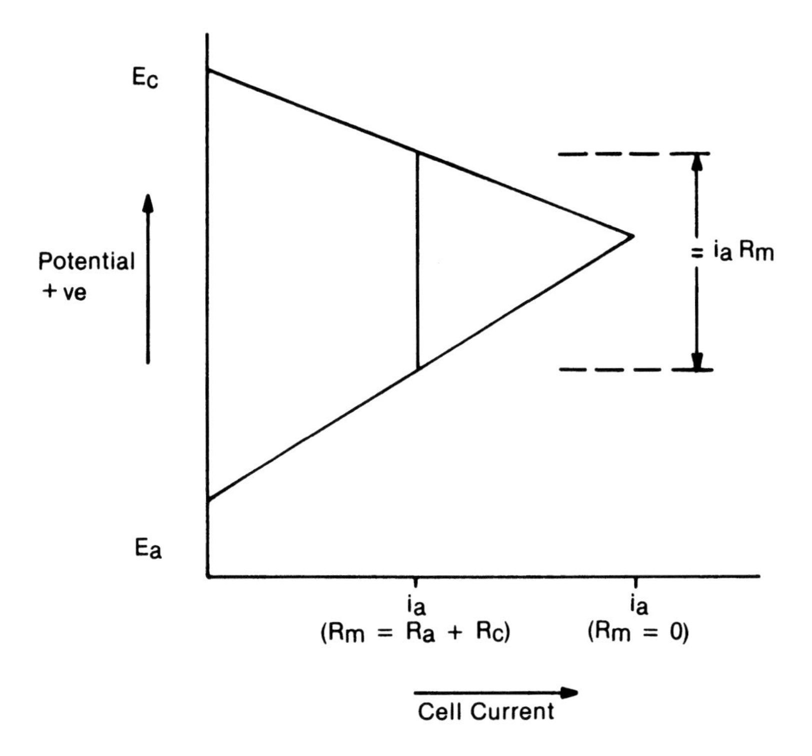

FIGURE 4 - Potential of anode and cathode plotted against the cell corroison current.

anolyte resistivity, then the corrosion would still have been evenly distributed over the anode surface and, being only 10% of that originally in the cell, would result in a tenfold decrease in the rate of penetration.

A similar decrease in the cell current may be achieved by increasing the cathode resistance, again by painting, but now, since the cathode is not corroding, an increase in the current density at one point on the cathode will be of no consequence while the rate of corrosion at the anode will decrease. There is no change in the current distribution at the anode so that corrosion failure will be greatly delayed.

Corrosion Cell Potentials

The potential readings between the metal and the electrolyte are made by connecting the electrolyte to a reference half cell through a salt bridge and then measuring the voltage between the half cell and the metal. This technique is essential and several suitable half cells complete with salt bridges are available.

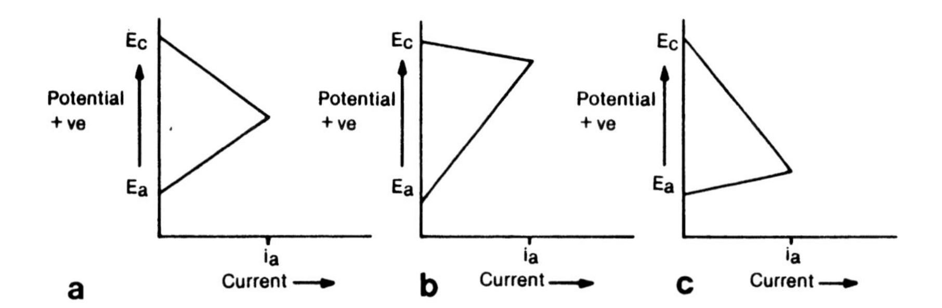

FIGURE 5 - Anode and cathode potentials in cells under (a) mixed (b) anodic and (c) cathodic control.

The use of a salt bridge introduces a considerable resistance into the circuit and so measurements are made with a high input impedance voltmeter or potentiometer. As can be seen from the electrical circuit diagram, different values of voltage can be measured by varying the point of connection to the metal or the point of contact of the salt bridge to the electrolyte. Potential measurements made between the electrodes and the electrolyte in their close vicinity show the anode has a more negative potential than the cathode. The exact value of this will depend upon the position of the salt bridge in the electrolyte which is the electrical equivalent to the point of connection on resistors R_a or R_c.

To measurements taken with the salt bridge a considerable distance from the anode and the cathode, the cell potential of both (assuming R_m to be negligible) will be that displayed by the junction of the resistors R_a and R_c. Where mixed control operates this potential will be about mid way between E_a and E_c, that is

$$\frac{E_a + E_c}{2}$$

A cell with anodic control will display almost the cathode potential while cathodic control will produce a cell potential close to the anode potential.

Corrosion Cell Current

By Ohm's law an increase in the total cell resistance will cause a decrease in cell current while a decrease in this resistance will have the opposite effect. Cell resistance will generally increase in high resistivity electrolytes and decrease in low resistivity electrolytes such as sea water: this fact is one of the principal reasons why low resistivity electrolytes enhance the corrosion of metals. As the cell current will vary with changes in the circuit resistance, so will it vary with the driving voltage, that is $E_a - E_c$. The value of this difference will depend upon the cell conditions such as the separation between the electrode metals in the electrochemical series (Table I) or the variations in the electrolyte or metal interfaces. Any means of reducing this potential difference will diminish the cell current and bring about a reduction in corrosion in the cell.

Polarization

Suppose two plates of metal are placed in a salt solution and current is caused to flow from one to the other through the electrolyte, then a driving voltage will be needed. This voltage will depend upon the current density and can be divided into two components; firstly a voltage that will be required to overcome the ohmic resistance associated with the electrode size and shape and the electrolyte resistivity, and secondly, a non-ohmic component that will apparently exist very close to the electrode/electrolyte surfaces of the anode and cathode. This latter voltage is termed polarization or overpotential.

The origin of this polarization can be traced to four sources: activation polarization, concentration polarization, ohmic and psuedo-ohmic polarization. The first may be described as that necessary to bring about the specific anode and cathode reactions and it depends upon the particular reaction and the current density. The activation polarization and current density are related by the equation due to Tafel, where η is the activation polarization, σ the current density, and a & b constants.

$$\eta = a + b \log \sigma$$

The values of a and b are characteristic for each electrode reaction. The activation polarization will vary with the physical characteristics of the half cell. The electrode surface roughness will influence the actual current density as opposed to the mean current density. Generally this type of polarization decreases a few millivolts per degree centigrade with increasing temperature. Activation polarization, being associated with a particular reaction, can be varied by introducing depolarizing chemicals. For example, hydrogen polarization can be decreased by the introduction of oxygen as a depolarizing agent.

Concentration polarization occurs where the anodic or cathodic products cause a change in the electrolytic environment at the metal interface. Fast flowing electrolytes or moving electrodes keep this to a minimum as do concentrated salt solutions where replacement of lost ions is rapid. This

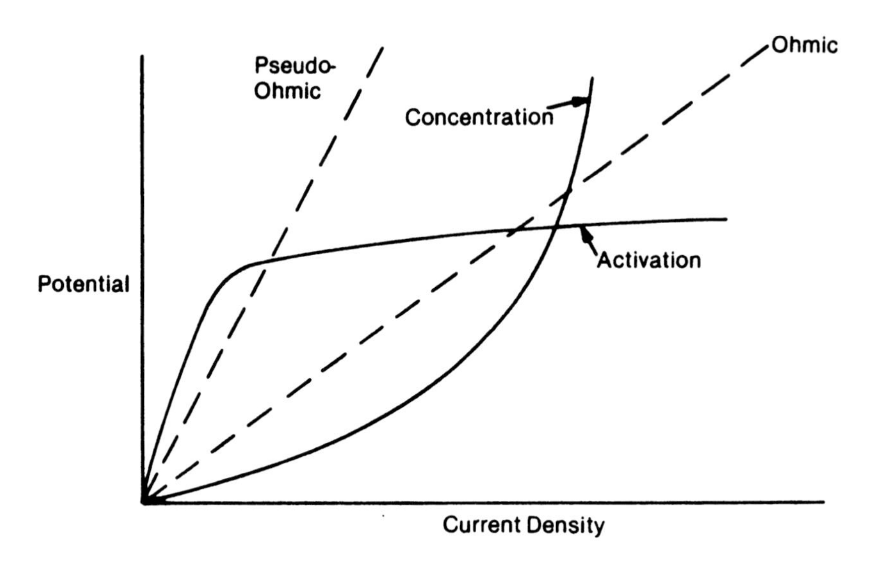

FIGURE 6 - Potential against current density of various types of polarization, anode or cathode.

type of polarization can be recognized by its persistence after the cell current has been switched off.

Ohmic and pseudo-ohmic polarization occur at the interface of the electrode and are purely resistive. The ohmic component is associated with a changed resistivity at the interface either by the loss of conducting ions, by drying out caused by electro-osmosis, or by the temperature change associated with the electrode energy dissipation.

Pseudo-ohmic polarization is associated with oxide or other films that form on the electrode. These films may not form immediately and will generally be time dependent; they are of great importance in corrosion resistance and cathodic protection.

The three types of polarization are plotted against current in Figure 6, which shows the activation, concentration and ohmic components. The last would change with time, increasing by the formation of a pseudo-ohmic film. The magnitude of each effect will vary with the particular cell.

These types of polarization exist in the simple cell and they can be represented electrically as components (not all of them ohmic) of the resistors R_a and R_c in the equivalent circuit diagram. Perhaps more accurately they could be introduced into the cell voltages E_a and E_c, making these functions of current density and time. The effect on the corrosion of the cell can be illustrated by redrawing Figures 5a, b and c to include these factors. The major effect, where film formation takes place, will be the

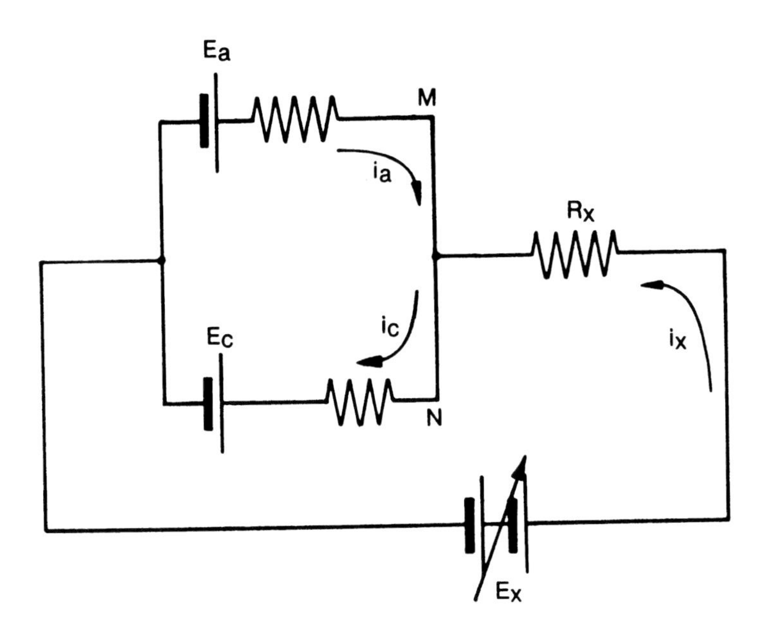

FIGURE 7 - Electrical circuit of simple cell with added external circuit.

marked reduction in the corrosion rate with time, cathodic and anodic films both decreasing the corrosion rate.

There are two further ways of decreasing the corrosion rate or current in a simple cell, both achieved by causing an increase in the polarization. Either chemicals called inhibitors may be added which will cause heavy polarization of one or both electrodes or the electrolyte may be treated to remove any depolarizing agents from it.

The methods of corrosion prevention so far considered have relied upon decreasing the cell current by increasing the electrical resistance or by decreasing the driving voltage of the cell. If, however, two facts are accepted, first that the corrosion process is analogous to that of an equivalent electrical circuit, and second, that the corrosion only occurs at the anode, then it is possible to prevent corrosion by the introduction of a current from an external electrical circuit.

Cathodic Protection

Simple Cell

Figure 7 shows the equivalent electrical circuit for a simple cell (Figure 3) redrawn with $R_m = 0$ and an external electrical circuit added. This circuit consists of a source of direct current comprising a cell E_x and a resistor R_x. Suppose the currents flowing are i_a the anode current, i_c the cathode current and i_x in the external circuit; then by Kirchoff's laws

$$i_a + i_x = i_c \tag{1.1}$$

and

$$E_a = i_a R_a + E_c + R_c i_c \tag{1.2}$$

by substitution

$$E_a = i_a R_a + E_c + R_c(i_x + i_a) \tag{1.3}$$

Now to prevent corrosion there must be no anodic reaction: that is i_a must be zero or negative. Making $i_a = 0$

$$E_a = E_c + R_c i_x \tag{1.4}$$

or the potential of point M is equal to that of the anode.

Thus in a simple cell it is possible to prevent the corrosion that would normally occur at the anode by an externally imposed current. The cathode of the cell will still act as a normal cell cathode though with an increased reaction rate. The imposed current may be increased beyond the value necessary to reduce i_a to zero when the anodic current will be reversed. This will mean that the former anodic area will now itself be acting as a cathode and no corrosion will take place.

It is possible, therefore, to prevent all the corrosion in a simple cell by the application of an external current; this method of corrosion control is called Cathodic Protection.

Simple Polarized Cell

If the conditions in the cell are such that polarization occurs then E_a and E_c become functions of the current density at the anode and cathode. E_a is the anode potential on open circuit and $E_a = E_a - \phi_a(i_a/A_a)$, where ϕ_a is the anode polarization function and A_a the anode area. Similarly $E_c = E_c - \phi_c(i_c/A_c)$. As i_c is in the opposite direction to i_a polarization will tend to reduce the potential difference $E_a - E_c$. Under these conditions equation (1.3) can be rewritten

$$E_a - \phi_a(i_a/A_a) = R_a i_a + E_c - \phi_c(i_c/A_c) + R_c(i_x + i_c) \tag{1.3a}$$

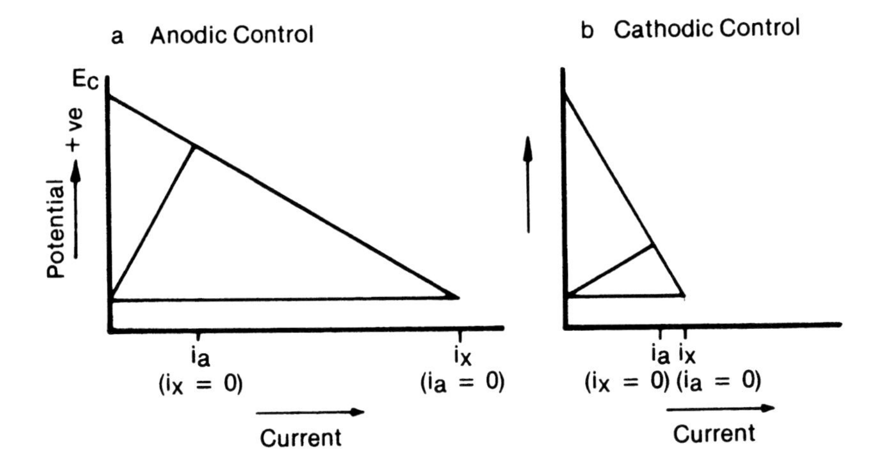

FIGURE 8 - Anode and cathode potentials aginst cell current under (a) anodic and (b) cathodic control.

and when the conditions of no corrosion, that is $i_a = 0$ and therefore $i_c = i_x$ is reached

$$E_a = E_c - \phi_c(i_x/A_c) + R_c i_x \tag{1.4a}$$

That is the point M is now held at the open circuit potential of the anode E_a and the sum of the cathodic polarization and the voltage developed across the cathode resistance is equal to this value: this argument holds irrespective of the shape of the polarization curve and so cathodic protection will be effective in any simple cell with polarization.

In both cases cited, the current necessary to achieve protection can be determined by a knowledge of the other constants in equations 1.4 and 1.4a. These may be rewritten as

$$E_a - E_c = R_c i_x - \phi_c(i_x/A_c) \tag{1.5a}$$

$$E_a - E_c = R_c i_x \tag{1.5}$$

In this form the equations indicate that sufficient current must flow to cause the total voltage drop associated with the cathode, through polarization, and with the catholyte resistance to equal the open circuit driving potential of the cell. Thus a large value of R_c and a highly polarizing

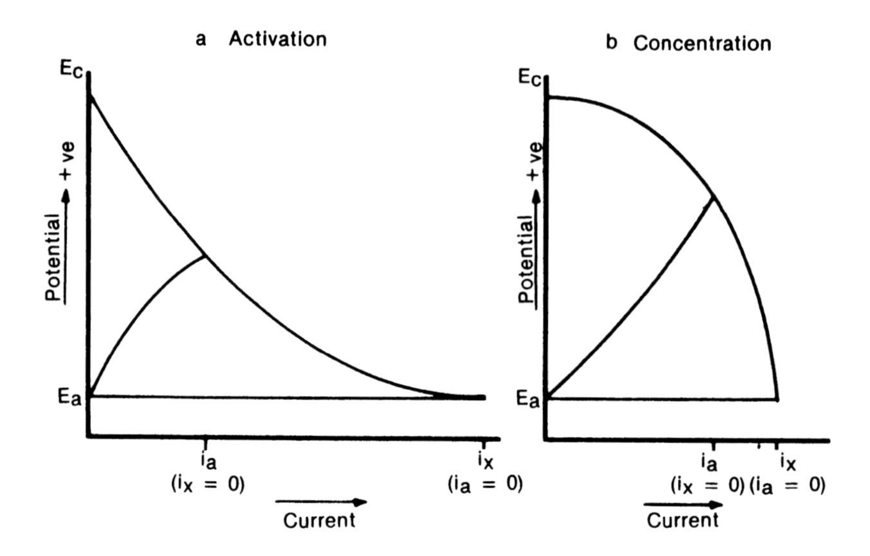

FIGURE 9 - Anode and cathode polarization curves with (a) activation and (b) concentration polarization.

cathode will decrease the value of i_x the current necessary to achieve protection.

This can be shown by an extension of Figs 5b & c to include the value of i_x as in Figs. 8 and 9.

Where the cathode polarizes at low current densities or there is a large ohmic resistance associated with it, protection will be achieved with little impressed current. Where the cathode fails to polarize and the catholyte resistance is low, a large external current will be necessary to bring about cathodic protection.

Practical Cathodic Protection

So far the arguments have been confined to a single simple cell. In practice the cell may be surrounded by an electrolyte of considerable resistivity and an equivalent circuit diagram will be as in Figure 10. It is apparent that to achieve protection the point M must assume the open circuit potential of the anode, E_a, or exceed it, so that i_a becomes zero or its direction of flow is reversed.

On a practical structure such as a buried metal plate, the surface will consist of a multitude of anodes and cathodes each joined electrically through the metal. Cathodic protection will be achieved if, at each anode, the point corresponding to M is at least at the open circuit potential of that anode. Measurements show that these conditions are easily met and so cathodic protection is a feasible means of preventing the corrosion of metal

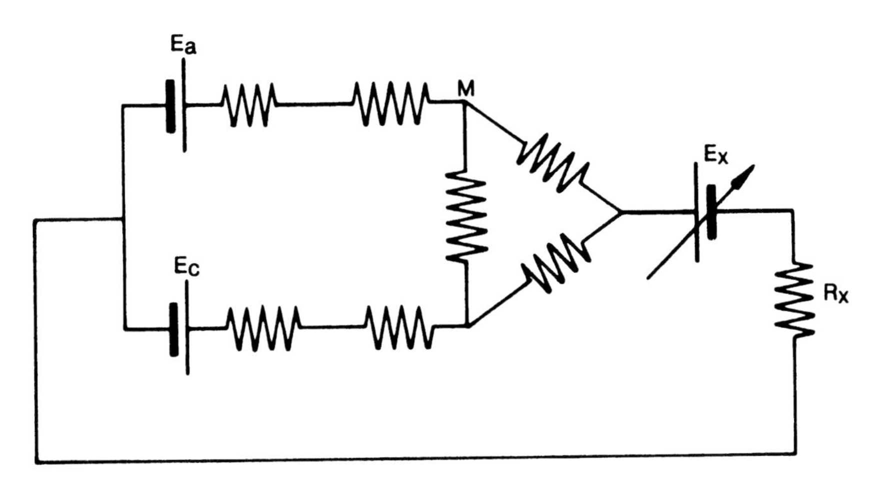

FIGURE 10 - Equivalent electrical circuit of practical cathodic protection.

structures when they are in contact with an electrolyte. While the techniques of galvanizing and base cladding provide cathodic protection to a surface which may only be covered with drops of moisture, the cathodic protection process will only be considered where the electrolyte is present in bulk. This may be as a liquid varying from the sea to the water contained in a small storage tank or, as a pseudo-solid, the soil surrounding a pipe-line or underlying a tank bottom.

Intermittent and Pulsed Current

The external current source has been considered as providing a direct constant current; in practice this current may arrive intermittently at the surface either deliberately or accidentally. In a tidal estuary there may be breaks in the protection when the tide recedes or under other circumstances there may be breaks in the continuity of the electrical supply.

Where the cathode has polarized there will be effective protection during periods of about one day in the absence of the external current depending upon the degree and type of polarization. Some overprotection may be necessary when the current supply is available to ensure the polarization lasts through the break period. On structures that are affected by the tide, protection is usually achieved to just above the mid-tide level.

With half wave and with full wave rectification of single phase a c the output will not be smooth and it has been found in practice that it is not necessary to smooth this output to achieve protection. Where phase angle control of a thyristor or similar device is used the output from the transformer rectifier will consist of a series of pulses. These will be of the order of 10 milliseconds apart and the pulse itself may be considerably less than 1 millisecond in width. In all practical experience, and particularly in sea water, this type of current output has provided excellent cathodic protection.

There has been considerable interest aroused by papers which suggest that short duration pulses of the order of 10 microseconds may provide adequate cathodic protection at much lower power consumption. There is evidence to suggest that a superior form of polarization occurs in sea water. Most practical experience comes from systems in which the output is pulsed fortuitously by the operation of solid state control. The widest range of closely inspected structures are ships' hulls and the results are discussed in the chapter on their protection.

The cathodic protection current can be automatically controlled and this is done by sensing the potential of the structure using an electrode. It is usual to use an average or integrated value of potential and the same criterion of protection has been applied that has been established with a smooth, direct current.

Bacteria

Much of the worst corrosion in the ground occurs in water-logged clays, particularly where these are anaerobic (oxygenless). This condition can be identified by the gray-blue color of the clay as opposed to the rich yellow brown of the aerobic clays. It would seem that corrosion should soon stifle itself under anaerobic conditions as the chief depolarizing agent, oxygen, is absent: this absence should lead to cathodic polarization and a reduction of the cell current almost to nothing. Under these conditions, however, rapid corrosion is encountered.

The paradox was first explained by the Dutch scientist, von Walzogen Kuhr, who identified at the metal interface a species of bacteria which can reduce sulfates. These are reduced to sulfides and the reaction provides an acceptor system for cathodic hydrogen and acts as a powerful cathodic depolarizer allowing the corrosion current to flow almost undiminished. The corroding metal becomes covered with a black slime of iron sulfide and this is recognized by its characteristic smell and taste.

The bacteria are widely scattered and extremely hardy; they have been found in most soils and waters throughout the world. Under aerobic conditions, that is in soil or water where oxygen is present, the bacteria are inactive though not dead and they will become active again on their return to an anaerobic site.

If two identical iron electrodes are immersed in a suitable electrolyte and one is innoculated with bacteria, then the inoculated electrode will be more base, that is more negative in potential, than the sterile one. Some

French workers have reported potential differences between such electrodes of 60 mV, but whether this can all be attributed to an increased anode activity is not clear. Most work has indicated that a larger negative potential change of a steel structure in anaerobic soil is necessary to achieve cathodic protection. Such a potential is initially difficult to obtain as the cathode is heavily depolarized.

Certain soils, notably some Roman remains near York, have been found sterile and iron buried in them has been found virtually free from corrosion; this is attributed to various tannates present in the ground which have acted as a poison. There has been some commercial application of this and other bactericides in tapes and coatings for underground pipelines. Highly alkaline conditions have also been found to inhibit the bacterial reaction and pipe-line trenches have been treated with lime as an alkalizer.

Cathodic protection is possible in the presence of these bacteria though generally a larger current density, particularly initially, is necessary and a different anode open circuit potential is found so that the metal potential has to be made more negative with respect to the point *M*.

The use of pulse techniques in the presence of bacteria has yet to be investigated fully.

Criteria

Cathodic protection can be used to prevent the corrosion in a bulk electrolyte of a metal that corrodes by the reaction:- metal to positive metal ion. This protection may be achieved under all practical conditions of polarization and environment including the presence of sulfate reducing bacteria. Protection is achieved when the potential of the structure relative to the corrosion cell boundary is polarized to be more negative than the open circuit potential of the anode.

In practice there are two forms of criteria that can be used to determine the degree of the cathodic protection by measurement of the potential of the structure. With certain metals there is a maximum anode open circuit potential in a particular electrolyte and the potential of the structure relative to a standard half cell placed at the corrosion cell boundary can be used to determine the adequacy of protection. As an alternative it is sometimes possible to ascribe a maximum potential difference that will exist between the anode open-circuit and the corrosion cell potentials. Protection is achieved by causing a potential change of the metal relative to the corrosion cell boundary, greater than this potential difference. In a practical case these criteria will be interchangeable.

As the change in potential is generally achieved by polarization it is often found that the sufficiency of protection can be determined from the immediate residual polarization on switching-off the cathodic protection current. Some workers prefer to use a swing criterion, that is the potential immediately on switching-off compared with that after some time has elapsed, while others seek to achieve the standard protection potential immediately on switching-off. These instant-off potentials are used on long structures where there is concern over the adequacy of protection at all points.

It is also possible to use as a criterion the change in the cathode electrical reaction as the cell becomes wholly a cathode. It is detected as a variation in the shape of the plot of the cathode potential against the logarithm of the cathodic protection current and is discussed later.

In all cases the criteria are met by causing an external uni-directional current to flow from an external source into the corrosion cell through the electrolyte and on to the structure receiving protection.

Practical Cathodic Protection Parameters

CHAPTER 2

Metal Potentials

Measurements

The measurement of the potential of a metal immersed in an electrolyte is, in practice, the measurement of the potential of a cell which consists of the metal and solution as one electrode system and a half cell as the other, the two being joined by an electrolyte bridge. In theoretical electrochemistry, it is usual to refer to the hydrogen electrode as the standard half cell. This is not a convenient instrument for use in the field and other, more rugged, electrodes are used.

The use of the half cell introduces a large resistance into the circuit and potential readings must be made using a high impedance instrument. The voltage measured usually has a value of between ½ V and 1 ½ V and an accuracy of ± 2 mV, or exceptionally ± 1 mV, is required. The total circuit resistance varies from below 20 ohms in low resistivity environments to more than 50,000 ohms in high resistivity soils. Moving-coil instruments are available with resistances of 100,000 ohm per volt and these are suitable for low resistance circuits. In high resistance circuits electronic solid state voltmeters or potentiometers are employed, one is shown in Figure 11.

Fig. 12 shows a typical potential measurement being made. Generally the metal will be connected to the negative of the meter and the half cell to the positive. Thus a reading of 600 mV between steel and a copper sulfate reference cell would mean that the copper half cell was 600 mV positive to the metal-soil half cell. In cathodic protection work it is general practice to refer to the potential of the metal-soil electrode system relative to the half cell, that is the metal has a potential of − 600 mV to the copper sulfate half cell. This convention, being universally adopted, will be used throughout the remainder of the text.

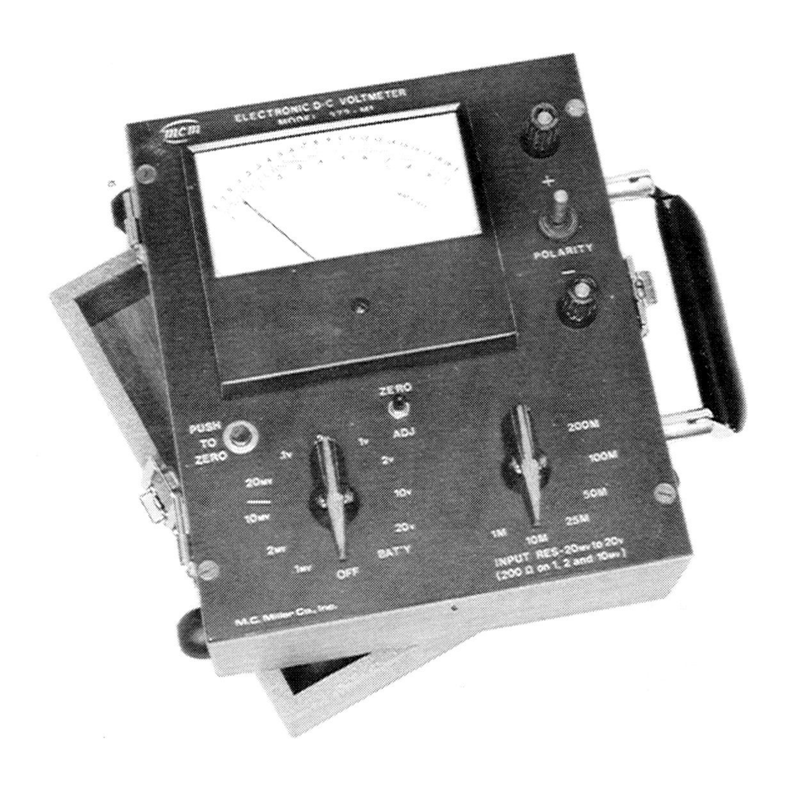

FIGURE 11 - Electronic dc voltmeter for using in potential measurement.

Consider Fig. 12, the metal may be acting as an anode or as a cathode. Suppose it is acting as an anode, then if the copper sulfate half cell is moved away from the metal the observed potential will become less negative, that is, it may now be −550 mV relative to the half cell. Similarly if the copper half cell were moved nearer to the metal its potential would become more negative, say −650 mV, relative to the half cell. If the copper half cell is returned to its original position where the metal displayed −600 mV potential and the current is increased, the metal potential will be less negative with respect to the half cell, say −500 mV, while if the current were decreased the metal would show a more negative potential, say −700 mV. All these properties are ascribed to the metal when it is an anode.

The change in potential caused by moving the half cell is due to the ohmic voltage drop in the electrolyte; this may be considerable in a soil that

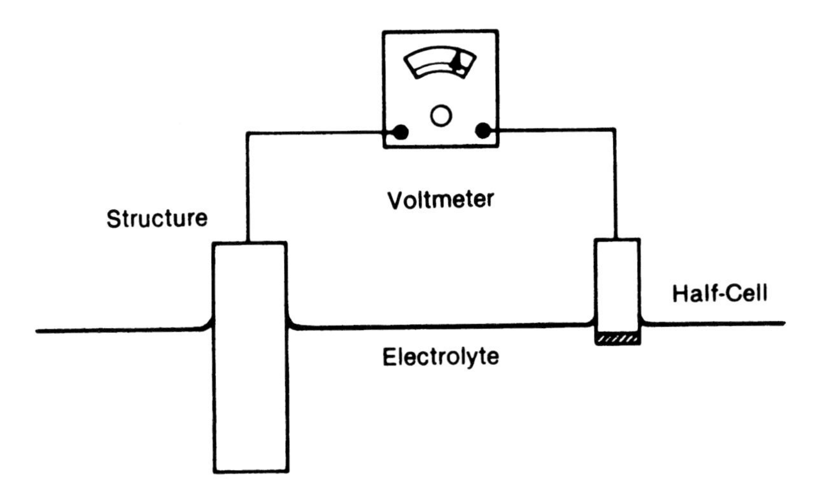

FIGURE 12 - Measurement of metal potential using half cell voltmeter.

has a high resistivity, especially if the current flux or density is high. The changes that occur on varying the current are those due both to polarization of the metal interface and the changed voltage drop through the ohmic resistance of the soil. The opposite of the above effects would occur if the metal were acting as a cathode; moving the half cell away would cause a more negative metal potential and bringing the half-cell nearer would cause the metal to display a less negative potential. Increasing the current flow towards the metal which is acting as a cathode would make it display a more negative potential, while decreasing the current will make the metal potential more positive.

These effects can be illustrated by considering an insulated box of electrolyte divided by a block of metal as in Fig. 13a with a direct current flowing from left to right through the electrolyte and metal. The metal potential relative to various half cell positions is shown in Fig. 13b for several values of the current.

If the current flowing into and out of a vertical cylinder of metal buried in the ground (or suspended in a liquid electrolyte) is considered, then the curves shown in Fig. 14 give the potential that the metal will display relative to a half cell at various positions. The left-hand side of the graph shows the relationship when the metal is a cathode and the right-hand side when the metal is an anode. The shape and magnitude of these curves will vary with the metal shape and the resistivity of the electrolyte. Changes in the current density will affect the curve both by the change in polarization at the metal interface and the ohmic voltage in the electrolyte.

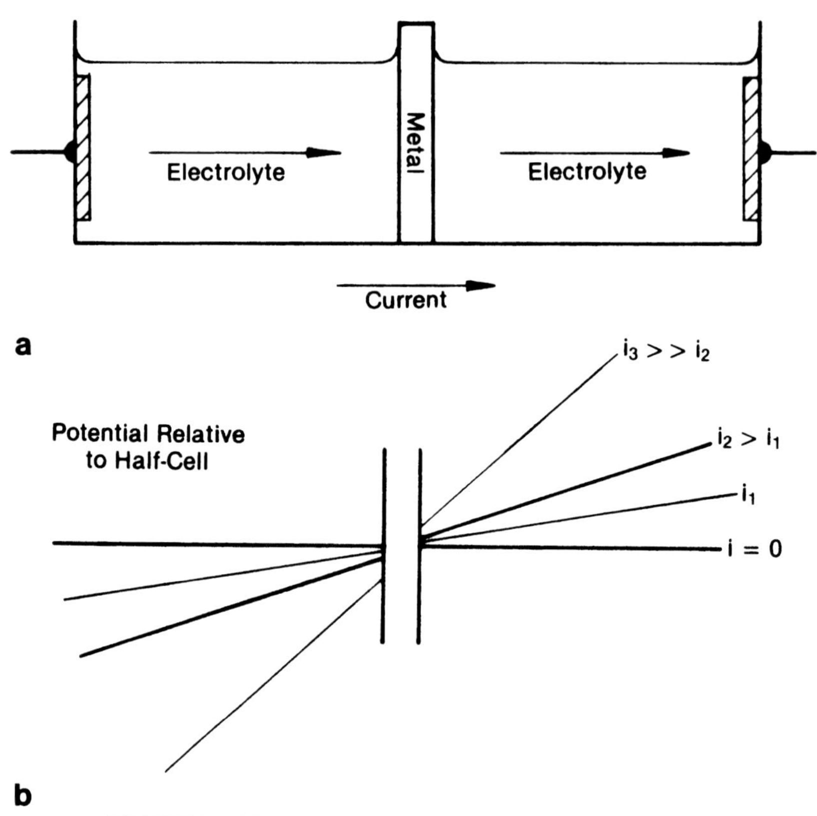

FIGURE 13 - Potential variations close to electrodes at various currents into and out of metal plate lying in electrolyte trough.

Electrodes

In the discussion on the equivalent circuit for a simple cell in Chapter 1 it was shown that the potential of the anode in the cell would be more negative than that of the cathode. In a practical corrosion cell the potential of the anodic area will be more negative to a half cell adjacent to it than is displayed by the cathodic area to a similar half cell placed close to it. This is illustrated in Fig. 15a.

If the current flowing through the cell is controlled by inserting a resistor in the electrical circuit (as opposed to the electrolytic circuit) then the potential of the anode and cathode relative to points x and y is shown in Fig. 15b. The difference between the curves relative to x and y being the voltage loss in the electrical circuit by virture of the introduced resistor.

It is possible to determine the anodic and cathodic areas on a structure

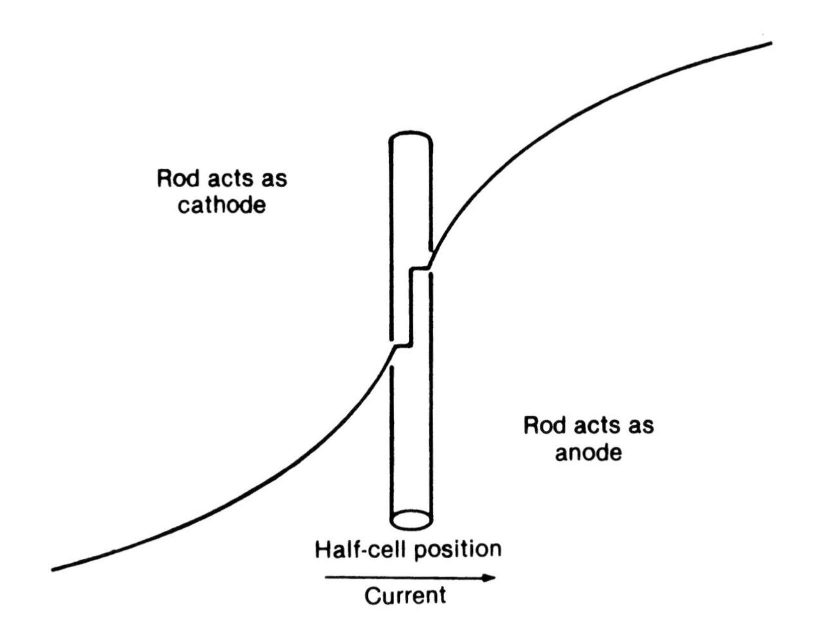

FIGURE 14 - Potential of metal rod relative to variously placed half cells.

by making a detailed potential survey; to do this the potential of successive small areas of the structure is measured relative to a local half cell. In the case of an extended or highly resistive structure both the half cell and the cathode lead have to be moved together. The half cell will give a mean potential over an area of the surface which is a circle whose diameter is approximately four times the structure to half cell distance; the central area will contribute more than the periphery. Small anodic and cathodic areas can be detected by a close half cell, say within a few inches from the surface, while larger anodes and cathodes will be detected from readings relative to a half cell further from the metal.

As the area contributing to the half cell potential depends upon the distance between it and the structure, it will be necessary to take a large number of closely spaced measurements in order to detect small anodic areas while large anodic areas may be mapped from more widely spaced readings. If measurements are taken beyond the 'suggested' distance from the metal, an integrated potential will be measured, the value of this depending upon the relative anode and cathode resistances. To determine the type of corrosive attack, it is usually necessary to carry out such a survey.

The anodic areas will display the more negative potentials, the cathodic areas showing the more positive potentials. Both areas will

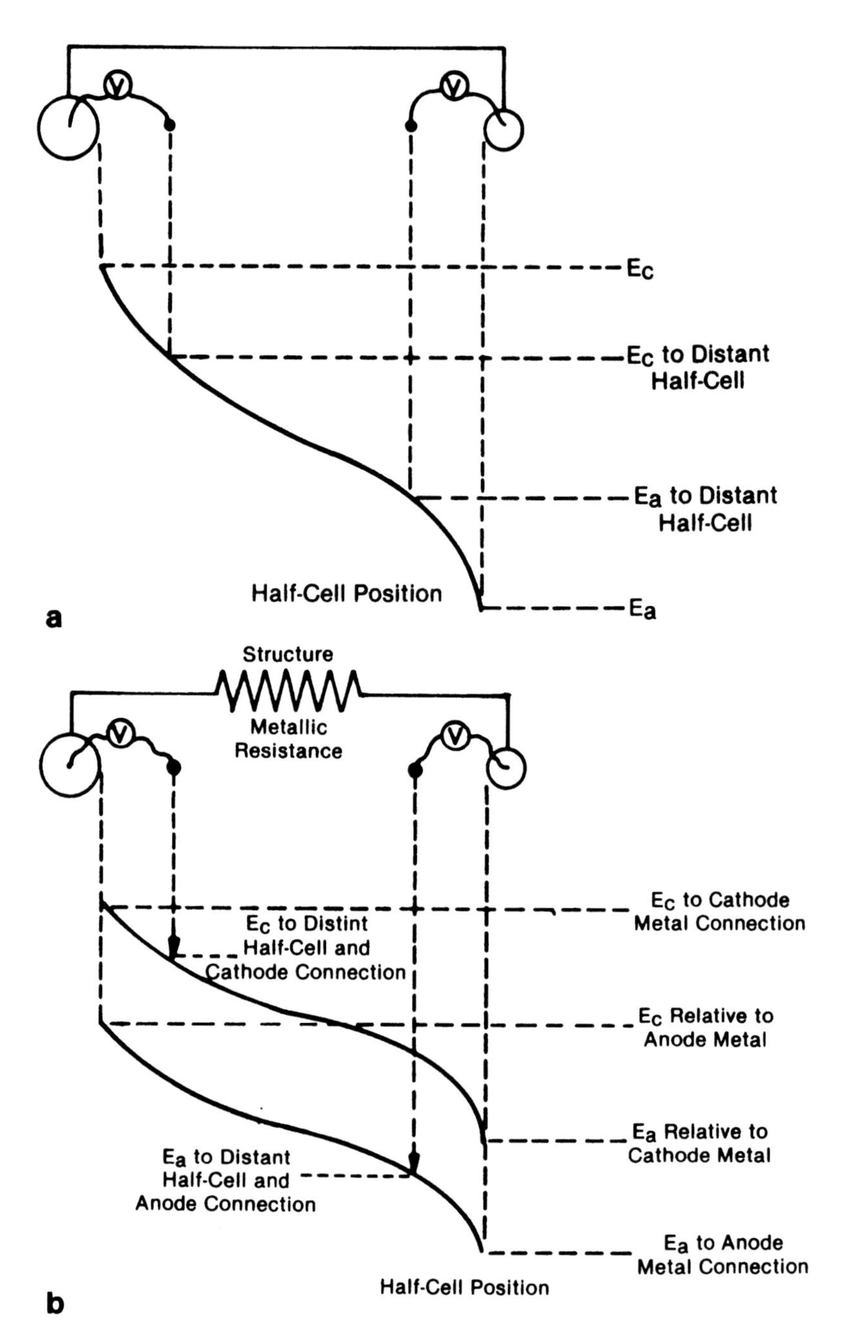

FIGURE 15 - Potential of anode and cathode in a simple cell against half cell position and electrode connection (a) with no metal resistance (b) with resistive structure.

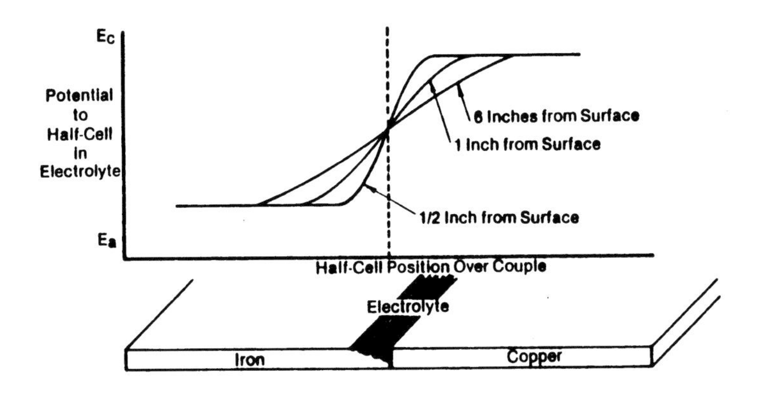

FIGURE 16 - Potential of a bi-metallic junction relative to three close anodes.

polarize and this polarization will decrease their extreme potential values; this means that if there is a steel anode and a copper cathode, then the steel potential will be the more negative. However, if these metals are close to each other along one edge as in Fig. 16, then the most rapid corrosion of the steel will occur immediately next to the copper sheet as illustrated.

A series of potential surveys taken at various distances from the two plates are shown in the graph. It can be seen that most corrosion occurs at the metal junction where the iron potential is most positive. This ennoblement is caused by the excessive polarization of the anode where the maximum density of current leaves the metal.

Should the copper be present only as a thin strip surrounded on both sides by iron, then a remote potential survey would reveal a slightly less negative band about the copper strip. Thus, though the worst corrosion will be occurring in this area, a remote potential survey would indicate a cathodic zone. Equally, if the iron were present as a thin strip surrounded by copper then a remote survey would indicate a slightly anodic zone in the vicinity of the iron though, in fact, the copper would be most cathodic and would be receiving a degree of cathodic protection.

This illustrates the problem of potential surveys. A microscopic survey may reveal small anodic areas and large cathodic areas developed on a section of steel plate. A remote survey might indicate that the section considered shows an overall cathodic potential relative to the remainder of the sheet. This section would then be considered free from rapid corrosion by a study of the remote readings whereas, had a microscopic survey been made, the presence of rapid pitting corrosion might have been predicted.

The anode will be the area displaying the more negative potential, however a piece of steel acting as an anode to a piece of copper, though still

TABLE III — Half Cells Commonly Used in Corrosion Practice and Cathodic Protection

Metal/solution	Metal Potential Volts
Copper/saturated copper sulfate	-0.85
Silver/silver chloride, in sea water	-0.80
Silver/silver chloride, in saturated salt water	-0.76
Calomel electrode	-0.78
Zinc in sea water	+0.25

more negative than the copper, will become polarized; this polarization will make the steel display a more positive potential than it normally would. If this same piece of steel is corroding by its surface having anodic and cathodic areas, then the anodic areas will display a more negative potential and this, despite some polarization, will be more negative than the potential that the steel plate would display to a remote electrode.

Half Cells

A practical half cell consists of a piece of metal, usually a rod, surrounded by a standard solution of one of its salts, connected to the corroding electrolyte via a salt bridge. In practice the bridge consists of a porous plug of wood, sintered glass or ceramic, soaked in the cell electrolyte. In the majority of half cells the bridge ends with a cross-sectional area of a few square centimeters and for most field purposes this acts as a point contact. For laboratory work it is general practice to make the bridge tip as small as possible in order to achieve a reading at a precise point, often by the use of an extension salt bridge.

The most common half cells used in corrosion practice and in cathodic protection are listed in Table III. Opposite each cell is indicated the potential that would be measured if they replaced each other in a particular measurement of steel at its protection potential. The copper sulfate electrode is used generally for non-marine work and the zinc and silver chloride cells for marine work. The calomel electrode finds limited research use in both fields.

Cathodic Protection Criteria

In the first chapter it was shown that corrosion in a simple cell could be prevented by an external current; this method is called cathodic protection. While protection is achieved by virture of the external current, corrosion ceases when the anode current is reduced to zero or is reversed. This condition can be achieved when the potential of the structure relative to a half cell at the corrosion cell boundary is made more negative than the open circuit potential of the anodic areas in the cell. The potential of the structure relative to this point can be the basis of a criterion for protection. It has been suggested that there may be preferable criteria but they are not as widely accepted.

Many metals, notably steel, are found to be adequately protected in a variety of environments by maintaining the potential of the structure more negative than a specific potential value. Some metal/electrolyte combinations can be protected when their potential is altered by a particular amount. Protection criteria have been established for the more common metals by either or both of these techniques. These criteria have been established from field practice, that is, these values have been found sufficient. They will tend to be conservative as the heavy economic losses from corrosion will inhibit experiments at lower levels of change. Cathodic protection has been used extensively for the past 50 years, and for steel the criterion has been well established. Other metals have not been cathodically protected so widely and with these some other method of establishing the appropriate criterion is often needed.

The majority of the measurements have been taken assuming a smooth direct current and potential. Where the external current is in the form of pulses, all the measurements are integrated to appear to be smooth, the integration period being of the order of one second. With moving coil meters the instrument itself acts as an integrator while digital meters have to be stabilized to a unique reading of the final digit. Empirical pulse techniques in natural waters have suggested a considerable reduction in the net charge density to achieve protection; this most markedly so where a 'calcareous' film has been formed. However, few of the tests have made a contemporary comparison of the pulses with the equivalent smooth current which might, in the same limited tests, have produced equal protection. There is a limited experience with pulse techniques and no criteria of protection other than the measurement of the smoothed dc equivalent has been established.

Where the structure has polarized to a stable level of protection and current it is possible to establish criteria of protection by measuring the potential of the structure immediately on switching-off. This can be used either as an absolute criterion relative to the half cell or it can be compared with the potential after the polarization has decayed. These are the equivalent of the absolute and swing potentials without the ohmic volts drop caused by the cathode current.

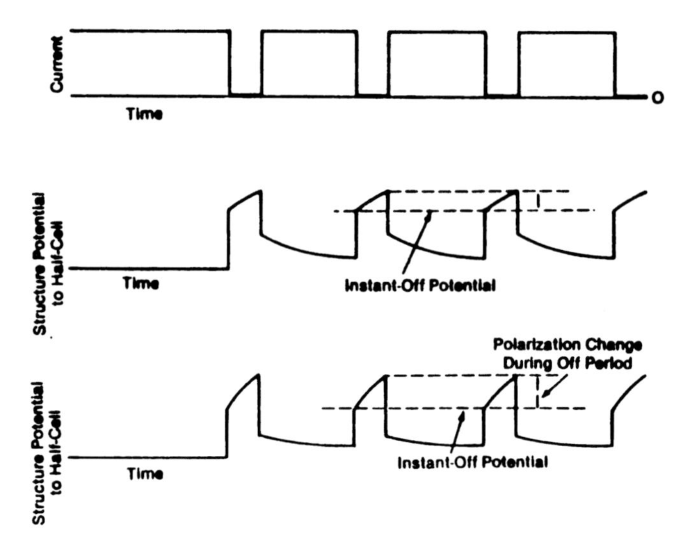

FIGURE 17 - Current and potential curves for instant-off criteria of protection for two structures.

In practice there are problems in undertaking the tests. On many structures there will be difficulty in achieving a sharp cut-off because of inductive and capacitive effects in the electrical circuits and in the structure itself. There will also be a differential rate of decay of the polarization as it will be affected by the equivalent of the normal corrosion currents. Where the current is interrupted for a considerable period of time a second measurement must be delayed until the polarization has been completely restored.

A field technique has been developed in which the current is interrupted for short periods of about one fifth to one third of the on/off cycle. This leads to a reduction in the polarization as the current density is not increased during the on period and the current is off for too short a time for the polarization to decay fully. The potential difference between the beginning and end of the off period is measured and used as a criterion fixed at between a half and a third of the normal swing criterion. This is illustrated in figure 17.

Instead of switching the cathodic current on and off, some workers prefer to make a step change in the current. This has many of the same objections, particularly in achieving an instant change in current. To some extent the measurement can be corrected by determining the potential gradient in the electrolyte which will change with exactly the same electrical profile as the current entering the structure.

The methods have tended to run into the techniques used for determining the ohmic volts drop around the structure by the use of a superimposed alternating current. The greatest application is in the protection of buried pipelines and these are discussed later in that chapter later in the book.

Iron and steel

Buried steel, particularly in the form of pipelines, has been cathodically protected for many years. The established potential criterion is −0.85 V relative to a copper sulfate half cell, or −0.80 V to a silver chloride cell when the corroding electrolyte is sea water. At high temperatures the criterion should be made more negative by about 2 mV/°C though positive proof of protection should always be sought in extreme conditions.

In anaerobic soils or waters that contain sulfate reducing bacteria steel is found to be protected when its potential is depressed a further 100 mV though some pipeline operators prefer a larger depression.

Many cathodic protection engineers use a swing criterion for the protection of steel. The accepted value for this under aerobic conditions is that the structure be made 300 mV more negative than its natural potential. Under anaerobic conditions this swing criterion becomes 400 mV.

In the case of buried structures, and particularly buried pipelines, there is a considerable variation in potential around the structure. This can give misleading readings. These ohmic voltages can be predicted from the soil resistivity and the current density and are discussed later in the book. They can be eliminated by the use of techniques which interrupt or change the cathodic protection current.

The use of 'instant-off' potentials on steel requires a minimum polarization change during the 'off' cycle of 100 mV and of 150 mV under anaerobic conditions.

Lead

Extensive cathodic protection installations have been made to lead cable sheaths, either buried in the ground or when they rest in ducts. The potential of a lead cable sheath to earth depends upon the earth conditions and can vary from −0.55 V to −0.65 V to a copper sulfate electrode.

This electrode is not always used and a piece of cable sheath alloy, usually an antimony alloy, or a lead/lead chloride half cell is employed. The lead alloy is either placed in a filled duct or at the bottom of a manhole and used as a zero reference. That is, when the lead sheath is receiving no protection it and the reference length of lead alloy display the same potential. The lead chloride half cell, usually lead 3 per cent antimony alloy in normal lead chloride (or sometimes saturated lead chloride) displays a potential of −0.58 V to −0.59 V to a copper sulfate half cell and so is generally about the potential of the lead cable sheath.

The swing potential, or potential to the coupon or half cell required to achieve protection, is quoted as varying between −0.05 volts and −0.50 volts. The majority of engineers show a preference for a potential of −0.10 volts to −0.25 volts. Some operators use a criterion relative to a copper sulfate half cell of −0.75 volts. A great deal of laboratory work has been performed to establish this criterion for lead, and Compton, who has been one of the major workers in the field, suggests as a potential device the use of bright lead in the corroding soil. To achieve protection he suggests that a 100 mV swing is necessary. This is based on tests where the cable potential begins to show over-voltage polarization and in specific cases it is suggested that smaller swings may be tolerated. The 100 mV negative swing will be achieved in most soils when the overall potential is −0.70 V to a copper sulfate electrode.

Lead and aluminum can suffer from corrosion by alkali attack and cathodic protection can cause considerable alkali build-up at the metal surface. The amount of alkali present will depend upon the soil, the rate of diffusion, any flushing by surface water, and upon the current density applied. In most soils current densities sufficient to produce a potential of −0.85 V to a copper sulfate half cell on a lead sheath will cause the electrolyte close to the metal surface to have a pH of about 10; at this concentration alkali attack could commence. There is some evidence that this does not occur while the cathodic protection current is flowing, but interruptions, caused either by failure of the electrical apparatus of by drying out of the soil or the duct, allow corrosion to proceed. Most companies operating lead cable networks flush their ducts regularly and in applying cathodic protection do not exceed 0.5 V negative swing or −1.0 V to a copper sulfate half cell, though this is influenced by local experience. Cables in asbestos cement ducts are particularly prone to cathodic corrosion and protection by a negative potential swing of more than 50 mV is to be avoided. The use of asbestos is now considered bad practice and, indeed, the use of exposed lead sheathed cables is disappearing with the introduction of plastic outer sheaths.

Aluminum

Aluminum is finding increasing use as a sheathing material for cables and considerable lengths of it are buried in the ground and in various waters. Similar developments are taking place in piping and—using modern welding techniques—aluminum pipe lines are being installed. A great weight of aluminum is being fabricated into chemical plants and many storage vessels are now made of it. Small ships and boats are being built with aluminum hulls and these techniques are being extended to light warships.

All these uses of aluminum involve its contact with bulk electrolytes. Coatings on aluminum have been tried with some success, but inevitably even the best coatings suffer damage and pitting corrosion occurs. Cathodic protection has been used successfully in a great number of applications and with a large number of aluminum alloys. In soils it has been found that a negative swing of 100 to 200 mV protects the aluminum and this makes its potential about −0.80 V to −0.85 V to a copper sulfate half cell. 99.5 per cent pure aluminum and some of the structural alloys have been protected in sea water and in various brackish waters at potentials of −0.85 V. The aluminum alloys used to clad the structural alloys often initially display potentials more negative than this in sea water. For cathodic protection in sea water a potential swing of −100 mV to −200 mV has been adopted. Where aluminum is used in chemical plant the cathodic protection criterion is best established in the particular electrolyte.

Aluminum suffers cathodic corrosion as the result of an excessive accumulation of alkali. In many soils potentials as negative as −1.3 V to copper sulfate can be tolerated and in moving sea water it is difficult to cause a cathodic attack. The presence of a second metal, such as steel, can cause alkali to accumulate by the cathodic reaction on the steel which will attack the aluminum at potentials more positive than those generally considered safe. Because of the diverse electrolytes and alloys that are found with aluminum careful use of swing potential criteria and site experiments will allow the successful use of cathodic protection.

Galvanizing

Zinc is rarely used structurally but there is widespread use of it in hot dip galvanizing and other similar processes. Steel that has been galvanized can be successfully cathodically protected: this should be achieved by the zinc which forms a sacrificial coating, but in practice this coating acts for a few days and then becomes passive. Galvanized tanks, pipes, ships' hulls, etc., can be protected and the potential required for protection seems to be about −1.0 V to copper sulfate. While the protection of the old galvanizing, or rather the underlying steel, can be achieved at less negative potentials than this, it is difficult to avoid potentials more negative than −1.0 V with new galvanizing when a swing criterion of 100 mV can be adopted.

Zinc dust in an inorganic binder used to coat steel in sea water can be protected at normal steel potentials. It is suggested that if the steel is protected by a further 100 mV, that is to about −0.95 V to −1.80 V to copper sulfate, then the coating will remain intact and the zinc not corrode or leach out. The coating integrity is retained but this may be caused by oxidation of the outer layer of zinc which then seals the surface.

Zinc metal coating, if it is in electrical contact with the steel, will act as a potential barrier which will display seemingly very high cathodic resistance at moderate potential swings.

It is suggested that zinc coatings on steel can be polarized for short periods and then in the absence of cathodic protection will continue to prevent corrosion. Proposals involving periods of about one minute application of current and one hour continuation of protection have been suggested.

Cathodic corrosion could conceivably occur with a galvanized structure, but since the corrosion of the galvanizing would reveal only steel which would be adequately protected, there is little information on this point.

Stainless Steel

Stainless steel relies for its corrosion resistance upon a film of oxide which passivates or protects the metal. To maintain this film it is essential that there is a supply of oxygen to the surface. This condition is achieved under anodic, not cathodic, conditions and there is considerable experience of successful anodic protection of stainless steel. Under cathodic conditions the supply of oxygen to the cathode is diminished. Small cathodic changes of 100 mV can destroy the oxide film on stainless steels and render them liable to corrosion. A further potential change, while destroying the natural protection of the oxide film, protects the steel by the normal cathodic protection processes.

Anodic protection of stainless steel and other passivating metals is achieved by holding the potential within the band at which the minimum corrosion occurs. This means that a miniscule anodic current is flowing and causing the continuous re-passivation of the surface. The corrosion rate under these conditions is the minimum and the system is designed to tolerate this.

Other Non-ferrous Metals

The yellow metals—copper, brass and bronze—can be cathodically protected. Copper was the subject of the famous investigation by Sir Humphrey Davy and his results are still applicable. When copper cable sheaths are used the operators have reported successful protection at a swing of −100 mV to −200 mV or potentials of −0.1 V to −0.25 V to the normal copper sulfate half cell. Dezincification of new brass can be prevented by cathodic protection; this has usually been achieved incidental to a wider protection scheme and the paucity of the experience forbids any reasonably established criterion. A negative swing of 500 mV seems to be of the order required to prevent dezincification.

Probably the worst bronze corrosion occurs on the propellers of ships. This is very often caused by the extreme turbulence and cavitation at the tip of the propeller and this is dealt with in detail in the chapter on Protection of Ships. Similar corrosion occurs on the tube plate of heat exchangers

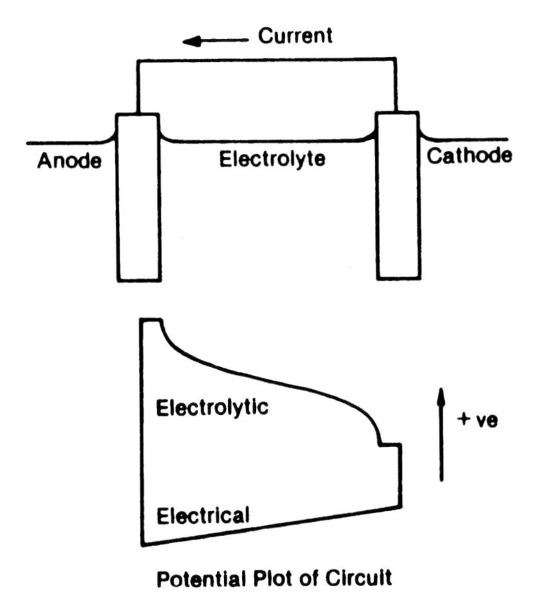

FIGURE 18 - Potential variation in a simple galvanic cell.

which can be prevented by cathodic protection, as can the corrosion of the alloys used in condenser and heat exchanger tubes at the same potentials that protect steel.

Half Cell Position

Protection Criteria

The criteria of protection are established as potential changes or particular potentials relative to a standard reference electrode. Equally essential is a definition of the position of the reference, or swing test, electrode during the potential measurements.

As indicated earlier, there will be a change in potential of any corroding structure relative to variously positioned half cells. Before making a study of the half cell position in respect of the cathodic protection criteria, it is first necessary to consider the geometry of a simple cell.

Suppose two metal rods of equal size and shape are immersed in a liquid electrolyte then the potentials along the two current paths, the electric and the electrolytic, will be as in Fig. 18. If, as is often the case in practice, the two rods of metal are physically joined together, then the electrical path will have zero resistance and the driving potential of the cell, $E_a - E_c$, will be used to drive the electrolytic current only. This will meet a resistance

R_a at the anolyte and R_c at the catholyte, so that the value of the cell current will be

$$i_0 = \frac{E_a - E_c}{R_a + R_c} \qquad (2.1)$$

The rate of corrosion per unit area will be proportional to the current density at the anode and this figure is of greater practical significance than the actual cell current. If the anode area is A_a then the rate of loss of metal per unit area will be proportional to

$$\frac{i_0}{A_a} = \frac{E_a - E_c}{A_a(R_a + R_c)} \qquad (2.2)$$

Electrode Size

If the dimension of the cell is increased by a factor m, then the resistances at the anode and cathode will decrease by this amount and the anode area will increase by its square, m^2. The cell current will increase because of the reduction in circuit resistance and the current density will be less because of the greater increase in area. The decrease in current density will mean there will be less polarization so the value $E_a - E_c$ will increase. Had there been no polarization the current density, and hence the rate of corrosion per unit area would be smaller by the factor m, the increase in dimensions of the cell. Where the electrodes polarize the decrease will not be as marked because the polarization will tend to maintain a constant current density.

If the metal rods are made unequal in size by increasing the size of only one, then the current will increase and the current density on the rod that remains the same size will be greater while the current density on the one that has increased in size will be less. The cathode will not be affected, but the anode will probably corrode more rapidly if its current density increases.

In a large corrosion cell the current density is smaller and polarization will have less effect, while in a small cell with high current density polarization will become important and can assume complete control. In a cell with electrodes of unequal size, the control—that is the factor that has the major influence on the cell current—may rest either with the cell resistance or the polarization of the electrodes. The rate of corrosion will almost invariably depend upon the size of the anode. As well as the influence of the size of the electrodes, considerable control can be exercised by the amount of electrolyte present. The current flowing in a cell formed by two flat metal electrodes lying on the bed of shallow water will be influenced by the depth of water over them until this depth exceeds the plate dimensions.

Corrosion Cell Boundary

The potential measuring circuit has been described earlier in this chapter. In a short circuited cell, the potential read will vary with the half cell position, the variation becoming vanishingly small as the half cell is placed further away from the couple. The general zone where the metal potential is influenced by small changes in the half cell position will be considered to be inside the electrical boundary of the cell.

The potential that the couple, that is the anode and the cathode, displays to a remote half cell will depend upon the polarized potentials of the electrodes and the resistance associated with each of these. In certain symmetrical electrode arrangements these resistances will be directly related to the square root of the surface areas of the anode and the cathode. A better understanding of this problem arises from a consideration of the potentials associated with a pair of infinite line electrodes spaced a constant distance apart.

Potential Survey in a Corrosion Cell

First consider the effect of causing a current to flow uniformly onto a long cylindrical electrode as in Fig. 19. The resistive volts drop can be

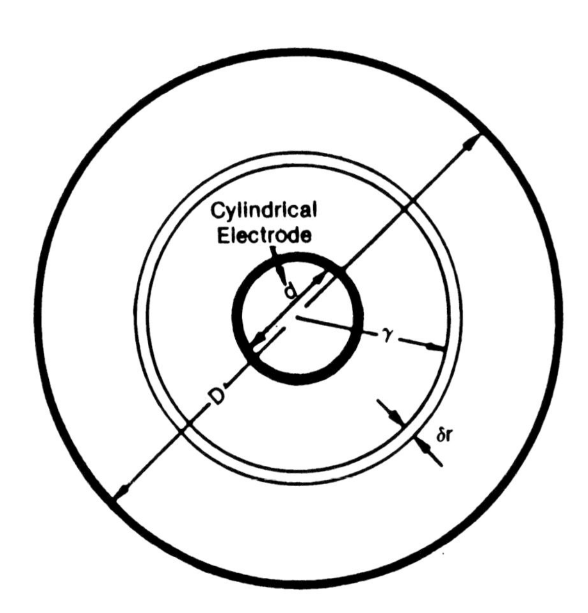

FIGURE 19 - Current/potential relationship around a cylindrical electrode.

calculated by considering the voltage drop in the small cylindrical shell of unit length, radius r, wall thickness δ_r; then the area through which the current flows will be $2\pi r$ the shell thickness δ_r, so that if I amp is flowing to the cylinder per unit length the volts drop in the shell will be

$$\delta V = \frac{I}{2\pi r} \delta_r \rho$$

where ρ is the electrolyte resistivity. If the cylinder has a diameter d then the volts drop between the cylinder and a point distance $D/2$ from its center will be

$$V = \frac{I}{2\pi} \rho \int_{d/2}^{D/2} \frac{dr}{r} \quad V = \frac{I\rho}{2\pi} \ln \frac{D}{d} \tag{2.3}$$

Two such electrodes are shown in Fig 20. These are separated by a constant distance in the electrolyte and are connected electrically outside it. A corrosion current flows in the cell and the two line electrodes are polarized to potentials E_a and E_c. The potential of the system to any point P will depend on the function log P_1/P_2, where P_1 and P_2 are the distances of the point P from the anode and cathode. Fig. 21 shows the plot of the potential

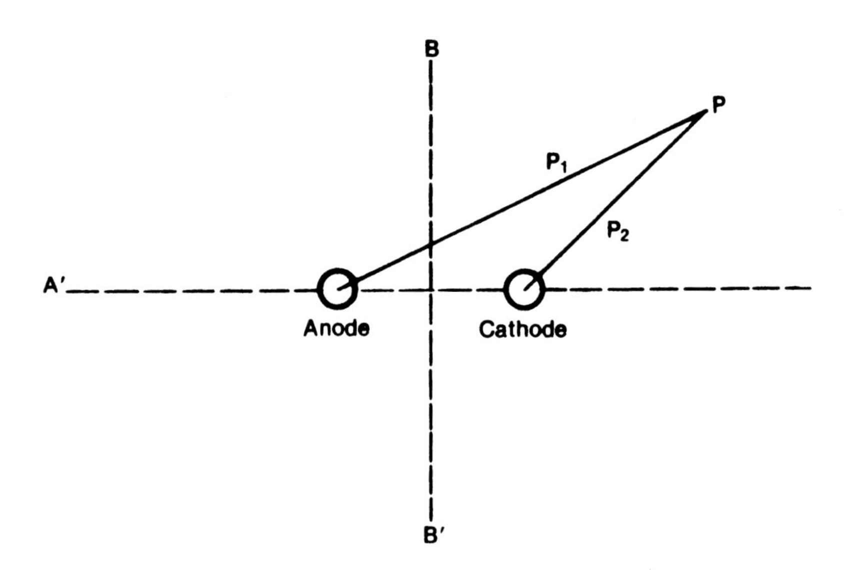

FIGURE 20 - Section through a pair of parallel cylindrical electrodes.

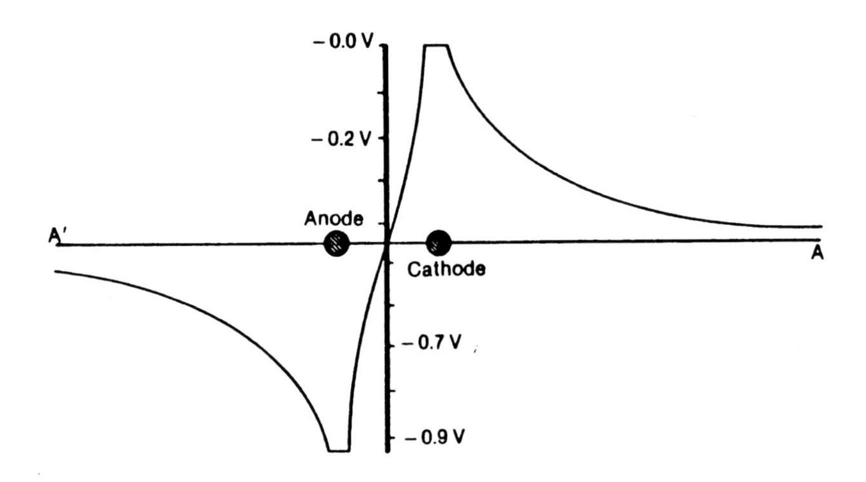

FIGURE 21 - Potential of the couple in Figure 20 relative to points along line AA'.

relative to a half cell moved along AA' where E_a and E_c are assumed to be −0.9 V and 0 V.

The effective cell boundary is not theoretically reached within the range of Fig. 21. The couple potential to a remote electrode is displayed by any point along the line BB'. At a point along AA' beyond 8 to 10 diameters from the couple the change of potential with distance is sufficiently small for these points to be considered outside the effective electrical boundary.

Potential Survey in a Corrosion Cell Receiving Cathodic Protection

When cathodic protection is applied to this couple the condition that no corrosion shall occur will be that there is no anodic current. In Chapter 1 this was shown to occur when the cathode was polarized to the anode open circuit potential and that this polarization could include a resistive component associated with the catholyte. This means that the cell current will become zero when the potential of the couple is equal to the anode open circuit potential and this measurement is made relative to a half cell placed to include the potential drop associated with the catholyte resistance.

Such a system has been examined experimentally and the couple subjected to cathodic protection. Fig. 21 shows a part of a potential along AA'. When an external current was applied, the potentials, as plotted in Fig. 22 show the condition when the anode current was reduced to zero. The geometry of the couple suggests that when no current is flowing to the anode there will be a considerable potential variation by virtue of the current that is flowing to the cathode; thus, to include the resistive drop of the

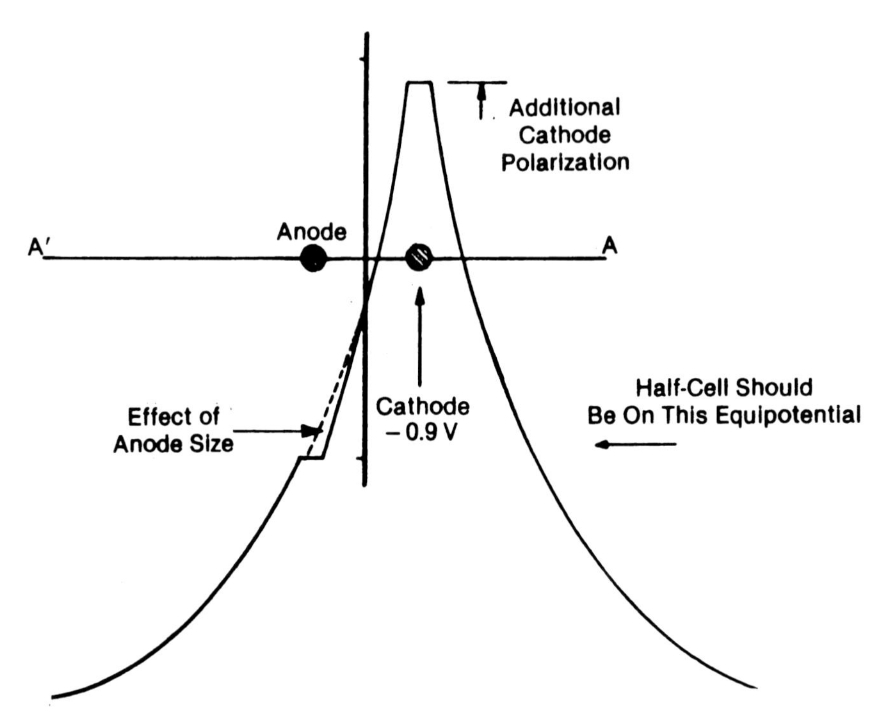

FIGURE 22 - Potential of the couple relative to line AA' when receiving cathodic protection.

catholyte the half cell should be placed on the equipotential which passes through the anode.

As can be seen from Fig. 22, the potential at the anode became the open circuit potential of the anode itself. Though this cell configuration is not typical of practical corrosion, the identity of the theory and practice indicates the importance of the location of the half cell. For example, if the half cell had been placed much closer to the cathode, then a very much higher external current would have been required to reduce that point to the potential of the anode, while if the half cell had been placed much further away, the potential of the anode would have been reached long before the anode current reduced to zero.

In practice it is essential to make a reliable estimate of the half cell location before using potential criteria: this location will depend upon the geometry of the corrosion cell, any resistive films including paints or coatings, and the resistivity of the electrolyte.

To indicate protection against the type of corrosion that forms anodic pits half-an-inch in diameter, the half cell must be placed half-an-inch from the metal surface. At this position the polarization of the cathode will include all the acceptable catholyte resistance potential drop. Placing the half cell remote from the structure would lead to a loss of protection as a potential drop caused by the resistance of the electrolyte would be included in the measured voltage. Equally, were corrosion caused by two different soil types over the length of a long pipeline, then applying protection relative to a half cell placed very close to the pipeline would demand more cathodic protection current than was necessary.

Practical Half Cell Location

Three practical corrosion problems might best illustrate the point. Fig. 23 shows a pipeline that is corroding in three ways: firstly by long line currents that flow from one geological region to the next; secondly by short line currents that flow from bottom to top of the pipe and between areas of slightly different soil, a matter of feet apart; thirdly by local cell or pitting corrosion caused by microscopic differences in the soils generally due to differential aeration.

If a copper sulfate half cell is placed at point 3 and the metal potential is depressed to -0.85 V, then the long line current corrosion is prevented and the two other types of corrosion are reduced. Protection at this criterion

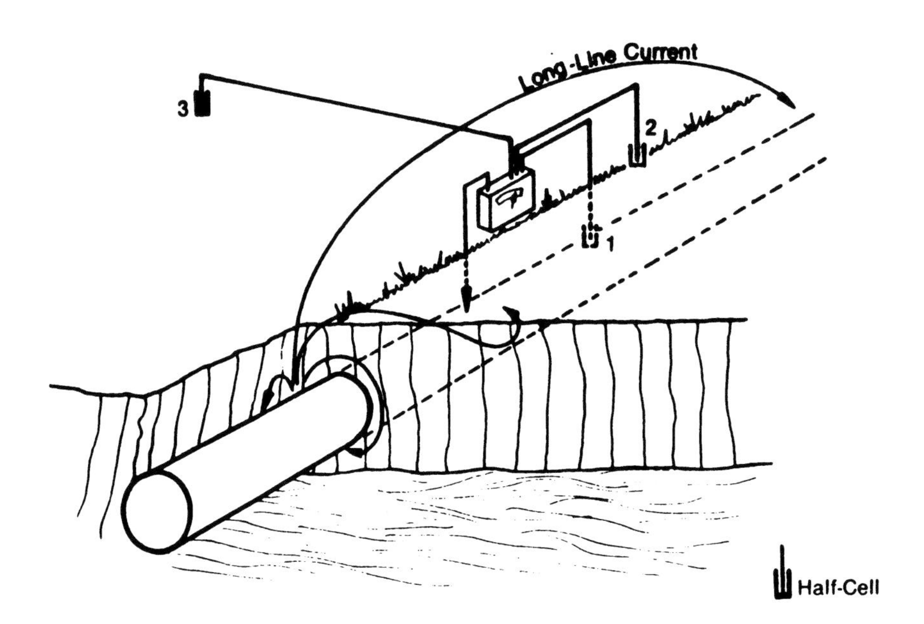

FIGURE 23 - Half cell location for buried pipe protection.

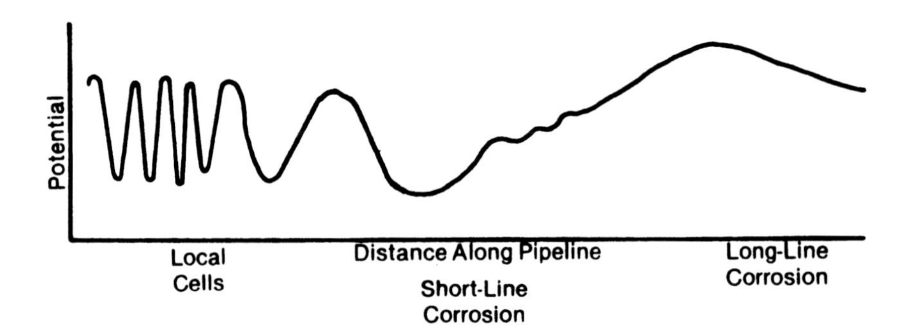

FIGURE 24 - Potential variation along pipeline.

relative to a half cell at position 2, that is, over the pipe at the soil surface, will prevent corrosion from short and long line currents and reduce local cell corrosion. Protection relative to a half cell at position 1, within an inch of the pipe metal, will prevent all three forms of corrosion.

Perhaps it would be useful to study a plot of potential along the pipe surface, assuming all three types of corrosion to be taking effect. Fig. 24 shows such a plot. In the curve, two features are important: the frequency of the change along the length of the pipeline, and the amplitude of these changes. A small ripple on the long line corrosion pattern will be suppressed by long line protection, but local cell corrosion with low frequency modulation of its amplitude will not be controlled by long line protection.

The second example illustrated in Fig. 25 shows a water storage tank being protected by a single anode. The tank base is covered with a layer of sediment from the water and a large differential aeration cell exists between this and the tank walls. The walls, though cathodic to the base, themselves suffer corrosion of an intense pitting character. The base-to-wall cell may be likened to the long line current corrosion and if protection potential is achieved relative to a half cell at position 1 this corrosion will cease and the pitting attack on the walls will be reduced. To control the pitting, or local cell action, at the walls, the protection potential must be achieved relative to a half cell within ½ inch of the wall.

The third example concerns the position of the half cell relative to a protected structure such as an H girder. This is illustrated in Figure 26. The half cell at position 1 will deal with corrosion from top to bottom of the girder that may occur with stratified sea water, river water and mud or by a temperature differential in an otherwise homogeneous liquid. The corrosion between the steel weld metal and the steel will be reduced, but in order that this is eliminated then the protective potential must be reached within the girder recess.

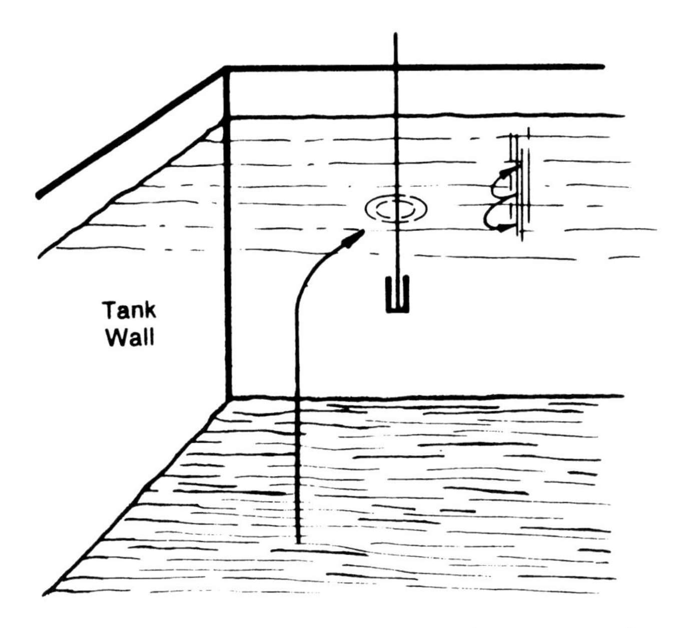

FIGURE 25 - Half cell location in water tank.

It is difficult to attain protection in recesses which have a depth greater than their diameter without considerable overprotection of the main body of the metal. In a later chapter, the spread of protection into such recesses will be discussed in detail and the half cell position considered then.

The position of the half cell decides the amount of catholyte ohmic volts drop included in the criterion voltage. The half cell position will be less critical where the change of potential with distance is least, that is, where either the current density onto the structure is small or where the resistivity is low. Coatings, whether caused by polarization, by paints or other organic coverings, will reduce the current density and hence the potential gradient in the electrolyte. Sea water has the lowest resistivity of practical electrolytes at about 20 ohm cms and generally electrolytes with resistivity below 100 ohm cms do not demand accurate positioning of the half cell except when associated with abnormally high current densities. In high resistivity electrolytes a quick check on the potential gradient will indicate the necessity for careful location of the half cell. If less than 20 mV is found between the positions suggested in the examples considered, then little attention need be paid to the accurate positioning of the electrode.

Switching Current On and Off

It is possible to eliminate the ohmic volts drop associated with the

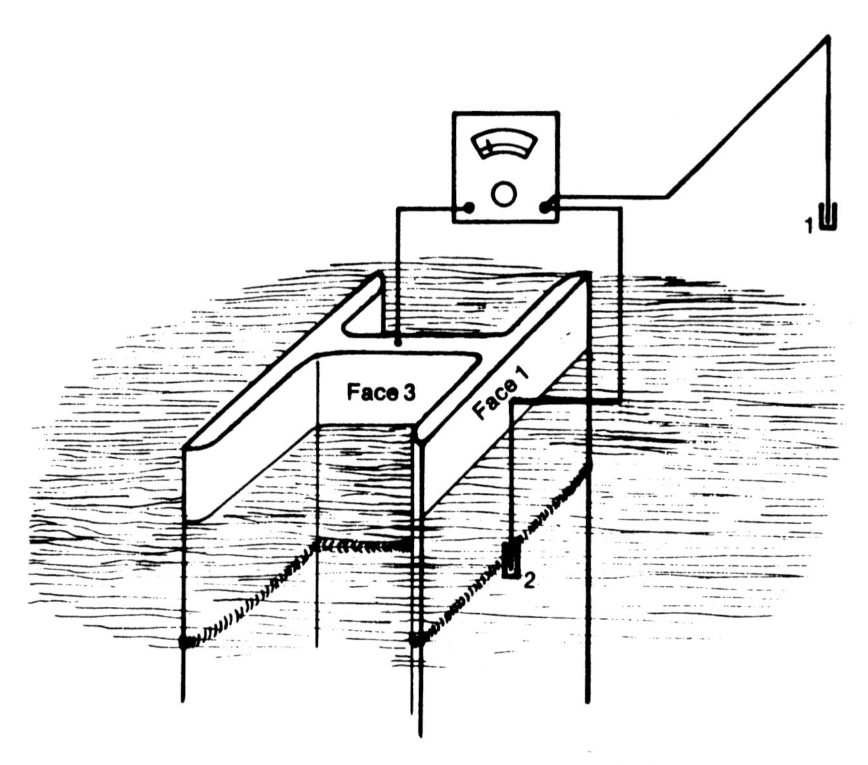

FIGURE 26 - Half cell location at 'H' pile.

catholyte by switching the cathodic protection system on and off and measuring the polarization potential. This will have different effects in different electrolytes and will depend on the degree of polarization that is sustained on the structure. It can usually only be used successfully only after a considerable period of operation of the cathodic protection, when a stable polarized condition has been achieved.

Current Density Required for Cathodic Protection

Having established the criteria for protection, the next practical parameter to discuss is the amount of current required to achieve protection. The protection is not influenced by the source of direct current, be it derived from normal generation or electrochemical reaction. The current density required to achieve protection can be estimated, but not with great accuracy, otherwise this method would be used to define protection criteria.

In most practical applications there will be two current densities; the mean current density over the whole surface—this will be the total current divided by the total area—and the absolute current density, which will be the minimum required to give protection to a particular area of the structure. Either of these quantities may be the larger, though generally the mean current density will be greater as it will include any overprotection of the system. Where one sector of the structure is subject to a depolarizing influence or perhaps part of its surface has a low coating resistance, the local absolute current density may exceed the mean current density. An efficiency figure could be calculated based on the ratio of the absolute current density to the mean current density, though this will have little meaning economically as the low efficiency in current density may be offset by a high efficiency in current generation.

It is usual to refer to the superficial area when the area of the cathode is being considered. A corroded or weathered surface will be rough and this roughness may increase the true surface area by 25% or more. Similarly, most metal surfaces will be covered by a layer of corrosion product which may be quite thin. Bright metal that has been freshly scraped will have a significantly different current demand from metal surfaces found in practice where both surface roughness and corrosion films will be present. The following considerations will include only the normally corroded and roughened surfaces of the metal. Mechanical and pneumatic erosion of the surface will be considered as it affects the normal current requirements. Temperature variations and other phenomena, such as relative movement of the metal and the electrolyte, will be considered where relevant.

Iron and Steel

The most common structural metal is iron, and either as steel or in one of its forms, it is extensively buried in the ground or immersed in water. The largest electrolyte is sea water and it is certainly the most corrosive of the common environments. Steel in the form of jetty piles, which are in tidal water, require five to six mA per sq ft for protection. At this current density, gradual polarization occurs, and after three to six months the current can be reduced to four to five mA per sq ft. In areas where there is no tide, such as sea-filled pools, static water storage tanks, etc., the current density is a little less and polarization more rapid.

If the water velocity is high, as around offshore structures, the curve of potential against current density will be logarithmic in form as Figure 27a which is a typical curve for temperate sea water. Figure 27b shows the variation with time of the current density needed to maintain the structure at −800 mV to Silver Chloride.

The curve reproduces the hydrogen overvoltage described in Fig 6, Chapter 1. In rapidly moving sea water bare steel may require 10 mA/sq ft to 20 mA/sq ft and if this velocity causes differential aeration or impingement then the current demand may be doubled. Sediment in the water can scour the surface and increase the current density required. The factor of importance is the rate of arrival of oxygen at the cathode, this seems to be

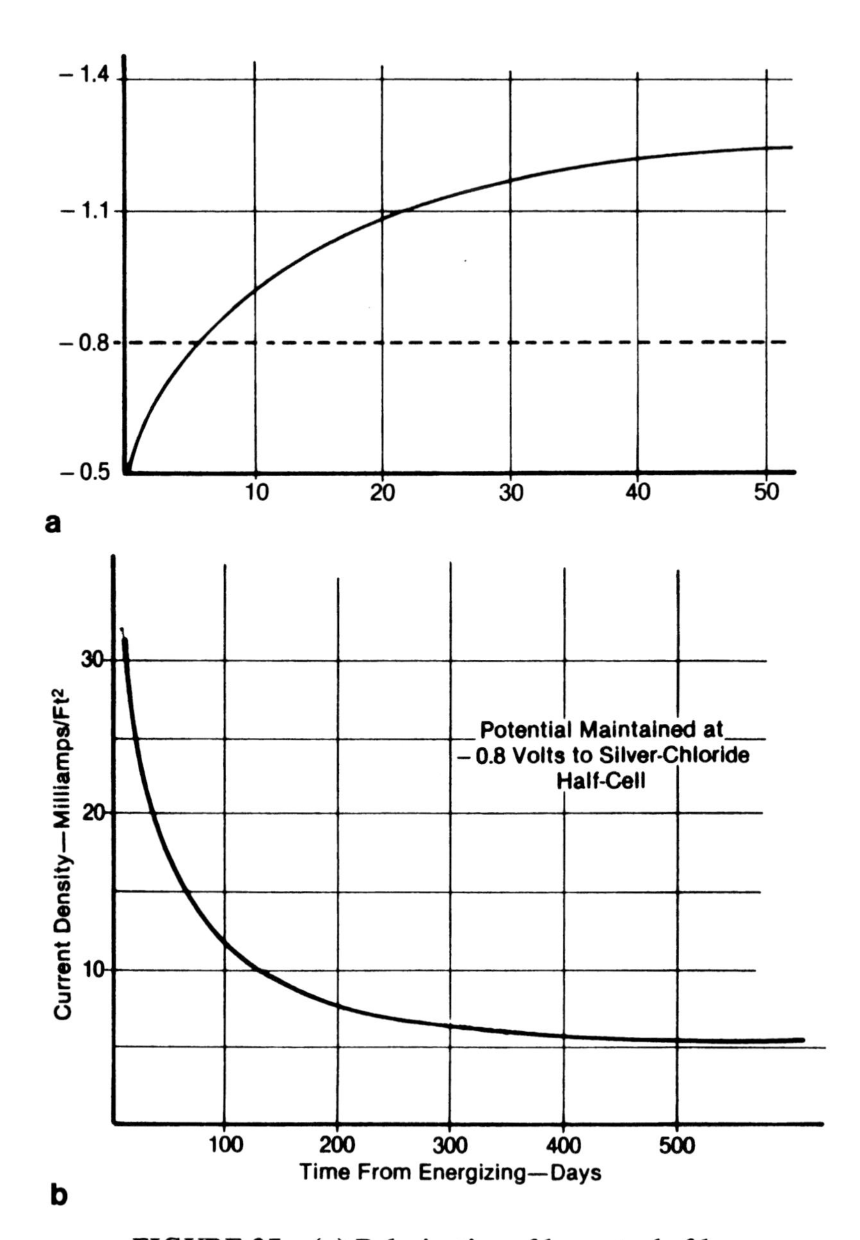

FIGURE 27 - (a) Polarization of bare steel of bare steel in slowly moving sea water. (b) Current density required to maintain polarization of 0.8V to silver chloride half cell.

dependent on velocity gradient rather than on velocity itself. The presence of surface coating on the cathode be it a corrosion product, a plated film of calcareous salts or paint will greatly reduce the current density required and its susceptibility to electrolyte velocity.

Steel in fresh water generally requires a lower current density of 1 to 3 mA per sq ft. The current demand is increased in hot fresh and hot sea water, and at 70 °C this about double that at 15 °C.

$$\sigma_T = \sigma_{15} \left(1 + \frac{T-15}{255}\right)$$

Present day practice in pipe laying is to use a highly insulating coating such as tapes or coal tar enamel. There are, however, many millions of square feet of bare steel buried in the ground. In aerobic soils this requires 1 to 3 mA/sq ft at which polarization occurs and reduces the on-going current density.

Under anaerobic conditions 0.5 to 3 mA/sq ft may be required depending upon the presence of sulfate-reducing bacteria. Excessive currents may be required to achieve the enhanced potential criterion in their presence but generally three to six months at 3 mA/sq ft will bring about polarization and the protective potential will be achieved. This delay seems to depend upon the bacterial activity and the total charge passed per sq ft of surface area. High current density, while causing the pipe to exhibit protective potentials, will not cause immediate sustained polarization. Steel will require a slightly higher current density in a low resistivity soil than in a high resistivity soil. Coatings on steel may reduce the required current density to 0.01 mA/sq ft and polythene tape, plastic extrusions and powder fusion coatings to even lower values. At about these values of coating resistance handling damage becomes significant and without the care comparable with that found in the laboratory lower current densities are not attainable.

Reinforcing bars in concrete are usually bare steel often with a keying indentation. Where the concrete is in good condition its alkalinity passivates the steel and corrosion is not a problem. In many areas carbonates and chlorides either naturally or by use of de-icing salt degrade the concrete and the steel rebar corrodes. Cathodic protection will prevent this corrosion and where the concrete is buried or immersed the current density is similar to that found with polarized steel. Where the structure is free standing the concrete is the electrolyte and can be treated as an extension of protection in soil. Both these applications are discussed as special cases in the appropriate chapters, together with the effects of coating and galvanizing the rebar.

Galvanizing and aluminizing reduce the required current density to about one third that of bare steel and polarization to more negative values is easier.

Non-ferrous Metals

Aluminum can be protected at current densities of 0.2 to 3 mA per sq ft, with the environment and the particular alloy influencing these figures. Heavily anodized aluminum requires 0.002 mA per sq ft to 0.01 mA per sq ft and from the results of limited tests this current density does not appear to destroy the anodic film.

With lead it is similarly difficult to predict the current density required, but between 0.5 and 5 mA per sq ft are usually required for protection. Lead cables are perhaps the main field for this type of work and here the current density often depends on the duct conditions.

Determination of Current Density

The above is an outline of the order of current density required to achieve protection; on many installations comparable experience will allow more reasonable estimates to be made. The criterion of protection is an absolute or swing potential and although this potential is realized by causing current to flow on to the protected structure, it is difficult to make a closer correlation.

It is possible to provide practical values by current density determinations. The absolute value of the current density may be determined by measuring the current required to protect a small uniform surface or the practical, or mean value, over the whole surface may be determined. The latter will include the often unavoidable over-protection near to the anodes or at promontories of the structure. Either of these determinations can be made by the simple expedient of applying a temporary source of cathodic protection and measuring the current required to achieve the desired potential. If an isolated section is selected, then it is possible to determine the absolute value at that point, while the value for the whole structure is readily obtained. This method, called a current drainage survey, is frequently used in cathodic protection engineering and a fuller description is included in the chapters concerned with the design of practical installations.

Swing Tests

Many structures exhibit an almost ohmic relationship between the applied current and the potential. This is either because of a particular polarization characteristic or because the potential is controlled by an ohmic coating. Where the current required to protect such a structure is low, below about 10 Amps, then a reliable estimate of this current can be made using a four terminal resistivity meter. In this type of instrument a current is caused to flow in the circuit connected to terminals C_1 and C_2 and

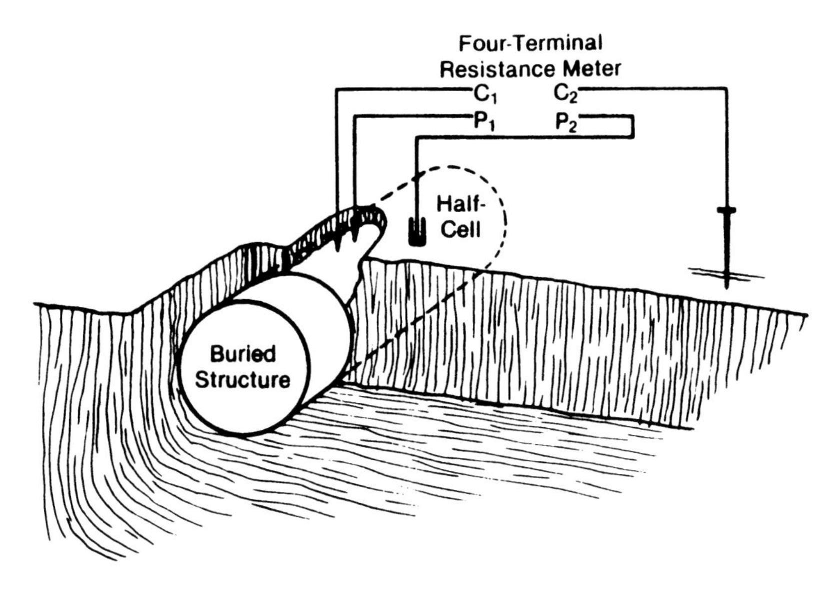

FIGURE 28 - Four pin resistance meter used to determine current required for cathodic protection.

the potential developed across terminals P_1 and P_2 is compared with it; this is generally indicated as ohms or mhos. To measure the current required for cathodic protection, the terminals C_1 and P_1 are separately connected to the structure, Fig. 28, and C_2 is connected to an earthing rod placed at the proposed cathodic protection anode location. P_2 is connected to a metal rod used in place of the half cell. The reading in ohms can now be interpreted by assuming the potential swing, for steel about 300 mV, required to achieve protection or this swing can be calculated from the known potential of the structure. Thus, if the resistance indicated is 0.03 ohms then 10 Amps will cause a 300 mV swing. This is about the maximum accuracy that can be obtained in the field with a four terminal resistance meter. The accuracy of the method is reasonable and it can be used to determine either the mean or absolute values of current density. It can also be used to show the effect of variously positioning both the anodes and the measuring electrodes.

Other Field Methods

Two absolute methods of determining protection have been described which do not rely upon any assumed criteria of protection. The first is generally attributed to Pearson and Ewing who introduced it as a practical method for the corrosion engineer. It had been observed in the laboratory that there was a change in the polarization curve of a piece of steel in a potassium chloride solution at a current just sufficient to protect the steel. The validity of this was tested in several solutions and it was found that if the potential was plotted against the log of the current the polarization curve resolved itself into two straight lines which intersected at the current required for protection.

The original work and most of that subsequently performed in the laboratory was in electrolytes of exceptionally low resistivity. Ewing realized the significance of this and attempted to measure the potential at the interface by calculating the ohmic drop through the electrolyte. Pearson devised an improved measuring technique and was able to demonstrate the break in the polarization curve in the field. Ewing suggested the break, or alternatively the point of intersection of the projection of the two straight sections of the plot of potential against the logarithm of the current indicates the current required for complete protection. This was found to be so in the laboratory where the electrolyte was of exceptionally low resistivity and very uniform, but in the field the soil will almost invariably be heterogeneous and of high resistivity, so this criterion will not apply.

The metal surface under field conditions will be far from equipotential, some areas being more positive and others more negative than the mean potential. On the application of the cathodic protection current, these areas will each, individually, act in a similar manner to the laboratory specimens, though each will start at its own potential and at some value of applied current a break in the potential/logarithm of current plot will occur.

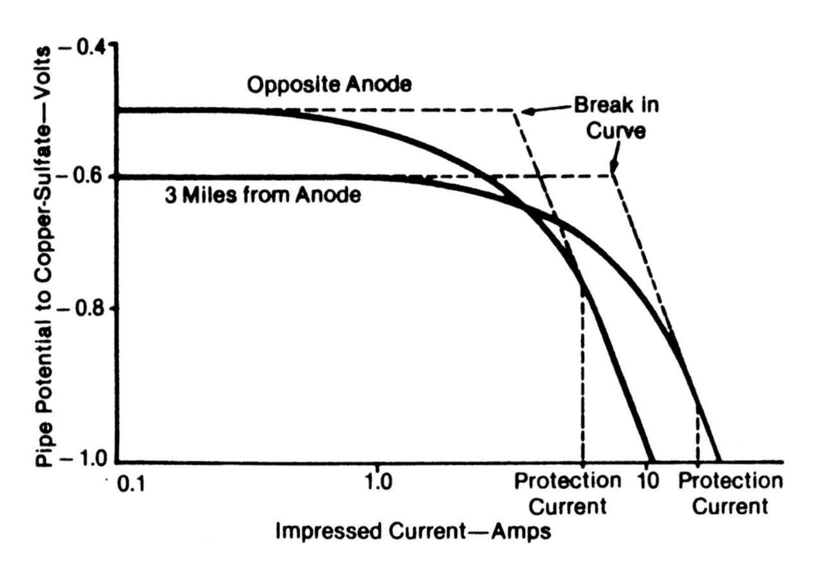

FIGURE 29 - Plot of potential against logarithm of current (E v log i) to show Ewing-Pearson criterion.

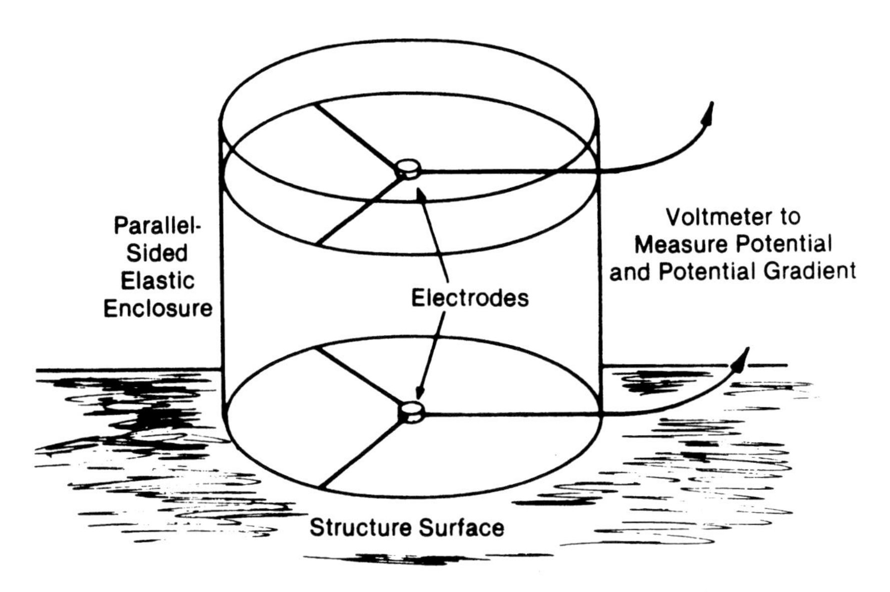

FIGURE 30 - Enclosure with twin electrodes to measure potential and potential gradient, hence current density onto surfaces.

The cathodic current will divide between the various small cathodic areas and this division will depend upon their relative potentials and the distribution of the current caused by the resistivity of the electrolyte, the cathode shape and the cathode resistance.

The combined effect of these many small areas will produce the type of curve found in a field determination. If the individual potential of each area could be plotted against the total applied current, then each would give a broken curve, though the 'flat' section of this would be at different potentials and the break would occur at different values of current. If the combined potential is measured—as indeed it is in practice—then initially the curve will be straight, and at the mean structure potential. As the current is increased this will be depressed as the individual areas become polarized. The process will continue until each of the areas is polarized, when the curve will again become a straight line.

Complete protection will occur when all the areas are polarized so that the current needed for protection will be that which causes the potential v logarithm of current plot to become a straight sloping line. As this will occur when the most negative, or anodic, areas are polarized, then the criterion will be very similar to that derived by considering the parameters of the electrical circuit. The potential measured by the two methods will not necessarily be the same, even if they indicate the same current requirement, as these potentials may be measured relative to differently placed electrodes.

Accepting this revised interpretation of the graph, the method has a great deal to commend it and doubtless is a powerful tool when new or different circumstances suggest a modified protection criterion. In the field the method suffers from two defects: firstly the large number of skilled measurements that have to be taken even to make the determination at a single point; and secondly, the method only works where a section of the structure receiving a current from a single source can be isolated. Thus, on a pipeline the point most remote from the cathodic protection should be subject to this type of test, but where this point falls between two installations the tests would have to be arranged by varying the current from both stations in the anticipated ratio that they will ideally have in practice.

Fig. 29 shows typical plots of installation close to and remote from the anode and how they will appear in practice. As can be seen, the interpretation of the break in the curve is comparatively easy, whereas the determination of the point where the practical curve deviates from a straight line is more difficult, particularly as the current, the factor that is to be determined, is on a log scale.

The introduction of electronic simple computing will revive interest in this technique, though it will still require considerable elegance and experience in the field in selecting the sites for the cathodic protection tests and in placing the reference electrodes.

A number of different types of enclosures can be used in sea water to determine the E v Log i curve break at any particular point. The ideal method seems to be one in which a sector of the electrolyte is shielded so that both the surface potential and the current density, by a measurement of the potential gradient, are determined. Such a unit is shown in Fig. 30 and it has the advantage of an open structure so that the electrolyte can flow freely next to the cathode. In using this technique the curve that results will either be the one obtained before polarization or it can be used where the structure is believed to be close to the protection potential and then the results will indicate the current density required after that period of operation.

The second series of methods of determining the minimum current required for cathodic protection is based on attempts to measure the net flow of current on to or off the metal surface. In a long pipeline, for example, it would be possible to determine that the current flowing in the pipe (towards the cathode return) is always increasing, showing that at all points there is a net gain of current and the pipe is cathodic. This might be difficult to interpret where the cause of corrosion is current flowing between top and bottom of the pipe. It is a most useful tool in surveying deep well protection.

Other techniques involve placing sample specimens isolated from the main structure and determining when there is a net cathodic flow. There are techniques in which simulated surfaces are arranged with no resistive components and these can either be used in a determination of the current density or to perform the potential v log current-break curve at that particular point, or instant-off criteria.

Experimental Techniques

The current density and potential criterion and the efficacy of the cathodic protection of a metal in a particular environment can be determined by laboratory experiment. Small samples of the metal can be placed in the electrolyte and the minimum current density required for protection can be determined. These results can be transposed to the field, though this has to be done with due regard to the change in dimensions.

In field trials, as opposed to the laboratory, the structure itself can be used and various parts of it subject to different degrees of protection. This can be done by techniques which cause different areas of the structure to receive different degrees of protection. For example, areas close to the anode may have a much larger potential swing than areas further away. Similarly, plastic insulating sheets can be used in certain electrolytes to cause a variation of potential in the electrolyte. The results of inspection will give the correct potential criteria in the field.

Alternatively metal coupons can be attached to the structure placed in close contact with and electrically bonded to the structure, either directly or through a zero potential measuring circuit. The coupons themselves can be accurately and rapidly monitored and they can be used to assess both the current density and the potential criteria. Care has to be taken that the coupon exactly matches the main cathode conditions, not receiving preferential or diminished protection. From these coupon tests the cathodic protection current density can be determined. This type of work has been carried out continuously for many years and in most environments the criteria are well known or a sensible interpolation can be made from the literature. In other cases, particularly where there is a bad history of corrosion, the application of cathodic protection—be it adequate or not—will reduce the rate of corrosion and from this practical application the best or sufficient criterion will be determined. There will be a reluctance to reduce the protection below the level that has been found to be adequate and so the criterion may be one which has shown itself to be sufficient though not necessarily the minimum.

Partial Protection

Sometimes it is found that rapid corrosion is taking place over a few limited zones. This is particularly true of structures that are exposed to a variety of electrolytes or conditions. For example, a pipeline may pass on part of its route through a highly corrosive area. In these cases it is often economical to apply local protection only to those areas worse affected. Protection is effective over this area and there is usually some negative swing of potential over the rest of the structure.

From the elementary circuit diagram of a corrosion cell, it can be seen that cell current can be reduced by a very much smaller external current than that required to make it zero. Under particular types of polarization, there will be a considerable reduction in the corrosion at only 25 per cent of the current density normally required for complete protection. This type of partial protection has found great use; it can often be justified economically where complete protection may not be necessary and an extension of the life of the structure all that is required. Equally, partial protection can achieve only a limited effect and in a system such as an electric cable, where one sheath penetration is sufficient to give complete breakdown, it would be a false economy. Partial protection by a change of potential of, say, 100 mV may be disastrous in the presence of sulfate reducing bacteria where a small cathodic change may stimulate their activity. It also is unwise to rely on partial protection where stress or other accelerating influences are present.

Protective Coatings

As the criterion of cathodic protection is a potential change, then the current density required to achieve this will depend upon the resistive state of the metal/electrolyte interface. This resistance can be greatly increased by coating the surface with an insulating coating such as a paint.

In a typical soil a pipeline may require 2 mA per sq ft to protect it, whereas a similar pipe, coated by modern techniques, would require only 20 micro amps per sq ft or 1 per cent of the bare steel value. This leads to a tremendous reduction in the cathodic protection components required, and in the case of a pipeline, greatly simplifies the engineering.

The current density that is required by a pipe covered with a modern coating will be the sum of two components, firstly that required to overcome the purely ohmic drop through the coating substance, and secondly, that required to polarize any pin holes, flaws or damaged areas. If no cathodic protection were applied the areas of damage would become the anodes and the coated areas the cathodes. The overall corrosion of the metal structure would be less, but the concentration of the corrosion current at the anodes would lead to deep pits and, where this is critical, corrosion failure. As a protective measure, this type of coating can cause more rapid failure than would occur in its absence. On the other hand, using such a coating will greatly reduce the cathodic protection costs. The total cost of the combined coating and cathodic protection may be more or less than that of the cathodic protection alone, according to the difficulty and cost of coating the structure. In most cases the minimum cost will be found where cathodic protection is applied as a complement to, not the substitute for, a coating.

The electrochemical reactions that take place at the cathode surface and the electrical circuit control of the cathodic protection scheme suggests a series of properties which a successful coating must possess. There will be an accumulation of alkali close to the cathode, so it is essential that the coating is not attacked by alkalies. When protection is just achieved a pH of about 9 will be found at the cathode but where some over-protection is inevitable and higher values of 10, 11 or 12 are possible.

The electrical potential which is established across the coating is capable of transmitting moisture by electro-osmosis. This phenomenon occurs in a variety of substances and the water generally travels in the same direction as the conventional current. The use of unsuitable fillers can greatly increase the rate of electro-osmotic penetration, particularly if the fillers are not well dispersed and some areas are over-filled. Water that travels through the coating can form bubbles behind the coating with breakdown of the coating to metal adhesion. Corrosion will occur and the cathodic protection from the outside will be of no effect in preventing it. A good coating to metal bond is essential, with coatings that are susceptible to electro-osmosis.

Electrically, it is essential that the coating maintains its high insulating property, otherwise constant adjustments of the electric generator and frequent detailed inspections must be made. Water absorption, whether by the influence of alkali, by electro-osmosis or purely as a property of the material, will reduce its electrical resistance.

The majority of coatings used on structures that are buried or immersed in water are organic and many of these are hot applied or tested for pin holes by high voltage breakdown. These techniques, unless carefully controlled, can burn the coating, causing the formation of a coke. As this will be in contact with the metal the cell so formed, metal-electrolyte-coke, can be very aggressive. This condition cannot be successfully be suppressed by normal cathodic protection techniques.

Pipe Coatings

The pipeline engineer has to use a coating that will stand up to soil stress found in the ground and one which will not be destroyed by contact with and movement through the earth. Most coatings rely on bulk for this type of protection, often with an outer shield wrap in bad conditions.

The earliest types of protectives were greases; these and petroleum based jellies and wases are still used though now reinforced with wrappers and tapes which often hold inhibitors. The electrical resistance of these is not high and about five per cent of the bare-steel current is required. Some

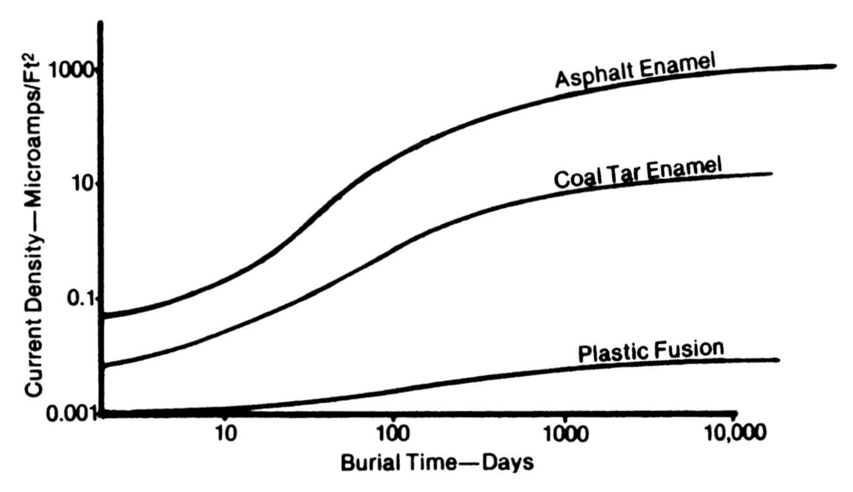

FIGURE 31 - Current density required to protect coated steel plotted against time.

of them have a tendency to absorb moisture and their resistance drops with time.

On long pipelines for site application over the trenches, asphalt, a product of the petroleum industry, coal tar enamels and plastic tapes are used. Asphalt is often filled with an inert powder and wrapped in an outer shield which protects it from damage. The coating may vary from 1/32'' minimum to a half inch or more. Asphalt has a high water absorption, 6 to 7 g per sq ft in 100 days, while a well formulated asphaltic enamel might absorb 2 g per sq ft in the same time. With this type of coating its water absorption can be considerably increased by electro-osmosis.

Coal-tar enamels provide a superior coating to the asphalts. Carefully applied, and this is essential, the right enamel reinforced with woven or felted glass fibers will have a high electrical resistance that will remain nearly constant with time. At low voltages, less than 1 volt negative swing on most practical coatings, osmosis is negligible, and even at higher swings of up to 2 V it is low. The current density required to provide protection to a metal surface covered with a well applied coal-tar enamel is shown in Fig. 31. Coal-tar enamel however, tends to have a very narrow working temperature range, softening above it and embrittling below it. Additions to the enamel to widen this range can be made with resins which do not significantly reduce its electrical properties.

Plastic tapes have been developed to a point where they provide excellent electrical properties, low water absorption, and only a small tendency to electro-osmosis. Their use is increasing and the early difficulties with adhesives and plasticizers causing delamination at the tape overlap have been developed out.

The extension of the plastic tape is a complete plastic coating which can be extruded or otherwise bonded to the pipe. This is usually factory or mill applied.

The fusion bonding of organic plastic coatings is a new development that is providing the equivalent of the best yard applied coaltar enamels. The film is thin and, although tough has to be handled with some care. Overwrapping is used to prevent soil stress damage.

Under-Water Paints

Paints used to protect marine structures have to stand the same electrical conditions as do coatings. The other properties required are slightly different. The coating used on submarine pipes will be similar to that used on land-based installations except that the pipe coating will be overlaid with a weight coating. This is usually one or two inches of concrete reinforced with a wire mesh. The pipeline will be constructed on a barge or on the shoreline by welding together short lengths of pipe. These may be single pipe lengths or two, three or four lengths previously welded together. Each individual length will have been coated, both with the anti-corrosive composition and with the concrete weight coat. The welded joint then has to be protected both against corrosion and mechanically. Where the pipe is constructed onshore and towed out to sea there will be time to make this joint by the same techniques that are used on the general length of pipe or even to set up a pipe mill and continuously coat the pipe before long lengths are floated out to sea. On the lay barge there will be very little time to make the joint, coat it and protect it. A number of techniques have been developed. The earliest was the use of an overfilled asphaltic mastic which was hot-applied to the joint to be flush with the concrete weight coating surface. The mastic provided only a low degree of corrosion control and principally reduced the amount of cathodic protection required at the joint by reducing the rate of arrival of oxygen at the pipe surface and preventing the dispersion of the inhibiting catholyte.

More recently the joint has been coated beneath the asphalt mastic, usually with a plastic tape. Coaltar enamels have been applied to the joint and mechanically protected by concrete or by resin/ballast mixes.

When a ship's hull is cathodically protected the paint will be required to stand higher potential variations in the comparatively small area close to impressed current anodes. The hull paint must remain smooth under the influence of cathodic protection and particularly not be roughened by bubbling. There will be an increase in the current required when the ship is underway. The increase in the cathode reaction will be compensated by the action of the moving water washing away any alkali which might otherwise accumulate. There is evidence that cathodic protection has adversely affected paint adhesion over certain primers. This technology is now well understood and the latest marine paints are capable of maintaining a smooth, pristine finish in the presence of cathodic protection over a wide range of potentials. Thick, self-polishing coatings restore hull smoothness when applied over microrough surfaces and retain this on a fully protected steel hull.

In many seawater applications the cathodic protection will have to supply massive amounts of current for long periods of time to protect a bare structure. There will also be problems with the distribution of this current within complex structures. Under these circumstances a coating—whether over the whole of the structure or only over part of it—that reduces the overall current requirement by a factor or two or three may be sufficient to make considerable difference to the cathodic protection engineering. A simple coating technique that could economically be applied to this type of structure and cause a reduction in current requirement would find a great deal of use on large sea-water structures.

Resistivity and Electrode Resistance

CHAPTER 3

Electrical Resistivity

Units and Magnitudes

One of the most important parameters in corrosion and cathodic protection is the electrical resistivity of the electrolyte: this is a property which is defined by measuring the resistance between the opposite faces of a specific cube of the material. The usual units are based on the ohm per cm cube and the ohm per ft cube, the former being the one most frequently adopted for corrosion measurements. The resistance of the volume of electrolyte will increase if the cube is distorted so that the distance between the measuring faces is increased and will decrease if the area of these faces is increased; thus the unit should be described as the ohm cm^2/cm or ohm cm. An ohm meter would be the equivalent of 100 ohm cm and an ohm ft the equivalent of 30.5 ohm cm. This property of the electrolyte is intrinsic; that is, it depends entirely upon the substance and not upon its dimensions; other intrinsic properties are color, temperature and potential.

The common electrolytes vary considerably in resistivity from sea water at 20 to 30 ohm cm to granite rock at 500,000 ohm cm. Water varies almost as much as any electrolyte. Pure water, or as near to it as can be obtained, has a resistivity of 20,000,000 ohm cm while the distilled water that is obtainable in most laboratories has a resistivity of about 500,000 ohm cm. Rain water that has collected in lakes, and melted ice, give resistivities of the order 20,000 ohm cm, though tap water varies from 1,000 to 5,000 ohm cm. If pure water, or near it, is contaminated with small quantities of salts, the change in conductivity, the inverse of resistivity, is as shown in Fig. 32: this indicates that a few parts per million of some salts greatly reduce the resistivity of pure water. At the other end of this curve lie the various estuary waters and sea waters. The Thames in the Pool of London has a resistivity of 200 ohm cm which varies with the state of the tide and

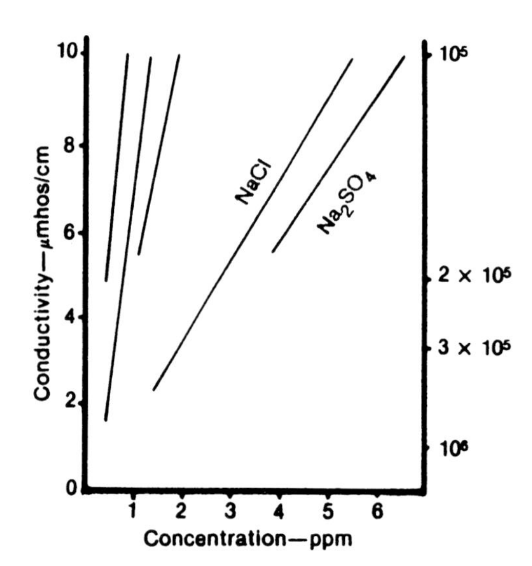

FIGURE 32 - Conductivity of water as a function of added salts.

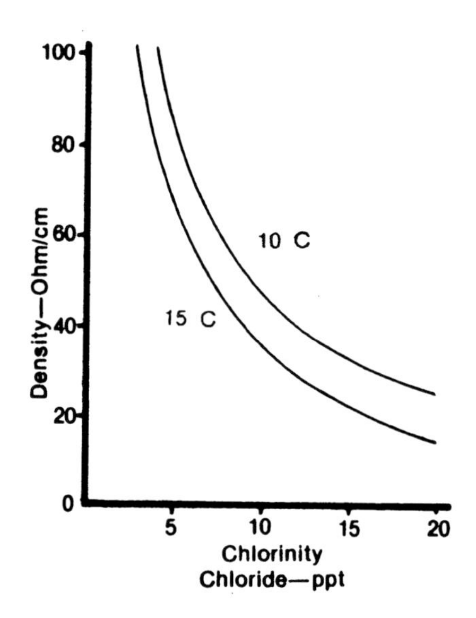

FIGURE 33 - Resistivity of seawater as a function of chlorinity.

the flow from the upper reaches. Sea water in an estuary mouth has a resistivity which varies about 30 ohm cm, while in the open sea the resistivity is as low as 20 to 25 ohm cm: this is the lowest resistivity bulk electrolyte met in nature. Solutions of higher concentration give resistivities of 1 ohm cm but are rare. The variation of resistivity with chlorinity for sea water is shown in Fig. 33. Ocean sea water has a chlorinity of 19 parts per thousand and its resistivity varies from 16 ohm cm in the tropics to 35 ohm cm in the Arctic regions.

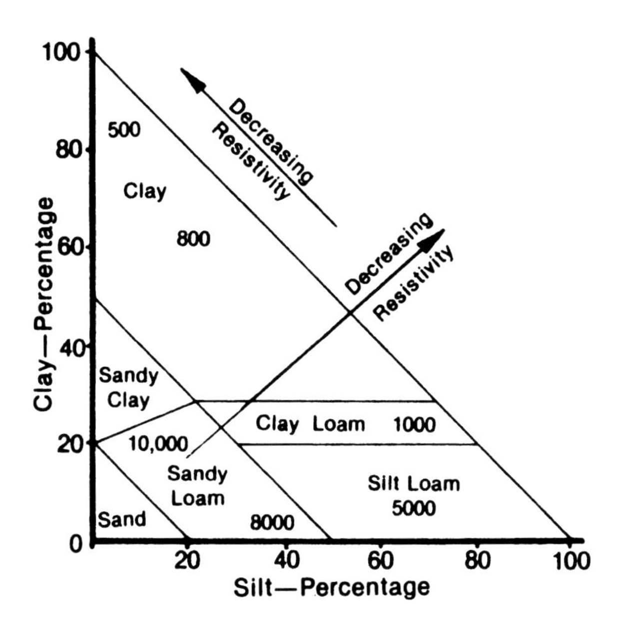

FIGURE 34 - Whitney diagram for soils with typical resistivities.

While sea water is a uniform electrolyte, the soil and rocks present a highly heterogeneous structure. Clays may have a resistivity below 1,000 ohm cm while clean gravel will have over 100,000 ohm cm resistivity. The soils can be divided by their content of silt and clay, sand being the remainder. The subdivision is illustrated in Fig. 34, which is the Whitney diagram for soils. Typical values of resistivity for these soils are given and the trends in resistivity are shown. Gravels and fine gravels with particles above 1 mm dia are not shown. The resistivity of these will be high, depending upon the amount of inter-particle filling and the resistivity of any included water.

The resistivity of the soil varies greatly with the water content and the resistivity of the contained water. Fig. 35 and 36 illustrate these properties. The resistivity of rocks is generally high depending, in the case of porous rocks, upon the water content. The formula for the resistivity of a porous rock is

$$\rho_R = \rho_w \frac{1}{\phi^2} \times \frac{1}{S_w^2} \qquad (3.1)$$

where ρ_R and ρ_w are the resistivities of the rock and included water; ϕ the rock porosity and S_w the water saturation.

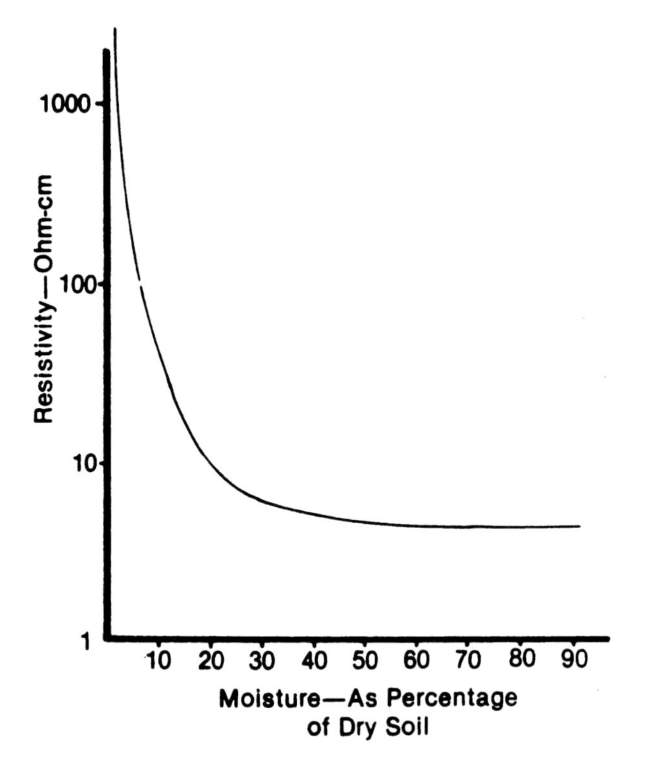

FIGURE 35 - Resistivity of soil as a function of moisture content.

Table IV shows the resistivities to be expected with various geological formations. Fig. 37 shows the change in sea water resistivity with temperature due to Crennell; similar variations take place in the soil. There is a marked rise in resistivity on freezing and this has been studied both in America and the Soviet Union. The results, correlated to air temperature, are shown in Fig. 38, which is the data supplied by Logan; he suggests that the resistivity of the unfrozen soil may be given by the equation

$$\rho_t = \rho_{15.5} \frac{40}{24.5 + t} \tag{3.2}$$

where ρ_t is the resistivity at temperature t °C.

Because of the dependence of soil resistivity upon these criteria there is considerable variation of resistivity with the seasons. The maximum resistance recorded through a 5 feet deep earthing system is 100 per cent greater than the minimum resistance, the highest resistance occurring in the winter

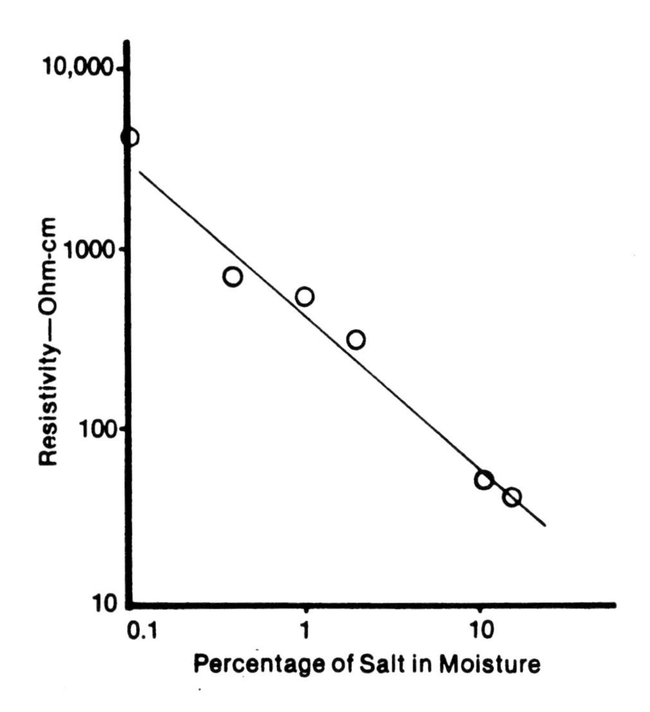

FIGURE 36 - Resistivity of soil against salt content of moisture.

TABLE IV

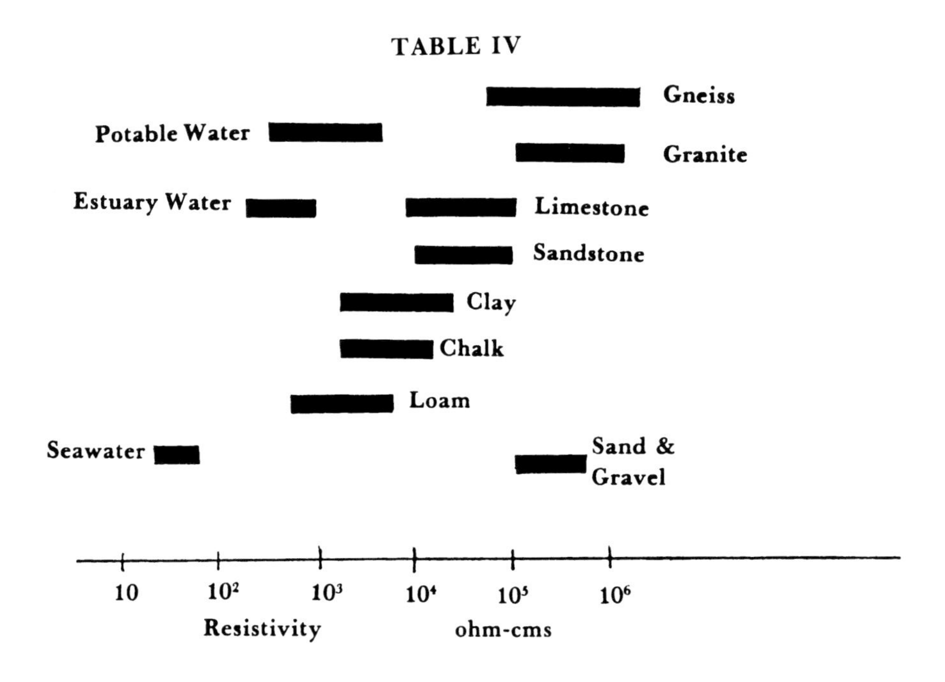

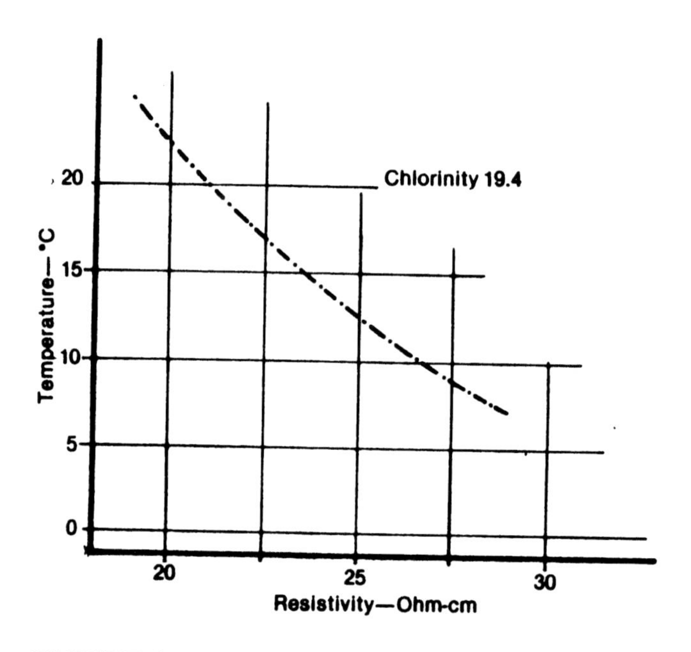

FIGURE 37 - Change in seawater resistivity with temperature.

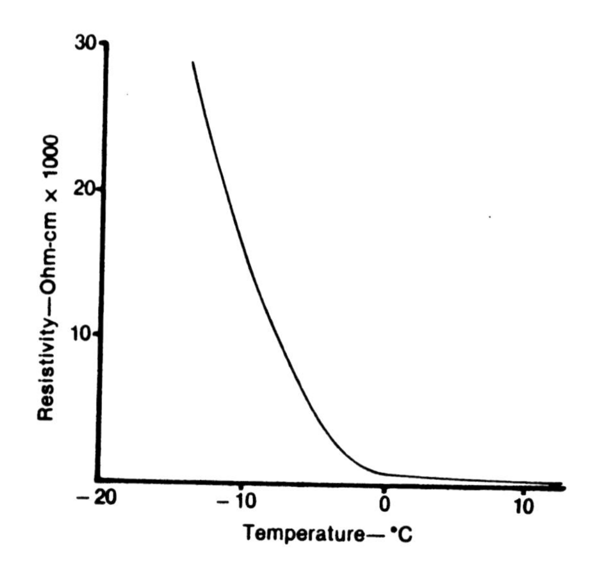

FIGURE 38 - Soil resistivity against temperature.

and the lowest in the summer. Measurements nearer the surface would indicate a very high resistance during freezing and during dry summers while the resistance of a deeply buried earth will depend upon the movement of salt-bearing water upwards and downwards through the soil. Variations of resistivity of 25 per cent and greater occur between the seasons in successive years.

Salt additions to the ground in the vicinity of electrodes show some tendency to stabilize these effects and the annual variation is greatly reduced.

Measurement

The resistivity of an electrolyte has already been defined as the resistance measured between the opposite faces of a cube of the material. To measure the resistivity of an electrolyte the resistance between the opposite faces of a cubic sample could be measured, and if the resistance is expressed in ohms and the cube has a 1 cm side, then the resistivity in ohm cm will be numerically the same as the resistance. As the resistance depends not upon the shape of the measuring faces but merely upon their area, then the resistance of a prism of any section and length could be used to derive the resistivity of the sample. The technique of measuring the resistance associated with a certain geometry of the electrolyte is the principal method used in practical determinations.

The simplest version is the soil box or tube, Fig. 39a; this is a plastic or other insulating box or tube, filled with the electrolyte and with two electrodes mounted at the ends. Current is made to flow between these and this causes a potential drop along the tube or box. If the box is sufficiently long—four or five times the maximum dimension of the cross section—then there will be a uniform distribution of current for a considerable distance near the center. The potential gradient in this area and its relation to the current are sufficient to determine the resistance per unit length and from the soil box dimensions the resistivity can be calculated.

A sampling auger model of this type of instrument is illustrated in Fig. 39(b). If the cross sectional area in sq cm of this is made equal to the length in cm between the potential measuring rings; the resistivity in ohm cm is then equal to the measured resistance in ohms. The resistance measurement can be made with a c or d c current and can either be read directly or calculated from the current and potential difference.

Where waters are the only electrolyte of concern then, using a c methods, a cell can be constructed of two electrodes at which both current and voltage are measured, there is a great variety of these measuring devices, or cells, each of which has refinements or methods of obtaining a geometric factor, particularly suited to the range of work that they perform: in some, temperature corrections are made automatically.

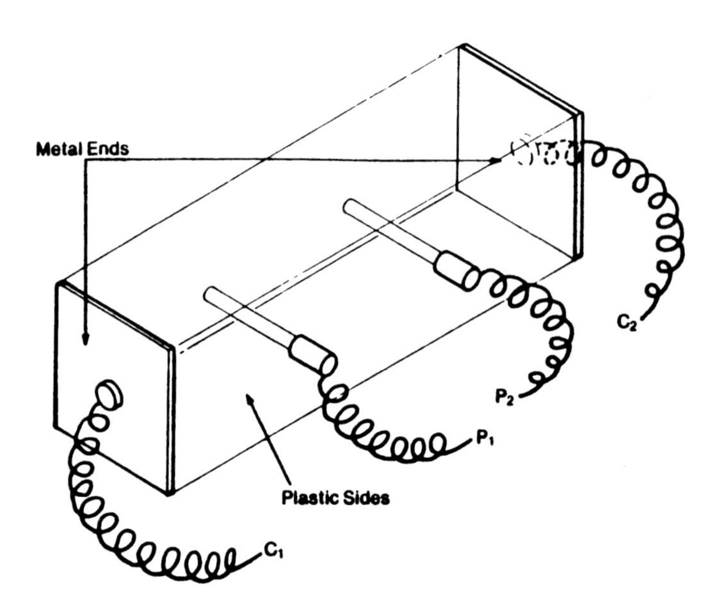

FIGURE 39(a) - Soil box for determining resistivity.

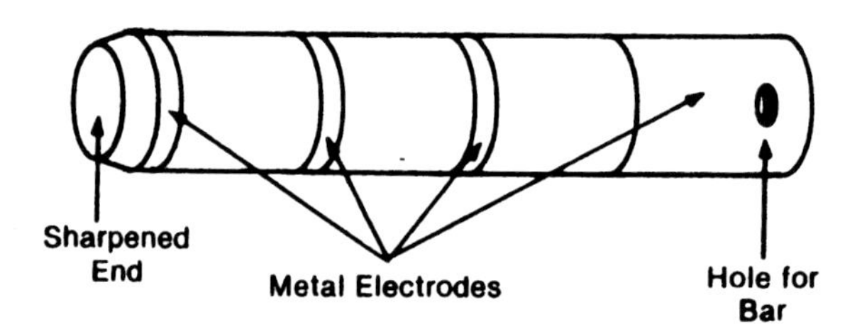

FIGURE 39(b) - Soil auger variation.

While the soil-box type of instrument is convenient, particularly for sample analysis, it is often necessary to measure the resistivity of the electrolyte in situ, particularly if the area is heterogeneous, when sampling techniques would be very difficult. Also the soil-box restricts the sample size, whereas a measurement taken in the ground can encompass a greater volume of electrolyte and a mean value can be obtained.

The simplest method of measuring the soil resistivity consists of inserting two metal rods for a specific distance into the ground at a set spacing and measuring the resistance between them. If the rods are replaced by metal tips on the end of a pair of insulated rods and they are spaced a considerable distance apart, then on insertion to a depth of 10 to 12 diameters into the ground the resistance measured will be a specific multiple of the resistivity. Such an instrument has been designed by Shepherd and is referred to as Shepherd Canes. The two electrodes are cones of about 1 in. dia base; one, the cathode, with a very acute, 20°, apex and the other, the anode, with an obtuse, 120°, apex. Direct current is caused to flow from a constant voltage battery and this current indicates the conductance of the circuit from which the instrument may be calibrated to read ohm cm directly. Polarization occurs and quick readings are necessary. The rods integrate the mean resistivity over a volume of about 1 cu ft in their close vicinity.

Other single, or walking-stick, probes have been used with a c measuring devices. The probes are made either with twin metal rings, metal tip and ring or tip and rod electrodes, the latter having two electrodes, one extremely large and the other much smaller which controls the resistance of the circuit. One instrument in which a bimetallic probe is used to provide the driving potential is particularly ineffective in determining the resistivity, as neither is constant voltage achieved nor uniform resistance measured as the corrosion product and polarization films have a large effect.

Wenner Method

The most useful method of measuring soil resistivity is that ascribed to Wenner and called the Wenner or four-pin method. In this four metal rods are driven into the ground equally spaced along a straight line as in Fig. 40.

Current is caused to flow between the outer pair and the potential developed between the inner pair is measured. The rods should be only driven a small distance (0.05a) into the ground and if the electrolyte is assumed to be uniform, then its resistivity is given by the equation

$$\rho = 2\pi a R \tag{3.3}$$

where R is the ratio of volts to amps in ohms, 'a' the inter-electrode spacing in centimeters, and ρ the resistivity in ohm cm. The derivation of this formula is simple: the potential of point P_1 relative to P_2 due to the current flowing from C_1 is

$$\frac{i\rho}{2\pi a} - \frac{i\rho}{2\pi \cdot 2a} = \frac{i\rho}{4\pi a}$$

where i is the current.

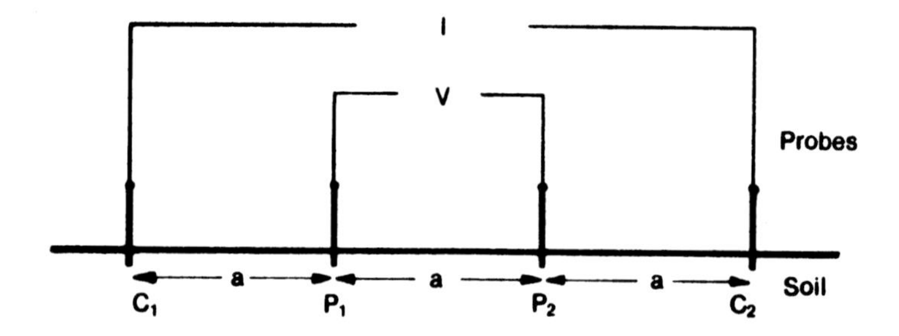

FIGURE 40 - Four-pin or Wenner method of determining resistivity.

Similarly the potential of P_1 relative to P_2 by virtue of the current in C_2

$$-\frac{i\rho}{2\pi \bullet 2a} + \frac{i\rho}{2\pi a} < \frac{i\rho}{4\pi a}$$

therefore, the total potential difference between P_1 and P_2 is

$$\frac{2i\rho}{4\pi a} = \frac{i\rho}{2\pi a}$$

and the value of R, the ratio of this potential difference to the current is

$$\frac{i\rho}{2\pi a} \bullet \frac{1}{i} = R \text{ or } \rho = 2\pi a R$$

The method sums the resistivity to a depth of approximately 'a' and so, by changing 'a,' samples of various sizes can be included in the measurement. Although the formula was derived for a homogeneous electrolyte, it is usual to use this method to measure the apparent resistivity under any circumstances. The apparent resistivity will vary with changes in 'a' and with changes in the location of the four pins. The simplest case of variation will occur when there are two layers of different resistivity with a horizontal discontinuity.

Suppose the top layer depth 'd' has a resistivity ρ_1 and the lower layer infinitely thick has a resistivity ρ_2. Then the reading when 'a' became vanishingly small would give the apparent resistivity ρ_a equal to the resistivity ρ_1, while at an infinite spacing ρ_a would approach ρ_2. If this resistivity model is considered and values of ρ_a are plotted against the separation 'a,' then a certain pattern will emerge which will depend on 'd,' ρ_1 and ρ_2. The ratio ρ_1 to ρ_2 will determine the shape of the curve while the

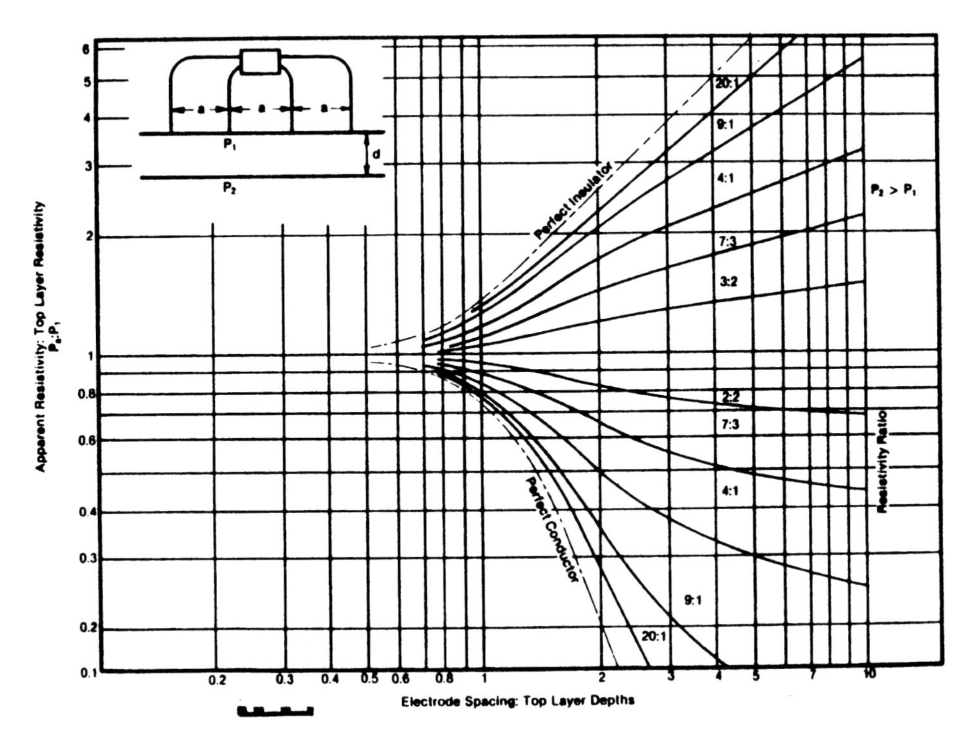

FIGURE 41 - Apparent resistivity found by Wenner method in two-layer system.

absolute values of ρ_1 and d will determine the size, that is a plot of ρ_a against 'a' on an enlarged model with d being 'nd' but with ρ_1 and ρ_2 being the same will look the same as the first graph when the electrode spacing corresponding to 'a' is 'na' and to a_1, na_1, etc.

Similarly, if the model is changed by increasing ρ_1 and ρ_2 to $q\rho_1$ and $q\rho_2$ then ρ_a will become $q\rho_a$ where the graphs correspond. If the plot is made on log/log paper, that is, the cartesian co-ordinates become log 'a' and log ρ_a, then the shape and size of the plots which have the same ratio ρ_1 to ρ_2 will be identical but translated from each other. Thus, by moving a tracing of the graph of the practical determination over that of a series of similar curves drawn for different values of the ratio ρ_1 to ρ_2 then when a fit occurs without any rotation of the axis the practical ratio ρ_1 to ρ_2 is found. From the translational movements the actual values of ρ_1 and ρ_2 and of the depth 'd' can be found: this is the principle of resistivity prospecting. Curves for various ρ_1 to ρ_2 ratios are shown in Fig. 41. By using these curves it is possible by the four-pin method to measure not only the surface resistivity, but also that of lower strata and to map accurately the depth of the change.

It is possible, instead of using the graphs as described, to program a computer to analyze the results and to produce the most probable resistivity configuration of the area. Some engineers may feel that immediate, if not very accurate, plotting in the field is useful as it indicates anomalies that might be expensive to check if a return visit is needed.

A variation of Wenner's method can be used to measure water resistivities. The four pins are replaced by four electrodes connected onto an insulated cable harness as in Fig. 42; this assembly is lowered into the liquid and the resistance measured in the normal manner, C_1 and C_2 being the current electrodes and P_1 and P_2 the potential electrodes; because these are now surrounded by an infinite, and not a semi-infinite electrolyte, the formula becomes

$$\rho = 4\pi aR \tag{3.4}$$

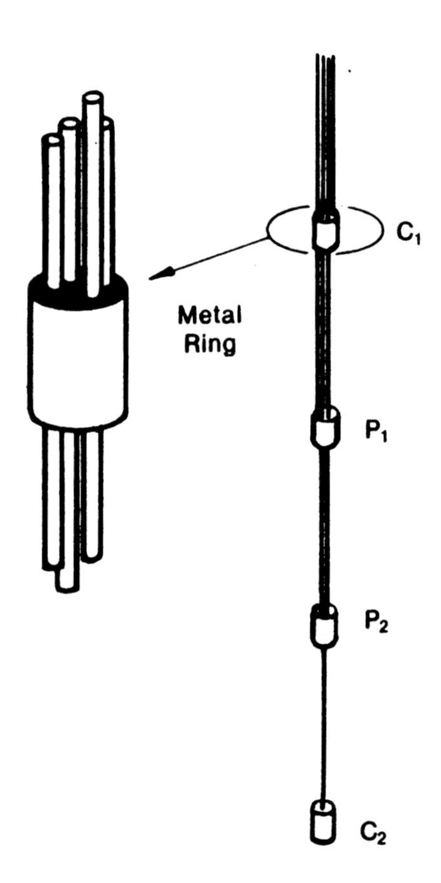

FIGURE 42 - Modification of four-pin method for use in water.

The four-pin method can be varied by moving the electrodes so that the distance between C_1 and C_2 is fixed, and the potential electrodes P_1 and P_2 are placed close together at a constant distance apart on the same straight line. The potential electrodes are moved from the vicinity of C_1 towards C_2, the four being kept in one line. The method is favored for geophysical work but not in corrosion studies. Other variations of the method are possible and simple geometric patterns can be evolved so that an accurate map may be made of the whole of a particular area.

It is often desirable to determine the area of lowest surface resistivity before evaluating the bulk resistivity accurately. This initial survey can be performed by the single or dual probe methods; a simple way is to make a sole plate, for either one or both shoes, which carries the electrodes leaving the hands free to make the measurements. A remote C_1 pin and C_2, P_1 and P_2 pins, carried by the investigator, say attached to the sole plate, gives good results.

The above methods have all involved measurement of resistance, either directly or indirectly, and the correlation of this with a geometric factor to determine the resistivity. It is possible to measure resistivity directly: this can be accomplished by several methods, most of which measure the effect of the conductor upon the linkage between two coils or determine the attenuation that occurs to radio waves; the former principle is employed in mine or pipe locators and these instruments can be used to give an indication of high or low resistivity. Greater accuracy can be obtained under stringently controlled conditions, but the rapid location of the lowest area of resistivity is of sufficient value to warrant the use of a simple instrument of field accuracy.

Inductive Methods

The inductive measurement of resistivity relies on a phenomenon found in soils that have low induction numbers, which are the majority. The technique is simple and has been developed as a replacement for the Wenner method over which it has advantages in convenience but against this it has some disadvantages in interpretation and conversion into groundbed resistance.

The principle of operation is to place a transmitter coil (as shown in Fig 43) which is energized with alternating current at audio frequency and whose magnetic field induces a current in a second coil which is placed some distance away lying in the same place, usually on the surface of the ground.

The primary field of the transmitting coil and the secondary field from the induced current flow in the ground are both detected by the second coil. The ratio of strength of these, that is the secondary to the primary field, is a function of the ground conductivity and the square of the inter-coil spacing. It also depends on the frequency of the transmission. If the distance and frequency are constant, then this ratio of the two fields can be used to give a

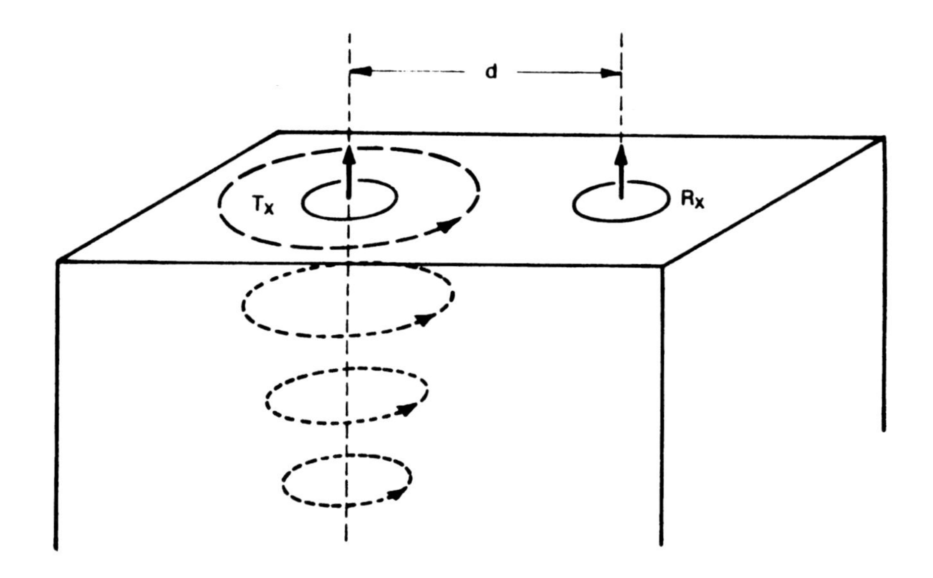

FIGURE 43 - Coil orientation and magnetic field with inductive resistivity meter.

direct indication of the resistivity. A portable instrument using coils at constant spacing is available, but it is limited in the depth to which it will work, usually about 10 ft.

Separate coils can be used, they can be carried easily and their data fed into the instruments to indicate resistivity (or, more usually, conductivity, the inverse of resistivity). With the independent coils two techniques can be used in which the orientation of the coils is varied: In the normal configuration the coils lie horizontally on the ground, that is with the axes of the coils vertical (haloes). The alternative technique is to use the coils so that they stand in a vertical plane, as does a wheel, with their axes horizontal. As there is now freedom to alter the spacing between the coils, this technique can be used to explore larger areas and encompass a greater depth of the ground. The two configurations have different relative responses from the coils and these are shown in Figs 44 & 45. The combined response from the two coil orientations gives a cumulative curve as in Fig. 46. The vertical axis coils (haloes) do not pick up, as can be seen, very much response from the surface layer, which can be a considerable advantage under dry top soil conditions, while the vertically oriented coils (wheels) pick their major response from close to the surface. However, the horizontal axis coils (wheels) are more usually employed as their exact alignment is less critical.

The response of a pair of coils at different separations compared to a two layer earth model has been calculated and proved in the field. These curves have different shapes from the Wenner curves but the commercially

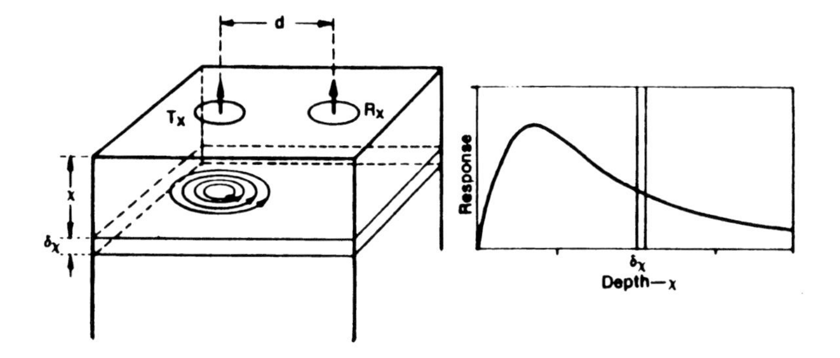

FIGURE 44 - Contribution of various small layers to overall response with vertical axis coils.

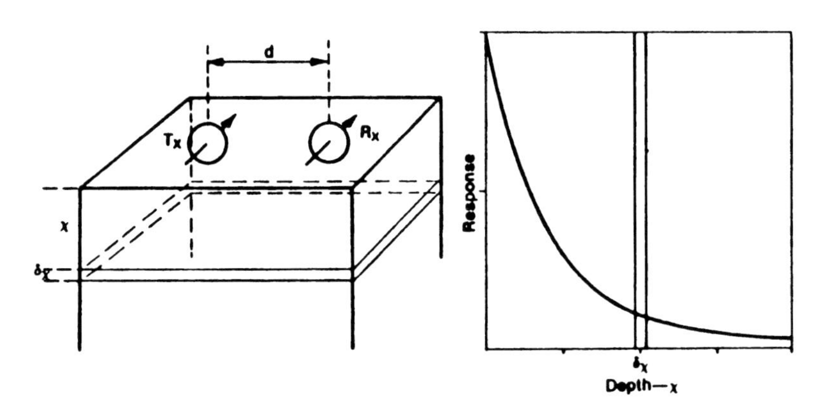

FIGURE 45 - Contribution of various small layers to overall response with horizontal axis coils.

available instruments are supplied with interpretation curves and the ground resistivity profile can be measured.

The ease with which the coils can be positioned and the fact that one does not have to place four pins in the ground make these ideal for rapid surveys; in dry surface conditions no water ring is required. However, the instruments are not useful below about 300 ohm cm resistivity and they have to be treated with considerable caution below about 2000 ohm cm. These errors are included in curves which link the measurements to the size, shape and resistance of the groundbed.

As an alternative method, transients from larger coils can be used, but this method is principally for exploring large areas, particularly to considerable depths. In the use of deep well groundbeds, this technique will have more value though its commercial use is in the detection of ore beds, etc.

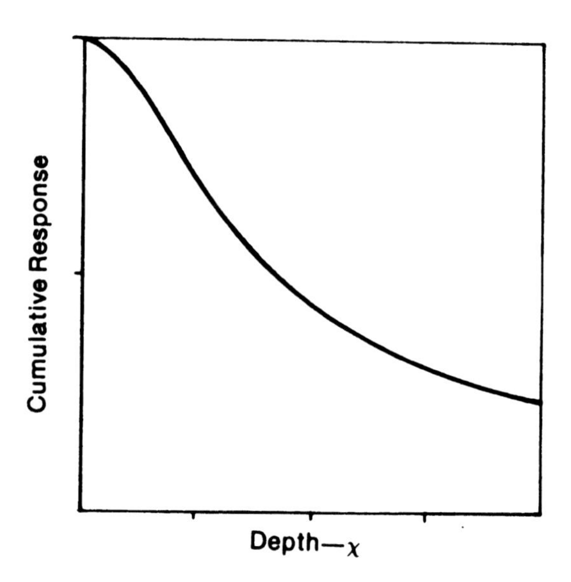

FIGURE 46 - Combined contributions to response from horizontal and vertical axes coils.

Radio wave attenuation will equally give a rapid indication of resistivity and as this equipment may be operated from an airplane, a quick survey over difficult country might suggest the least corrosive route for a pipeline.

Resistance of Ground Connections

If two pieces of metal are placed in the ground or any other electrolyte, a resistance may be measured between them; this will be the sum of three components: firstly, the metallic resistance of the pieces of metal which is so small that it can generally be ignored, secondly, the interface resistance of the metal/electrolyte boundary, this may be increased by the presence of scale, paint or grease to be quite substantial or it may be very low as is the case with bright steel rods; thirdly, there is a resistance associated wholly with the electrolyte and its resistivity. This last resistance will now be considered at length.

By dimensional arguments the resistance must be a function of the resistivity and the reciprocal of length, that is ohm cm $\times$ cm^{-1} = ohms, so there will be a factor associated with each size and shape of earth rod which will have the dimension length^{-1}; that is its resistance will decrease with increasing size of electrode.

First consider the case of a square based hollow prism with conducting ends and insulating sides as Fig. 47, filled with an electrolyte of resistivity ρ. The resistance measured between the metal plates will be

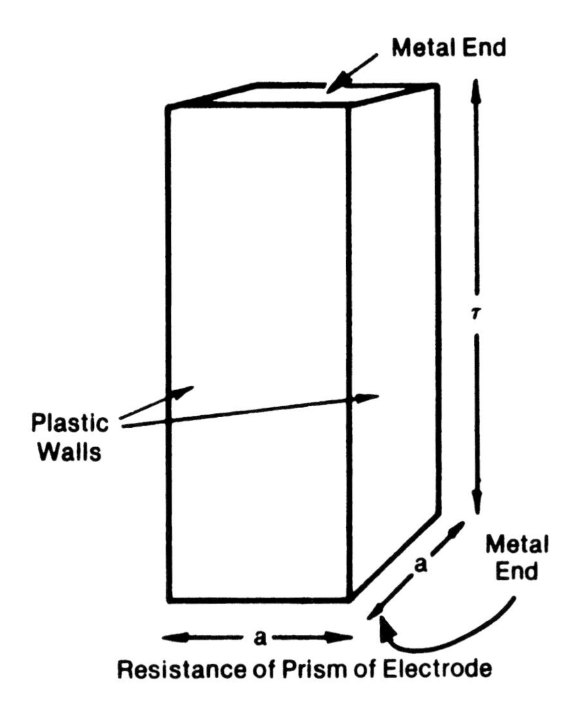

FIGURE 47 - Relationship between resistance and resistivity in rectangular prism.

$$R = \rho \frac{l}{a^2}$$

if the interface resistance is ignored.

Second, consider the case of a sphere of metal resting inside a concentric metal sphere of larger radius, with the space between being filled with an electrolyte of resistivity ρ, as in Fig. 48, then at a radius r, there is a thin shell δr whose resistance is:

$$\delta R = \rho \frac{\delta r}{4\pi r^2} \quad (3.5)$$

and the total resistance between the two spheres will be:

$$R = \int_{r_1}^{r_2} \rho \frac{dr}{4\pi r^2} = \frac{\rho}{4\pi} \left(\frac{1}{r_2} - \frac{1}{r_1} \right) \quad (3.6)$$

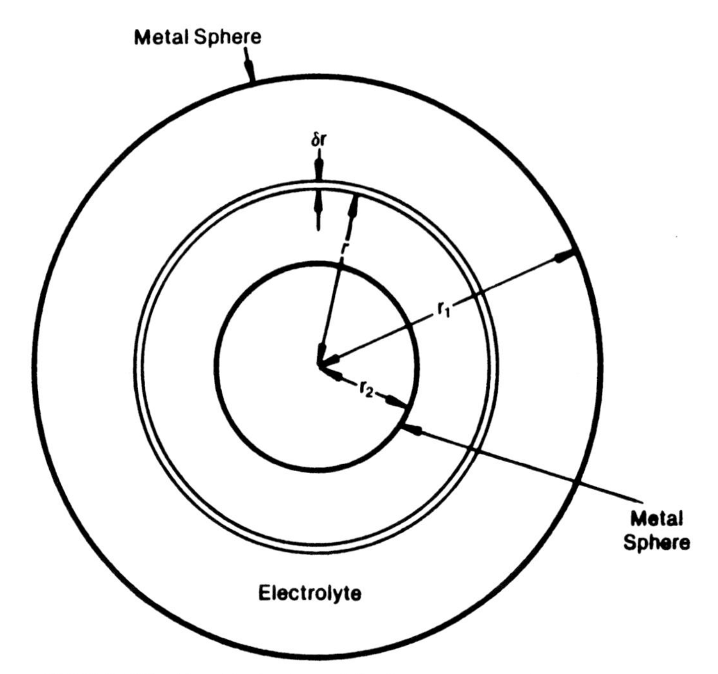

FIGURE 48 - Relationship between resistance and resistivity with concentric spheres.

where r_1 is the radius of the larger sphere and r_2 that of the smaller sphere. Similarly a pair of concentric metal cylinders of the same radii as the spheres, length l separated by the same electrolyte and with insulating ends will have a resistance between their cylindrical surfaces that will be

$$R = \int_{r_1}^{r_2} \frac{dr}{2\pi l r} = \frac{\rho}{2\pi l} \ln \frac{r_2}{r_1} \tag{3.7}$$

Consider two metallic bodies lying in an infinite electrolyte of resistivity ρ, then there will be a finite resistance between them such that if they are maintained at a potential difference V, then current $i = V/R$ will flow between them. An electrical potential will exist at all points in the electrolyte and lines of equipotential and lines of current flow can be drawn between them.

The same conditions would exist if one body were a shell and totally enclosed the other. If the enclosing shell is considered to have an infinite radius then R will be the resistance between the enclosed body and infinity. This would correspond, for example, to the cause of a sphere of metal suspended in deep water with current flowing to a distant object.

$$R = \frac{\rho}{4\pi r} \tag{3.8}$$

Now consider the particular case of two bodies, one of which is a mirror image of the other in a plane. The resistance of these two bodies to infinity (when they are electrically connected together by an insulated conductor) can be determined. Suppose their total resistance is

$$R = \frac{\rho}{4\pi r_1}$$

where r_1 is the size and shape factor. The resistance of each calculated from their mutual potential and the current flowing in each one separately will be

$$R = \frac{\rho}{2\pi r_1} \tag{3.9}$$

The arrangement is symmetrical about the image plane and so all of that one side of the plane could be removed without altering the shape and value of the equipotential lines or the lines of current flow and hence the resistance of one in the semi infinite electrolyte.

Groundbeds (as cathodic protection electrodes are called) set in the earth can be considered to lie in a semi-infinite electrolyte. The argument that applies to two symmetrical pieces of metal will apply equally to any shape that has a line of symmetry going through it. A particular case would be that of a hemispherical groundbed electrode with a radius r. The resistance of a sphere, as can be seen from Equation 3.6 when r_1 is made infinite, is given by

$$R = \frac{\rho}{4\pi r} \tag{3.8}$$

and so a hemisphere set in the ground would have a resistance of

$$R = \frac{\rho}{2\pi r} \tag{3.9}$$

The result can be obtained by a calculation similar to that which led to equation 3.6 where:

$$R = \int_{\infty}^{r} \rho \frac{1}{2\pi r^2} \bullet dr \; R = \frac{1}{2\pi r} \rho \tag{3.9}$$

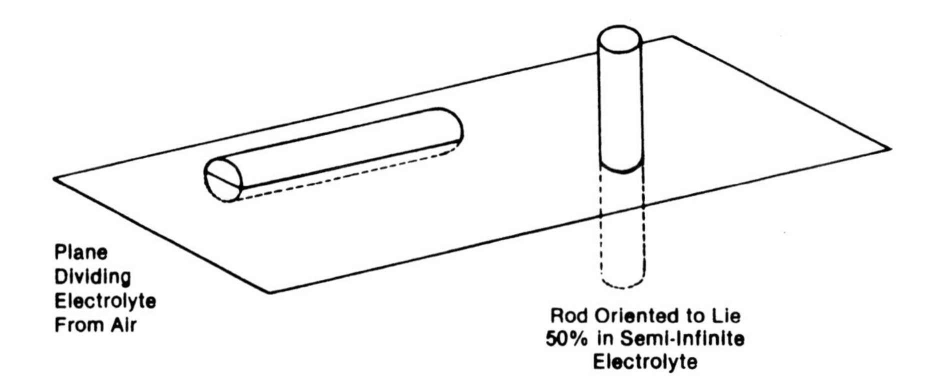

FIGURE 49 - Resistance in infinite and semi-infinite electrolytes.

As the resistance is principally associated with the area close to the electrode, then the extent of the 'semi-infinite' electrolyte will not need to be great.

Other symmetrical shapes can be treated similarly and Fig 49 shows two planes of symmetry in a cylindrical rod.

Rods and Cylinders

The formulae for the resistance of most elementary shapes have been derived and there are a variety of means of doing this, most of which have been proved experimentally. The resistance to infinity of a long thin rod of length $2L$ and radius a is given by

$$R = \frac{\rho}{4\pi L}\left(\ln\frac{4L}{a} - 1\right) \tag{3.10}$$

so that the resistance of the rod of half that length driven vertically into the ground would be twice the value in equation 3.10.

$$R = \frac{\rho}{2\pi L}\left(\ln\frac{4L}{a} - 1\right) \tag{3.11}$$

where ρ is expressed in ohm cm, L is in cm, a the radius, in cm, then R is in ohms.

The same formula will apply to a semi-cylinder whose flat surface coincides with that of the earth and whose length is $2L$ as a plane of symmetry exists along this surface and the earth can be assumed to be a semi-infinite conductor on one side of this plane (as in Fig 48). The formula becomes

$$R = \frac{\rho}{\pi l}\left(\ln \frac{2l}{a} - 1\right) \tag{3.12}$$

where $l = 2L$ the length of the rod.

If two vertical rods are placed parallel to each other and connected together then their resistance will depend upon their separation; at large separations it will be half that of each one singly while when they are close together it will be only slightly less than the resistance of a single rod. If the rods are separated by a distance 's' then the resistance of the pair of rods will be

$$R = \frac{\rho}{4\pi L}\left(\ln \frac{4L}{a} - 1 + \ln \frac{2L + \sqrt{s^2 + 4L^2}}{s} + \frac{s}{2L} - \frac{\sqrt{s^2 + 4L^2}}{2L}\right) \tag{3.13}$$

and this can be simplified for large values of s/L to

$$R = \frac{\rho}{4\pi L}\left(\ln \frac{4L}{a} - 1\right) + \frac{\rho}{4\pi s}\left(1 - \frac{L^2}{3s^2} + \frac{2}{5}\frac{L^4}{s^4}\right) \tag{3.14}$$

which as $s \rightarrow \infty$ gives the formula

$$R = \frac{\rho}{4\pi L}\left(\ln \frac{4L}{a} - 1\right)$$

which shows that their resistance is half that for a single rod (3.11).

For values of s/L which are small

$$R = \frac{\rho}{4\pi L}\left(\ln \frac{4L}{a} + \ln \frac{4L}{s} - 2 + \frac{s}{2L} - \frac{s^2}{16L^2} + \frac{s^4}{512L^4}\right) \tag{3.15}$$

or for very small values of s as $s \rightarrow a$

$$R = \frac{\rho}{2\pi L}\left(\ln \frac{4L}{\sqrt{as}} - 1 + \frac{s}{4L} \ldots\right) \tag{3.16}$$

An alternative calculation for the resistance of two rods in parallel which is accurate for rods separated by distance great compared with their diameter and of the order of their length can be made by replacing each of the rods by half buried spheres of equal resistance. Then if these spheres have a radius q and are separated by a distance s then their total resistance R will be

$$R = \frac{\rho}{4\pi}\left(\frac{1}{q} + \frac{1}{s}\right) \tag{3.17}$$

or this is equivalent to including the first term of the second part of equation 3.14.

A rod buried in the ground horizontally will form the same geometric arrangement as a pair of parallel vertical rods when they are considered with their images in the earth's plane as Fig 50. The depth of burial will be $s/2$ and the length of the rod $2L$. Thus for small values of s/L the resistance will be given as in equation 3.16

$$R = \frac{\rho}{2\pi L}\left(\ln \frac{4L}{\sqrt{as}} - 1 + \frac{s}{4L} + \ldots\right) \tag{3.16}$$

where $2L$ is the length of the rod and $s/2$ the depth of burial.

A similar mutual interference effect occurs when a vertical rod is buried so that it is wholly below the surface. Equation 3.11 gives the resistance of a vertical rod whose end reaches the electrolyte surface, and considering the method of obtaining this, the resistance of a rod buried infinitely deeply will be

$$R = \frac{\rho}{4\pi L}\left(\ln \frac{4L}{a} - 1\right) \text{ or } R = \frac{\rho}{2\pi l}\left(\ln \frac{2l}{a} - 1\right) \tag{3.18}$$

where $2L = l =$ length.

Applying the concept of two spheres similar to that used earlier then the resistance of a rod buried in the ground will be

$$R = \frac{\rho}{4\pi}\left(\frac{1}{q} + \frac{1}{s}\right) =$$

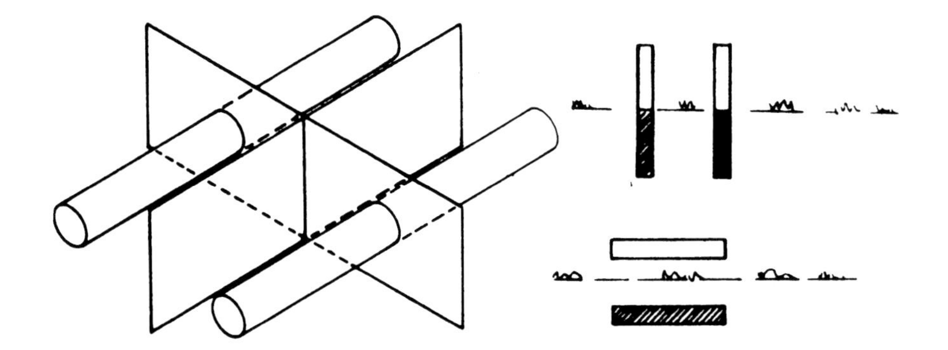

FIGURE 50 - Planes of symmetry in a pair of rod electrodes.

$$\frac{\rho}{4\pi}\left(\frac{1}{q}+\frac{1}{2t}\right)$$

where t is the depth of burial to the center, and q is the radius of the equivalent sphere; that is

$$\frac{1}{4\pi q} = \frac{1}{2\pi l}\left(\ln\frac{2l}{a} - 1\right)$$

$$\frac{1}{q} = \frac{2}{l}\left(\ln\frac{2l}{a} - 1\right)$$

$$\therefore R = \frac{\rho}{4\pi l}\bullet 2\left(\ln\frac{2l}{a} - 1\right) + \frac{\rho}{8\pi t} \tag{3.19}$$

this relationship only holds good for deep burial. If the resistance of such a completely buried rod is plotted as a function of the depth of burial of the rod center, then a curve will result as in Fig. 51.

Other Shapes

Formulae have been derived for the resistance of a sphere, hemisphere, vertical rod, completely buried vertical rod, horizontal rod and a pair of vertical rods. The resistance of a shape such as a tetrahedron is best found by assuming the existence of an equivalent sphere whose radius will be less than that of encircling sphere and more than that of the enclosed sphere.

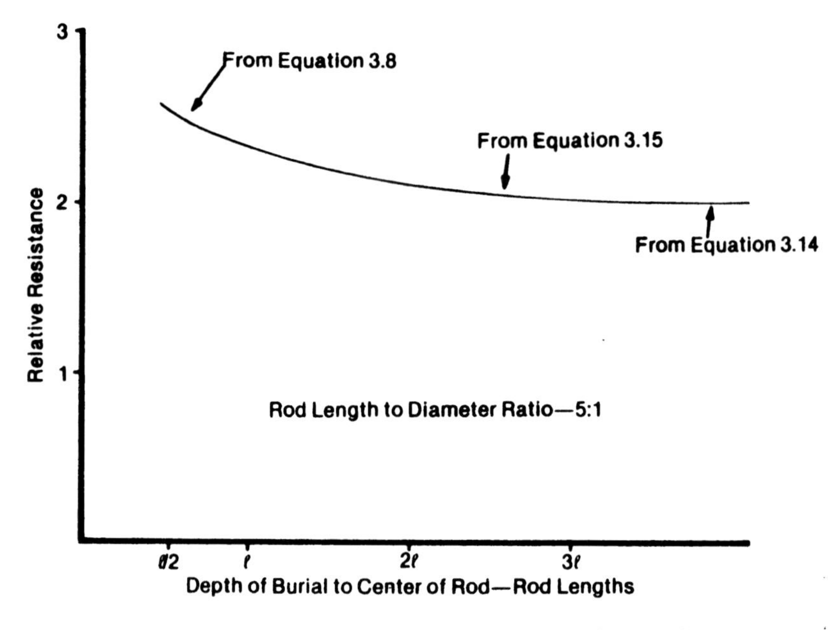

FIGURE 51 - Resistance of buried vertical rod as a function of depth.

If the mean of these two values is used then the calculated resistance will be approximately correct. For example, consider the case of a buried rod whose resistance is given by equation 3.18.

$$R = \frac{\rho}{2\pi l}\left(\ln \frac{2l}{a} - 1\right)$$

Then, if the above rule were applied, its resistance would be found by considering a sphere of diameter equal to the length and another of diameter equal to the cylinder diameter, and calculating the resistance of a sphere whose radius was the mean of the two considered. That is

$$R = \frac{2\rho}{2\pi (l + \mathrm{d})} \tag{3.20}$$

The ratio of this resistance to that derived from equation 3.18 would be

$$\frac{2}{l + \mathrm{d}} : \frac{1}{1}\left(\ln \frac{4l}{\mathrm{d}} - 1\right)$$

and assuming that the cylinder's length is five times its diameter this ratio will be

$$\frac{2}{6} : \frac{1}{5} \text{ (ln 20-1) or } \frac{1}{3} : \frac{1}{5}(3\text{-}1)$$
$$= 5{:}6$$

or the approximation will have given a value 17 per cent too small. The same calculation can be used to convert a rectangular rod to its equivalent radius by taking the mean of the enclosed cylinder and the enclosing cylinder. In the case of square section, sides, the equivalent radius will then be

$$r = \frac{1 + \sqrt{2}}{4} \times a \tag{3.21}$$

A slightly closer approximation would be found in this case if the geometric mean, as opposed to the arithmetic mean, were used.

The estimate of the resistance of an anode is most important and the value will depend on the size and shape of the electrode. Most sacrificial anodes are cast in some regular form and their resistance can be estimated with good accuracy if they are considered to be the equivalent of a sphere whose diameter is equal to the mean of the length, breadth and thickness of the anode.

The use of the length plus breadth plus thickness formulae can be compared with the Dwight formula for a rod. This is done on the basis of a circular rod whose formula will now become

$$R = \frac{\rho}{2\pi} \times \frac{3}{L + B + D} \tag{3.22}$$

and where the diameter of the equivalent sphere is one third of (length plus twice the rod diameter), and the formula for a square rod, where the equivalent diameter becomes one third of (length plus twice the side). As can be seen in Fig. 52, there is a good correlation in the general range of offshore anodes, that is where the length to diameter ratio varies between 5 and 12.

These calculations can be converted into easy-to-read tables in which the sum of the length, breadth and thickness required to protect 100 square meters of steel can be given relative to the current density requirement and the resistivity. Such a series of curves for aluminum-mercury and zinc anodes at 0.25 V cell or driving voltage and aluminum-indium at 0.30 V driving voltage and magnesium at 0.75 V driving voltage are shown in Fig. 53.

If the anode is at the surface of a semi-infinite electrolyte, and this would be the case of an anode attached to a painted hull, then the line of symmetry rule would apply and the resistance would be found by doubling

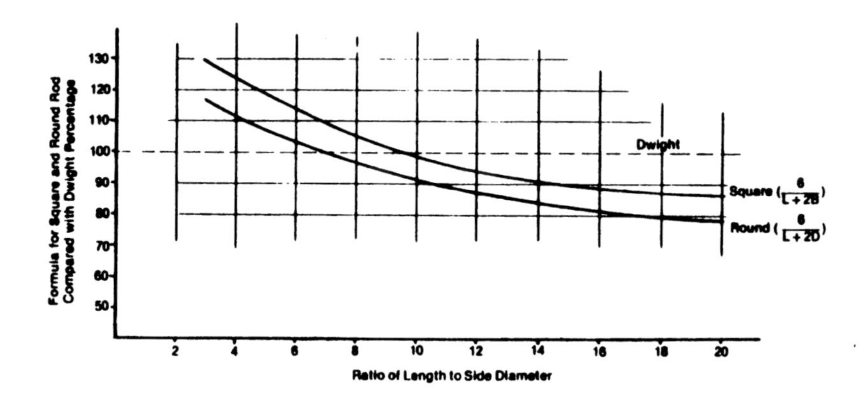

FIGURE 52 - Comparison of Dwight and Morgan formulae for rods.

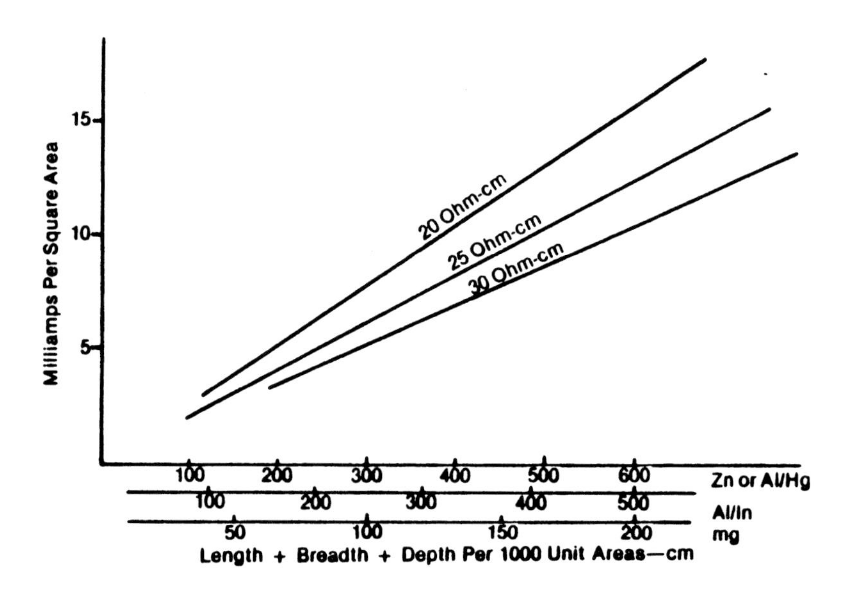

FIGURE 53 - Morgan's formula used to design anodes for protection in various sea waters.

the thickness of the anode and then doubling the resistance calculated as though the thickened anode were in an infinite electrolyte.

A more rigorous mathematical treatment, particularly that of Prof. Dwight, has led to equations for a number of simple geometric shapes and these are listed in Table V. The accuracy with which the earth resistivity may be measured, an assumed constant, is low, an overall 10% error being

TABLE V

One vertical ground rod: length L, radius a	$R = \frac{\rho}{2\pi L}\left(\log_n \frac{4L}{a} - 1\right)$
Two vertical ground rods separation s, $s > L$	$R = \frac{\rho}{4\pi L}\left(\log_n \frac{4L}{a} - 1\right) + \frac{\rho}{4\pi s}\left(1 - \frac{L^2}{3s^2} + \frac{2}{5} \cdot \frac{L^4}{s^4} \ldots\right)$
$s < L$	$R = \frac{\rho}{4\pi L}\left(\log_n \frac{4L}{a} + \log_n \frac{4L}{s} - 2 + \frac{s}{2L} - \frac{s^2}{16L^2} + \frac{s^4}{512L^4} \ldots\right)$
Buried horizontal wire, length $2L$, depth $s/2$	$R = \frac{\rho}{4\pi L}\left(\log_n \frac{4L}{a} + \log_n \frac{4L}{s} - 2 + \frac{s}{2L} - \frac{s^2}{16L^2} + \frac{s^4}{512L^4} \ldots\right)$
Right-angle turn of wire: length of arm L, depth $s/2$	$R = \frac{\rho}{4\pi L}\left(\log_n \frac{2L}{a} + \log_n \frac{2L}{s} - 0{\cdot}24 + 0{\cdot}2\frac{s}{L} \ldots\right)$
Three-point star	$R = \frac{\rho}{6\pi L}\left(\log_n \frac{2L}{a} + \log_n \frac{2L}{s} + 1{\cdot}1 - 0{\cdot}2\frac{s}{L} \ldots\right)$
Four-point star	$R = \frac{\rho}{8\pi L}\left(\log_n \frac{2L}{a} + \log_n \frac{2L}{s} + 3 - \frac{s}{L} \ldots\right)$
Six-point star	$R = \frac{\rho}{12\pi L}\left(\log_n \frac{2L}{a} + \log_n \frac{2L}{s} + 6{\cdot}9 - \frac{3s}{L} \ldots\right)$
Eight-point star	$R = \frac{\rho}{16\pi L}\left(\log_n \frac{2L}{a} + \log_n \frac{2L}{s} + 11 - 5{\cdot}5\frac{s}{L} \ldots\right)$
Ring of wire. Diameter D of ring, diameter of wire, a depth $s/2$	$R = \frac{\rho}{2\pi^2 D}\left(\log_n \frac{8D}{d} + \log_n \frac{4D}{s}\right)$
Buried horizontal strip: length $2L$, section a by b depth $s/2$, $b < a/8$	$R = \frac{\rho}{4\pi L}\left(\log_n \frac{4L}{a} + \frac{a^2 - \pi ab}{2(a+b)^2} + \log_n \frac{4L}{s} - 1 + \frac{s}{2L} - \frac{s^2}{16L^2}\right)$
Buried horizontal round plate radius a, depth $s/2$	$R = \frac{\rho}{8a} + \frac{\rho}{4\pi s}\left(1 - \frac{7a^2}{12s^2} + \frac{33a^4}{40s^4} \ldots\right)$
Buried vertical round plate	$R = \frac{\rho}{8a} + \frac{\rho}{4\pi s}\left(1 + \frac{7a^2}{24s^2} + \frac{99a^4}{320s^4} \ldots\right)$

the practical value. Constructions of groundbeds should lie within these limits of accuracy but the equations quoted will generally give accuracy beyond those required in the field.

Anodes that are of peculiar shapes that do not lend themselves to this treatment can be considered by making the best estimate of the equivalent sphere.

A resistance formula based on anode area has been developed by Crennell for what he describes as 'squarish' anodes and is extensively used for sacrificial anodes offshore,

$$R = \frac{0.3}{\sqrt{A}}\rho \qquad (3.23)$$

The constant 0.3 is the mean of those for a sphere, 0.282 ($R = \rho/4\pi a$, $A = 4\pi a^2$) and a disc, 0.313 ($R = \rho/8a$, $A = 2\pi a^2$). There is sometimes a greater computation in measuring the area of an anode than in determining its weight or length.

Once the resistance has been measured for a shape of a particular size then the resistance of a larger or smaller body of similar shape may be determined by a purely dimensional treatment. This can be done by comparing the scale factor, say the length, or the square root of the area or the cube root of the weight.

The above calculations assume the anode is freely suspended in water. There will be an effect if the anode is placed close to a structure depending on whether this structure is painted or not. With a bare structure there will be less increase in resistance as the anode is taken closer to the cathodic structure.

As an approximation it can be taken that if the anode is stood off by more than one half of its length, then the general formulae will apply. If it is between one half and one quarter of its length from the structure then the calculated resistance should be increased by 10 per cent on a bare structure and 20 per cent on a coated structure; if it is closer than one quarter of its length then these figures should be doubled. Fig. 54 shows the variation in resistance as a rod whose length is ten times its diameter approaches a plane insulated surface.

An anode that is immediately next to a coated structure will be influenced by the shape of the cathode surface. For example, an anode placed

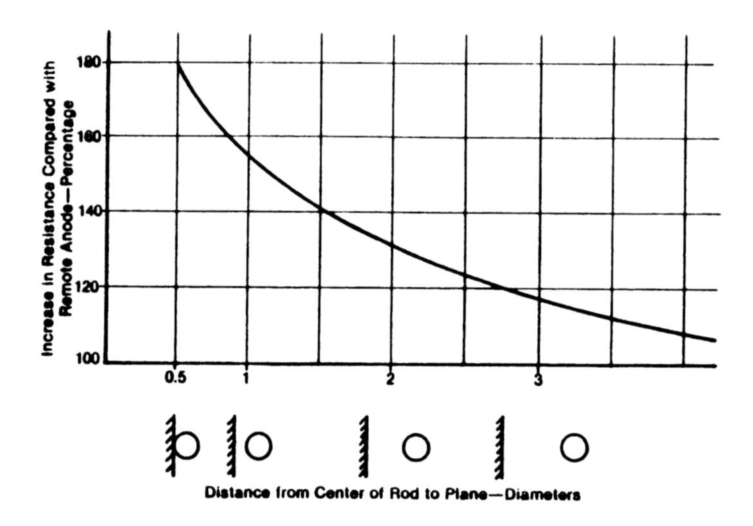

FIGURE 54 - Change in resistance of a rod anode as it approaches a planar-insulated surface.

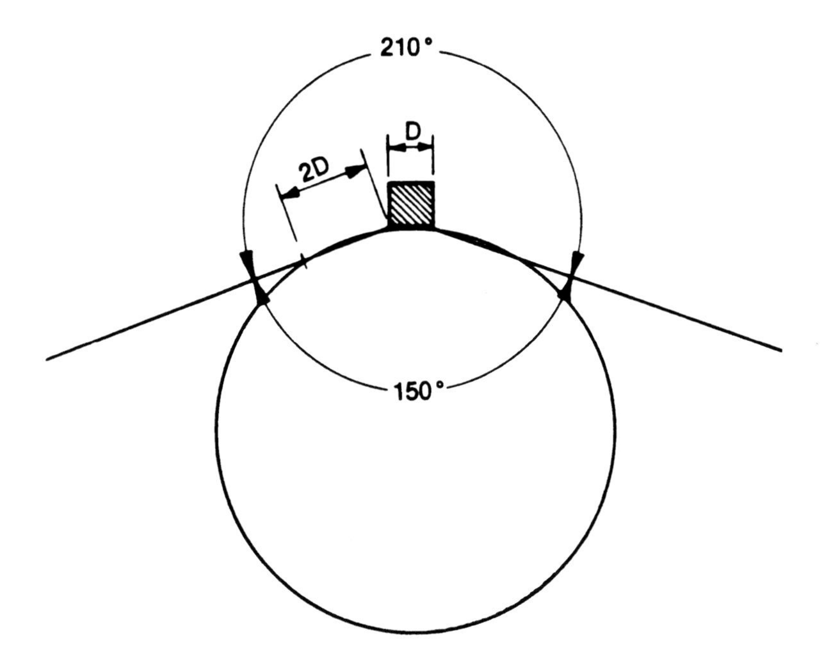

FIGURE 55 - Resistance of an anode on a tubular-insulated offshore member.

on a tubular platform leg will obviously have a lower resistance than one placed on a flat plane surface. A good approximation can be made by looking at the point which is two mean diameters from the center of the anode and calculating the ratio of sea water to insulated steel at that point. For example, if the sea water at that point occupies an arc of 210° against steel's 150°, then the anode resistance will be 5/7ths (i.e. 150°/210°) of that which it would have on a plane surface. This is illustrated in Fig. 55 and a similar approximation can be made on a three dimensional figure.

Composite Shapes

It is often most convenient to construct a groundbed from a multiplicity of small rods or other pieces rather than from one large rod. When these are buried in the ground their combined resistance will be greater than the sum of their electrical resistances if they were assumed to be electrically in parallel, the total resistance of the group being increased by a correcting factor called the spacing factor. This factor will vary with the distance between the units, being greatest when they are close together. If rods are used as the individual elements then the factor for a series of similar rods will depend upon the ratio of their spacing to length and there will be less effect with thin rods, that is rods whose length is very much greater than their diameter, than there will be with thick rods.

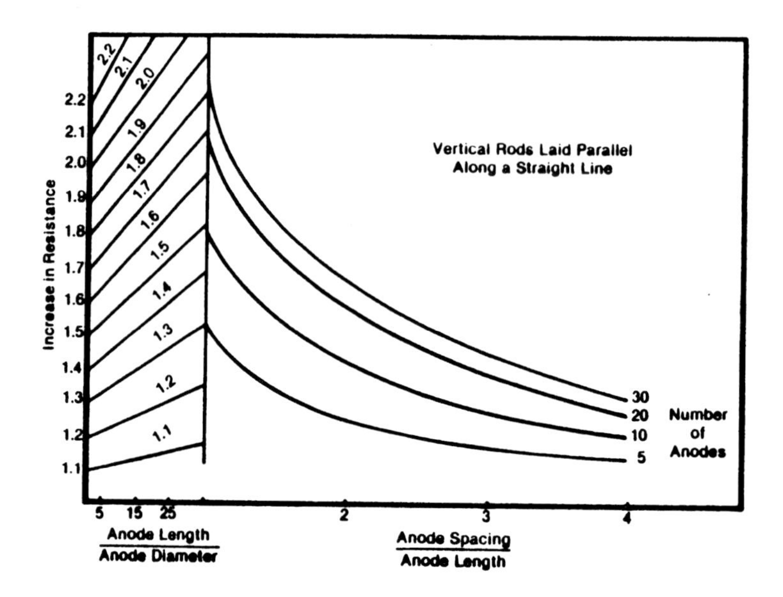

FIGURE 56 - Increase in resistance of groundbed rods over widely spaced anodes.

Several curves and equations have been suggested to give this spacing factor and one of these sets of curves is re-drawn in Fig. 56 in which the two variables, the inter-rod spacing and the ratio of the rod length to diameter, can be selected. There has been considerable difficulty in comparing the results obtained in practice and the curves given by the various authorities as the uniform conditions that are assumed in the mathematics are never found in the field. If a large series of electrodes are considered then a reasonable approach may be made by calculating their resistance when they are represented by a group of hemispheres.

If vertical rods are used to construct a groundbed then the electrical connection to them will probably be made by a buried cable and as this will require trenching it is common practice to place further rods in this trench. The Russians seem to favor this method even to the extent of prefabricating such massive 'comb' electrodes out of scrap pipe.

Coated Anodes

If an anode is coated over part of its surface then its resistance will increase. A long wire may be coated so that equal lengths of wire are alternately coated and bare: this is covered by the formulae for spaced anodes.

The coating of parts of a block anode will have a small effect on the resistance with about eight to 10 per cent increase if half is coated in strips, and about 15 per cent if larger areas, such as the sides are coated. This method is often used to control the shape of the consumption of the anode and the use of devices that limit the high current density at the end of an anode equally increases the resistance compared with that of a freely suspended rod. An increase in resistance of about eight per cent is found when a disc is attached to one end and 15% to both ends; a bracelet anode inset into the weight coating on a pipeline being an example.

Non-Uniform Electrolytes

The resistances derived so far have been those to infinity that is the resistance between the ground electrode and a conducting shell at infinity. Also for simplicity it has been assumed that the ground or liquid electrolyte is uniform in resistivity. In a deep lake or the open sea this condition may exist but the ground is rarely homogeneous, most frequently consists of a layer of top soil overlying some rock, clay or other formation. This lower formation is often of higher resistivity, sometimes to the extent of being 100 times greater and consequently the effective resistance of buried rods is higher than their calculated resistance to infinity.

In other cases, especially in desert areas and in places where the top cover is highly porous, the uppermost layer is dry and of high resistivity while at some depth there is a water table which causes a considerable drop in resistivity. In some areas this change may be from gravelly soils of 100,000 ohm cm to a lower stratum of 5,000 ohm cm.

When constructing a groundbed it is an advantage to be able to locate it in low resistivity soil and to have that soil saturated with water, the reason for this latter condition being explained later in the chapter. Wenner's four pin resistivity method can be used to determine the change in resistivity with depth and a method suitable for a two layer problem has already been indicated. If the apparent resistivity against pin spacing is plotted, then when the resistivity decreases as the spacing increases, the lower stratum is a better conductor while the reverse is true when the resistivity increases at large spacings. If the engineer has a complete freedom of choice in the method of construction of the groundbed he could locate the rods in the lower resistivity stratum. If this layer were reasonably thick then the resistance calculation would be relatively easy; either the surface of the ground could be considered to be lower in the case of a high resistivity top layer, or the problem considered as being similar to an additional groundbed buried deeply in the ground interfering with the constructed groundbed when this is laid in low resistivity top soil.

Often, however, the engineer has no choice and is compelled by the equipment or labor at his disposal to select the best site where a groundbed within his capabilities can be built. The problem that now arises is which of a series of resistivity readings is the correct one for a particular groundbed design calculation. For example, the Wenner resistivity readings may be made at pin spacings of 2 ft 6 in., 5 ft, 7 ft 6 in., 10 ft and 15 ft and the results obtained may be as below:-

ohms	Pin Spacing ft	in.	Apparent Resisivity ohm cm
80	2	6	3,000
160	5	0	2,500
240	7	6	2,800
320	10	0	3,000
480	15	0	3,500

If a groundbed is constructed from calculations based on a resistivity of 3,500 ohm cm, the reading at 15 ft, it may be too large while constructing one on the basis of a resistivity of 2,500 ohm cm would lead to the groundbed being to small. Probably the most popular of the easily constructed groundbeds is the long horizontal rod type.

In high resistivity soils, groundbeds of 100 ft may be used and it is possible to calculate the resistance it would have if it were buried near the surface of a simple two layer geological configuration. First, however, consider the case of a groundbed 100 ft long, 1 ft diameter buried so that its center is 3 ft deep. In a uniform soil its resistance to infinity would be given by

$$R_1 = \frac{\rho}{nl}\left(\ln \frac{2l}{\sqrt{sd}} - 1\right) \tag{3.24}$$

where l is its length, d its diameter and s its depth. The resistance of a semi-cylindrical groundbed, the flat face of which coincides with the ground level, relative to a larger concentric semi-cylindrical shell would be:

$$R_2 = \frac{\rho}{\pi l} \ln \frac{d_2}{d_1} \tag{3.25}$$

where d_1 and d_2 are the diameters of the two cylinders as in Fig. 57.

Now d_1 can be assumed to be of such a value that the resistance of this semi-cylindrical groundbed would be the same as the horizontal buried rod being considered. If the resistance of this semi-cylindrical rod to a shell of 15 ft diameter is considered, then the ratio of this resistance and its resistance to infinity will be 1.5 to 3.8. Thus 40 per cent of the resistance of the buried horizontal rod is derived from the electrolyte within 7 ft 6 in. of the center line of the groundbed and this will indicate the importance of the Wenner reading at spacings of 7 ft 6 in. and 10 ft. This method can be used

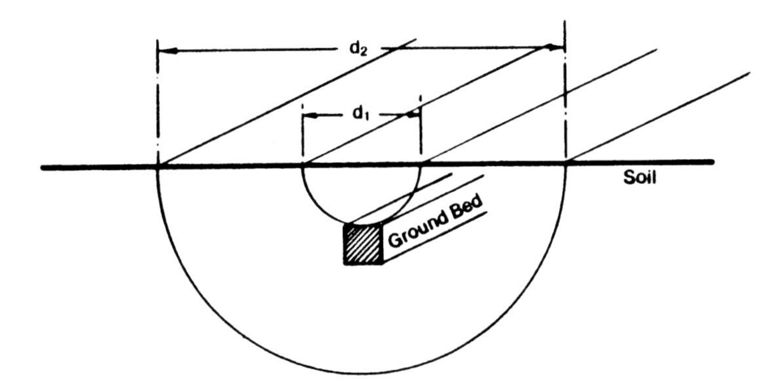

FIGURE 57 - Contribution of volume of soil closest to anode to total anode resistance.

as an aid to guessing the appropriate or apparent resistivity that will give the correct resistance for the groundbed when it is applied to the formulae given in Table V.

Further approximations with horizontal groundbeds are helped if it is remembered that the groundbed appears to the electrolyte to be a hemisphere when considered from distances greater than its length away, as a semi-cylindrical structure at distances intermediate between this and twice its burial depth, and as a cylinder buried in the ground only when measurements are taken close to or directly above the backfill. This can be verified by measuring the potential of the ground around the groundbed and away from it. The potential pattern as the measuring electrode is taken further from the groundbed will change from that expected from a buried rod to that expected from a semi-cylindrical groundbed and finally to the same pattern that would be obtained from a point earth electrode when measurements are made remote from the groundbed.

A series of computer programs can be used to determine groundbed resistance with great accuracy which can be linked directly to the inductive or Wenner technique and to the particular favored geometry of the groundbeds most expediently constructed by the particular contractor.

Two-Layer Configuration

For horizontal groundbeds in a two layer geological structure Sunde has suggested methods of determining the resistance to infinity and these values can be compared with the resistance that would be calculated from Wenner method resistivity measurements taken at various spacings under the same conditions. Figs 58 show the resistance of a 150 ft groundbed of 1 ft diameter buried 3 ft deep in three two-layer resistivity configurations. The resistance that would have been calculated from the Wenner readings

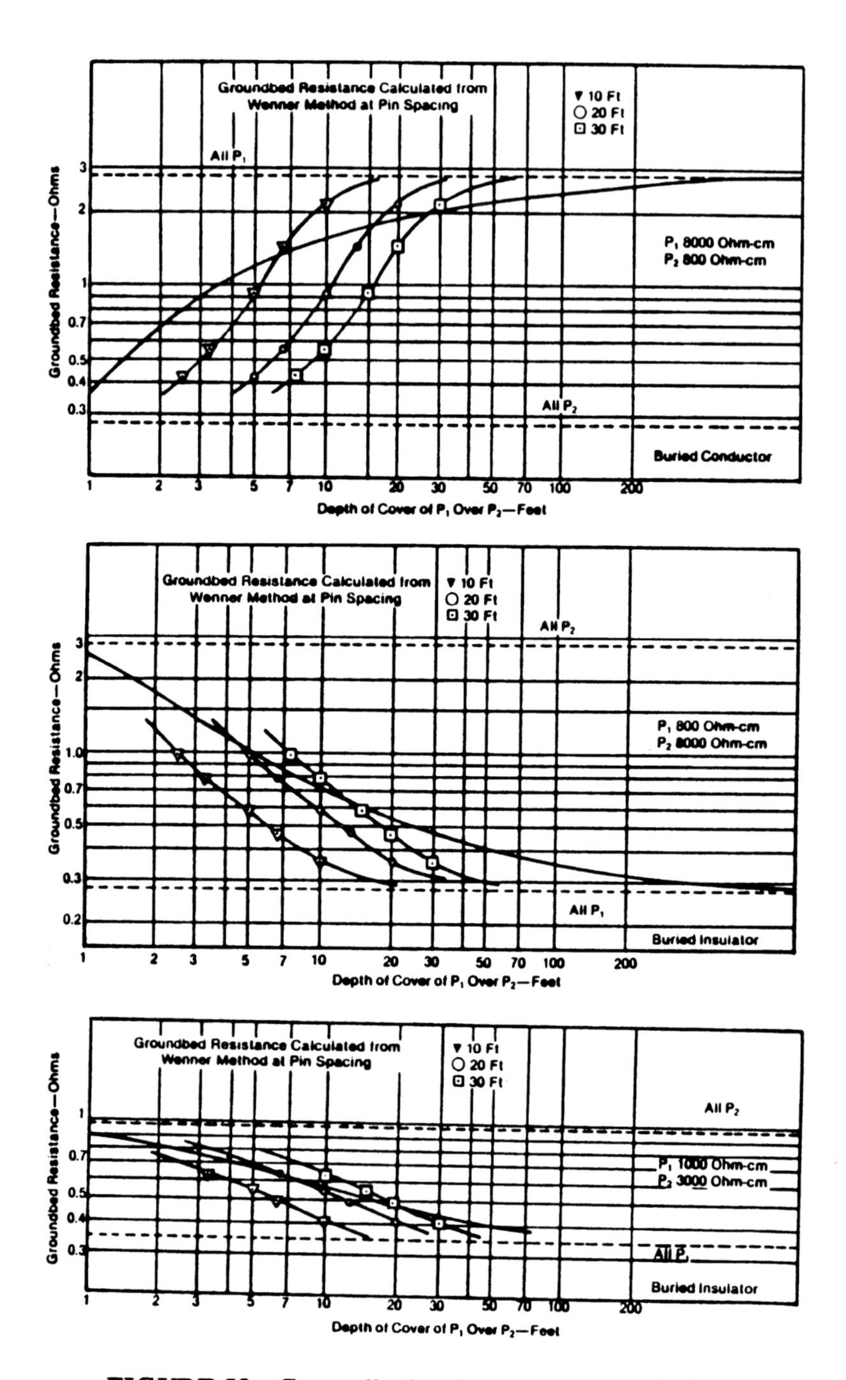

FIGURE 58 - Groundbed resistance compared with that calculated from the Wenner method in three two-layer configurations.

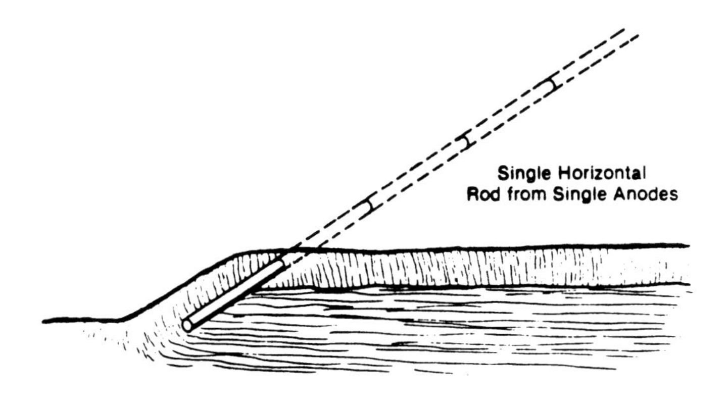

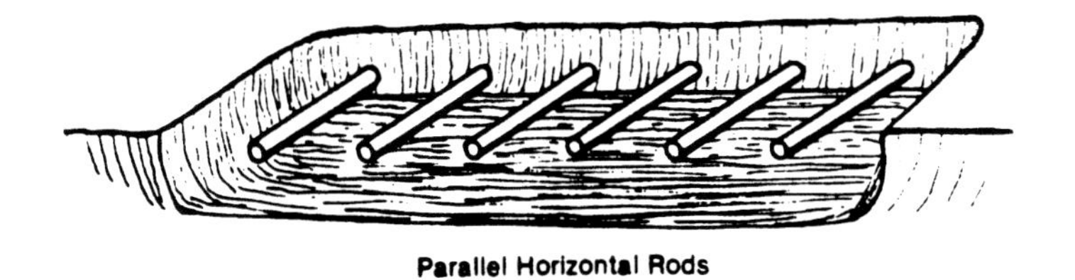

FIGURE 59 - Arrangements of horizontal rods to form groundbeds.

is also shown. It is suggested that the pin spacing in the Wenner method should be 20 per cent of the groundbed length and this seems as reasonable as any other general method of obtaining the appropriate value of resistivity to be used in the calculation.

Groundbed Construction

Groundbeds are generally constructed from a series of standard rods. These may be long and of small diameter as would be the case with electrodes of platinized titanium, magnesium or lead or they may be of much thicker cross section and quite short, graphite, zinc and silicon iron rods would fall into this category. The groundbed configurations that can be fabricated from these rods are innumerable though practically they can be divided into horizontal and vertical configurations. This terminology only applies to groundbeds close to the surface as in the body of the electrolyte their orientation does not matter. In both configurations the groundbed can be made into one continuous rod or into a series of parallel rods and these are illustrated in Figs. 59 and 60.

Generally the resistance of a long rod will be less than that of a series of

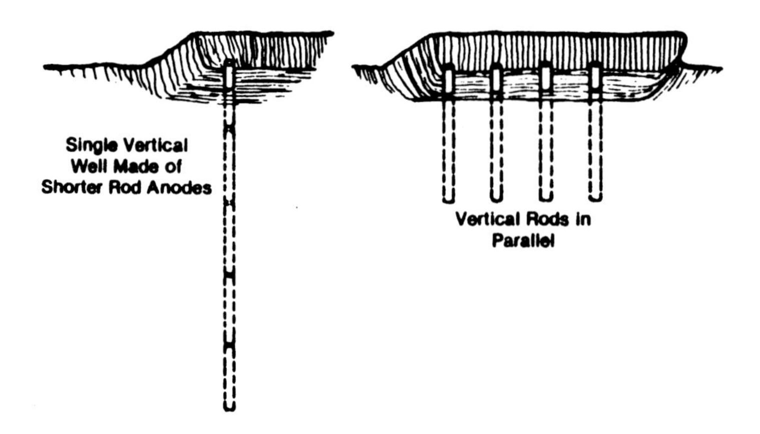

FIGURE 60 - Arrangements of vertical rods to form groundbeds.

closely spaced parallel rods, and parallel rods lying in one plane, or along a straight line, will have a smaller resistance than they would as a group. Vertical anodes will have lower resistance than their horizontal counterparts, so that it would seem that one long vertical rod would be the ideal anode and this is so. However, the great depth required for its construction may bring the groundbed into high resistivity rock and the expense of sinking a hole 12 in. diameter and 100 ft deep is large unless a very high degree of utilization can be made of the drilling equipment. Thinner rods driven to these depths will have other disadvantages in cathodic protection engineering. It is relatively easy to drill a hole only 8 to 9 ft deep and so a parallel array of shorter vertical anodes finds a great deal of favor.

All of the vertical rods have to be electrically connected together either by a cable which is laid in a trench, or mole-plowed into the ground or by locating the vertical electrodes at the feet of the poles of an overhead conductor system. If hand excavation or mechanical trenching are the only methods available the single horizontal rod is preferable, because this requires the minimum of excavation as the cable connecting the anodes and the anodes lie in the same trench.

This is also the case where there is only a thin cover of low resistivity soil when it has an advantage over vertical anodes of lying solely within the top stratum. A groundbed of horizontal parallel rods has little to recommend it as extra work is necessary to lay the connecting cable in its separate trench. Some considerable saving of material can be effected by removing part of the center of a long horizontal single rod groundbed and the resistance is little affected, for example, if a 200 ft long groundbed is split into two 80 ft lengths separated by 80 ft their combined resistance will probably be less than the original anode.

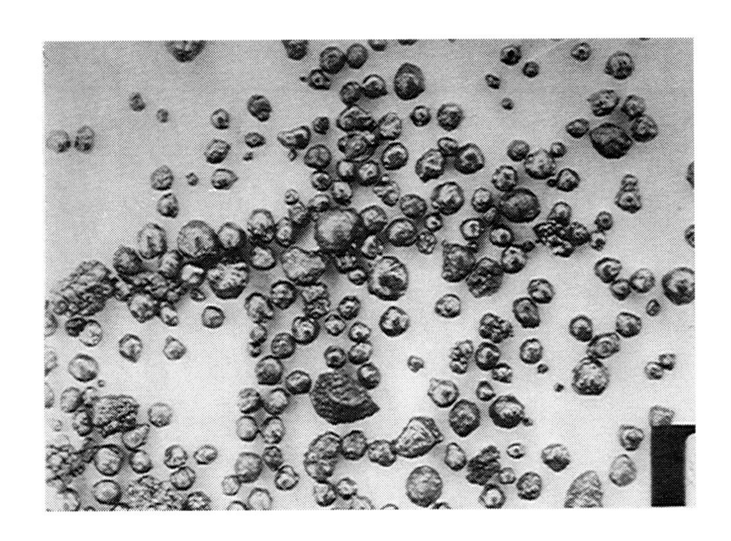

FIGURE 61 - Spherical noodles of petroleum coke used in backfill.

Horizontal parallel rods are generally used in muds or on the bottoms of ponds where the anodes can be fabricated into a 'ladder-like' formation, the anodes lying in the rung positions with two parallel feeder cables either side. End to end stringing has the disadvantage that fragile anodes, such as graphite, magnetite and high silicon iron, may fracture and isolate part of the groundbed. Generally horizontal groundbeds have to be laid 3 ft deep so that they are below plow level and constructing the anode at this depth also means that the ground around it will not freeze in temperate zones.

Backfill

One method of reducing the groundbed resistance is to surround the rod electrodes with low resistivity soil and this is done in three ways. In the first method graphite, magnetite, platinum and high silicon iron anodes may be surrounded by a mixture of coke breeze and graphite and this mixture, called a backfill, can have a resistivity as low as 2 or 3 ohm cm which, compared with the soil, is an excellent conductor. The resistivity of such a mixture depends greatly upon the grading and simple experiments with local coke breeze will indicate the best mixtures; for example, using a number 10 and ¼ in. sieve it was possible to reduce some London gas works breeze from 25 ohm cm to 12 ohm cm. Petroleum coke can be specifically manufactured to provide much lower resistivities and by careful grading and selection as low as 1 ohm cm can be obtained, Figs 61 and 62. Equally important is the compacting or tamping that the breeze receives

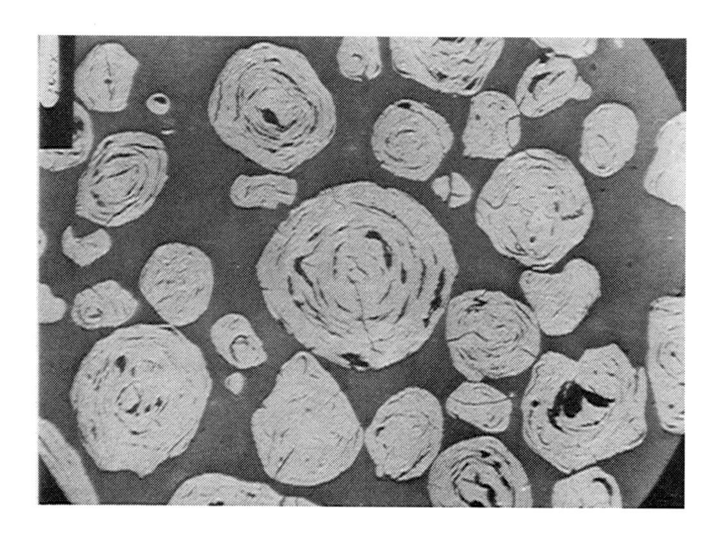

FIGURE 62 - Section through spheres of petroleum coke. (Photo reprinted with permission from Cathodic Engineering Equipment Co., Inc., Hattiesburg, Mississippi.)

and as this is more easily achieved in horizontal groundbeds these are often used with this type of backfilled anode.

The coke breeze and anode rod can be compacted in the workshop and sealed into a metal cannister, allowing groundbeds to be made by this method in muddy soils or deep holes, Fig 63. The nature of the coke breeze causes excessive drainage of the soil water and this can be a disadvantage as the ground surrounding the anode may dry out quickly. The construction of long horizontal groundbeds must be done most carefully to avoid creating land drains which can remove the coke breeze from the anode surface by the water running in them. Precious metal anodes can be used with the better quality coke breeze and graphite backfills in which these act as point contacts to what is virtually a massive coke anode. Both wire and mesh anodes have been used and from experiments begun 20 years ago, the author's experience suggests that these form an excellent groundbed. Contact must be maintained with the mesh or rod. For this reason the mesh or expanded metal anode appears to have advantages, fig. 64. In deep well groundbeds it is possible to use a slurry of coke which can be placed more easily and by mud and jetting techniques the anode can be removed for replacement or inspection.

A second type of backfill can be made by mixing chemicals with clean clays usually common salt or gypsum is mixed with bentonite and this is

FIGURE 63 - Canned anode assembly in spiral can. (Photo reprinted with permission from P.I. Corrosion Engineering, Ltd., Hants, Alresford, England.)

either poured as a slurry around the electrode or is packed around the anode by placing both in a cotton bag and lowering the whole into the ground, Fig 65. The resistivity of these mixtures varies with the clay used and the amount and kind of salt added; resistivities below 100 ohm cm can

FIGURE 64 - End connected anode of LIDA type of metal-oxide coated titanium mesh for burial in conducting backfill. (Photo reprinted with permission from Impalloy, Ltd., Bloxwich, Walsall, England.)

be obtained with sodium chloride additions while 300 ohm cm is the minimum usually obtainable with mixtures of gypsum and clays.

The third method of reducing the groundbed resistance is to salt the area in which it is laid. This is usually engineered by applying common salt and gypsum to the soil surface or to just below this level. The common salt has an immediate effect while the gypsum, because of its lower solubility, has a long term value. If crushed gypsum is spread as a 1 in layer in a 3 ft deep horizontal groundbed it should last for at least 50 years and a longer life could be ensured by mixing additional layers with the trench backfill. The gypsum and salt reduces the resistivity close to the rods and may effect a decrease in anode resistance varying from 10 per cent to 50 per cent. The treatment seems most effective in clean washed gravels and sands; sand gypsum mixtures give resistivities of about 500 ohm cm when the pretreat-

FIGURE 65 - Cast magnesium cylindrical anodes bare and packaged. (Photo reprinted with permission from Corrintec/USA, Cathodic Protection Services, Inc., Houston, Texas.)

ment resistivity of the sand was 10,000 to 20,000 ohm cm. Calcium chloride would seem to be an attractive salt for this purpose as it has a high deliquescence.

Electro-Osmosis

Close to the anode groundbed there will be a large potential gradient in the soil and this will cause electro-osmosis, which is the movement of water through the soil by an applied potential gradient. The movement takes place in the direction of flow of conventional current so that this process dries out the area near to the anode and wets the cathode. Clays and fine silts are most prone to this osmotic flow which is considerably reduced in the presence of salts, and salting even a low resistance groundbed may be of considerable use. Where the groundbed is laid well below the water table it is unlikely that complete drying out will occur; water drainage into the anodes can often easily be encouraged and this will prevent any increase in resistance through loss of moisture by osmosis. High current density anodes in fine muds can be dried out.

Soil Heating

Another factor that may cause drying out of the soil in the vicinity of the groundbed is the heating effect which occurs in high resistivity soils. The drying out of the soil by evaporation will increase the power consumed at the groundbed, assuming the current to be constant, and so the soil temperature will rise progressively. The ratio of the thermal and electrical conductivities is approximately the same for most soils and this constant is

$$c = 10^{-2}\,^{\circ}\mathrm{C\ Volts}^{-2}$$

When this ratio is assumed, the temperature rise of the soil adjacent to the groundbed will be determined by the following expression irrespective of the size or shape of the groundbed or the nature of the soil.

$$T = 1/2\, c\, V^2 \tag{3.26}$$

where V is the driving voltage of the groundbed to a remote electrode. Thus a groundbed operating at 50 V would cause a temperature increase of approximately 12 °C or (22 °F) which would occur close to the anode and would be too low in temperate climates to initiate any serious evaporation. In hot climates, however, this, coupled with electro-osmosis, could cause a serious increase in groundbed resistance.

Perma Frost

In areas of perma frost, such as the tundra, the soil will change markedly in resistivity with the seasons. In particular, where the oil in the pipeline is heated to reduce its viscosity, there will be a thaw bubble around the pipe. This will vary in magnitude with the seasons, and possibly with the rate of flow of the oil, and there will be a thaw/melt zone in which it has been noted that there will be a decrease in resistivity in the thawed area close to the interface. This may be because the successive ice crystallizations of the soil moisture causes a concentration of salt. Anodes placed near such an area may possibly lie sometimes within the thaw bubble and at other times outside it, and occasionally straddle it in the low resistivity zone.

It may be possible under specific circumstances or by elegant engineering to use the soil heating effect around an impressed current anode to maintain a small thawed zone in otherwise frozen ground.

Variations in Anode Potential

In the arguments so far it has been assumed that the rods or anodes, including the conducting backfills, have been electrically at the same potential. With extended groundbeds and anodes where the resistance to earth is small there may be a significant potential change along the length of the anode structure. Consider a 1/4 in. diameter aluminum wire used in sea water as an anode to protect a moving ship. If this is streamed in the sea at a mean depth of 2 ft and the sea water has a resitivity of 20 ohm cm 100 ft of anode will have a resistance of

$$R = \rho \bullet \frac{1}{\pi l} \left(\ln \frac{2l}{\sqrt{ds}} - 1 \right)$$

$$= \frac{20}{\pi \times 100 \times 30} \left(\ln \frac{2 \times 100}{\sqrt{1/24}} - 1 \right)$$

$$= 12 \times 10^{-3} \text{ ohm at } 1.2 \text{ ohm per foot}$$

A 1/4 in. diameter aluminum conductor will have a resistance of 0.8 ohm per 1,000 yds or 0.27×10^{-3} ohm per ft, and the wire can be considered as a leaky transmission line when the attenuation factor would be

$$a = \sqrt{\frac{0.27 \times 10^{-3}}{1.2}} = 1.5 \times 10^{-2} \text{ ft}^{-1}$$

and the voltage at the feed end of the anode would be given by the expression

$$\begin{aligned} E_a &= E_m \cosh al \\ &= E_m \cosh 1.5 \\ &= E_m \times 2.4 \end{aligned}$$

where E_m is the voltage at the remote end. The resistance of the 100 ft of anode will be

$$\frac{E_a}{I_a} = \sqrt{\frac{r}{g}} \coth al$$

$$= \sqrt{0.27 \times 10^{-3} \times 1.2} \bullet \coth al$$

$$= 2 \times 10^{-2} \text{ ohms}$$

This resistance is almost twice that calculated for the equipotential wire and this increase is caused by the resistance of the metal itself. The essential feature for a significant increase to occur is that the resistance of the anode metal should be of the same order as the resistance of the anode shape through the electrolyte.

A second example of this phenomenon is afforded by the resistance of the coke breeze backfill surrounding graphite anodes. Suppose the anodes and backfill are laid to form one long horizontal rod groundbed with a continuous bed of coke breeze the graphite anodes being placed at intervals in it; each graphite anode is 5 ft long and they are placed at 15 ft centers. The coke breeze has a 1 sq ft section and this could have a 'metallic' resistance of 1/2 ohm per foot. In low resistivity soils, say 1,000 ohm cm, such a 100 ft groundbed would have a resistance of approximately 0.40 ohms so that the resistance per foot would be 40 ohm. The attenuation-factor in this case would be

$$a = \sqrt{\frac{0.5}{40}}$$

$$= 0.11 \text{ ft}^{-1}$$

so that over the distance of 5 ft the potential at the anode, that is E_a, would be related to the minimum potential between anodes E_m, by the equation

$$\begin{aligned} E_a &= E_m \cosh 0.55 \\ &= E_m \times 1.16 \end{aligned}$$

and the resistance of this 5 ft length of backfill will be

$$\begin{aligned} R &= \sqrt{0.5 \times 40} \coth 0.55 \\ &= 9 \text{ ohms} \end{aligned}$$

The resistance of each 15 ft length which includes one anode and 5 ft. each side of backfill, will be

$$\frac{1}{R_{15}} = \frac{1}{9} + \frac{1}{9} + \frac{5}{40} = \frac{25}{72}$$

$$R_{15} = 2.88 \text{ ohms}$$

whereas this would have had a resistance of 2.67 ohm if the groundbed were at equipotential.

Thus there is an 8 per cent increase in resistance of the groundbed by virtue of the metallic resistance of the coke breeze. A decrease in the soil resistivity, an increase in the coke breeze resistivity or an increase in the inter-anode spacing would cause a greater percentage increase, for example at an anode spacing of 25 ft with four anodes in the 100 ft groundbed the resistance is increased by nearly 30%. A reduction in the resistivity of the backfill below 5 ohm cm would mean an imperceptible change in groundbed resistance, but where precious metal anodes are to be used the criterion of 3 ohm cm might be more suitable.

Groundbed to Structure Distance

These arguments and equations apply to the resistance of the anode to an infinite shell electrode or infinite earth. While this might be considered reasonable for calculations where the groundbed is remote from the cathode it is not true for anodes that are placed close to the protected structure. The nearness of an anode is a difficult parameter to define or to determine although some idea of the effect can be judged if a typical cathodic protection installation is considered. Suppose a long pipeline is protected and at the mid-point between groundbeds the anode to pipe distance is 5 miles while the pipe approaches within 50 yds of the groundbed at its nearest point. The nearness, or rather the remoteness, of the groundbed may be estimated by considering the voltage drop that would occur in the vicinity of the cathode were this structure not present and the groundbed output was flowing to an infinite cathode. At a point 50 yards from such a 20 amp groundbed in 2,000 ohm cm soil the potential of the ground to infinity would be

$$V = \rho \frac{i}{2\pi d} = \frac{20 \times 20}{2\pi \times 50} = 1.25 \text{ V}$$

so that a pipeline that was laid to run within 50 yards of that groundbed could expect a considerable cathodic swing caused by the positive swing of the earth near the groundbed. If the pipe to groundbed distance had been 200 yds then the positive swing of the ground relative to an infinite cathode would be

$$\frac{20 \times 20}{2\pi \times 200} = 0.3 \text{ V}$$

In a practical installation this would be a negligible swing.

The change in the resistance between the anodes and cathodes in these two cases will not be great and any reduction in driving voltage will be less than the soil potential swings that have been calculated. The effect of the proximity of the anode upon the cathodic protection of the structure will be discussed later in the book.

One of the fields of cathodic protection where anodes are often mounted unavoidably close to the cathode is in the protection of marine structures. Sometimes, as in the case of ships' hulls, the demands of streamlining mean that the anode has to be mounted directly onto the structure and the calculation of the resistance of the anode in such cases is difficult. It is not possible to dissociate the effects of the two electrodes and the anode, or rather anode to cathode, resistance will depend as much upon the cathode parameters as upon those of the anode. Usually the effects of

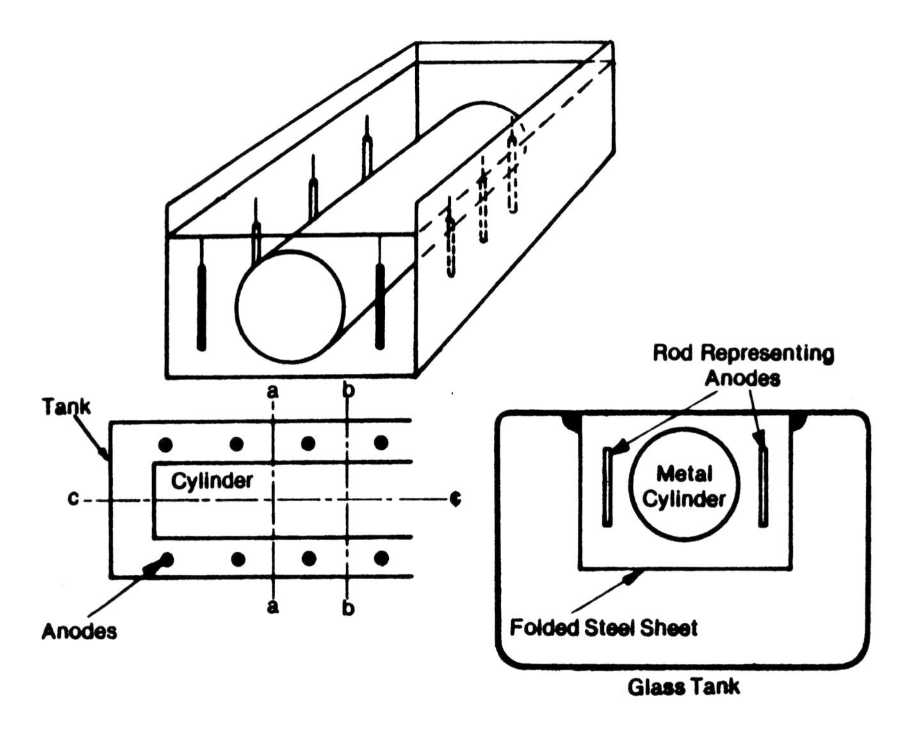

FIGURE 66 - Use of a model to solve a practical resistance problem.

polarization and of a resistive film upon the cathode, which have so far been ignored, will be most important. The engineering in this field is discussed in Chapter 7.

In certain cases a compromise calculation can be made as, for example, in the case of an anode in the water box of a condenser. Here an equivalent radius can be given to the anode and to the shell of the box and the resistance between two spheres of these radii calculated. The spheres can be assumed to be concentric for the calculation, eccentricity decreasing the calculated value of the resistance. Usually in cases of this kind calculations using various approximations to simple shapes and intelligent guesswork will give answers within the limits of the equipment available.

Models

In many problems it is possible to construct models of the anode and cathode and one advantage of these is that the size, shape and location of the anode can easily be altered. If such a model is constructed and surrounded by an electrolyte of the same resistivity as the actual electrolyte then the resistance of the model will be greater than that of the practical installation by the scale factor; a model constructed where 1 in. represents 1 ft will give a resistance 12 times the full scale value.

As an example of this technique, consider the protection of a steel cylindrical tank which is submerged in high resistivity water contained in a rectangular open top tank as illustrated in Fig. 66. It was decided to protect both the tanks, by suspending cylindrical anodes as shown in the figure; what would be the circuit resistance? By symmetry there would be no current flowing across planes *a* to *a'*, *b* to *b'* and *c* to *c'* so that the small section bounded by these planes could be considered. As a cylinder of metal was readily available it was decided that it would be as easy to deal with the section bounded by planes *a* to *a'* and *b* to *b'*. The water level in the tank would not be varied but the anode spacing could be varied by the designer so it seemed that the model might be more useful if it were turned on its side, the metal tank then being represented by a hollow box section with three metal and one insulated side. This was made by placing a piece of bent steel into a glass tank and sealing it against the bottom and one wall with a plastic compound. The metal cylinder was placed inside this and two small steel rods to represent the anodes were suspended by thread in the position they might occupy. The model was filled with water to a predetermined depth and the resistance between anode and cathode measured when the anode was at half the water depth. This arrangement was repeated with various depths of water, the anodes being moved in sympathy so that different anode spacings were represented.

In the particular case the water used in the model had a resistivity of 1,800 ohm cm while that used in the actual tank was 2,500 ohm cm. The model was made to a scale 1 in. to 2 ft so that the ratio between the resistance of the model and that found in practice would be

$$24 \times \frac{1,800}{2,500} = 17$$

The resistance found from the model was divided by 17 and the circuit resistance of the full scale installation determined. From the estimated current requirements and the back e m f's of the materials, the rectifier voltage was calculated. When the installation was completed, an accuracy of about 10 per cent had been achieved in the resistance model.

The use of such models, therefore, seems justified and as simple models can be made quickly, and anode shapes, sizes and positions altered, the method has much to recommend it.

From the experience gained with models and practical installations, and by the use of simple formulae the circuit resistance of an installation can be accurately estimated.

Computer Modeling

While there is a considerable satisfaction to the engineer to be able to use a model and to plot the potential around the anode and to make other determinations, the same results can be obtained using a computer. In the more sophisticated programs the surface polarization of the cathode can be built-in to the system, which is difficult in most modeling because the change in resistivity of the electrolyte to procure a good scaling factor will alter the polarization characteristics of the surface. The computer requires a considerable time to set up various programs so that while it will be used where there are multiplicities of applications which fall within its program, modeling for the unique cases may be the more economic method.

Sacrificial Anodes

CHAPTER 4

Mechanism

History

The earliest experiments on cathodic protection were performed with zinc anodes which were electrically connected to copper plates, and the whole immersed in sea water. In the large galvanic cell so formed, the zinc was the anode and corroded while the copper was the cathode; when sufficient current flowed the copper was cathodically protected. The zinc corroded to provide this current and is called a sacrificial anode. This method of cathodic protection can be used with other combinations of metals providing the electrical parameters are such that sufficient current can be generated. This necessary condition was demonstrated very simply by Faraday who experimented with zinc, or rather zinc-coated, nails and small iron plates. Placing both in small vessels containing fresh and salt water he showed that when the zinc and steel were in electrical contact in the sea water the steel did not corrode. When the zinc was not in electrical contact with the steel and also when there was electrical contact between them in fresh water, the steel corroded.

Anode Properties

The simple arrangement of anode and cathode in Fig 67 will have an equivalent circuit diagram as in Fig. 68. The voltage V_d will be the driving voltage between the polarized anode and protected cathode and the current will be given by the Ohm's law relationship

$$I = \frac{V_d}{R_a + R_m + R_c}$$

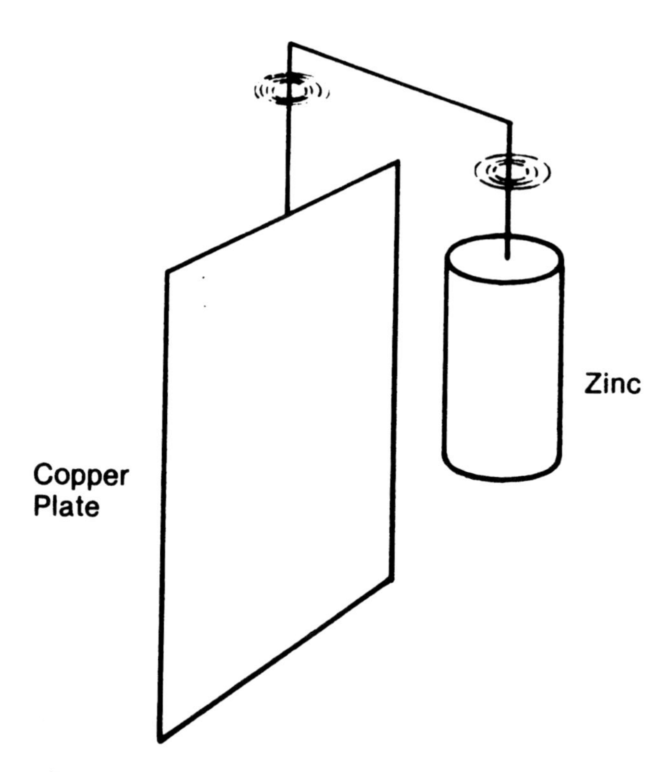

FIGURE 67 - Basic sacrificial anode protection.

where R_a is the anode resistance, R_c the cathode resistance, excluding those components of polarization, and R_m is the electronic resistance of the metal circuit. From this simple picture several necessary properties of the sacrificial anode material and of the anode itself emerge.

The anode must exhibit a driving voltage to the protected cathode, usually steel; that is the voltage of the anode when delivering current must be more negative than -0.85 V (or -0.95 V under anaerobic conditions) relative to a copper sulfate half cell. This driving voltage must be sufficient to overcome the ohmic resistances associated with the cathode and the anode.

For example consider the protection of the inside surface of a large steel pipeline containing water of 2,000 ohm cm resistivity where the pipe has a diameter of 3 ft so that its circumference will be 10 ft: the current required to protect the pipe metal may be 3 mA per sq ft or 30 mA per linear foot. Suppose that the pipe is to be protected by a continuous rod of sacrificial anode metal 1 ft diameter and concentric with the pipe. The resistance per foot between the concentric 1 ft diameter rod and the pipe wall will be given by the equation

$$R = \frac{2000}{30 \times 2\pi} \ln 3 = 11.6 \text{ ohms}$$

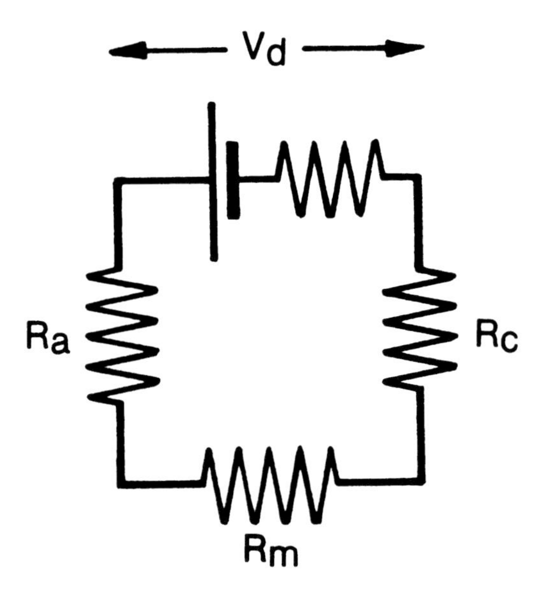

FIGURE 68 - Equivalent electrical circuit of Figure 67.

and at a current of 30 mA per foot there would be a voltage drop of 350 mV between the rod and the pipe. Therefore, a sacrificial anode system of this type would require an anode metal with at least 350 mV driving voltage. If the anode were constructed as a rod of 2 in. diameter then the resistance per foot run between the anode and the pipe wall would be

$$R = \frac{2000}{30 \times 2\pi} \ln 18 = 31 \text{ ohms}$$

and the driving voltage required of the anode metal would be almost 1 V. Similar calculations can be made with other sacrificial anode installations. The anode resistance may be varied independently of the cathode and so the anode shape and size can control the useful current output in a particular electrolyte.

While the anode metal must have sufficient driving voltage this, or rather the anode potential, must be predictable. Preferably it should be constant irrespective of the anode's environment and the current density off the anode surface.

If the anode potential varies with its environment then the anode becomes restricted in its use. The only consistent electrolyte is sea water, soils vary greatly from one area to another. The relationship between current density and driving potential is usually that the driving potential falls slightly with increase in current density. The driving voltage of the anode may vary with anode and electrolyte temperature. There may be some temperature hysterisis but both these properties, if they can be predicted, are acceptable.

While the anode material must have a predictable driving voltage initially, it is equally important that this remains constant throughout the anode's life, an initial rapid polarization to a subsequently constant potential is acceptable but a gradual decrease in driving potential is incompatible with a useful anode material. The consumption of the anode will cause an increase in anode resistance, but this is a regular, predictable decrease and can be overcome by suitable anode design or by an anode replacement program; a stub factor being included in the calculation.

Anode Life

The sacrificial anode must maintain its electro-chemical properties and must be consumed sufficiently slowly to give a reasonable life. The rate of consumption will depend upon the dissolution of the metal, as given by Faraday's Laws, and the efficiency with which this process occurs. The latter may be defined as the useful amp hours charge that is derived from the metal in practice compared with that which should be theoretically obtainable. Low efficiency of the anode metal may be caused by parasitic corrosion, that is the anode corrodes without delivering useful current, and if this tendency is great then the anode material may be unsuitable for practical use. When parasitic corrosion occurs the efficiency is usually a function of current density and will improve at high current densities.

An important factor in determining the useful life of an anode is the rate of volume consumption of the anode metal as it is the shape and size of the anode, rather than its mass, that affects its resistance in the electrolyte. In a particular metal the ratio between the anode resistance and the anode weight will determine the life that can be expected. In the same environment, the life of a sphere of anode material will be greater than that of a rod of the same weight.

As the resistivity of the electrolytes in which the anode may be used will vary from sea water at 20 ohm cm to soils at more than 2,000 ohm cm then the preferred shape and size of an anode will vary considerably. Practical difficulties in casting or forming the anode metals will limit the range of shapes and economic considerations will limit the number of sizes of anodes that can be stocked. In general, bulky anodes will be used in low resistivity environments and long, thin anode rods in high resistivity electrolytes.

The consumption of the anode will be proportional to the total charge delivered, but the rate of consumption at any point on the anode surface will depend upon its current density: at edges, corners and other sharp points the current density will be greatest and the anode will corrode most rapidly there. If the anode starts life as a right circular cylinder 9 in. diameter, 20 in. long, then it will corrode to an elongated ellipsoid about the size and shape of a rugby football and thence approach a more spherical form as corrosion proceeds.

The ratio of the surface area to the enclosed volume will depend not only upon the shape but the size of the anode. If a sphere of radius r is considered then its surface area will be $4\pi r^2$ and its volume $4/3\ \pi r^3$, thus the ratio of its surface area to its volume will be $3/r$ and so by altering r different volumes of metal can be enclosed per unit area of surface. Similarly the life of an anode in a particular electrolyte will depend upon the anode resistance and the weight of its metal. Again considering the sphere its resistance will be proportional to the inverse of its diameter and its weight proportional to the cube of this so that a longer life can be obtained by increasing the anode diameter and a much smaller life obtained by decreasing it. For example, 10 spherical anodes 3 in. in diameter connected in parallel and set some distance apart in the electrolyte, will give the same current as one 30 in. diameter sphere but will only have 1 per cent of its life.

A similar analysis can be applied to the use of a sacrificial anode to protect a structure, say a small anode protecting a water storage tank. If the dimensions of the tank and the anode are doubled then the area of the protected metal and hence the current required for protection will be increased fourfold. The anode resistance, which will be inversely proportional to its length, will only decrease to half its previous value; in order to maintain protection at the same current density the driving voltage of the anode would need to be doubled. If this were possible, say by an impressed voltage, the anode weight would have increased eightfold so that the anode would have a life of twice that of the original model. This problem will be discussed in greater detail later in the book when the application of sacrificial anodes to the protection of specific structures is considered.

Backfill

To achieve the maximum performance from sacrificial anodes they are occasionally surrounded by a special chemical mixture or backfill. This has two functions: firstly, it provides a uniform environment so that the anode output is predictable and constant and, as far as economically possible, the best that can be achieved from the anode metal; secondly, the backfill is of lower resistivity than the surrounding soil and so reduces the anode resistance. In sea water anodes require no backfill and in fresh water the anode is rarely treated with backfills, which are usually soluble, though sometimes

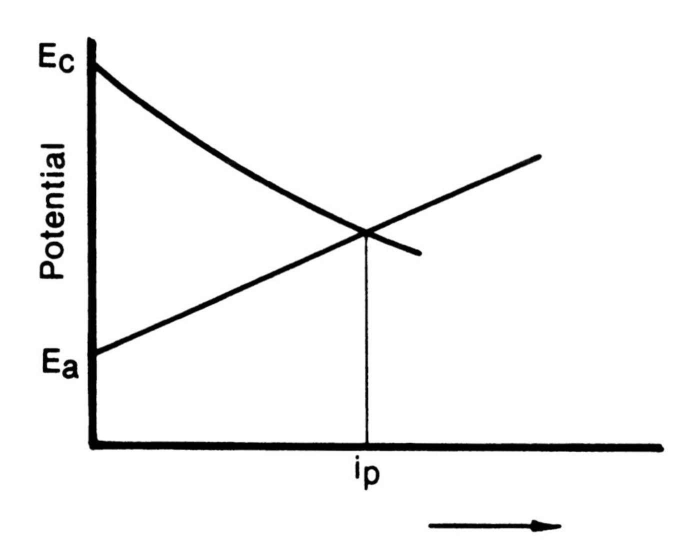

FIGURE 69 - Anode and cathode potentials as a function of protection current.

anodes are enclosed in a porous bag to prevent their corrosion product entering the water system.

Current Output

The current output of a sacrificial output will be determined by its electrical parameters. If the potentials of the anode and cathode relative to a remote electrode are plotted against the delivered current, then the behavior of the simple circuit shown in Fig. 68 will be as displayed in Fig. 69. The output of the anode will depend upon the slope of the anode curve and the open circuit potential of the anode. As there should be minimum polarization the major component in this slope will be the resistance of the anode in the electrolyte. To provide sufficient current to give protection the two curves must pass at the point i_p, the minimum protection current, in figure 69.

If the anode metal has a low driving voltage then the anode polarization curve must be flat and the anode constructed so that its resistance in the electrolyte will be low. Alternatively, if the anode metal has a high driving voltage, the anode polarization curve will be steeper. The resistance of the anode in the electrolyte will be determined by its shape and size and the electrolyte resistivity so that when the driving voltage is known the output of one or many anodes can be determined.

Often it is desirable to restrict some of the anode parameters. For example, the current density which, with certain metals, is kept low to avoid

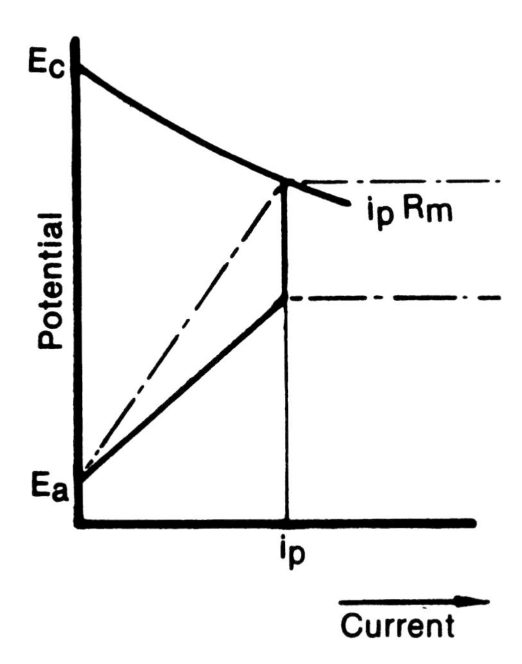

FIGURE 70 - Anode and cathode potentials against protection current in presence of resistor.

polarization, while with others it is maintained at a high level to achieve maximum efficiency.

When designing for a long life in a protective system there may be a problem that the anode will be too large and hence give out more than sufficient current for protection. To overcome this it is common practice to insert a resistor in the anode to cathode connection or, exceptionally, in the electrolyte path. This has the effect of increasing the anode polarization and potential measurements taken in the cell will now look as Fig. 70 when there is a metallic resistance, or as in Fig. 69 if the resistance is increased within the electrolyte. This may be by restricting the current path from the anode through part of the electrolyte or by partly blanketing or coating the anode itself.

Change in Protection Requirements

Within the above limitations it is possible to design a cathodic protection installation with sacrificial anodes. While such a design is relatively simple the natural parameters upon which it is based may not remain constant. The resistivity of the soil varies with the seasons and the consumption of the anode material brings about a change in the resistance of the anode.

The structure may equally be affected and its current requirement varied. This could be because the alkali nature of the cathodic reaction causes a polarization of calcium and magnesium salts or a high resistance coating, such as an asphalt, will deteriorate naturally, or even more rapidly, under the influence of electro-osmosis.

It is essential to study the effect of these changes on the cathodic protection by the sacrificial anodes and three systems will be compared; these are the low driving voltage system, the high driving voltage system, and a high driving voltage system with a resistor in the electrical circuit.

The first system employing anodes of low resistance is easily obtained in sea water with conventional ingot shapes, while in high resistivity soils long ribbons of metal might be employed. The second case would occur where a short life is required in a low resistivity environment or where the anode shape and size and the electrolyte resistivity demanded a high driving voltage. The third system would be employed where the resistance of a sufficient mass of anode metal to give the required life would be too low and give more than the required current at its driving voltage. These polarization curves are illustrated in Figs. 71 (a), (b) and (c).

Consumption of the Anode Metal

The consumption of the anode will reduce its metallic volume and will tend to convert the anode shape into a sphere. Where the anode shape was approximately spherical, then the resistance will increase as the inverse of the anode diameter, that is, for the anode resistance to double the volume must decrease to only one eighth of its original value, anode replacement would be advisable earlier. If the anodes are elongated then the increase in resistance with consumption would be more rapid and the anode resistance will have doubled when about 20 per cent of the original metal remained. At about this time replacement would be essential. The effect of this increase in anode resistance upon the cathodic protection system will be the same as in the more general cases discussed below. It is, of course, possible to design anodes, either by their shape or by the use of shields, so that their consumption has the minimum effect on the anode resistance.

Change in Electrolyte Resistivity

Variations in the resistivity in the area of the anode will affect the anode resistance directly; a 10 per cent increase in resistivity causing a 10 per cent increase in anode resistance. Any change in anode resistance will affect the cathodic protection system and the effect of halving and doubling the anode resistance is shown in Fig. 72, which reproduces the polarization curves on Fig. 71 with the modifications caused by the change in resistance.

Changes in anode resistance affect the second case most. The anode resistance almost completely controls the current in the circuit, while in the low driving voltage system the cathode polarization has a great influence on

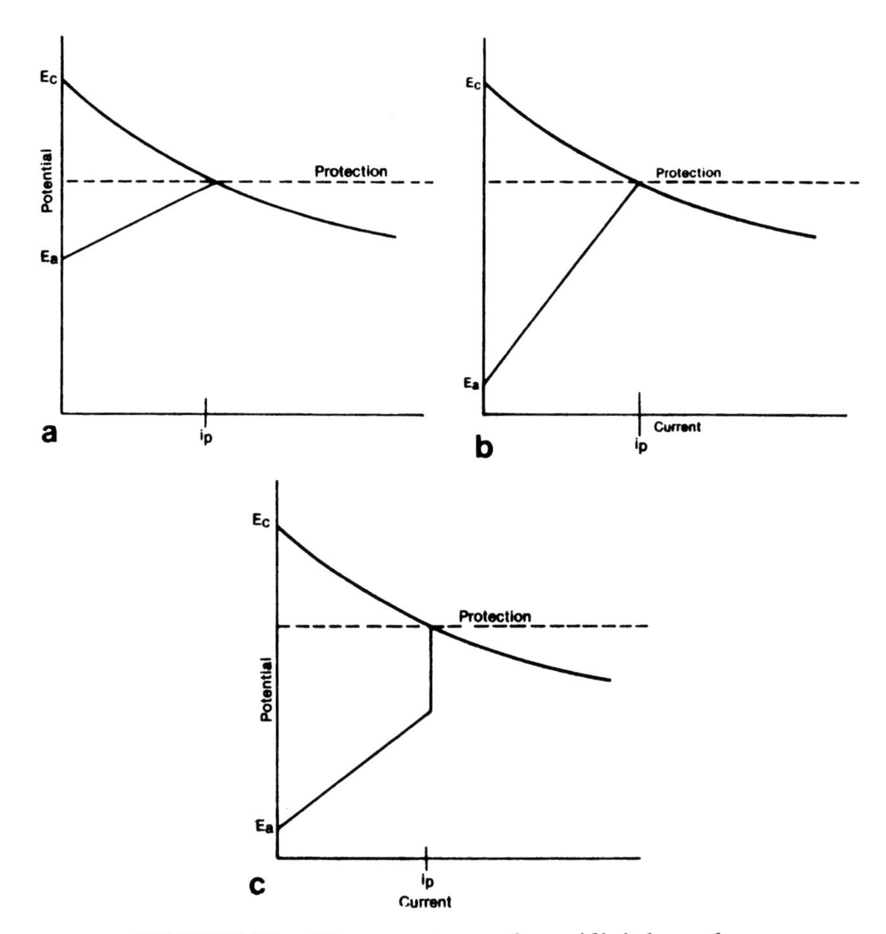

FIGURE 71 - Three systems of sacrificial anodes: (a) low driving voltage, (b) high driving voltage, and (c) high driving voltage with anode resistor.

total current, and in the third case it is the sum of the anode resistance and that of the resistor which affects the overall circuit.

Where there are regular, frequent variations in the electrolyte caused, for example, by the tide in an estuary, a degree of overprotection at the periods of greatest salinity, matched by a lower protection current at other times, can be used to overcome this problem.

Change in Current Required for Protection

Probably the most important variation in the conditions that can affect the cathodic protection system is a change in the current required to

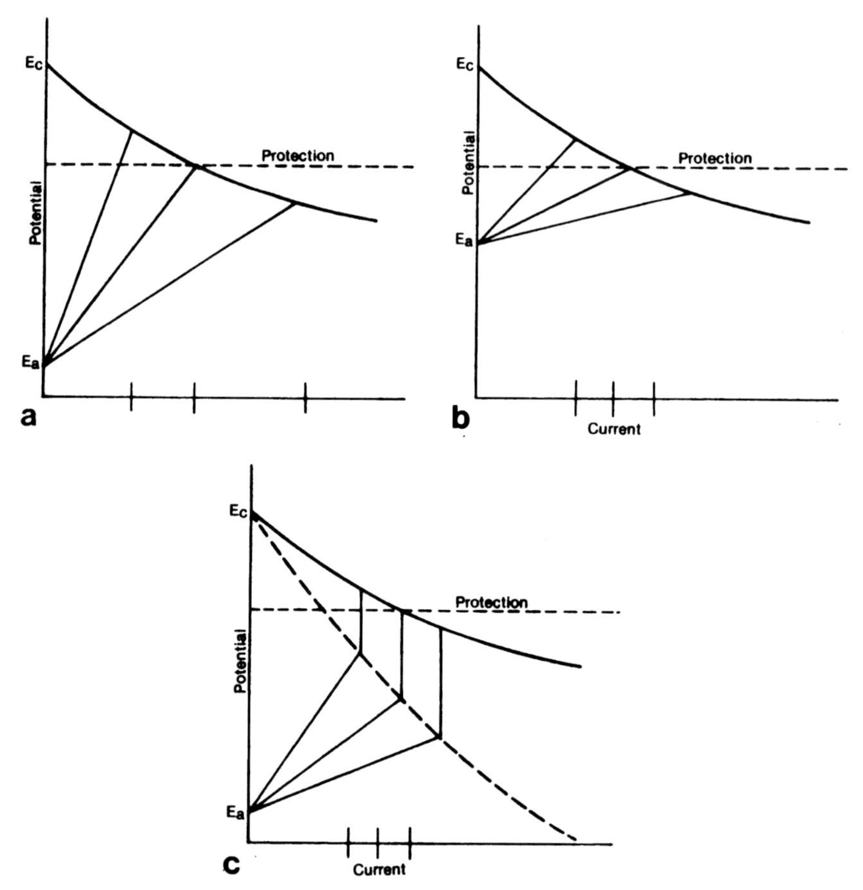

FIGURE 72 - The three systems in Figure 71 with variations in anode polarization.

achieve complete protection. There are a variety of causes that may bring this about and the variation in the current may be a 50 per cent decrease as in the case of marine chalking, or the current required may incease twenty-fold because of coating deterioration, or damage, on the cathode surface. The three systems considered are compared in Fig. 73 which shows the polarization curves with the cathode polarization slope of Figs. 71 and 72 doubled and halved. Again, the dual criteria of decrease in potential with underprotection and excessive current with overprotection are applied. The low driving potential source is clearly the best while both the high driving voltage sources are poor, the fixed resistance in the electrical lead causing case (c) to be identical with the high resistance anode polarization of case (b).

From these observations on the changes most likely to occur either

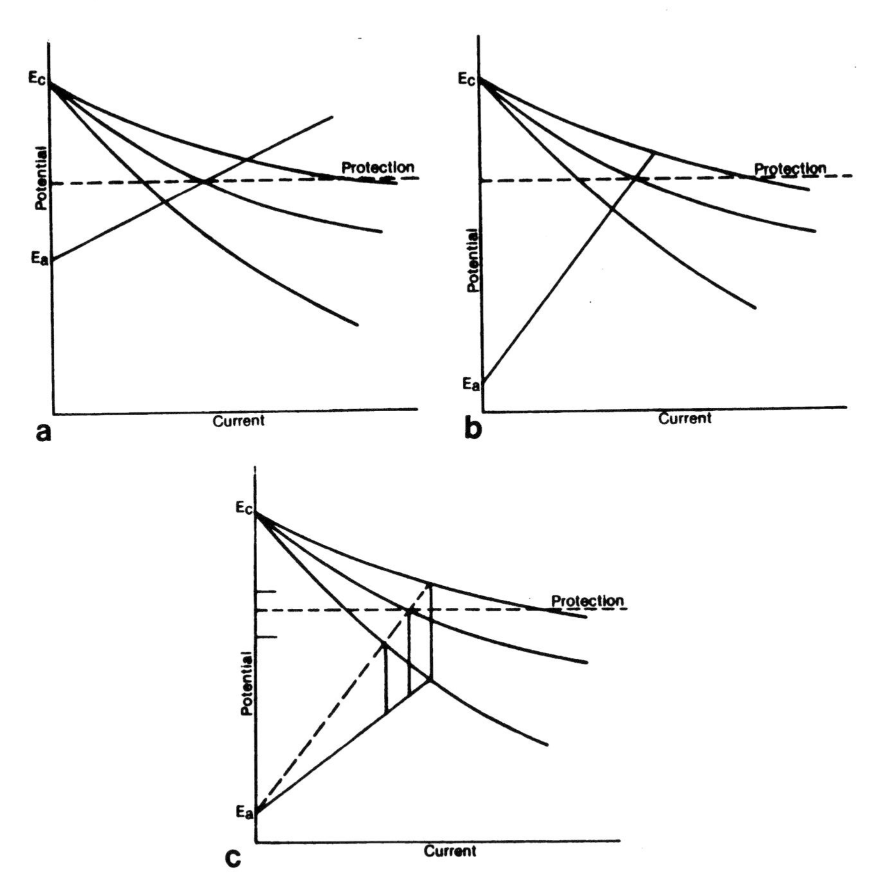

FIGURE 73 - The three systems in Figure 71 with variations in cathode polarization.

singly, or together, it appears that the low driving voltage anode makes the best sacrificial anode system, achieving a large degree of automatic control. This system has the added advantage that not only are variations in the conditions compensated but a similar compensation is enjoyed by the badly designed installation. It also has the widest variation in anode current so that its life is most affected.

Unfortunately, the bulk of the material required to operate this system in high resistivity soils will often rule it out. In this class of electrolyte a high anode resistance will be unavoidable and a high driving voltage anode metal must be used. However, in some cases of high resistivity environments the cathodic polarization measured to a remote point will be as high so that while protection will be achieved by a potential swing of 300 mV relative to a close half cell this may entail a swing of 600 mV to a remote

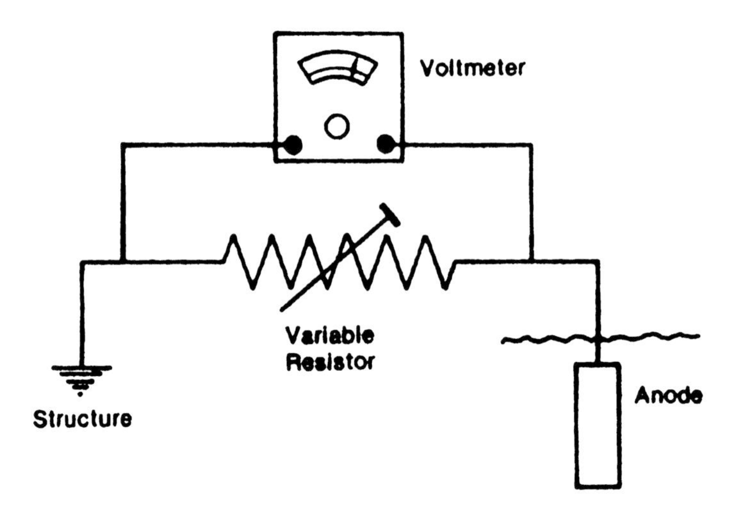

FIGURE 74 - Arrangement for variable resistance control.

electrode, thus an anode metal with a driving potential of 600 mV relative to protected steel will only have 300 mV drive to the polarized cathode and will act as a low driving voltage anode under such conditions. Each case will need to be considered on its own merits and the simple relationships established in Figs. 71, 72 and 73 determined to assess the degree and type of protection required.

Variable Resistor Control

The third case considered had a fixed resistor placed between the anode and the cathode. This could be replaced by a variable resistor which either manually or automatically reduced the driving potential of the anode by a fixed amount. Fig. 74 shows a simple circuit that would achieve this and the result of maintaining a constant potential means that a high driving voltage anode can be used with a low polarization curve, Fig 75. The advantage of this system is that the high driving voltage anodes have the ability to operate in a wider range of electrolytes than do some of the low driving voltage anodes. While manual control, except in a few cases, is labor consuming, modern solid state devices could achieve this automatically.

Economics

The economics of sacrificial anode protection are not difficult. The cost of the metal and its rate of consumption are known so that the cost can be formulated in terms of dollars per amp year. The installation is not difficult but it may be tedious where a large number of anodes are involved or

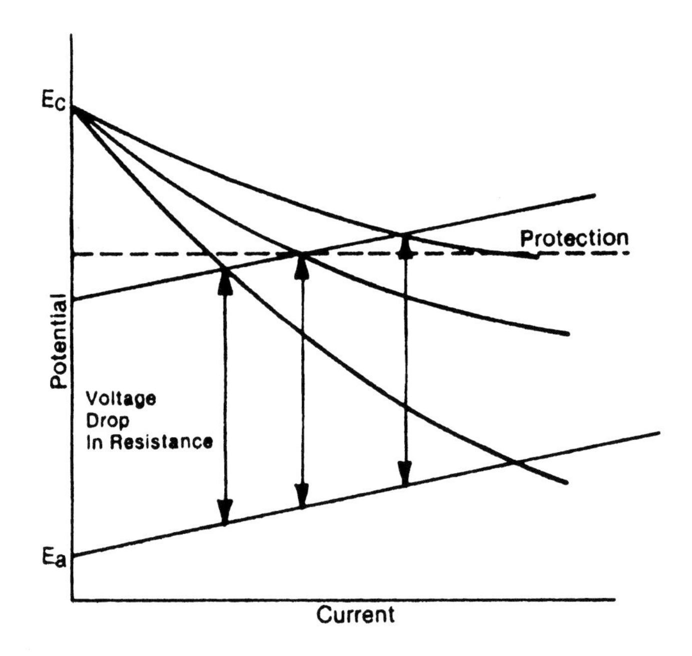

FIGURE 75 - Cathode polarization variations with control as in Figure 74.

where staging or other forms of access are expensive. Perhaps the greatest cost occurs in replacing and inspecting the system. The inspection is required to ensure that the structure is receiving adequate protection and would be necessary irrespective of the cathodic protection technique employed. The multiplicity of the sacrificial anodes can make such an inspection very costly.

Replacement of the consumed anodes following inspection will be comparatively easy and in many designs anodes can be added without the original anodes being removed. Other designs, however, require complete replacement of the anode and this leads to the loss of the anode 'stub' and a high engineering cost, often more than that of the original installation as anodes have to be removed and replaced.

If all these expenses are computed and included in the total then the cost can be calculated on the basis of dollars per ampere year. This figure will vary depending on the effectiveness of the system, the evenness with which the system is designed to give current, both in relation to the space around the anodes and over the period of protection.

The sacrificial anode system alternatively can be costed on the basis of an initial cost to cover a definite period and the maintenance and inspection costs during that time. In comparing a sacrificial anode system with an impressed current system there is a major distinction that has to be drawn. The sacrificial anode system will provide a capacity of so many ampere hours of protection. The impressed current system, on the other hand, will provide a specific maximum current. This will exceed the mean current required but will be independent to a very large extent of the time that the system is operating.

Anode Metals

The most widely used construction metal, and hence of most importance in cathodic protection, is steel. Three common metals are sufficiently base to be suitable for use as sacrificial anodes: magnesium, aluminum and zinc. All three metals or alloys of them will be discussed below.

Magnesium

The early marine application of zincs to ships' steel hulls was not successful and, on the revival of cathodic protection interest in the late forties, magnesium was investigated as a possible sacrificial anode metal. Pure magnesium should have a driving potential of 850 mV to protected steel but experiments with this showed that it corroded very rapidly with a very low efficiency. Some commercially available alloys with about 150 mV less driving voltage were tried and with some refinements proved to be satisfactory in a variety of electrolytes. These alloys have found extensive use as sacrificial anode metal. Magnesium is an extremely light metal and its properties are summarized in Table VI.

TABLE VI — Properties of Magnesium

Atomic weight	24.3
Valency	2 or possibly 1
Relative density	1.74
Melting point	650 C
Ampere hours per lb at valency 2	1000

Pure magnesium and most of its low alloys corrode rapidly in aqueous solutions, the pure metal dissolving in dilute acids with rapid evolution of hydrogen and generation of considerable heat; the metal is not amphoteric and is not vigorously attacked by alkalies. In aqueous electrolytes the dissolution is assumed to be by the direct displacement of hydrogen ions

$$Mg + 2H^{+} \rightarrow Mg^{++} + H_2\uparrow$$

This reaction proceeds irreversibly at a less negative potential than that suggested by theory and because of this the concentration of metallic salts in the electrolyte has little effect on the metal potential. Three electrochemical properties of magnesium suggest its usefulness as a sacrificial anode: it possesses a high driving voltage to protected steel, it has a low electrochemical equivalent and it has good anodic polarization properties. The potential of a piece of pure magnesium immersed in a dilute aqueous salt solution is – 1.70 V to a copper sulfate half cell; thus it is anodic to the common metals, steel, lead, aluminum, copper and zinc (both as the metal and as galvanizing) and with these metals magnesium has a considerable driving voltage even when they are fully cathodically protected.

If magnesium is considered to corrode by a divalent reaction then its theoretical electrochemical equivalent is 1,000 ampere hours per lb or 9 lb per ampere year. There is some doubt on this point and magnesium may corrode by a univalent reaction giving 500 ampere hours per lb or 17.5 lb per ampere year. Efficiencies of more than 50 per cent are seldom reported though several reliable cases of efficiencies up to 60 per cent tend to detract from the univalent theory. The subsequent oxidation of the univalent salts to the divalent form must also occur, but this reaction would not contribute to the useful anode current.

The products of the anodic reaction generally produce a highly soluble chloride and sulfate and in the presence of chloride or sulfate ions it suffers no polarization. In fresh water or electrolytes which contain none of these ions, the hydroxide and carbonate may form, but these do not seriously polarize the anode, the introduction of very small quantities of sulfate or chloride ions immediately causing depolarization. These ions are usually artificially introduced into the electrolyte as a backfill when a deficiency is expected; the hydroxide which is preferentially formed because of its low solubility, becomes enriched with the backfill anions and itself functions as a backfill. The metal is consumed uniformly and so the anode shape degenerates to a sphere and with well designed inserts all of the anode metal is available for sacrificial consumption. Three factors influence the efficiency of magnesium as an anode metal: the current density, the composition and the environment.

Current Density

The anode efficiency, that is the useful ampere hours per lb, increases at high current densities; composition has some effect upon this, but generally maximum efficiency is obtained at current densities above 100 mA per sq ft, Fig. 76 shows this relationship. The curve shown suggests a certain amount of parasitic corrosion, that is the metal corrodes at a certain rate even when it is not acting as an anode; the evidence for this is conflicting and it appears that a relatively high efficiency can be maintained at low current densities in electrolytes rich in chloride or sulfate ions. While a magnesium anode corrodes uniformly on a macroscopic scale at all current densities, the low current density attack is characterized by a considerable amount of pitting.

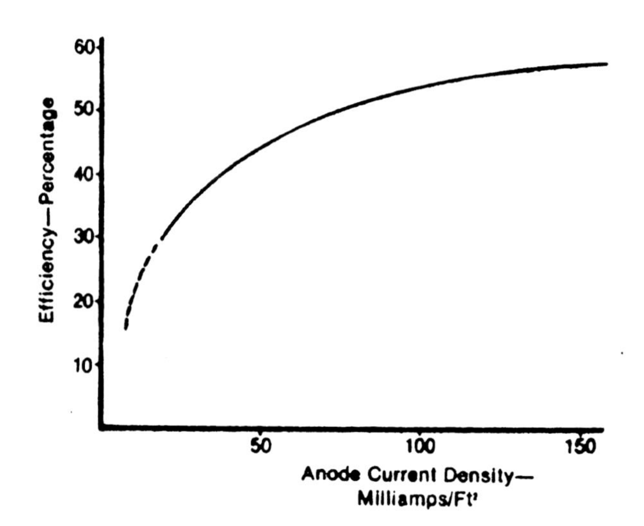

FIGURE 76 - Efficiency as a function of current density.

This process has been attributed to a parasitic corrosion, though selective corrosion would give the same effect. At high current densities the anode surface becomes smooth and this is most pronounced in sea water. The rate of corrosion, or rather the current density off a metal, usually influences its polarization, but with magnesium no such polarization occurs in the presence of chloride or sulfate ions. In solutions devoid of these ions but containing large amounts of phosphate, carbonate or hydroxyl ions, the working potential of the anode is considerably reduced, and in some environments the anode driving voltage may be halved at a current density of 500 mA per sq. ft.

Anode Composition

The early experience with magnesium and its alloys showed that while the pure metal had a higher driving voltage, some 150 mV greater, certain alloys, particularly the 6 per cent aluminum 3 per cent zinc type, gave much greater efficiencies. The reason for this appeared to be the very much lower rate of parasitic corrosion in the alloys; considerable attention was, therefore, paid to those impurities which would encourage parasitic corrosion. It seemed probable that the more noble metals, particularly those of low hydrogen overvoltage, would cause this type of attack and copper, nickel and iron are impurities of this class found in commercial magnesium. The noble metals with a high hydrogen overvoltage, that is lead, tin, cadmium and zinc appear to have little effect upon the anode performance.

As the alloy chosen as a starting point for these experiments contained 6 per cent aluminum and 3 per cent zinc, the effect of variations in the amounts of these was first investigated. No difference within the accuracy of the experiments could be found between alloys with an aluminum content in the range 5 per cent to 8 per cent or between alloys within the range 2 per cent to 4 per cent of zinc and the standard 6 per cent aluminum, 3 per cent zinc alloy was, therefore, adopted.

Iron was feared to be one of the major causes of parasitic corrosion but it was known that manganese had a very powerful scavenger effect upon it. This is found to be particularly high in magnesium and two processes are involved in reducing the effect of the iron; its content is reduced by the settling of the iron from the melt and the iron remaining in the alloy is surrounded by manganese so that it is replaced in its role as a cathodic impurity. Manganese is much less cathodic than iron and so is not dangerous as an impurity. In commercial anodes the iron content of the alloy is restricted to 0.003 per cent and the manganese is added to a minimum of 0.15 per cent. The improvement achieved with manganese depends upon the surrounding electrolyte and is most marked in sea water. There is considerable evidence to show that nickel can be tolerated up to 0.003 per cent before it has any effect upon anode efficiency although to some extent this property, as with several other minor impurities, depends upon the total amount of all the minor impurities. It seems prudent, therefore, to restrict the nickel content to below 0.003 per cent and preferably below 0.002 per cent.

The effect of copper on anode efficiency is difficult to assess as it is usually present in magnesium in company with silicon and sometimes with lead and tin. Manganese has a beneficial effect upon copper impurities somewhat similar to that found with iron and, with the manganese content required to counter the effect of iron, copper can be tolerated up to 0.1 per cent in sea water and to 0.03 per cent elsewhere. The other impurities, tin, silicon and lead show little detrimental effect at their normal levels of impurity; silicon should be restricted to 0.1 per cent, lead to 0.04 per cent and tin to less than 0.005 per cent.

Two series of anode alloys can be developed: firstly, the high purity anode for use in soils, and particularly high resistivity soils, and the lower purity anode—hence cheaper as more scrap metal can be used in its manufacture—for use in sea water. The compositions of the two alloys might well be as in Table VII.

Later work resulted in the production of a high purity magnesium anode with about 1 per cent of manganese in place of the aluminum and zinc. The anode has a higher driving voltage, about 200 mV more than the 6 per cent aluminum 3 per cent zinc alloys and so gives about 25 per cent higher driving voltage to polarized steel. The efficiency of the material is slightly lower than the alloys but as the anode is primarily intended for use in high resistivity soils when the increased current density off the anode will

TABLE VII — Magnesium Alloys

	Low Impurity (per cent)	Sea Water (per cent)
Copper	0.02 maximum	0.03 maximum
Iron	0.003 maximum	0.003 maximum
Nickel	0.002 maximum	
Manganese	0.15 minimum	0.10 minimum
Silicon	0.10 maximum	

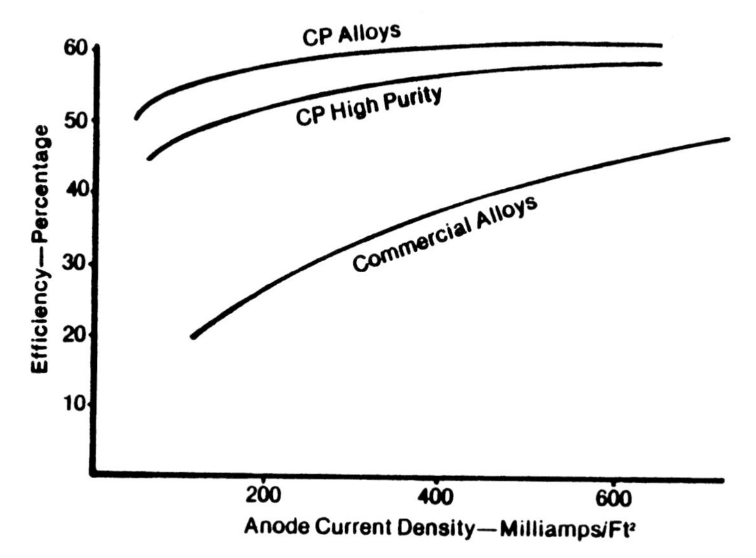

FIGURE 77 - Efficiency versus current density for three magnesium alloys.

enable it to work at efficiencies comparable with the best of the alloys. Fig. 77 shows the efficiencies generally reported for the various anode metals in saturated calcium sulfate solution.

The success of the high voltage alloy lies in its micro-structure and the addition of considerable manganese to the metal. This is in large enough quantities to be retained in the metal during cooling and not be precipitated. Considerable reduction of the effect of iron is, therefore, achieved by transfer of the iron into solution in the manganese. The properties achieved in this high voltage anode depend upon the cooling history of the anode.

Environment

Magnesium anodes operate at peak performance in sea water, the low resistivity high chloride content and the ease with which the corrosion product is washed off by movement of the water contribute to this. In soils which contain sulfate ions, most agricultural land does in the form of gypsum, the anode operates well; good drainage into the anode is necessary and this is usually obtained automatically where the anode is buried below the water table. In dry soils and those denuded of sulfates and chlorides the anode tends to polarize and as these soils will display a high resistivity, anode efficiency is reduced by the low current density achieved. Similar conditions exist in fresh water though polarization is reduced by sedimentation of the corrosion product. This is not harmful in drinking water, indeed magnesium and aluminum salts achieve some success as stomach powders.

Where the soil conditions are poor it is customary to surround magnesium anodes with a mixture of chemicals or backfill. This has two functions: it provides a suitable environment for the anode so that the best performance is obtained from it, and it reduces the soil resistivity in the immediate neighborhood of the anode, hence reducing the anode resistance and allowing it to work at a higher current density, both of which effects are beneficial to magnesium. The backfills recommended for this purpose are numerous and are usually mixtures of chemicals with the natural soil from the anode hole or with bentonite, a high water absorption class of clay. The chemicals used are gypsum, sodium sulfate and common salt, the last, having a detrimental effect on plants, is used sparingly while gypsum is common and cheap, so mixtures of this with a smaller amount of sodium sulfate are popular. There are two methods of applying the backfill: the chemicals and bentonite may surround the anode in a porous cotton bag and the whole be lowered into the ground as a unit, or the chemicals may be mixed on site with local soil or imported clay and poured as a slurry into the anode hole around the magnesium. The results of the two methods are similar, fig 78.

Magnesium is used as a high driving voltage anode; it is universally applicable and is little affected by its environment or, where it is, packaging and backfilling techniques overcome this. The metal is light and so has a high volume consumption and this, coupled with a poor efficiency at low current densities, means that the anode cannot generally be used economically to give a long life. The consumption of the metal will be about 17 lb per ampere year. Wastage of the anode metal on replacement of the anode stub will increase this consumption by 3 or 4 lb per ampere year. Magnesium is a comparatively expensive metal and at present-day prices its cost per ampere year is much greater than the other sacrificial anode metals.

In massive applications in sea water, such as are sometimes required in the offshore industry, the high driving voltage of magnesium can be used to great advantage. Anodes can be engineered that are simpler, and hence

FIGURE 78 - Cast magnesium anode with connecting cable. (Photo reprinted with permission from Impalloy, Ltd., Bloxwich, Walsall, England.)

very much cheaper to fit, either as replacements or for augmenting an inadequate cathodic protection system and the bulk of the anode means that a sensible current density is maintained off the surface coupled with a reasonably long life.

The overall cost of these systems can be several times cheaper than the equivalent zinc or aluminum anodes, which are limited in their scope and must generally be applied in very large numbers.

Zinc

The early experiments with zinc on copper-sheathed hulls or warships were made over 150 years ago, and the efficacy of this cathodic protection caused the zincs to be used on steel hulls when they replaced wooden vessels. The zincs were placed in the stern region near to the propeller and close to other areas of yellow metal; they failed, however, to produce cathodic protection on the steel hulls or even to protect the yellow metal propellers. Zinc should give a potential of − 1.10 V to copper sulfate and so should have a driving voltage of 0.25 V to cathodically protected steel. It is a fairly common metal and some of its properties are listed below in Table VIII.

TABLE VIII — Properties of Zinc

Atomic weight	65.4
Valency	2
Relative density	7.1
Melting point	420 °C
Theoretical ampere hours/pound	372

The failure of the marine zincs or protectors was due to the formation of a dense adherent corrosion product on the metal, and the advantages that could be gained by the development of a non-polarizing zinc anode are very considerable. Several groups of scientists worked to achieve this end and as a result zinc alloys are available for use as sacrificial anodes. The potential of the zinc anodes is 250 mV more base than protected steel, that is −1.10 V to a copper sulfate half cell. This is the no-current potential of the anode, which is unaffected by environment, special backfills being used to prevent heavy current polarization of the anode in the soil. The anode polarization film is very loosely adherent in sea water and can be removed from a suspended anode by lifting it out of the water or equally by the effect of a tide stream, specific alloys being particularly good in this respect.

The corrosion of the zinc is by a divalent reaction at nearly 100 per cent efficiency: the theoretical anode consumption is 23 lb per ampere year and in practice 25 lb per ampere yere are consumed. Zinc suspended in sea water or buried in the ground does not corrode rapidly by parasitic reactions and the anode alloys maintain their high efficiency at very low current densities. The corrosion pattern of the zinc anodes is very uniform though ultra high purity zinc anodes corrode leaving a large grain structure visible and this is likely to cause additional loss of metal towards the end of the anodes' life by electrical insulation occurring between the grains, by grains falling out of the matrix or breaking away from the insert. The large-scale corrosion pattern is uniform and anode shapes tend to degenerate to ellipsoids and spheres. Steel inserts can be bonded to the metal if they are galvanized or sherardized.

Composition and Environment

The performance of a zinc anode depends on two parameters, the particular zinc alloy, or rather its impurities, used to make the anode and the environment into which the anode is placed.

If a pure zinc rod, that is pure to within a few parts per million of any impurity, is used as an anode, it polarizes by about 20 mV and then gives an excellent performance. The consumption is low and this is not affected by variations in the current density. Commercially, 99.99 per cent zinc is available and it would be of considerable advantage if this could be used, even with some modification, as an anode material. The properties and composition of the zinc alloy for use in sacrificial anodes have been resolved into two distinct requirements according to whether the anode is to find application in the marine field or in the non-marine field. In the former it is generally agreed that the major harmful impurity in commercially pure zinc is iron even when it is present in minute quantities, and an acceptable performance is obtained when the iron content is less than 15 ppm though improved performance occurs when even less iron is present.

The addition of aluminum has proved to be of doubtful advantage but certainly it is not detrimental; the iron content of the aluminum addition must be kept exceptionally low to avoid contamination of the zinc. Two metals, both in the presence of aluminum, have proved to have a large beneficial effect on pure zinc of comparatively high iron content. Silicon, if added with aluminum, has an effect similar to that of manganese in the magnesium anodes, the iron being partly lost from the melt and that which remains being entrapped by the silicon, which results in an alloy with an almost iron-free zinc performance. The technique of alloying is not difficult and the small grain size of the resulting alloy gives a better corrosion pattern and a structurally stronger metal. This work has been extensively reported by Crennell and Wheeler. In America cadmium was found to confer equally beneficial properties as silicon when added with aluminum. Other metals, manganese, magnesium and calcium have been suggested though no successful use of them is reported, while techniques such as amalgamation of the anode surface have been successful but do not lend themselves to practical application.

In soils, or rather backfills, the standard 99.99 per cent pure zinc is satisfactory. The best backfill material, as with magnesium, is a mixture of gypsum, sodium sulfate and bentonite clay, while phosphates, sodium carbonate or hydroxide and ammonium salts are to be avoided. The addition of zinc sulfate to some backfills during field trials has a slight beneficial effect but this was probably caused by the lower resistivity of the backfill rather than an electrochemical effect. With the very low driving voltage available it is important that the backfill surrounding a zinc anode should be free of voids and holes, and care in making up and filling the anode hole is necessary. It is essential that the anode hole remains wet, and besides ensuring that the anode is below the water table, a hygroscopic wetting agent could be added with advantage.

Zinc can be used to protect installations in fresh water where the 99.99 per cent pure zinc is suitable though a low iron content gives some improvement. The current required to protect bare steel will generally preclude the use of zinc where the water resistivity is greater than 500 ohm cm, while galvanized, aluminized or painted steel structures can usually be economically protected in most fresh waters. Potable water may be tasted by as little as fifteen parts per million of zinc salts dissolved in it and vomiting may result from drinking higher concentrations. Neither of these levels is likely to occur in normal practice when water storage or header tanks are cathodically protected. The toxicity of the corrosion product is of use in sea water where the anode does not foul, and it acts as a bactericide in anaerobic conditions.

Zinc can be used as an anode material with a low driving voltage. The high efficiency means that only 24 to 25 lb. of zinc are consumed to give 1 ampere year of charge and with anode stub wastage this figure is increased to 28 or 30 lb. The material is dense, has a low volume consumption and the efficiency remains high at all practical current densities so that long life anodes are easily designed. Zinc at present prices is about half the cost of magnesium when considered on an ampere year basis and to this in many cases must be added the extra efficiency of the low driving voltage system over that of the high driving potential metal.

The low driving voltage available limits the use of the metal to low resistivity soils except when it is used to protect very well coated structures. Zinc is also temperature-sensitive and inversion of the driving potential relative to steel has been found to occur above 70 °C in hot water and this phenomenon has been reported in muds containing small amounts of chloride and sulfate ions. Zinc cannot, therefore, be used to protect galvanized hot water storage cylinders.

One use of zinc metal anodes that will find increasing application is the cathodic protection of aluminum. Cathodic and anodic corrosion of this metal can occur and experiments on 99 per cent and 99.5 per cent aluminum and some structural alloys show that zinc provides adequate protection without the possibility of an all-aluminum structure suffering cathodic corrosion. This intrinsic safety of zinc will be much more acceptable than either anode or resistor control of magnesium anodes or manually controlled impressed current cathodic protection. Similarly, it can be used to protect reinforcing bars in marine concrete without the possibility of hydrogen debonding, Figure 79.

Aluminum

The third common metal that should act as a sacrificial anode to steel is aluminum and its position in the electrochemical series suggests a high driving potential to steel as it lies between magnesium and zinc. Structural aluminum displays potentials which, for the unalloyed metal, that is 99.5 per cent pure, lie in the range −0.6 V to −0.8 V and varies with its environments, while aluminum alloys exhibit a variety of potentials, some sufficiently base that the alloy may be used as a cladding on the more noble alloys, the function of which parallels the use of galvanizing on steel. Aluminum has attractive electrochemical and mechanical properties and these are reproduced in Table IX.

TABLE IX — Properties of Aluminum

Atomic weight	27.0
Valency	3.0
Relative density	2.7
Melting point	660 C
Theoretical ampere hours per lb	1,350

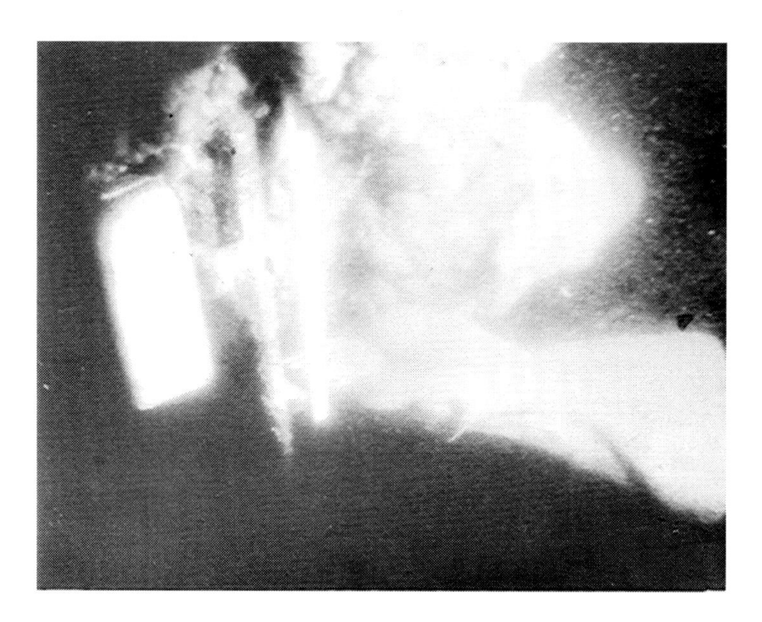

FIGURE 79 - Underwater attachment of zinc anode clamped to exposed rebar in concrete pile. (Photo reprinted with permission from Federated Metals Corp., Somerville, New Jersey.)

In most galvanic couples aluminum corrodes being anodic to steel, copper, brass and bronze. This corrosion in water and soil usually occurs as small pits on the metal surface which reach a maximum depth and are then stifled, though failure of many thin wall aluminum structures would occur by this process. The cladding technique has been developed to prevent pitting and by a sacrificial action distribute the corrosion which in terms of loss of metal is very small. One of the chief causes of aluminum corrosion is stagnation or small areas of differential aeration caused either by sedimentation or intrinsic in the structure design.

Early Alloys

A great deal of work was done in the 1950's on the development of an aluminum anode for use in sea water. The possibility of producing a very attractive anode material is great since at 100 per cent efficiency only 6.5 lb. of aluminum would be required to provide 1 amp year of charge and aluminum is comparatively cheap and abundant.

Tests were completed on three groups of alloys but they all proved to be no more than substitutes for zinc, providing no better economy and virtually no better performance.

The earliest of these was based on an aluminum zinc containing about 3 per cent zinc. The anode was improved by some forms of heat treatment and it was suggested that high purity aluminum produced a superior anode. It obtained an efficiency not much better than 50 per cent, at which point it broke even with zinc. Practical application of the anode on ships' sterns and in some water tanks confirmed this. For use in the soil, special backfills were required, but apart from experimental use it appeared to have little application.

A series of alloys of aluminum and tin, and aluminum, zinc, and tin were developed which showed a more base potential of -1.3 V to copper sulfate which is intermediate between magnesium and zinc in driving voltage. The alloys were improved by heat treatment and less than 0.5 per cent of tin was used to make a working anode metal. The driving voltage varied with concentration of sodium chloride and with current densities. The efficiency of the anodes was variable, but did provide practical cathodic protection, though the additional driving voltage was of no consequence as the anodes only operated marginally economically when they were in sea water.

The third group of anodes which were developed used mercury as the alloy ingredient or contained in the backfill. A number of anodes were developed, none of which seemed to show improvements noticeably better than the 3 per cent zinc anodes.

Present Anodes

In the 1960's there was a considerable study of binary, ternary and quaternary alloys of aluminum. It was found that manganese and copper caused a positive shift in the potential of pure aluminum, while mercury, tin and indium caused a negative change, with cadmium, magnesium and zinc bringing about only a slight negative change in potential. Two ternary alloys—aluminum zinc mercury and aluminum zinc indium—showed high capacities at potentials about the zinc potential. The former alloy had an immediate impact on the market and is extensively used offshore.

Problems occurred when the anode was used in environments that were not as encouraging as the Gulf of Mexico and a second generation of these alloys was produced with a higher purity and a larger quantity of zinc which operated in a wider range of electrolytes but with a small decrease in efficiency. A variation of the alloy was capable of operating in saline muds, though the density of the mud has a considerable effect on the operating efficiency. In certain types of compacted mud there may be a drop to 60% efficiency.

The basic feedstock of the aluminum was found to have a considerable effect on the anode performance. In the United States virgin aluminum has a lower iron content than European production, though silicon levels are often higher. Silicon is capable of sequestering the free iron and a careful balance of the silicon to iron ratio can give galvanic efficiencies of better than 90%. In Europe a series of alloys containing significant amounts of magnesium, upwards of 8%, were found to give high strength alloys with better casting characteristics, though with some loss of efficiency.

Copper has a dramatic effect on the anode alloy and it is suggested that a minimum of 0.02% mercury is necessary for reasonable anode activity. With low iron and copper levels consistently good performance can be achieved with 0.4% zinc and 0.04% mercury. A higher level of zinc is of benefit in anaerobic mud.

Environmentally there was great concern at the use of mercury in cathodic protection anodes, and it was feared that mercury may pollute sea life, particularly crustaces, in the area around an offshore platform. There was evidence of this in a report made off the coast of California where zinc was found in very heavy concentrations, though it was not, remarkably, attributed to the sacrificial anodes. The second group of alloys developed are all of a generic strain, which use indium as the principal alloying material, usually with high purity aluminum and some zinc; the capacity of the anodes is high and they operate at close to 90 per cent efficiency. They exhibit higher driving voltage than the mercury anodes, usually at −1.10 V to silver chloride or 0.3 volt to protected steel. This represents a 20 per cent increase in the driving voltage, or where the steel is overprotected, as it initially would be in a long life offshore structure, it may increase the driving voltage by one third. These anodes, again, are capable of operating in saline mud, which they do with less loss of efficiency than the mercury alloys. Japanese workers have suggested the addition of cadmium to the alloy which improves the corrosion pattern on the anode and softens the corrosion product.

The increase in the efficiency of aluminum has given it a considerable economic margin over zinc. There is some reluctance to use these alloys in all applications; in anaerobic conditions, particularly where mud such as drilling mud may cover the anodes, the use of zinc with its corrosion product acting as a bactericide, is favored.

The indium and mercury alloys of aluminum tend to lose their efficiency at low current densities and to lose both efficiency and driving voltage in brackish water. The efficiency is effected by up to 20 per cent in an estuary which has 5 per cent sea water. The anode potential is reduced, but only markedly when operating at high current densities which it would be difficult to achieve in that resistivity of water.

In hot saline mud, usually when used as bracelet anodes on hot oil lines, many of the aluminum anodes lose considerably in efficiency and also suffer a small decrease in driving voltage. The indium alloys suffer less in

TABLE X — Typical Aluminum Anode Alloy Additions
Percentage by weight

Mercury	0.03-0.05	0.03-0.05	0.03-0.05	—	—	—
Indium	—	—	—	0.01-0.03	0.01-0.02	0.005-0.05
Zinc	1.3-2.7	0.35-1.00	3.0-6.0	4.0-6.0	2.0-4.0	0.5-5.0
Iron	0.10 max	0.13 max	0.13 max	0.1 max	0.12 max	0.13 max
Silicon	0.10 max	0.11-0.21	0.11-0.21	0.1 max	0.08-0.2	0.10 max
Copper Nickel	0.005 max	0.006 max	0.006	0.005 max	0.006	0.01 max
Mg, Mn	0.01 max	—	—	0.01 max	—	—
Others each	—	0.02 max	0.02 max	—	0.02 max	0.02 max
Capacity						
lbs/Amp year	6.8	6.8	7.1	7.3	7.6	7.6
Kg/Amp year	3.1	3.1	3.2	3.3	3.4	3.4

this respect than do the mercury alloys. The mercury alloys initially have a higher efficiency and the transition point is reached between 40 and 60 degrees centigrade.

The general composition of these alloys is given in Table X. In the general design of a cathodic protection anode, particularly those used offshore, the problems of casting the anode tend to restrict them to rectangular shapes about 8 ft long and 12 in × 10 in. The current density off the anode surface is about half the mean at the center points of the major faces, and about twice the mean at the ends of the anode. This could lead to a decrease in the average efficiency of the anode and a reduction in its useful life.

Some aluminum anodes are produced by continuous casting. This leads to an anode free of voids and cracks. The anode insert has to be continuous through the center of the anode and so the shapes and the insert configuration are considerably limited.

Aluminum anodes require some skill in casting and surface cracking, particularly on the open mold surface, can penetrate a considerable distance into the anode, Figure 80. In general, these voids are of little consequence to the anode in terms of its operational efficiency unless they reach through to the insert. When they do they can mechanically weaken the anode and sea water entering a crack can cause corrosion to take place between the anode and the insert. The corrosion at the foot of a void can be prevented by filling the crack with insulating compound, either a resin or a hot setting compound such as a bitumen.

Much of the information on sacrificial anodes is derived from laboratory or control tests which are carried out under laboratory-type conditions. Reports from companies using these anodes in the North Sea suggest that they may only be operating at 60 per cent efficiency. This is revealed on an analysis of the results which are often reported as indicating that there is an increased current requirement in the North Sea based on the fact that the anodes must be operating at almost 100 per cent efficiency or that there is

FIGURE 80 - Casting aluminum anodes for offshore protection. (Photo reprinted with permission from CORRINTEC/USA, Cathodic Protection Services, Inc., Houston, Texas.)

an error in the techniques of calculating anode resistance which is corrected to support a high efficiency of the anode.

The techniques of monitoring anode output, though fully developed for a number of years, have not found even minimum application and without this practical evidence there must be some concern over the anode efficiency.

Use of aluminum anodes to protect structural aluminum requires considerable care in order to avoid smears of the anode alloy onto the protected aluminum where it may promote rapid corrosion.

Anode Design

Shapes and Sizes

Sacrificial anodes are designed to give a particular current output for a definite time and so are usually designed to be multiples or fractions of the weight of metal consumed in 1 ampere year; zinc anodes are usually made in a series of multiples of 30 lb and magnesium in multiples of either 17 lbs or 22 lb. Both these anode metals can be cast easily and the simplest technique produces a D section ingot from an open top mold. Anodes produced by this method are usually several times as long as they are wide and are suitable for a large number of applications including packaged anodes. Anodes for high resistivity soils are made into very long thin rods while anode shapes for use in sea water and other low resistivity environments are short and thick and tend to be almost spherical, a hemispherical anode being popular for long life marine applications.

The long thin anodes of magnesium are made by extruding the metal as a rectangular or circular rod over a galvanized or aluminized steel core, the core or insert being used to give the ribbon strength and to maintain continuity after the sacrificial metal has been consumed. Zinc may be obtained in similar shapes though a continuous casting technique is often employed. Most aluminum anodes are being cast into rods or other simple shapes; continuous casting and extrusion of the alloys is simple if a little more costly. The widest use of aluminum anodes is in offshore structures where they are generally made as large as possible within the scope of casting and handling or are formed into special shapes, such as bracelet anodes described later.

Inserts

Inserts for the anodes are usually made of steel and are galvanized or aluminized to give a better bond to the anode metal. This bond is not usually sufficient and keying of the anode to the insert is necessary as plain rods or bars used as inserts have been found to fall out. Several simple designs of insert have proved to be successful and these may involve making holes or indentations in a strip or tube or the use of spiral or parallel wires or strips. During the cooling of the anode the insert will be subject to the contracting stress of the metal and so must be robust.

With magnesium anodes and to a lesser degree with aluminum and zinc the anode metal nearest to the exposed part of the insert will corrode rapidly, particularly in low resistivity soils and water because of the local bimetallic effect. This can be prevented by insulating the insert where it leaves the metal either by casting a hole close to the insert and filling with a non-conducting compound or by coating the insert and possibly part of the anode with an insulating material. Some of the failures of magnesium

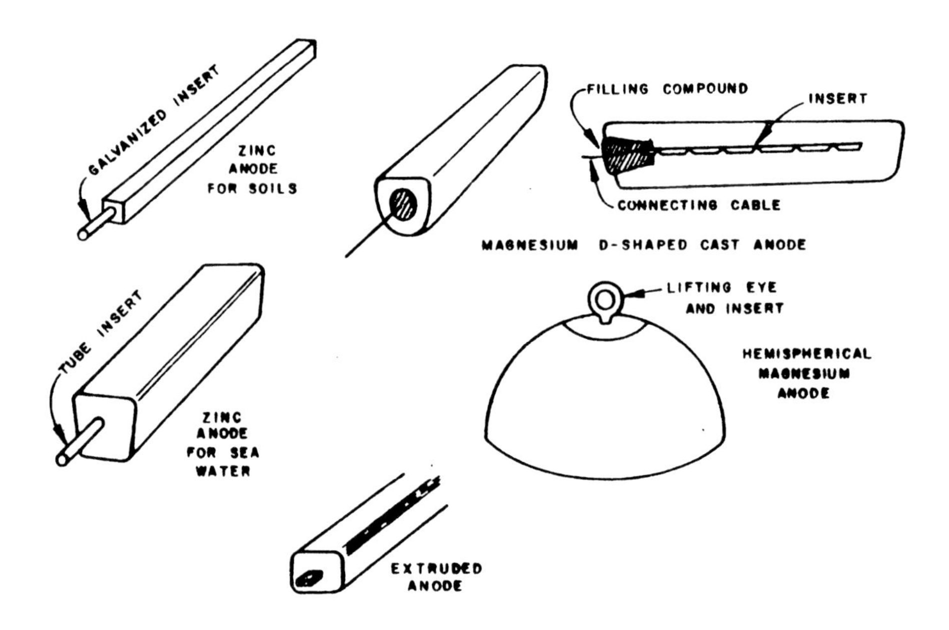

FIGURE 81 - Selection of sacrificial anodes.

FIGURE 82 - 'D' shaped cast sac. anodes. (Photo reprinted with permission from CORRINTEC/USA, Cathodic Protection Services, Inc., Houston, Texas.)

anodes, particularly those that were suspended from deckheads in tankers, suffered from this type of failure where during the non-ballast period sea water was retained on top of the anode next to the insert and the bi-metallic reaction continued.

Many anodes will be specifically designed for the installation in which they are used. These anodes will be described later in the book when the protection of particular installations is discussed. Anodes of a more general character are illustrated in Fig. 81, which shows cast magnesium, zinc and aluminum anodes.

FIGURE 83 - CORRINTEC zinc condenser anode. (Photo reprinted with permission from CORRINTEC/USA, Cathodic Protection Services, Inc., Houston, Texas.)

FIGURE 84 - Stand off Galvulum III anode for offshore platforms. (Photo reprinted with permission from CORRINTEC/USA, Cathodic Protection Services, Inc., Houston, Texas.)

FIGURE 85 - Cast zinc anodes with rod insert. (Photo reprinted with permission from Impalloy, Ltd., Bloxwich, Walsall, England.)

FIGURE 86 - Streamline-shaped zinc anode for hull attachment by welded lugs. (Photo reprinted with permission from CORRINTEC/USA, Cathodic Protection Services, Inc., Houston, Texas.)

Impressed Current Cathodic Protection

CHAPTER 5

Mechanism

In the last chapter the means of providing the cathodic protection current by sacrificial anode corrosion was explained. The resulting protection is achieved by virtue of the current and not by an intrinsic property of the sacrificial metal itself, be it magnesium, aluminum or zinc. As a supply of direct current is the essential requirement this can be obtained from more conventional generating equipment, for example, by a transformer rectifier operating from the ubiquitous a c power supply.

This current can be caused to flow through an electrode, the anode, into the electrolyte and thence to the protected structure. The anode, if similar in shape and size to the sacrificial anode which it might replace, is now powered by a d c source in place of the metal potential. There is no reason for the anode to corrode and it would be of particular advantage if it were permanent or had a long life.

The current required by the structure will be the same as with sacrificial anodes, if both are of a similar location and disposition, and the voltage required at the generator can be calculated from the ohmic components in the electrical circuit, the electrolytic anode resistance and the back e m f. This e m f will be similar to the driving voltage of the sacrificial system, though it will be in opposition to the external driving voltage when the anodes are permanent. From these two parameters, the circuit voltage and current, the power required to achieve protection can be determined. This method of cathodic protection is known as the impressed current system.

Sacrificial anodes were limited by their driving voltage and had to be engineered to suit this limitation under the particular field conditions. Impressed current engineering is freed from this restriction and different design techniques are employed to take advantage of it. To protect a large structure, say a pipeline, with sacrificial anodes a large number of them would be distributed along it, involving a multiplicity of electrical connections and considerable installation work. With impressed current protection such a structure could be protected by a very much smaller number of anodes, each fed with a greater current, although this would lead to the use of somewhat higher driving voltage, say 20V. One undesirable feature of such unit design is that certain parts of the structure may be overprotected in achieving the complete protection of other parts. This will be so on a pipeline where the relative nearness of the anode or groundbed and the build up of voltage in the pipe metal will cause local overprotection. Often a structure is deliberately overprotected from one anode position, as overall protection is more easily obtained from it than from a number of smaller anodes which would give more even protection.

It is not always possible to lump together the anodes into one large group and the multiplicity is unavoidable. In such cases the impressed current may either be generated from a single source and fed by a cable network to the anodes, or alternatively, local generation of d c can be used, though this often requires an a c distribution system: a c control from a single point can be used in this case.

The cathodic protection current can be controlled by varying the output of the d c source. On a transformer rectifier this is by control of the output voltage by tapping the transformer or can be done either automatically or manually by using controlled rectification. The control can be exercised with sacrificial anodes is similar but the disadvantage is in the number of anodes and consequently of the adjustments that would be required. These adjustments are time-consuming as each will tend to affect the other and there is a considerable time delay between the adjustment and the effect. Indeed, in many cases the best form of control is a step-wise approach in which a specific adjustment is made and the structure is examined a day or so later to see whether an additional step is required.

The same arguments would apply to impressed current when a large number of individual anodes fed by a cable network form a single d c source are adjusted using series variable resistors. It is very difficult to justify anything other than a simple variation of the current and adjustment of individual anodes has never, in the writer's opinion, been operated successfully. Changeover links between anodes and standby anodes, adjustment of current outputs between individual impressed current anodes within a group all seem to be unnecessary encumbrances on the operator brought on by the designer's inadequacy.

As well as the generation of the direct current from a c mains it is possible to use other current sources. A direct current generator may be driven by a gas turbine or some other power source and there is increasing interest in wind driven generators. Solar power and thermal generators are used though intermittent sources of energy create problems where continuous protection is required.

Increasingly automatic control will be used with impressed current. This may be a constant current device that will overcome the variations in electrolyte resistivity or it may be complete control from a variety of reference potential points which monitor the structure. Because of this increasing use of automatic devices the various criteria of protection that are recommended will nearly all be translated into a simple and direct measurement which can be used to influence the control.

Many of the simple control devices that were proposed say switching from low to high linked with either the movement of a ship or the start-up of a pump are now replaced with fully automatic equipment.

Economics

The costs of impressed current cathodic protection are assessed differently from those of sacrificial systems. The cost will have two major components: an initial or capital cost, and added to this will be a running cost which will include the power consumed by the system and any inspection or maintenance found necessary. The former cost will include the anodes, which will probably be permanent or have a very long life, the electric cables, the d c generator and any potential monitoring or control system.

The cost of the anodes will vary greatly and may be the major factor in the cost of the system. Recent anode developments, however, have meant that the anode itself has become comparatively inexpensive but its mounting and the cabling costs have not been significantly reduced. The electric power unit may be a simple transformer rectifier but there may be considerable expense in bringing an a c supply to it. The cost of the transformer rectifier will depend upon the power as well as the current and will be influenced both by the resistance of the anode and its back e m f. Generation from gas or oil driven machines is more expensive and these have to be housed in a suitable building. Thermo-electric generators, solar generators and wind generators are all expensive in terms of the power that they produce and those that are intermittent may require battery storage for the d c. It is a considerable advantage with these to be able to reduce the power required to a minimum.

In reducing the power requirements it may be necessary to extend the size of the anodes, and this in itself will add to their cost. A detailed analysis of these engineering costs is discussed later.

There are innumerable variations open to the designer of an impressed current system as well as many constraints. These will vary from the type of anode to be used to the distribution of the anodes or the concentration of anodes which may well affect the current required. a c distribution to small power units will have to be balanced against d c distribution using heavier cables but from a more economical large unit. Added to these are the problems of maintenance, reliability and redundancy of equipment.

The engineering will be the prime consideration in many of these spheres, but in the overall context the engineer will be expected to provide the system that can most readily be justified economically as well as technically.

Consumable Anodes

Since the current is being supplied from an external source, the anode material used for impressed current requires different properties from those of the sacrificial metals; preferably the anode system should be permanent, or rather should have a low overall life cost. The cost of the anode will depend on several features of the design: the life required, the driving voltage that is most economical, the method of mounting the anode, and possibly, the ease of anode replacement.

If only a 10 year life is required from the anode the cost of a truly permanent anode may be greater than the cost of an anode that would be consumed during that period and this might be particularly true where the consumable anode is made of some scrap metal. The criterion for the selection of the anode material will chiefly be economic, though since the cathodic protection system as a whole is aimed at achieving even greater economies, the reliability of the cathodic protection equipment is the prime consideration.

Earlier the mechanism of anodic corrosion was explained; impressed current anodes are subject to these reactions and so careful material selection is important. Many metals exhibit excellent corrosion resistance but cannot be used as anodes either because this resistance is not maintained under the anodic conditions or a highly resistive film is formed which reduces the flow of current from the anode. An increase in the circuit voltage would overcome this latter effect but it will not be economical to work at high voltages where a more suitable anode material would allow low voltage operation. A similar limitation exists on the almost complete consumption of the anodes, which will require a higher voltage. Though it is easy to build spare capacity into the d c generator an increase in the overall power of 20 to 30 per cent, that is, when the anodes are about half consumed, is the maximum that can be tolerated.

Anodes for use with impressed current protection must be cheaply consumed, or equally cheap if they are permanent, and the resistance of the anode must not change greatly either by consumption or by polarization. There will be two main types of anode materials, consumable and permanent, and because of the general replacement and inspection difficulties the latter will be of most value, though cases will arise where consumable anodes prove to be most economical.

Steel

The pioneers of impressed current protection used scrap iron and steel as anodes; the metal is theoretically consumed at about 20 lb per ampere year and so where ample scrap exists it makes a cheap consumable anode. The corrosion product that builds up on the iron or steel polarizes the anode and this is particularly large when the anode is buried directly in the ground. This trouble can be remedied by using a carbonaceous backfill around the anode although in such a case, the cost of the backfill becomes a major item unless anodes of particular shapes, usually long thin cylinders, are used. Such scrap is not always available and, even when it is, may be expensive. The consumption of the metal under these conditions has been reported as being exceptionally low but the significant feature of the attack is its distribution over the anode surface. Where the coke backfill does not remain properly consolidated and in close contact with the steel, then rapid corrosion occurs. The consumption of the metal can take place at a few points and anode electrical continuity may be lost. Very careful engineering is required if the anode life is to be greatly extended by the use of coke breeze.

In liquid electrolytes steel anodes have a considerably lower polarization. In fresh potable or process water the red color imparted to it by the anode consumption is usually undesirable and steel anodes are rarely employed in these electrolytes. In river, brackish or sea water, steel anodes are acceptable and the degree of polarization that they suffer decreases with increasing salinity.

Scrap seal anodes are very heavy and will be expensive to support so they usually rest on the sea or river bed. In designing such anodes, care has to be taken that they do not fail through rapid corrosion at or near the electrical connection isolating large pieces of anode. This type of failure may be avoided by welding shields over the electrical connections and close to any joint as shown in Fig 87. The shields are consumed more rapidly than the joint which is not attacked until the shield has almost completely disappeared. This technique is superior to coating the joint and connection areas as this leads to the rapid corrosion of the metal at the edge of the coated section. The corrosion pattern of steel anodes is similar to that of the sacrificial anodes and a long anode of steel such as a rail or a length of hawser will be consumed from its ends.

A special problem occurs where the anode is in a stratified electrolyte, as, for example, in a tank containing some hot and some cold water: the hot water will rise to the top of the tank and, having a lower resistivity, the part of the anode in the hottest water will be most rapidly consumed so that the lower end of the suspended anode will drop off. This type of premature failure can be prevented by three methods: the anodes can be arranged so that they have a higher resistance near the surface of the water by an in-

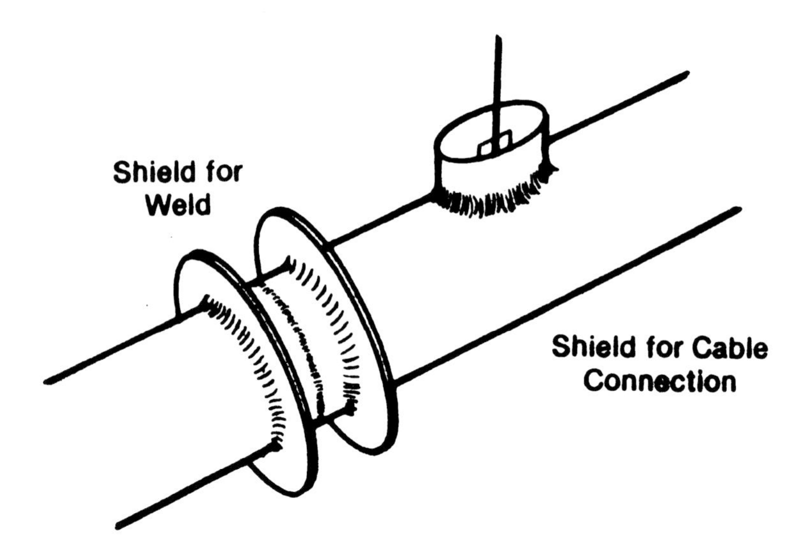

FIGURE 87 - Shields to prolong life of welds and cable connection on scrap steel anode.

terference or similar technique; secondly, extra anodes can be introduced into the top layer of the electrolyte to compensate for the more rapid corrosion loss there; or a third alternative is to lower the anode slowly into the electrolyte so that a constant length is always in the electrolyte and each part has passed through the region of highest consumption. This might not be practical in the case of a hot water tank but will certainly have advantages in other applications, particularly where the rate of anode consumption is high.

As all these conditions and modifications to the anode are easily carried out with steel, it makes an ideal consumable anode when it is available as cheap scrap. The consumption rate of 20 lb per ampere year will mean that in a practical installation 60 ampere-years of useful output would be obtained from each ton of scrap. The steel will polarize in the ground unless surrounded by a carbonaceous backfill and lose most of the economic advantages that a cheap source of scrap metal might provide.

Iron anodes are used to condition aluminum brass and other copper based condenser tubes in sea water by the electrolytic injection of iron into the cooling water stream.

Aluminum

Because steel has for many purposes such an undesirable corrosion product aluminum has been used as a consumable anode. In fresh water two types of alloys are recommended: these are the commercially pure aluminum and the heat treated alloys H 14 and H 15. Both retain the good mechanical and electrical properties of aluminum, making mounting and fabrication easy and the connection of the electrical feeder cable simple. The practical rate of consumption of the metal is about 10 to 12 lb per ampere year, and initially the anodes have no polarization and little back e m f. In fresh water there is a gradual build up of corrosion product which polarizes the anode and increases its resistance threefold, this process taking about a month under anodic conditions where the current output is between a quarter and a half amp per square foot. The initial high output of the anode aids the polarization of the cathode but this can hardly be claimed as an advantage. The ease with which a very low circuit resistance is obtained is a considerable help and in this respect aluminum has a distinct advantage over more brittle materials. The anode consumption in most water including sea water, is even and little or no pitting occurs though this property is only found in these alloys. The corrosion product as was the case with aluminum sacrifical anodes is non-toxic and it is suggested might retard the formation of hot water scale.

Aluminum anodes can be used successfully in sea water where the degree of polarization is less than in fresh water and the pure aluminum anodes are usually preferred to the alloys.

Anti-Fouling Anodes

The corrosion of a consumable anode means that the metal used in it goes into solution in the corroding electrolyte. It has been suggested that as copper, arsenic and lead salts are poisonous to marine fouling organisms these metals might be used as consumable anodes in sea water and provide cathodic protection with complementary anti-fouling. This has only a limited practical application including copper cladding on platinized anodes to prevent the fouling that may otherwise occur before they are energized.

Sacrificial Anode Metals

The metals used as sacrificial anodes can equally be used as consumable impressed current anodes. The operation of the anode metal with an increased driving voltage would allow the use of higher current densities and this would probably result in maximum anode efficiency. Impressed driving voltage can be used to boost their natural driving potentials so that their use could be extended to higher resistivity electrolytes or the driving voltage could be used to achieve an initial high current density on the cathode. The advantages that can be gained by this use of sacrificial metals is small.

Where the cost of d c power generation from alternative energy sources is high then there is a particular advantage in using sacrificial anodes. This results in a very low power consumption which is aided by the low ohmic resistance of the bulk of the sacrificial anode used. There may be a particular advantage in the protection of remote pipelines where an intermittent source, such as solar or wind power, could be used to boost the protection from the anodes which would revert to a sacrificial role during breaks in the power supply.

Permanent Anodes

Graphite

The first permanent anodes were made from graphite and carbon. These materials do not possess ideal mechanical or electrochemical properties as they are brittle and porous. Improved forms are available impregnated with inert resins and waxes. Unimpregnated carbon and graphite allows liquid electrolytes to reach the metallic connector to the anode and corrode it; almost all anodes used at the present time are impregnated. The graphite content of the available anodes varies and there is some evidence that this may be a critical quality. High purity graphite, such as that used for spectroscopic electrodes, has a greatly improved performance.

In the ground graphite is an extremely good anode material. It is cushioned against severe mechanical shock and the general high resistivity means that it will usually be operated at very low current densities. Buried directly in the ground graphite should not be run at current densities in excess of one third amp per sq ft on any part of its surface which usually imposes a restriction of a mean current density of a quarter amp per sq ft. At these outputs its consumption will be less than 1 lb per ampere year. Graphite is usually made as rods of between 2½ in. and 6 in. diameter and in lengths that vary from 3 ft to 7 ft. The electrical connection is buried a short distance into one end or both ends and is made by pressure contact of copper. Because the end of the anode is consumed first some contacts are buried deeply in the anode and, indeed, some anodes are tubular. The pressure contact of copper is made either by an expansion of the connection or by some form of tapping device. Both of these hold the anode in tension and as the anode is consumed, particularly if the impregnant is destroyed, the anode cracks and the connection is lost.

The superior connection can be made to the anode holding it in compression. This can be made at the mid point to the anode and as well as being cheaper it allows the whole of both ends of the anode to be available for consumption. The type of connector designed by the writer is shown in Fig. 88.

Greatly improved performance can be obtained from a graphite anode by surrounding it in the ground with a carbonaceous backfill. The object of this is to lower the anode resistance by metallic contact between the backfill particles and the anode, hence increasing the effective anode size, a process

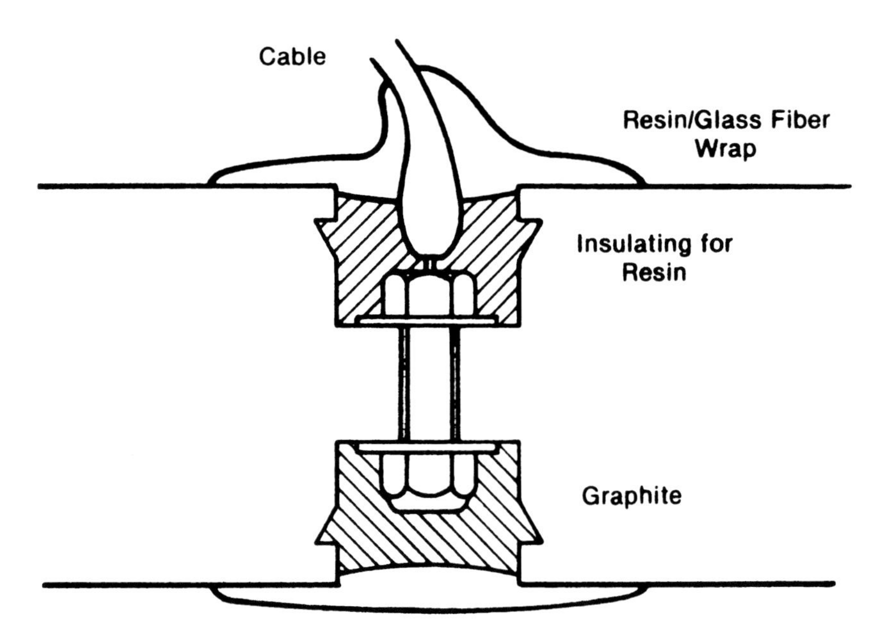

FIGURE 88 - Central compression connector for graphite anode.

which also transfers the anodic reaction to the backfill particles and so the anode is not consumed. The properties of the backfill are most important and for the anode to be completely preserved these properties should be carefully controlled. The backfill should possess a low resistivity of less than 10 ohm cm and this can usually be obtained from steel works coke if it is first screened and then remixed to a particular grading. The most suitable grading can be determined by experiment, the coke breeze resistivity being determined by soil box techniques as described in Chapter 3. An ideal soil box or tube can be made from a 6 in. diameter plastic tube which has the particular advantage that the backfill can be compressed to reproduce the compaction it would have in the field. If the desired resistivity cannot be obtained from mixtures of the local coke then graphite or special coke particles can be used and mixed with the coke. These will greatly reduce its resistivity. Some results obtained with London Gas Works coke and graphite dust sweepings are given in Table XI.

It is possible to make superior grades of coke specifically from petroleum for use as a backfill. In this the particles are not only graded, but are spheroidal so that they pack more closely and form better contact one with the other. By careful selection a reduction in the resistivity to less than one ohm cm is possible. Additions to the coke by way of lubrication allow better settlement and compacting of the mixture.

TABLE XI

Material		Resistivity—Ohm cms		
Coke Breeze	*Graphite*	*Dried*	*Damp*	*Wet*
100	—	52		13
86	14			10
75	25	Not Compressed		2.4
67	33			2.0
57	43			2.5
—	100	150		18
Sand	*Gypsum*			
100	—		27,500	5,800
89	11		10,000	2,000
80	20			1,150
—	100		400	
Coke Breeze				
40% passed 1/8 in. mesh				14
Large >3/16 in. mesh				24
Small <3/16 in. mesh				50
Coke	*Gypsum*			
5	1			21
5	2			22
Coke				
Fine	*Coarse*			
6	1			26
1	5			11
Wet includes London water				2,000

To ensure that good metallic contact is made between the anode and the coke particles the backfill must be tightly packed around the anode. The effect of compression on the resistance between two sheets of steel separated by coke particles is shown in Fig. 89 and a compression of 2 lb per sq in. should be obtained in practice.

While the resistive properties of the backfill are important it is equally essential that the gases formed at the anode or on the backfill particles are free to escape through the backfill. Very fine graphite powder or the addition of powdered lime to a backfill will prevent the escape of these gases. The backfill has a lower resistivity when wet and backfill should be installed below the water table. Salts and lime can be added to the surrounding soil to reduce any electro-osmotic tendency, the lime having the additional advantage that it reduces the acidity in the vicinity of the anode.

When the anode is suitably buried in backfill the current density should not exceed 4 amp per sq ft at any point, this value being governed by the backfill resistivity and the contact resistance between the graphite and the coke particles. At high current densities the anode surface will conduct current electrolytically, as well as through the direct contact with the backfill particles: at the limiting current density electrolytic conduction will reach that normally allowed when the anode is buried directly in the soil.

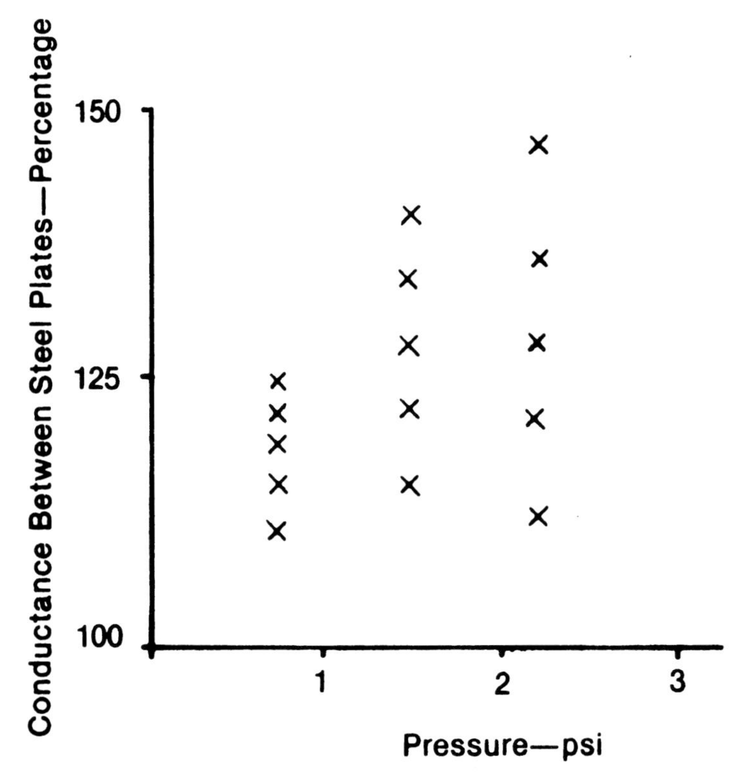

FIGURE 89 - The effect of pressure on coke contact conductance.

As the backfill is a good conductor it is contemporary practice to reduce the resistance of a groundbed by extending it with coke breeze backfill, making a long cylinder of the backfill and inserting into it a series of graphite anodes which, act as electrical connections to the coke breeze. This practice tends to drain excessive current off the ends of the anodes causing their rapid deterioration as the anode current will flow along the backfill so that equal amounts of current leave each part of the coke breeze cylinder. This has been discussed in Chapter 3. The maximum spacing that should occur between anodes can be calculated and generally should not exceed 10 ft between the anode ends when they are laid in a 1 sq ft section of backfill. Corrosion of the end of the graphite anode will be serious and will destroy the electrical connection. The consumption of the connector end can be prevented by placing the anodes so that the connector ends are separated by only a foot or so as shown in Fig. 90. The use of center compression-connected anodes has a great advantage.

The construction of groundbeds of this type is difficult in marshy or

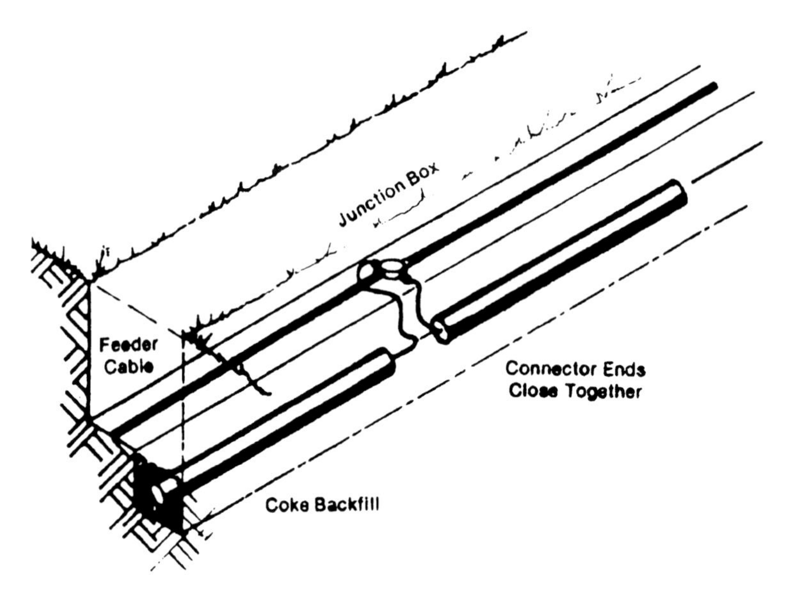

FIGURE 90 - Horizontal anodes arranged to reduce connector-end consumption.

boggy ground, or where there is excessive water, as the backfill cannot be consolidated nor can any reasonable shape be maintained in the trench. In these conditions a pre-packaged graphite anode surrounded by carbonaceous backfill and enclosed in a metal cannister can be used.

The correct backfill can be tightly compacted around the anode before the can is sealed. The can is electrically connected to the anode and is first consumed. As there is no limit to the current density that can be used with the can metal, the packaged anode can be run at a very high current initially to provide a polarizing charge for the cathode. Properly compacted the backfill and cannister allow the anode to be shipped without the careful packing that is normally required with a graphite anode. By using this technique graphite anodes can be successfully used, at the current densities permissible only with a properly backfilled anode, in muds, bogs, etc., where often the lowest resistivity ground is to be found.

Two types of graphite coke breeze groundbeds are shown in Figs. 91 and 92. In the vertical groundbed configurations the gassing of the anode makes it necessary to cover the top of the anode hole with stone chips or coarse gravel to a depth of 2½ ft with short anodes and to greater depths with long vertical anodes.

Deep well anodes have become popular, particularly in areas either of low water table or where there is considerable congestion. The anodes are placed deep in a drilled hole and in their vertical configuration can meet

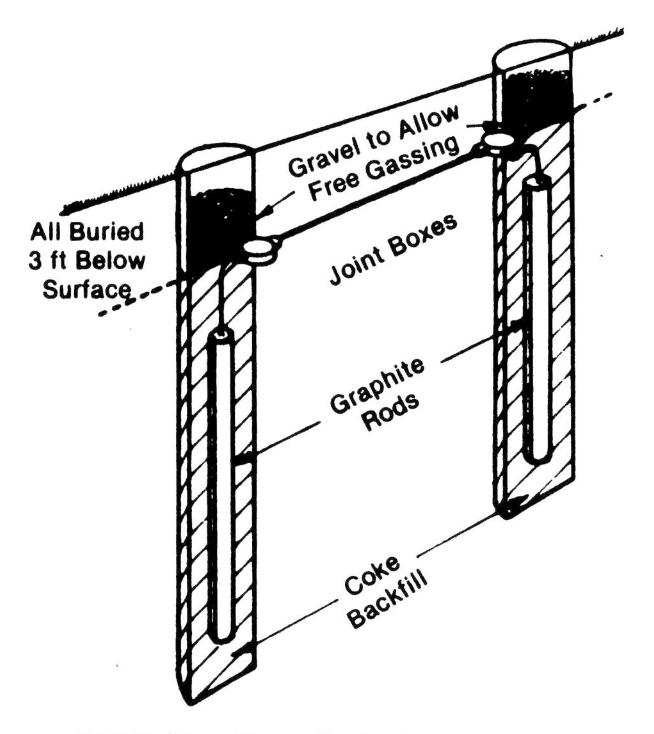

FIGURE 91 - Groundbed with vertical anodes.

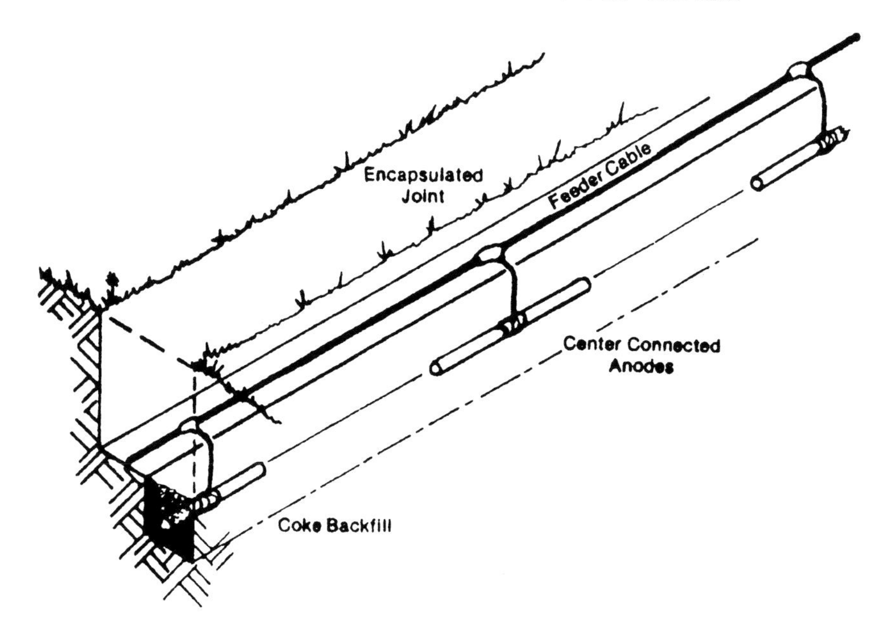

FIGURE 92 - Horizontal anode groundbed.

with a number of problems. Firstly, the anodes have to be carefully installed so that the weight of backfill does not rest upon one anode initially or after consumption of some of the backfill, secondly, that the anode products, particularly the gases, are free to escape and do not damage the insulation on the cable.

A number of techniques have been developed which vary from the attachment of the anodes to a rigid or flexible structure which is lowered into the hole and the coke backfill then pumped in as a slurry around them, to other techniques where anode and backfill are successively lowered into the hole Fig 93. Replacement of anodes in deep groundbeds is difficult but there have been several successful developments of recovery of anodes using fluidizing techniques. Special backfills have been developed that are particularly susceptible to this. There is some move away from graphite anodes for this application.

In fresh water, graphite can be used and will remain permanent at low current densities which are restricted to the same values, a mean 1/4 amp per sq ft, that are used when the anode is buried bare in the soil. The brittle

FIGURE 93 - Installation of deep well graphite anode groundbed. (Photo reprinted with permission from P.I. Corrosion Engineers, Ltd., Hants, Alresford, England.)

nature of graphite and its consumption make mounting the anode difficult and several systems are suggested including suspension from the anode electrical feed cable, strapping by an insulator to a suitable rack or mounting point on the structure, or suspension of the anode from holes drilled through it. The particular method chosen will depend upon the application and the size and shape of anode used.

At high temperatures above 50 °C the anode deteriorates more rapidly but this seems to be a failure of the impregnant rather than of the graphite. The anode consumption varies, in cold fresh water there should be no measurable loss of graphite while in hot water up to 2 lb per ampere year is consumed; again, as with the consumable anodes, most corrosion occurs at the anode extremities.

In sea water the anodic reaction is the evolution of chlorine and oxygen; both these have a detrimental effect upon the anode, though again, it is the impregnant that seems to suffer. If the current density of the graphite is restricted below 1 amp per sq ft everywhere, little or no consumption of the graphite occurs. As in fresh water the corrosion of the graphite increases at high temperatures though this can be somewhat offset by using a lower current density. In protecting marine structures the poor mechanical properties of the anode are a disadvantage and considerable care must be given to the details of the mounting; insulated holders set in steel racks have been used with some success though they are expensive. In certain areas where the sea bed is suitable, anodes can be laid directly onto it; large diameter anodes should be used and a separate electrical connection made to either end of the anode so that should an anode break there will be no great loss of graphite. These results are rather clouded by the high percentage of failures from anode fractures.

Economically graphite is relatively cheap particularly when considered as a cost per unit volume. Used in carbonaceous backfill in the ground, the anode, if not the backfill, is permanent and at low current densities the graphite will not be consumed in fresh or sea water though its mechanical properties make its use in water particularly difficult when there is a tide or other stream flowing. Consumption of the graphite often occurs as small particles spalling off the surface and when graphite is used for the internal protection of tanks or in water circulating systems these particles may accumulate in various parts of the plant, possibly remote from the protection. As graphite is noble and, unlike most corrosion products, has a high conductivity, an electrolytic cell is formed between these particles and the plant metal which may be particularly aggressive and rapid corrosion may result. This is an undesirable feature of the anodes. The back e m f of graphite relative to protected steel is about 2 V and this must be included in any calculation of the d c circuit voltage required.

High Silicon Iron

High silicon iron anodes with approximately 14% silicon are unfortu-

FIGURE 94 - Silicon iron anodes with end-connection. (Photo reprinted with permission from Impalloy, Ltd., Bloxwich, Walsall, England.)

nate in mechanical properties which are inferior to graphite and it cannot be machined except by grinding. It is extremely brittle and is transported with great difficulty. Electrical connections are made either to a cast insert by driving a soft metal plug into a cast hole or by pressure devices. The material is dense and resembles cast iron in appearance (Fig 94).

It displays a noble potential and has a small interface resistance giving a total back e m f of the same order as graphite. Anodes have been cast in a variety of shapes and long thin rods seem to be preferred although this shape produces an extremely fragile anode which is difficult to use except by burial or free suspension in the electrolyte. Experiments and practical tests have shown that the metal has a remarkable corrosion resistance when used as an anode in fresh water. The results of consumption tests show some scatter but they indicate that the corrosion rate is less than 4 oz per ampere year at current densities below 2 amp per sq ft while the rate of attack increases to 2 lb per ampere year at 10 amp per sq ft.

The attack is characteristic of the metal, a brown film being formed over the majority of the surface with pitting occurring at a number of points. These pits are a disturbing feature of the consumption and can lead to premature failure of the anode. In fresh water high silicon iron anodes should have a very long life though it might be advisable to protect the metal in the vicinity of the electrical connection to avoid the possibility of failure by pitting corrosion.

Similar performance can be expected from the material when buried bare in the ground. Jetting techniques are suggested for burial and a low groundbed resistance can be achieved by using a large number of rods in place of fewer backfilled anodes. Several failures of the anode are reported from soils that have a high chloride content and common salt should not be used to reduce the resistivity of the ground in their vicinity. The use of silicon iron anodes in carbonaceous backfill is entirely parallel to the use of graphite anodes.

In sea water high silicon iron anodes are an enigma. An anode may be in pristine condition over 90 per cent of its surface while the other 10 per cent will be badly pitted, the pits being up to an inch in diameter and equally deep with some undercutting. Gas blocking has been suggested as the cause of these because it has been found that various inert attachments to the anode have caused accelerated corrosion in their vicinity. If this is so the corrosion product which forms even in a shallow pit would probably augment any tendency to gas blocking locally so aggravating the condition, and this may be the mechanism of deep pit formation.

A modified alloy which includes molybdenum is superior, the rate of consumption of molybdenum silicon iron anodes in sea water is a function of the current density and is shown graphically in Fig. 95. At elevated

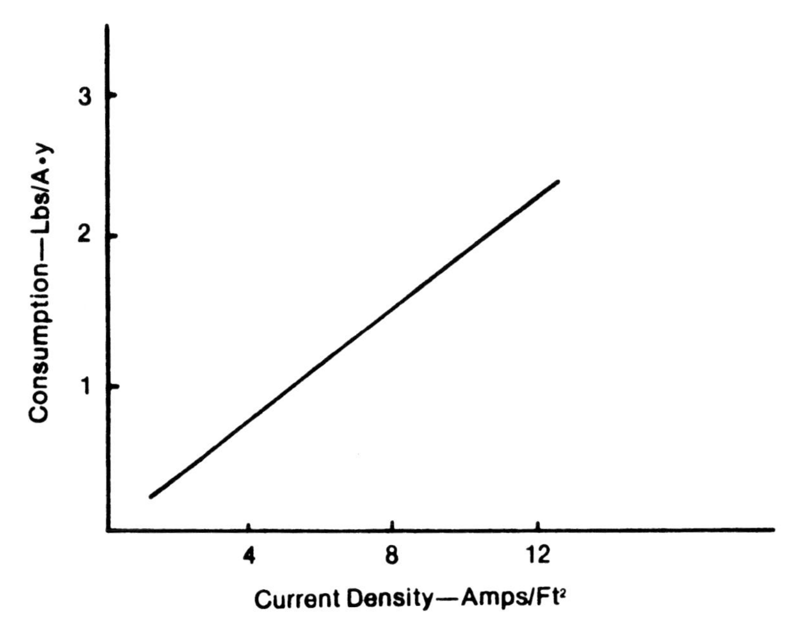

FIGURE 95 - Consumption of molybdenum/silicon iron anodes in seawater.

temperatures the metal corrodes more rapidly and it is not recommended for use above 140 °F, the molybdenum addition again conferring superior high temperature consumption and is suggested for use up to boiling water temperatures.

Economically silicon iron anodes are equal to graphite and equal difficulty is found in mounting them. They are efficient in fresh water and seem to be ideally suited to that electrolyte. The higher current densities that are economical for reasonable installation costs increase the power consumption and a balance between these factors must be made. In soils they offer no advantage over graphite anodes except where both are used bare. The pre-packaging or canning techniques used with graphite and coke breeze backfill will allow it always to be used with a carbonaceous backfill. In sea water the anode material is mechanically unsuitable and although the molybdenum addition improves its electrochemical properties it still remains unattractive to handle and use.

Magnetite Anodes

For many years there has been considerable use made of anodes of magnetite in various forms, generally processed in some way onto a supporting rod. The anode has similar properties to lead in sea water and to graphite and silicon iron in the soil. The early anodes were not attractive in that they were new and did not have sufficient advantage to promote their use in competition with the established anodes. Improvements in the techniques of anode manufacture have materially improved the anode's performance and particularly its susceptibility to breaking under thermal cycling. The anodes are made by casting the magnetite, often with an added flux metal, into a hollow cylinder closed at one end. This is often of the order of 2½ in. diameter with a ½ in. wall thickness (Fig. 96). Because of the low conductivity of the magnetite this is backed by an internal coating of copper to which the cable is attached. The cable connection itself is encapsulated in resin but the anode is filled with a soft plastic core which absorbs the differential rates of expansion of the resin and the magnetite. These anodes now operate under many circumstances five to 10 times as effectively as graphite or silicon iron and at a much lower cost than lead 2% silver. However, to ensure long life their current density must be considerably reduced to the order of 1 amp per square foot and this puts them at a disadvantage in sea water.

There has been an increase in their use but they are now against formidable opposition as anodes are expected to last in excess of 20 years. This not only makes the task of developing a new anode more difficult but it also extends the period over which the anode has to be proved. The material is very promising. It has already demonstrated remarkable properties, particularly in relation to its first cost. No doubt there will be further

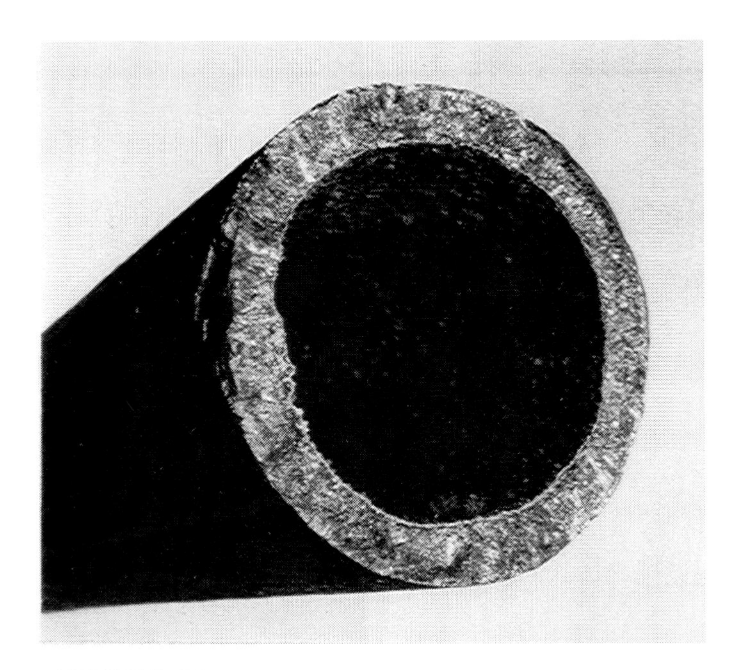

FIGURE 96 - Magnetite anode section. (Photo reprinted from P.I. Corrosion Engineers, Ltd., Hants, Alresford, England.)

refinements to the techniques of manufacture which will improve it. When more practical experience is available it will be seen how closely the low consumption rate of the anode, quoted at less than 1 oz per amp year, is met and whether the fragility of the very early anodes has been successfully overcome.

Magnetite has a formula of approximately Fe_3O_4 and a similar oxide closer to ferrite has been used to form anodes. These have properties akin to the magnetite anode and generically will be referred to as magnetite. An anode called ebonite, which is a ceramic titanium dioxide, has been announced and has approximately the same electrolytic characteristics as the magnetite series but is claimed to have better mechanical properties.

Platinum and Platinum Alloys

Platinum has been employed as an anode in electrochemistry for many years and its use as an anode in cathodic protection is equally established. The metal is workable and electrical connections are readily made to it by welding or other techniques. As an anode it is truly permanent except in sea water when a 10 per cent addition of rhodium or palladium makes it so.

Because of this property thin wires or films of the metal are used as an anode or it may be plated as a thin film onto another metal.

The metal has a high back e m f and the general small size of the anodes means that a high driving voltage will be required. When the metal is formed into a thin wire or foil it is susceptible to mechanical damage and when used in bulk its cost is exceptionally high and there is a considerable risk of theft in accessible installations. In many cases, particularly in the protection of chemical plant, it can be used with economy and its trouble-free existence will be a great advantage.

Platinized Titanium

If a thin layer of platinum is coated onto a substrate of titanium then the platinum can act as an anode and the titanium will form an electrolytically insulating film by anodizing. This gives the platinum an inert but conducting base and so it can be used as an anode. Because of the very low consumption of platinum when acting as an anode in most electrolytes, a very thin film of platinum so deposited will give a long life. In practice, one tenth of a thousandth of an inch is sufficient to provide a life of about ten years at a current density 70 amps per sq ft, on ½ amp per sq in.

The titanium anodizes in most electrolytes. In electrolytes that contain chloride ions it will break down and corrode if the potential across the interface exceeds approximately 8 V. In waters that have sulfate ions or that are virtually pure water, the breakdown voltage is much increased. This is a minor limitation in so far as the anode surface is concerned as the platinum in acting as the anode proper reduces the voltage across the titanium surface at those points that are bare of platinum to a very low value, in most cases to the order of 1 V. This allows for mechanical damage to the anode and in many cases for partial cladding of the surface without the risk of the titanium breaking down.

Titanium as a substrate is reasonably expensive and it has the disadvantage of high electrical resistivity. The metal is extremely strong, can be welded and formed—though this is a skilled operation—and electrical connection can be made to it, usually by compression onto a threaded rod. It is not easy to solder to it but other metals can be welded to it by friction welding techniques.

The platinum surface itself is capable of emitting very large current densities of the order of hundreds of amps per sq ft. The rate of corrosion of the platinum film, which is usually electrolytically deposited is initially high and then greatly reduced so that probably one half of the anode platinum is sufficient to provide 1000 amp years of charge off each sq ft.

The platinum itself is plated onto a pre-etched surface of titanium and there appeared in the early anodes to be problems with adhesion. Some disc anodes showed sign of lifting in which the platinum became detached rather in the manner of miniature birch bark. This often occurred where a stud had been welded to the back of the plate and there is some suspicion that detachment was aggravated by this. The problem had never appeared, to the author's knowledge, on rod anodes, particularly those of ½ in. diameter and less. There has also been concern that the anode would fail if operated in water below 4 °C, as many parts of the northern oceans reach this temperature, particularly on platform or harbor installations serving some of the more northerly oil fields.

If titanium is made a cathode then it will form a hydride and titanium will, on its surface at least, lose much of its mechanical strength. It is suggested that much of the difficulty may lie in the fact that the hydride is formed during the plating and that the bond between the platinum and the titanium is, therefore, very weak. A new technique of bonding has been developed which uses pre-treatment with a thermally decomposed noble metal primer. This allows the deposited platinum to grow from a number of sites, thereby reducing the problems associated with hydrogen absorbed brittleness in the platinum film when only plated from a few sites. The mechanical bonding of the platinum to the titanium can be enhanced by thermal treatment in an inert atmosphere, though this is not used for cathodic protection unless an abnormally thick deposit of platinum is required when the treatment is an intermediate stage in the plating. Platinum coating can be from fused salt baths instead of aqueous baths but the cost limits the size and complexity of the anodes though a much better platinum surface results. Failures of the actual platinized titanium anode, as opposed to the failures of the connection or support or its wearing out, have been so few that a proper comparison between the two surfaces is difficult. In a jetty in Oslo fiord the author many years ago placed half of the anodes with the new surface and the other half of the older anodes in the installation and none of the anodes failed though they are all operating beyond their predicted life.

Platinum itself will corrode in the presence of certain sugars and there have been one or two disastrous failures in plant where these sugars were used. It also will dissolve under a c conditions. Some types of supposedly d c have a large a c component and it appears that low frequency reversal is required to cause this corrosion. Very occasional reversal, such as might occur during switching-on surges, do not appear to detract from the platinum life, which probably will stand the order of 10^5 reversals before impinging on the life expectancy of the average anode.

Rapid consumption of platinum has been reported at low frequencies. The upper limit of rapid consumption is below 50 Hz. This means that rectified mains will have a ripple frequency of 100 or 120 Hz and so will not be subject to this. At low frequencies a number of the experiments have been carried out using interrupted d c. One of the problems with these experiments is that interrupted d c has a large spectrum of frequencies which, in a practical cathodic protection circuit, can lead to a current reversal during part of the cycle. It is possibly this technique is more to blame for the platinum consumption than the frequency of the modulation itself. Because of this limit, it is unusual to see mark space modulated control with a platinized anode, but rather to use phase angle control where, although the wave form is distorted, there is no reversal of the current.

The electrochemical, mechanical and electrical properties of the anode, as well as its cost, have dictated two principal forms of anode. The majority are rod anodes, varying in diameter from 1/4 to 1 in., and plate anodes which are usually about 1/8 to 1/4 in. thick. Both these anodes have connections made in the form of a screw thread either by a stud welded to the plate or by machining the end of the rod. A mechanical connection is then made between the anode and the electric cable. Techniques such as crimping and swaging can be used equally successfully.

These two basic anode shapes can be extended principally for use in high resistivities. The rod can be extended into a wire and the plate extended into a mesh or other form of expanded metal. Connection is still by a threaded titanium rod. Because of the high current density that the anodes can emit their principal use is in sea water or other low resistivity electrolytes. This includes their use in coke breeze. There a wire may be used but the author prefers the use of a mesh to which a rod has been welded. The rod can then be taken to a plastic box where it can be interconnected with the cable. The anodes can then be used in the same manner as a graphite anode, the same space in between anodes can be employed, but where one 5 ft graphite anode will be placed in every 15 ft of trench, leaving a spacing of 10 ft between the ends of the anodes, the platinized titanium mesh would need to be placed at 10 ft intervals. The anode arrangement is shown in Fig. 97 and this type of anode pre-packaged with suitable backfill in a can is shown in Fig. 98.

The anode breaks up sea water providing copious amounts of chlorine, which is generally recombined to form hypochorite, and some oxygen. This has a deleterious effect on the insulators that are used to support the anode and the resins used to encapsulate the electrical connection. Fortunately, developments in these insulators have run parallel with the development of the anode and it is possible, by careful engineering, suitably to hold and insulate the anode. The anode does not operate happily in estuary muds, particularly where these are of a fine silt variety, as there is considerable electro-osmosis around the anode which can dry out by this process and produce a very high resistance.

Because the titanium anodizes it can be used as a support for the anode itself. This allows the anode surface to stop short of the insulator and reduce the hypochlorite attack upon it. Titanium has a low breakdown potential

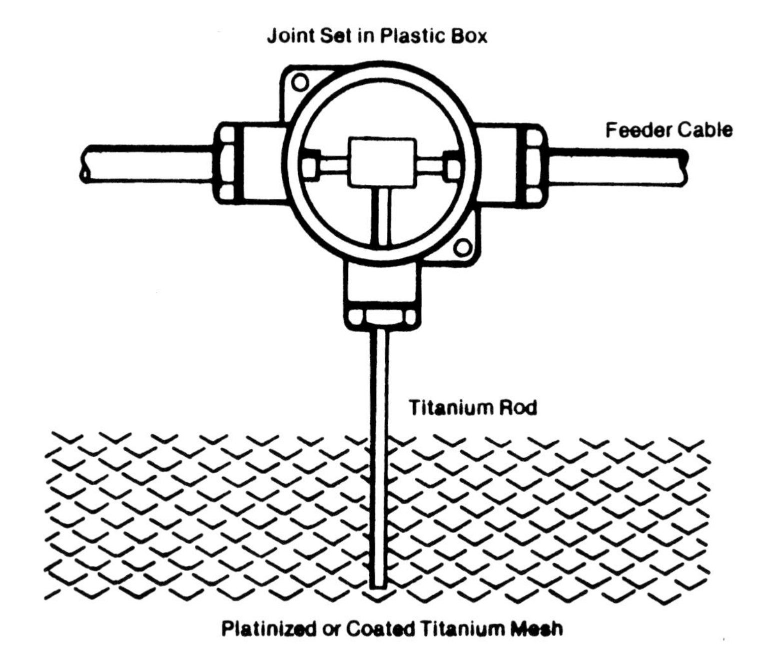

FIGURE 97 - Mesh platinized titanium anode, connector rod and cable connection.

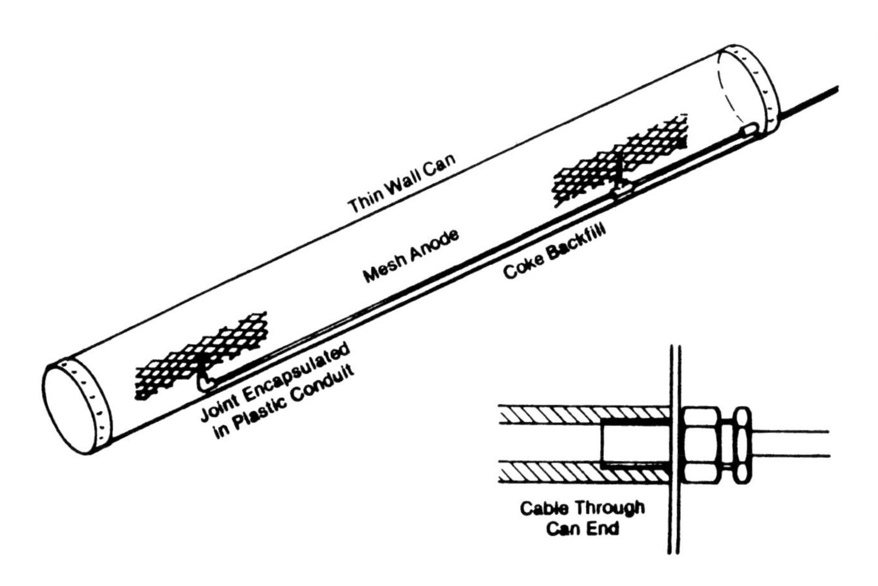

FIGURE 98 - Canned anode assembly with twin mesh platinized titanium anodes.

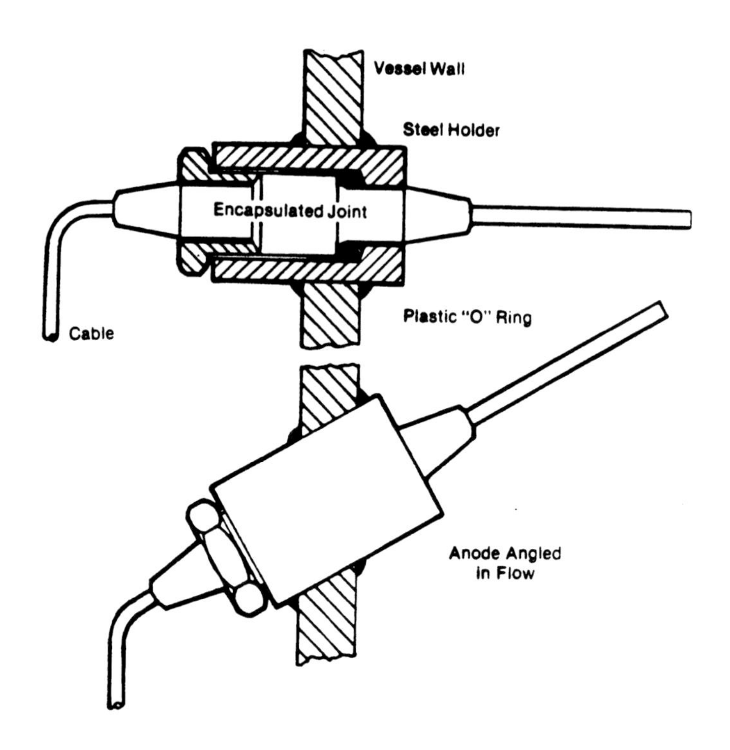

FIGURE 99 - Cantilever anode with molded holder.

and so the anode has to be engineered with some care. Rod anodes—say 3/8 in. diameter and 2 ft long—could be platinized over 8 in. of one end of the rod and the anode supported by the anodized titanium over the center 10 in. using the end 6 in. for support and connection. This cantilever type of anode is very popular, but great care has to be taken that no area of the titanium has a potential greater than 8 V between it and the electrolyte. There a number of techniques that can be used to prevent this, including the use of shielding, and one particular type of shielding in which an increased electrolytic path is involved by a surrounding polythene shield, say 1 in. diameter when the anode is 1/2 in. diameter.

Alternatively, the more resistant resins can be used to cover this area and these add to the rigidity of the anode; this is shown in Figs 99 and 100. There has been some problem where the anodes had been held in a clamp-like attachment, particularly where these are subject to vibration. The anodizing on the titanium has been destroyed and this has led to failure and breaking of the anode rod.

When long anodes are used, whether in the form of wire or a long rod, a difficulty arises with the high electrical resistivity of the titanium itself.

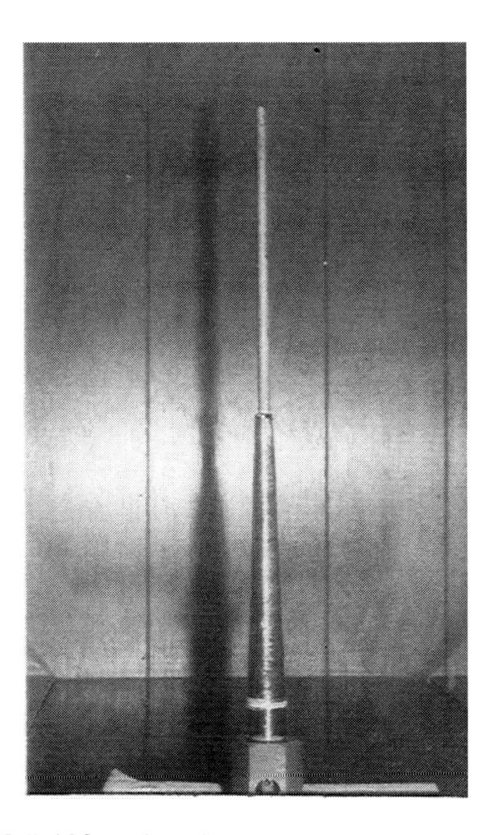

FIGURE 100 - Platinized titanium rod, resin glass fiber wound for insulation and support. (Photo reprinted with permission from Norton Corrosion, Ltd., Woodinville, Washington.)

There is a considerable voltage drop along the wire and there are two techniques by which this can be overcome. Either a number of connections are made to the wire, and this can be done by using titanium connectors, or the center of the platinized titanium wire can have a copper core. This technique allows for better distribution of current and if coupled with skip plating, that is, where areas—say 6 in. long— are plated every 1 ft, the anode current can flow for a considerable distance down the wire anode.

Care has to be taken in the ratio of the copper core to the titanium thickness; when thin walled anodes have been bent copper has leached through the titanium and broken the continuity.

The detailed engineering of platinized titanium anodes, which have played a major role in the development of almost every field of cathodic protection, will be described in the applications chapters. However, the main effect of the anode has been to reduce the anode cost to be small compared with the cost of the cables and the electrical equipment. The anode is difficult to destroy but operating at high current densities it makes the support and insulation of the connection extremely difficult. The majority of failures have been when the support and/or connection have not been well engineered. The best properties of the anode are only available to the cathodic protection system when the anode is properly engineered.

Platinized Niobium

Niobium is similar to titanium but it has a higher breakdown voltage when anodized and is a considerably better electrical conductor. It is also very much more expensive and much weaker metal.

Because of the higher breakdown voltage it is suggested that it can be operated at higher current densities. This is only a marginal advantage as the high current densities mean more difficulty with support and connections and higher costs of power and of the power-providing equipment.

There are applications where the high breakdown voltage is an advantage but it is as an alternative anode material rather than as a better anode material that this will find use. The fact that there is a higher breakdown voltage in no way excuses or inculpates the designer who badly engineers the connection or the support. The idea that a better electrochemical anode is the panacea for bad engineering is one of cathodic protection's myths.

Platinum Alloys

As well as platinum coated niobium and titanium it is possible to coat tantalum. This provides an even higher overvoltage but at greater expense. Platinum can be replaced as the surface metal by alloys of the platinum group and platinum-palladium and platinum-rhodium are used. It is also possible to use oxide anodes such as Ruthenium oxide. All these are variations on the basic anode and find application in particular environments generally associated with the chemical industry.

The use of some of the alloys improves the cold water performance, others are less readily attacked by sugars or provide a better source of hypochlorite. There are marginal differences in the overvoltage and hence the power consumed. None of these affect the cathodic protection system markedly.

Metal Oxide Anodes

An anode has been developed for the electrochemical production of chlorine based on a layer of metal oxides, principally titanium oxide, on titanium. The process consists of applying the metal oxide as a liquid coating and subsequently fusing or baking it on to the titanium substrate.

The anode has good acid resistance, that is when it is placed in an environment such as seawater mud or in the ground, the high acidity generated in the vicinity of the anode does not have a detrimental effect on the anode performance. This is unlike platinized titanium where the anode does not perform well in mud, having an increased consumption rate of up to five times. Anodes in the ground are usually surrounded by a conducting carbonaceous backfill and the reaction to a large extent is transferred to the surface of the backfill away from the anode. Because there is a finite resistance between the backfill and the anode proper, some anode reaction takes place at the anode/backfill interface and an acid anolyte is developed at this point. The oxide anode does not suffer an increase in the rate of consumption because of this. In terms of consumption of the outer oxide layer, the anode is consumed at 0.3 micrograms per amp hour in the chlorine generating mode and 0.2 micrograms in the oxygen generating mode, that is in the soil. It is suggested that at low current densities, at 5 amps per sq ft, the consumption rate in soil may be ten times lower. This compares with the figures for platinized titanium of 4 micrograms per amp hour in chlorine production and 25 micrograms per amp hour buried in the soil. The oxide has a density approximately one third of that of platinum so the volume consumption figures will be closer. The metal oxide anodes have a rate of consumption that is dependent on current density. In sea water the anode has a lower rate of consumption, that is in micrograms per amp hour, at high current densities than it does at low current densities. This can be expressed by the formula:

$$\log \text{life} = a - b \log \text{current density}$$

When 'b' is 1 the consumption will be constant at all current densities. The metal oxide anode has a 'b' of less than 0.5 in sea water.

The anodes' low rate of consumption coupled with its acid resistance makes it an ideal anode for direct burial in seabed muds or in low resistivity ground where it performs as a longlife medium current density anode with the workability of its titanium substrate.

The developers of the anode have produced a special connection in which the tubular titanium anode is compressed by crimping on to the anode feed cable and forms an effective seal. The cable has a heavy insulation of ethylene/propylene copolymer and an outer sheathing of hypalon which retains the seal during the anode's life, Fig 101.

Lead Alloys

The impressed current anodes so far discussed have been ideal for use in the ground, in the case of graphite and silicon iron, or in backfill where all four types of anode can be used, or in the case of the platinum anodes, at high current densities in sea water.

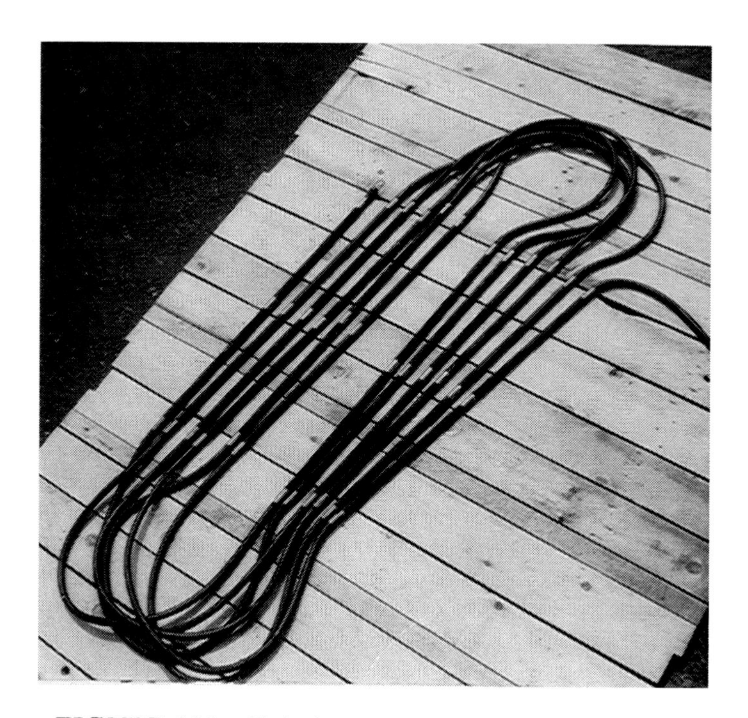

FIGURE 101 - Tubular anodes of metal oxide surface; titanium crimp connected over the anode cable. (Photo reprinted with permission from Impalloy, Ltd., Bloxwich, Walsall, England.)

Prior to the invention of platinized titanium a series of lead alloys had been developed for use in sea water. The early alloys were made from chemical lead with an addition of silver. Whereas Fink had suggested high silver contents, it was discovered that alloys with silver levels of between 1 and 2 per cent made an ideal anode for low current densities in sea water. The anode is cheap, is very easy to use, form and support, and operates in a medium range of current densities, at between 1 and 50 amps per sq ft, the major range for its operation being between 5 and 20 amps per sq ft.

The initial experiments were carried out on 1 per cent silver by the British Navy and 2 per cent silver by the Canadian Navy. The author developed an anode which had 6 per cent antimony as well as 1 per cent silver which had an intermediate performance.

The lead alloy anode works by the formation of a conductive peroxide film on its surface. This film itself acts as the anode and is not destroyed but very slowly grows with time. The effect of excessive current density is to cause the peroxide/lead interface to develop lower chlorides and oxychlorides of lead which are not conducting and which tend to isolate the peroxide from the lead. This often occurs in the form of bubbles or nodules on the surface. At low current densities and in fresh water the peroxide may not be formed and lead then dissolves by a normal anodic process.

The role of the silver appears to be to promote the formation of the peroxide. The use of 2 per cent silver, or possibly silver in excess of 1.7 per cent, gives an anode with a wide range of formation current density and an even wider operating range. The anode can operate in fresh water if formed in sea water and should the peroxide film be damaged the anode will reform a new surface film. The fine layer of peroxide that is formed on the 2 per cent silver anode is flexible and is not cracked or damaged by normal usage of the anode.

Because of these properties the anode is most suitable for use where a low current density anode or a distributed anode is required. The principal use of lead silver anodes is the protection of ships' hulls. The anode emits a low concentration of hypochlorite. The anode can be extruded in a variety of shapes and these can be used to key the anode into a plastic holder. Electrical connections are easily made to the anode and it is a good conductor of electricity.

In place of silver, or occasionally in addition to silver, being used to promote the lead peroxide surface a similar film can be produced by the use of small pins of platinum which act as nuclei for its formation and interconnect the peroxide with the lead base. This means that the anode can be used at higher current densities and properly formed under conditions that might make the formation of the peroxide film on 1 per cent silver lead difficult. However, each platinum pin is only effective over a short distance and so in a long anode, a large number of pins have to be provided as the failure or loss of any one pin could be catastrophic. On plate anodes the pins can form a pattern in which the loss of one or more pins can more readily be tolerated. Where a high current density anode is required the platinum group anodes have taken over with their advantages and the platinum pin, or platinum bi-electrode anode, is not as popular.

Alternative nuclei for the promotion of the lead peroxide film have been proposed, including carbon fiber, and recently composite electrodes of lead dioxide and magnetite. This latter is reported as having favorable consumption and operating characteristics in estuary waters of the order of 1,000 ohm cms.

All the lead alloy anodes seem to have a low consumption, initially losing about 2 lb per amp year but this tailing off to a level of approximately 1 to 2 oz per amp year. Less than 1/2 in. of lead alloy acting on one surface as an anode can be expected to have a life somewhat in excess of 40 years:2 per cent silver anodes which have been in service now for more than 20 years have shown this to be the case.

Lead anodes can be used for protection in sea water other than on ships' hulls and the weight of the anode makes it an ideal cheap, and to some extent, expendable anode for suspension in marine environments. The anode has ideal properties in this respect; it has to be in free water as if buried in mud it tends to corrode more rapidly. In heavy muds, or if the anode is deliberately buried, then it can fail within a very short period of time.

Small castings of lead, as well as extruded anodes, are used in chemical plant and heat exchangers but they are in general superseded by the smaller high current density platinum coated anodes.

Carbon Fiber Anodes

It is suggested that carbon fiber would make an ideal anode material and also that this could be used in conjunction with lead to form an anode in which the carbon fibers replaced the platinum pins. Some individual use has been made of both these principles but neither seem to have gained in favor except in conjunction with a current dispersing layer in concrete bridge decks.

Anode Consumption

Most of the anodes that are used in impressed current are very slowly consumed. This may be, in the case of the graphite anode in sea water, over a matter of 5 to 10 years as opposed to a consumable anode of the same size which could disappear in a matter of months. Many of the anodes have a limited range of current densities within which they operate most successfully. The power used by the anode is considerable so there is a tendency to make anodes that have low resistivity shapes, that is in the form of long rods or plates. These shapes, as well as conferring a low resistance on the anode, tend to increase the current density at the ends and the mean current density may differ considerably from the current density at these points. It is possible to design anodes so that the current density off their surface is more uniform without losing all the advantages of low resistance. For example, an anode that is a long cylinder has a more even consumption if insulating discs are placed at either end of the anode. Equally, the anode can be shaped so that the greater rate of consumption at its ends is accommodated. This gives rise to the "dog-bone" anode which has a longer life and a more constant resistance than a plain cylinder. Some of the early silicon iron anodes were, in fact, short lengths of silicon iron pipe with cone flanged ends and these have a dog bone effect fortuitously and also provided for a center-connection inside the tube.

Equally, supporting an anode using an insulator can cause an end effect next to the support. This often is the point of maximum bending moment and the combination can cause premature anode failure. Techniques of overcoming this are described more clearly in the applications section.

Power Sources

The source of current and voltage for impressed current cathodic protection is very important. The maximum current from a power unit will be limited by practical engineering considerations and it probably will be that units will not exceed 500 amps in capacity, it being better to provide a multiplicity of units. On the minimum current side, impressed current will be in competition with the current derivable from sacrificial anodes in some applications providing less than 1 amp. The maximum current output, therefore, from a particular d c source might be expected to lie within the range 5 amps to 500 amps.

Similarly, circuit and power considerations will limit the output on the voltage side. The minimum voltage required will probably be 5 V as the back e m f of the anode and the polarization of the cathode may well produce 3½ V of this, the maximum voltage of a power unit being probably 50 V as beyond this it will be economic to enlarge the anode rather than pay the power costs.

Within these ranges the power unit will need to have some mechanism for adjusting its output, both because the cathodic protection parameters may vary and because it will be more sensible to manufacture a limited range of power unit sizes. The accuracy with which the current and circuit resistance of a particular installation can be determined should be better than +33 per cent −25 per cent and if a particular error occurs in both to the maximum suggested then the positive and negative errors give a 3:1 ratio in voltage, that is, a unit of 32 amp output should be adjustable from 32 amps 42 V to 18 amps 14 V when the calculated output is 24 amps 24 V. This, the maximum required in selecting a single unit does not take account of the adjustment to the unit once it has been selected. It does indicate the steps in which power units might be manufactured and shows that these can be sensibly placed in a ratio of 2:1 or 5:2 in current output, that is, for example, 10 amp, 25 amp, 50 amp, etc. Similar steps in voltage may be used, though here the use of the transformer rectifiers will fall into two main categories, those used in sea water and those used in other installations. Two voltages, one at about 12 and one at about 30, might then be sufficient.

Within each unit, in addition to its universality, the unit must be controllable in its particular application. Because there will be a back e m f as well as the other voltages in the circuit, the range of voltage variation will be less on low voltage units than on high voltage ones. Generally, a 5:1 ratio is required within the particular unit in addition to the variation that may be required to make the unit fit the particular application. In some applications there will be very large variations in current and these may be caused for example, on a ship, when the difference in the current required when the ship is new and light in the water, will be very different from when the ship is old, fully ladened and traveling at speed. Control down to 1 per cent of the power would seem to be necessary.

As well as being able to vary the output from the power source it is convenient to be able to measure it and in-built meters are preferable. On some equipment this can be a single switched meter whilst others, both ammeter and voltmeter, will be fitted. The power source must be reliable and unless considerable attention can be given to the machinery, non-moving units and solid state devices will be most suitable. The equipment will probably be situated in an area where either its body or its casing will be exposed and it is essential that this is properly designed to give a trouble-free life and to be completely tamperproof, or enclosed in a suitable housing (Fig 102).

Corrosion protection is particularly important as there can be no worse advertisement for a cathodic protection system than a corroded power unit. Life expectancy of the power unit must match the other equipment and be in the order of 30 years.

Transformer Rectifiers

The a c electricity supply is the most convenient and popular source of power for cathodic protection. The a c supply is transformed and rectified to give a d c output. The transformer is usually single phase and steps down the voltage. In many cases a primary autotransformer is used and this may be continuously variable or variable in steps. It is usual for the switching to be on the high voltage, low current side, but it is possible to use low voltage, high current switching or to use links.

The rectification is either full wave or bridge; half wave rectification is rarely used. Even so, the output will have a considerable ripple on it and occasionally this can be harmful or noisy to other communication and signalling equipment. When this is the case capacitor smoothing can be used or three phase equipment employed. At large currents the equipment is often three phase and this leads to a smoother output and less imbalance in an enclosed remote supply system.

In addition to the manual control systems transformer rectifiers can be controlled automatically. This may be by replacing the primary transformer, with a saturable reactor, which will limit the power into the main transformer rectifier. As the current will vary as approximately the square root of the variation in the power, the saturable reactor will be limited in the control it can exercise. The alternative is to use a silicon controlled rectifier in the form of a thyristor or in the back-to-back device called a Triac. In these the method of control is usually by phase angle; that is, the control switches on during part of the cycle and allows a pulse of a c current through either the transformer or into the d c circuit. Where the control is

a

b

FIGURE 102 - (a) Resin fiber glass switch and transformer rectifier housing. (b) Inert gas purged resin glass fiber instrument and control building.

exercised in the primary then distorted a c of lower power is fed into the transformer rectifier. This will have a very increased heating effect and the transformer rectifier is designed to cope with it.

When the rectifiers themselves are controlled, similar de-rating of the transformer is required. As these control devices are costed on their current performance larger units will operate with a c control and smaller units with d c control. Three phase units are treated similarly; there is a problem in some of them with feedback at low levels of power. As an alternative, on a three phase supply, units of single phase construction can be used one placed in each of the phases and a control signal from a single controller used to regulate the level of power. At small currents d c control can be exercised and this can be done by power transistor.

Alternative methods of control using pulse instead of phase angle are possible. The variation in output will be by the ratio of the number of pulses gated through to the number stopped. Thus in a twenty cycle spell the control could be from one half cycle to twenty full cycles or 1 to 40 in power. Some problems of very rapid consumption of platinum anodes may occur with this mark space control. Pulsed protection current may be achieved by a similar wave form and the same techniques can be used. Special power units are needed to provide pulses outside this range.

The rectifiers used in cathodic protection are principally silicon because of their sealed nature, but are sometimes germanium and occasionally selenium. Improvements in the impregnation of the transformer and in the construction of switches, or more particularly in the use of solid state control devices, mean that a larger number of transformer rectifiers are air cooled. Many of these are base mounted and intended primarily to be used inside some form of building or shelter. Others, which are usually pole mounted or mounted on a short support pole of their own, are built in a substantial case which is almost completely weatherproof and often of plastic, such as glass fiber reinforced polyester.

Large units, and particularly some of them that are subject to salt spray, are oil-immersed; this improves the cooling and protects most of the equipment from atmospheric degradation. By placing the switches and the meter connections under oil an instrinsically safe unit can be made. A unit that is flame-proof to British Standards or explosion proof can be supplied, though these have to be housed in special tested enclosures. Where instrinsically safe equipment is required digital meters can be used and a digital display may be shown through a glass panel in the side of the tank, providing coded readings of current voltage and potential. Alternatively a conventional air-cooled unit can be placed in controlled atmosphere housing (Fig. 102b). For use in the tropics, the transformer rectifier should have a sun shade which will keep the majority of the tank in shade during most of the day. The unit must be tamper-proof so that unauthorized access to the switches or other adjustments is not possible (Fig. 103).

FIGURE 103 - Zone 2 hazardous area certified transformer rectifier with sunshade. (Photo reprinted with permission from Hockway, Ltd., Croydon, Surrey, England.)

Cathodic protection of a large structure can consume a considerable amount of power, and though as a constant load this is welcomed by the generating authority, great economy may be achieved by reducing or switching off the protection during short peak-load periods. Switching off in many installations, even for a few hours, will not affect the protection over a 24-hour period; this switch off can be automatically controlled from a peak demand device.

Other Generators

When no a c power is available, the d c source can be a rotating generator. The generator may be driven by d c power if that is available, or

FIGURE 104 - Vertical axis wind turbine generating about 1 kVA in average wind conditions. (Photo reprinted with permission from P.I. Corrosion Engineers, Ltd., Hants, Alresford, England.)

by some form of engine. This can be a diesel engine, a steam engine or a gas turbine, and will depend upon the fuel available for it possibly from the pipeline product. In cases where there is d c power available solid state devices can be used to reduce the voltage from storage batteries to provide the cathodic protection. Alternatively, thermoelectric generators can be used and these are more popular where there is a cheap supply of energy, as on a gas pipeline. The thermoelectric generators provide a continuous d c current, though they are not particularly efficient.

Solar generators can be used, though naturally only during hours of sunlight, and these provide a d c current which can either be stored or used as an additional polarizing current to a structure that is protected by sacrificial anodes.

Wind generators in many areas will provide a longer period of power than can be achieved from solar systems (Fig. 104). However, the sun rises daily while there can be longer periods of calm. The efficiency of the wind generator and its capacity is generally greater than that of the solar collector, but it does have moving parts and requires maintenance. Both systems can be used to augment other protection techniques or to charge a storage system to give continuous protection.

Automatic Control

Increasing use is being made of automatic control for cathodic protection. Usually this operates from a potential measurement on the structure. This point of measurement has to be carefully selected so that it gives an accurate indication of the protection without calling for an excessive reaction from the generator during periods of polarization or following de-polarization as would be the case in after an ocean storm or an oil platform. The power unit has itself to be protected against excessive demand so that it is not burnt out. The structure may also need protection where, in order to polarize the zone selected, other areas are subject to excessive protection and the possibility, at least with high tensile steels, of hydrogen embrittlement or of coating damage.

The problems of automatic protection vary from structure to structure as does the need for this type of control. It will be discussed with the applications of cathodic protection.

In general, continuous manual control of cathodic protection is difficult chiefly because of the slow response of the majority of structures to the cathodic protection current. This response may be over a period of minutes, hours, or sometimes days. Devices that can be continuously varied tend to be over-adjusted and the system chases the desired level. With step-wise control it is easier to make a single adjustment and a subsequent adjustment. It is usual that step control in cathodic protection is on an arithmetic basis not a geometric basis, that is, that if the control is divided into 63 steps then these may be approximately 1/2 volt adjustments between 5 V and 40 V. There is much to be said for the use of geometric control in which each step represents a 3 or 4 per cent change in output voltage. Continuously variable auto-transformers can be used but they can, in certain atmospheres, present problems with contact contamination and corrosion. Alternatively, the automatic control device can be manually activated and these provide continuous variation over a very wide range of power.

Cables and Connections

While the anode of an impressed current system may be made from the materials that are permanent under the anodic conditions the cable conductor to the anode must be made of the conventional electrical metals, copper or aluminum, and these would suffer rapid corrosion if they were exposed to the electrolyte. The conductor must be extremely well insulated and this insulation must extend right up to the anode material itself. Additionally, in the anode vicinity the insulating material must be able to withstand the oxidizing and acidic conditions met there which will include, in certain environments, heavy gassing of chlorine and oxygen.

The insulation materials will be buried in water or soils and so must retain their electrical properties when wet. Natural rubbers will be attacked by the anode environment and the only suitable common cable insulation materials that can be used are plastics and synthetic rubbers. The best results are obtained with polythene and cross linked polythene cable insulation sheathed with PVC synthetic rubber to give it protection against mechanical damage; at elevated temperatures only irradiated polythene may be used.

Some anodes are recessed around the electrical connection and after the conductors have been soldered, or otherwise connected, the recess is filled and stoppered, a typical anode section being shown in Fig. 105. The initial filling is made with a setting compound which has a good adhesion both to the cable sheath and to the anoe. If this is also mechanically strong and resists the attack of the oxidizing conditions at the anode it may be used alone, otherwise a plug of a suitable resistant material, such as polythene, should be used to close the mouth of the recess as shown in the drawing. Polyester and epoxy resins have been used to seal the anodes with good results.

With the platinized anodes the connection is usually either to the end of a rod or to a stud. The connection is mechanical and is usually characterized by a flexible anode cable and a comparatively rigid anode. The connection is encapsulated and this is usually done in a polyester or epoxy resin. The polyester resins are better, not because of their improved chemical resistance, but because in general they can be made to shrink on to the connection and cable, whereas epoxy resins sometimes contract away from the cable. It is usual with many cables to pretreat the cable insulation itself so that it more readily adheres to the resin. It is also essential that the resin is reinforced to ensure that flexing of the cable will not break the brittle resin. Glass reinforced resin is generally employed and winding techniques have proved to be the most durable under highly stressed conditions.

In many cases the anode is brought into a container which may be steel or plastic, and the cable to anode joint is made within this container, which is then either sealed mechanically to enclose it or it is sealed by the container being completely filled. In many of these containers the cable can be brought out of it by the use of a conventional cable gland. This anchors the cable, and its armouring if present, and transfers the stress away from the jointing compound.

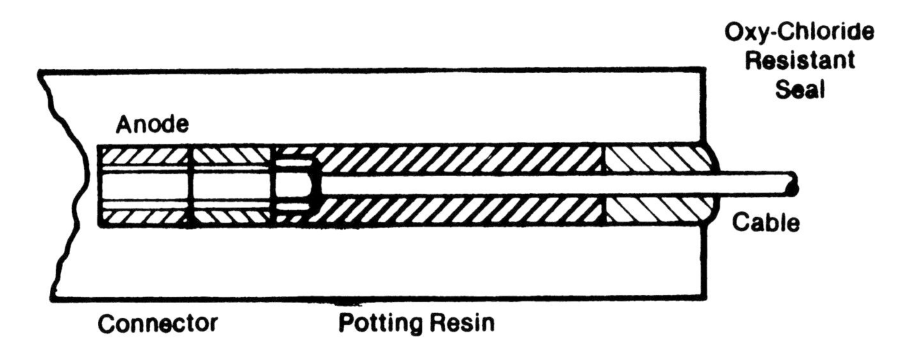

FIGURE 105 = Anode to cable connection encapsulated in resin with oxidation resistant plug.

Where these joints are integral with the anode construction they will be described in more detail in the chapters on application. One type of graphite anode with a compression connection, was illustrated earlier in the chapter. This shows the type of cable encapsulation that can be built up around the anode and has proved to be extremely reliable in service.

Any joint or junction of the anode conductor cable which is in the soil or under the water must be made with equal care, though most of these will not be in the close vicinity of the anodes and so will not suffer the anodic chemical attack. Present standard practice is to use a plastic box around the joint and to fill this with an insulating compound, cold setting resins, other plastics or bitumens being suitable for this type of filling.

Where a large d c network of current is employed then the voltage drops along the cables may become quite large, for example, a groundbed connected by 200 yds of 0.1 sq in. copper cable will suffer a voltage drop of 2 V at a current of 40 amp and similar voltage drops will occur in the cathode leads. Because there is a considerable voltage between the anode and the earth the drop of potential in the anode lead will be of little consequence, particularly in electrolytes of resistivities of the order of 1,000 ohm cm. The change in potential of the cathode, however, will be greater than the swing required to achieve protection and any differential voltage drop along a series of cathode cables will influence the spread of protection. The cathode leads must be carefully balanced in resistance so that the current that is required to be drained from any part of the structure causes the same potential drop in its cathode lead as occurs in each of the other cathode cables.

It has become general practice to pole bare d c conductors and bare power cable an arrangement prone to be struck by lightning. The two ends of the conductor are attached to the groundbed and structure, both of which

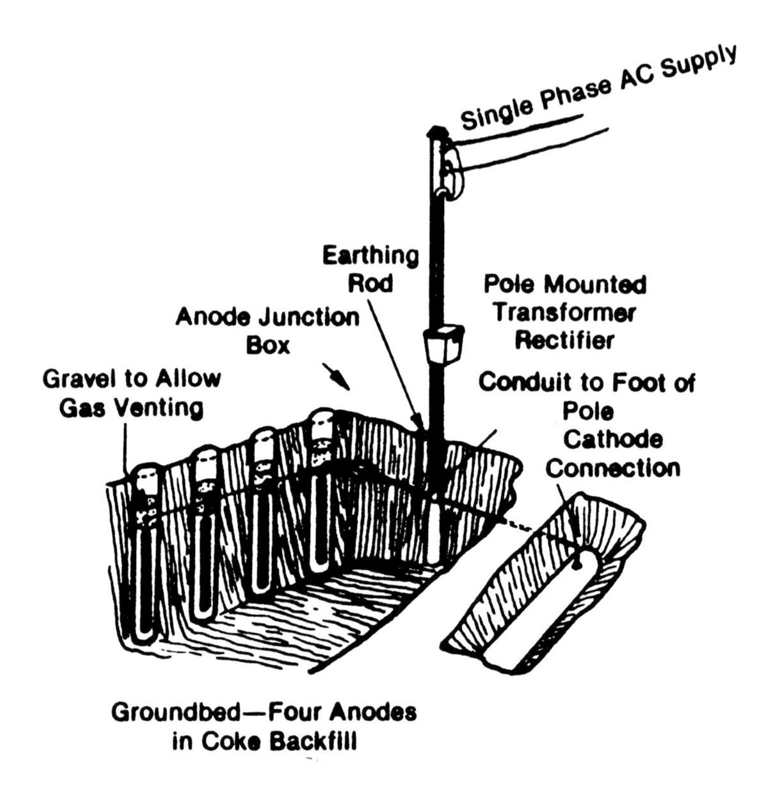

FIGURE 106 - Typical small impressed current installations.

have low resistances to earth and should suffer no damage; it is advisable to fit a lightning arrestor across the output of the power generator. The major effect of the current surge will be upon the meters and switching these out of circuit between inspections is a sensible safety precaution. A typical pipe protection system is shown in Figure 106.

Cathodic Protection of Buried Structures

Chapter 6

The structures that receive cathodic protection can be divided into three groups; those that are buried in the ground constitute the first and this includes pipelines, tanks and foundations; the second group comprises structures that are immersed in water, both marine and fresh, including stationary objects such as jetties, dock gates, wharves etc, moving vessels of all sizes and offshore structures. The third group consists of those applications where the electrolyte is contained within the structure and its members vary from a household cold water tank to an oil tanker in ballast.

The present chapter is concerned with providing cathodic protection to the first group; the buried structures.

Pipelines

Pipelines can be divided for the purpose of cathodic protection engineering into two classes: short pipes in which the longitudinal electrical resistance is sufficiently small for them to be considered at equipotential—to within 50 mV—while being protected, and long pipes, or pipelines, where their longitudinal resistance will play an important part in the design and engineering of the cathodic protection installation. The classes may be sub-divided further into small and large diameter pipes. The small diameter pipes are those in which the ratio of the diameter to depth of burial is small so that there is no appreciable effect of the surface of the ground upon the electrical potential distribution around the pipes; large pipes, on the other hand, are those laid so that the earth surface has a considerable effect upon the current and potential distribution around the pipe.

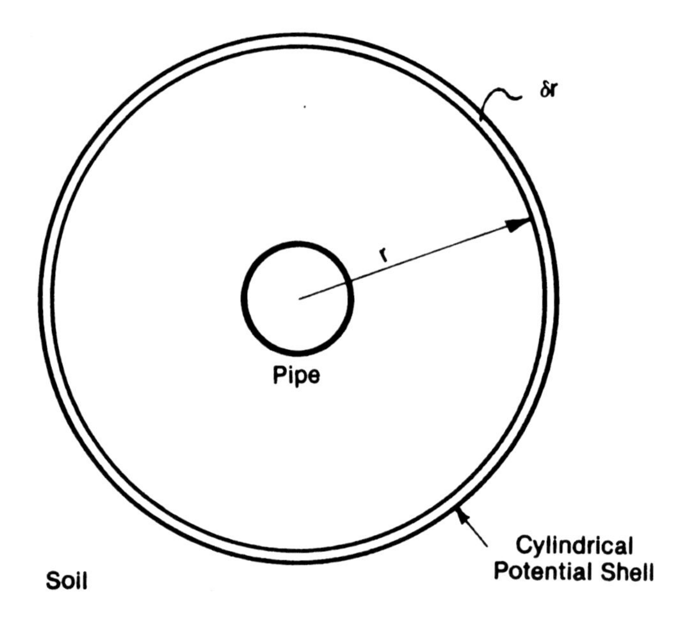

FIGURE 107 - Pipe section in infinite homogenous soil.

Short Small Pipes

Small short pipes are usually found as services conveying some product to a consumer and may carry gas, water, oil or be electric cables. In general these pipes will be of steel and a large proportion of them will be galvanized. The effect of galvanizing will be similar to that of a coating and will reduce the current required for cathodic protection. Metals such as aluminum, lead and copper, which may be used in the manufacture of this type of pipe, will be discussed as variations of the steel pipe which will be dealt with in detail.

Suppose there is a steel pipe of diameter d inches and it is buried in a soil of uniform resistivity ρ ohm cm in which the pipe requires 4 mA per sq ft for protection. Then if a 1 ft section is considered the resistance between it and some distant point can be calculated. Figure 107 shows the pipe in section. The resistance of the small cylindrical shell surrounding the 1 ft section of pipe will be

$$\frac{1}{30}\,\frac{\delta r}{2\pi r}\,\rho \text{ ohms} \tag{6.1}$$

and so the resistance from the pipe to some enclosing cylindrical surface at a distance R inches will be

$$\frac{\rho}{60\pi}\int_{d/2}^{R}\frac{dr}{r} = \frac{\rho}{60\pi}\ln\frac{R}{d/2} \tag{6.2}$$

or the voltage drop between this cylindrical shell and the pipe surface will be

$$\frac{4\pi d}{12} \times \frac{\rho}{60\pi}\ln\frac{2R}{d}\ \text{mV} \tag{6.3}$$

If reasonable values are inserted into this equation, say a 2 in. diameter pipe, ρ, 2,000 ohm cm and R, 4 ft, the potential drop will be

$$\frac{8\pi}{12} \times \frac{2000}{60\pi}\ln 48 = 86\ \text{mV}$$

This voltage drop will have two effects: it will enhance any potential reading made with the half cell 4 ft from the pipe by 86 mV compared with a reading relative to the pipe surface, and it will reduce the driving voltage of a sacrificial anode system by the same amount. An increase in the degree of cathodic control of the system will occur as the voltage drop associated with the cathode will be greater.

Now assume that there is a particular length of 2 in. service pipe, as described above, 100 ft long, to be protected. The surface area of the pipe is

$$\frac{100 \times \pi \times 2}{12} = 52\ \text{sq ft}$$

so that if the current density is uniform 208 mA will be required to protect it. The linear resistance of the pipe will be about 0.01 to 0.02 ohm for each 50 ft length, assuming a centrally located installation, and as 104 mA are needed to protect each half, then the voltage drop in the pipe metal will be

$$\frac{104 \times 0.015}{2} = 780\ \mu\text{V}$$

This would be undetectable by normal pipeline potential measuring techniques and so the line can be considered to be short.

Sacrificial Anode Protection. A protection current of 210 mA can be derived from sacrificial anodes or a very small impressed current system.

Magnesium or zinc anodes could be used and these should be surrounded by a backfill. The pipe to be protected is assumed to require a potential of -0.85 V to a close copper sulfate electrode so that its potential relative to an electrode 4 ft from the pipe would be -0.936 V which allows a driving voltage of 164 mV with zinc and 600 mV with magnesium. In practice the current would be obtained by selection from a number of stock anodes; amongst those available might be a 30 lb. zinc anode and a 22 lb. magnesium anode which, complete with backfill, might have resistances of 3 ohm and 6 ohm respectively in the particular soil. These values could be calculated from the anode shapes, sizes and data in Chapter 3. The total resistance of the zinc anode installation should not exceed 0.5 ohm and of the magnesium installation 2.5 ohm. Therefore, six zinc anodes distributed along the line or three magnesium anodes in a group would be sufficient to provide the protection.

The anodes will be required to provide protection for a definite period or until they are partly consumed when they will be replaced. In the latter case the anode might need replacement when it is half consumed; its resistance will then have increased by about 25 per cent so that its initial resistance should be 20 per cent less than those quoted, i.e. 0.4 ohm and 2 ohm. This will cause an increase in the mean current of about 10 per cent. Now eight zinc anodes would be required or a group of four magnesium anodes.

The life of each system can be calculated from the anode weigh and rate of consumption. The zinc anodes weight 240 lb and at a consumption of 30 lb per amp year would have a life of 19 years at 210 mA before they are half consumed. The magnesium would weigh 88 lb in a group and at a consumption of 20 lb per amp year would have a life of 10 years before they were half consumed.

Usually the anode replacement would not be by a percentage consumption, but by making additions to maintain the protective potential; replacements might, therefore, take place at more frequent intervals with some saving in current output from the anodes as they would only be required to provide the minimum current for protection. If the anodes used were expected to give a definite life, say 15 years, then either two extra magnesium anodes could be used and grouped to reduce their output or the zinc anodes described above could be employed. The greater weight of zinc, 240 lb. as compared with the weight of magnesium, 132 lb., would make the costs of the two systems comparable.

It has been assumed so far that the anodes have been located remote from the pipe. With the distributed anodes these would be placed at short intervals along the pipe and almost even protection could be assumed if the anodes are located so that they are separated from the structure by half the distance between themselves. If they are closer to the pipe than this then there will be an increase in current density opposite them which will effect the overall current demand. The anodes may be located within the right of way of the pipeline when they can either be buried much deeper than the pipe or long thin anodes can be laid close to and parallel to the pipe so that there is only a very small distance between anode ends. It was assumed earlier that only one set of stock anodes, one zinc and one magnesium were available, but most engineers will carry a stock of anodes in which the ratio of anode weight to resistance in the electrolyte will vary. To protect a simple pipe as described above, a long thin anode could be installed to give an economical short life or a more massive cylindrical anode to give long life protection. Grouping or widely distributing the anodes will give further variations.

The pipe described, while being typical of a bare steel pipe, required a large current to achieve protection in sterile soil while at this current density in anaerobic soil containing bacteria the protection criterion of $-0.95V$ to copper sulfate would have precluded the use of zinc anodes.

Coating or galvanizing would reduce the current demand and hence the associated cathode volts drop; a tape coating may reduce the current required to 1 mA when a single zinc anode would provide adequate long life protection. Galvanizing would reduce the protection current to below 100 mA when either anode system could be used. As equation 6.3 suggests, the voltage drop at the cathode reduces with decreasing size so that small galvanized pipe, when used for household services, is readily protected.

Impressed Current Protection. Impressed current could be used to protect such pipes, though generally it will rarely be economical for current demands of less than 500 mA. A small air-cooled rectifiers of 8V to 12V, 1 amp capacity, similar in size to a small computer battery charger, could be used to drive current through a single graphite rod surrounded by backfill preferably pre-packaged. The resistance of the circuit would be about 10 ohm and a back e m f would have to be overcome so that 8 V drive would be adequate; the anode would be truly permanent and the power consumption only some 8 to 10 W which would be negligible. Such a system would be more costly to install but might compete economically with sacrificial anodes if used over a long period, say more than 20 years, or in higher resistivity ground. The system finally adopted will depend upon the particular site conditions and the life and current demanded for protection.

Short Large Pipes

The discussion so far has been concerned with small pipes but there are many short lengths of large diameter pipe in existence. These may include a short buried length of a pipeline which is otherwise above ground, an isolated short line or short section of a long pipeline between insulating couplings. The same criterion of shortness will apply. An 8 in. pipe line may have a resistance of 0.02 milliohm per ft. so that 100 ft of this pipe which may require 3/4 amp for protection would have a potential drop of

$$\frac{0.002 \times 0.75}{2} \text{ volt} = 0.75 \text{ mV}$$

along its length if protected from one end. Therefore, considerable lengths may be protected from a single cathode or drainage lead. The voltage drop along a pipe will increase as the square of the length as both the current and resistance increase linearly. Thus, if a limit of 50 mV potential drop is adopted then 1630 ft of pipe may be protected from a central installation.

This length will vary with the size and resistance of the pipe and with the current density required to achieve protection, for example, a pipeline with a high quality coal tar enamel coating may require 0.05 mA per sq ft or 0.1 mA per linear foot if 8 in. diameter. The resistance per foot of the pipe would be the same so the length that could be protected would be

$$\frac{1630}{2} \sqrt{\frac{7.5}{0.1}} = 7{,}000 \text{ ft} \left(\begin{matrix}\text{ratio of current}\\ \text{per linear foot}\end{matrix}\right)$$

$$100 \sqrt{\frac{50}{0.01}} = 7{,}000 \text{ ft} \left(\frac{\text{permissible mV}}{\text{mV to protect 100 ft}}\right)$$

either side of the drainage point and this protection would require 1.4 amp. Such a calculation would have been equally true for a small diameter pipe.

The difference to the cathodic protection engineer between the small and large diameter pipes lies in the current and potential distribution in the ground around the pipe. The earth surface acts as a semi-infinite conductor and its effect can be reproduced by considering an infinite conductor with the pipe and its image in the earth's surface enclosed in it. The arrangement is then as shown in Figure 108.

If the earth were perfectly uniform and the pipe metal, or rather the outer surface of the pipe-metal/soil interface, were at the same potential, then it would be possible to calculate the field which surrounds the pipe. As neither of these conditions is likely to be strictly true in practice the exercise must be an approximation. Pearson has suggested a modified formula for the calculation of the potential of any point relative to the pipe and represents the pipe/soil interface as having purely resistive properties.

This will be true of pipes that have coatings which display on an ohmic conductance and will illustrate the type of potential distribution to be expected with other pipes. The significance of these potentials on bare pipes is decreasing as very few uncoated pipes are buried nowadays. If the pipeline shown in Fig. 108 is considered to receive i amp current per foot from the

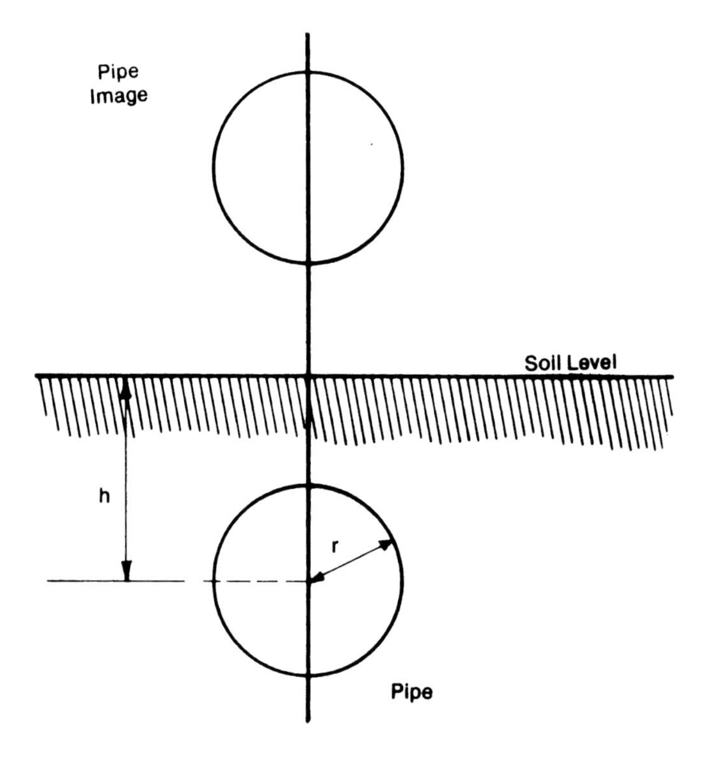

FIGURE 108 - Pipe and pipe image in section.

groundbed then a point which is distant R_1 and R_2 from the centers of the pipe and its image will have a potential V which will be the swing potential

$$V = \frac{i\rho}{2\pi} \ln \frac{R_1}{r} + \frac{i\rho}{2\pi} \ln \frac{R_2}{r} + C \qquad (6.4)$$

where ρ is the resistivity in ohm ft and C a constant of integration. If the coating or interface has a resistance of K ohm ft then the potential swing of the soil next to the pipe relative to the pipe potential will be iK. Now this potential can be derived from equation 6.4 giving

$$iK = \frac{i\rho}{2\pi} \ln \frac{r}{r} + \frac{i\rho}{2\pi} \ln \frac{2\,h}{r} + C \qquad (6.5)$$

as the point where R_1 equal r and R_2 equals 2 h will lie on the pipe surface. The method is not exact but equation 6.5 gives a solution which is a reasonably close approximately for most practical cases.

Now it is convenient mathematically to allow K to be:

$$K = \frac{\rho}{2\pi} \ln M$$

so that:

$$C = \frac{i\rho}{2\pi} \ln \frac{M_r}{2h} \tag{6.6}$$

and V now becomes:

$$V = \frac{i\rho}{2\pi} \left(\ln \frac{R_1 R_2}{r^2} + \ln \frac{M_r}{2h} \right) \tag{6.7}$$

At the surface of the earth R_1 equals R_2 and both are equal to $\sqrt{h^2 + x^2}$, where x is the distance from the center line of the cylinders to the point considered.

Thus as the earth's surface

$$V = \frac{i\rho}{2\pi} \left[\ln \frac{(h^2 + x^2)}{r^2} + \ln \frac{M_r}{2h} \right] \tag{6.8}$$

Similarly on the vertical center line the value of V will be given by

$$V = \frac{i\rho}{2\pi} \left[\ln \frac{(h + s)(h - s)}{r^2} + \ln \frac{M_r}{2h} \right] \tag{6.8a}$$

where s is the depth of the measurement.

Assuming some practical values for a pipeline, the potential along the earth's surface and its variation with depth over the pipe can be plotted as in Figs 109 and 110. The pipe may have a diameter of 12 in., be buried 2 ft 6 in. deep in soil of 3,000 ohm cm resistivity (100 ohm ft) and require a current of 2 mA per sq ft or 6.3 mA per linear ft to give a swing of 300 mV across the coating. Fig. 109 shows the variation of the measured potential with measurements taken on the earth's surface at right angles to the pipe route.

Fig. 110 shows on a larger scale the variation of potential with depth and compares it with the variation shown in Fig. 109. Fig. 110 also includes a similar curve for a pipeline requiring 1 mA per sq ft in 1,500 ohm cm soil or where the product of the current demand and soil resistivity is a quarter

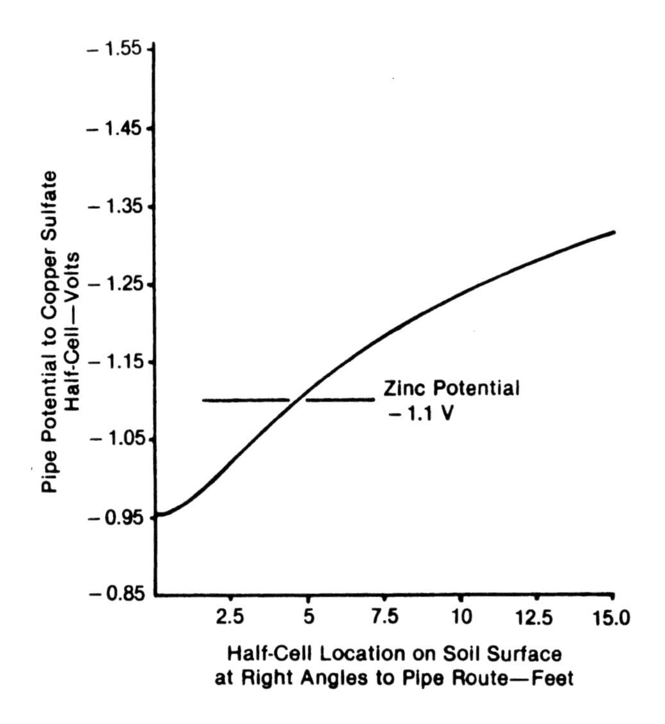

FIGURE 109 - Potential variation along soil surface at right angle to pipe.

of that in the case cited in detail. The potential variation will decrease with a reduction in the current density required for protection and with the soil resistivity. As the accuracy of the potential measurement will only be ± 4 mV then a twentyfold reduction in the current required $\times$ resistivity product will make any potential changes in the ground insignificant. The pipeline considered is not very large and is laid reasonably deep: larger pipes laid closer to the surface will cause greater potential variations in the soil.

Potential Measurement. There are two important effects of this phenomenon. Firstly, it will influence the position of the half cell when it is used to determine the efficiency of the cathodic protection, and secondly, the overall voltage drop to a point near the anode location will influence the amount of driving voltage available.

As the reference electrode potential and the driving voltage are not known to better than 4 mV and 10 mV respectively only bare or poorly wrapped pipes in high resistivity soils will be affected. The half cell position has been discussed in Chapter 2 but it will not always be easy to measure the potential of the pipe relative to a closely placed electrode without excavation. The volts drop in the soil associated with the pipe can be analyzed

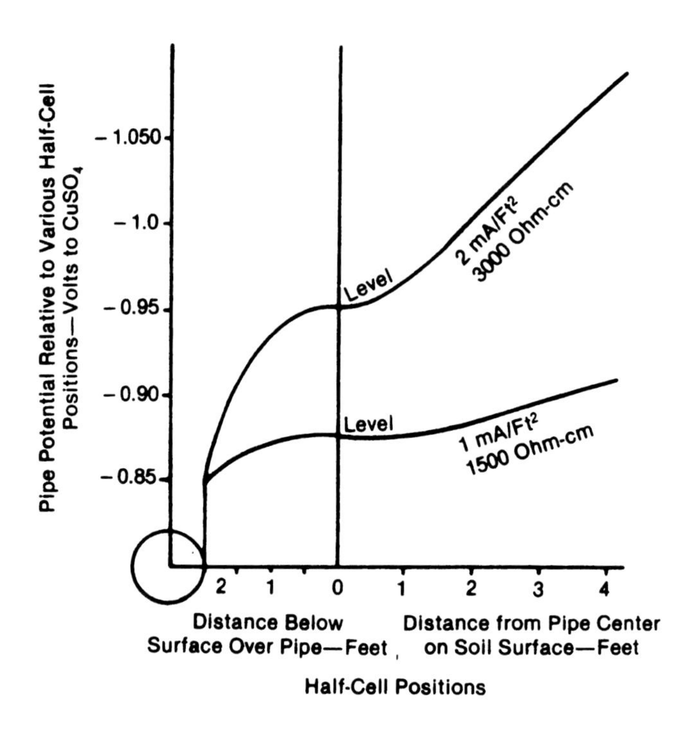

FIGURE 110 - Potential variation in the soil vertically above the pipe and along the soil surface.

so that the potential at the metal/soil interface may be determined from measurements made at the soil surface. If the potential relative to a half cell placed on the surface of the ground immediately over the pipe is determined and if this is n mV more negative than the potential that would have been measured relative to a half cell close to the interface it will be possible to find two points on the surface of the ground, one either side of the pipe, at which a half cell would display a reading n mV more negative than the potential measured with half cell on the soil surface over the pipe. If the current density or soil resistivity changes, the value of n will change equally relative to both positions. Thus for a particular pipe configuration, the difference in potential of the metal relative to a close half cell and to a half cell on the ground surface above the pipe can be found from two measurements of the pipe potential with surface electrodes. The distance between the surface positions of the half cells will be a function of the pipe diameter and its depth of burial; Fig. 111 shows a set of curves which give this distance. The method suffers several inaccuracies but these are small and the method generally gives the answer to better than ± 5 mV which is adequate.

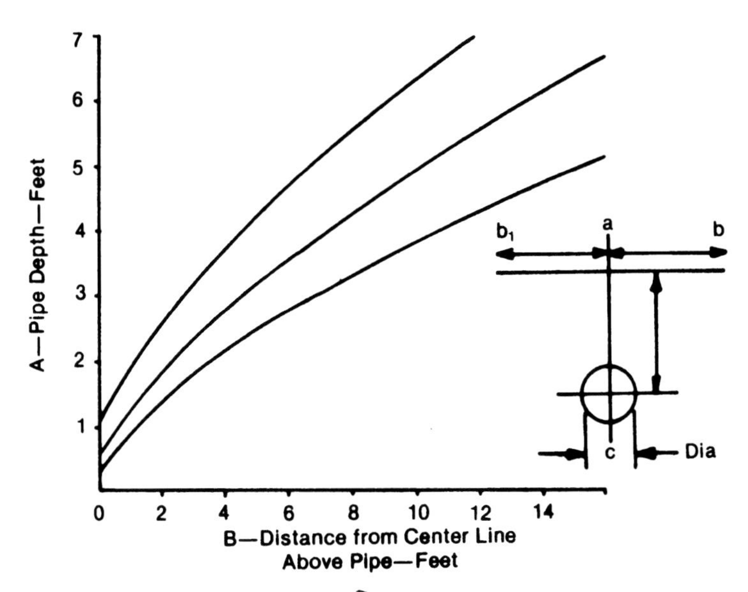

FIGURE 111 - Curves for determining volts drop between pipe surface and soil surface for measurements on soil surface only.

Alternatively, a computer program can be used that will give measurement of the pipe potential when the cell is placed over the line and a specific distance either side of it. This will mean that the graph will not need to be referred at on each reading and the measurement will automatically be corrected. The information, however, would need to be fed into the computer. The advantage of the technique may lie in that the computer could analyze from the series of 4 or 5 measurements taken at set distances apart and locate the pipe as well as determine the error in the measurement.

The volts drop in the soil can be eliminated by measuring the immediate polarization on switching off and comparing this with the potential after decay and using this measurement as a criterion. The polarization of the pipe immediately on switching off, however, excludes the ohmic components associated with the coating. These components are essentially part of the 300 mV swing or the 0.85 V potential.

It is difficult to switch off both transfer rectifiers so that a sharp cut-off in current from each occurs simultaneously at the point of measurement. It is possible to switch off one transformer rectifier and to compute the effect of each. It is also possible to make an abrupt step in the protection current and to use the change associated with this step to compute the ohmic drop in the soil. These techniques need a great deal of skill, expensive and difficult-to-use equipment, and the pipe must not have any effect on the switching transients. The advantage of the multi-electrode computation is that it is free from these constraints. A multi-electrode technique can be used and the pipe located accurately by the supposition of a c on the d c signal.

Sacrificial Anode Protection. The voltage drop that can be tolerated before low driving voltage sacrificial anodes cannot be used is not easily determined. In the case discussed earlier the zinc open circuit potential is reached at about 5 ft from the pipe metal, and for zinc anodes to deliver current for cathodic protection they would need to be placed closer to the pipe than this and be fairly well distributed around it. Such an installation would be impractical except for the installation of continuous ribbons of the anode and these would need to be placed at least either side of the pipeline. They would need to be connected frequently to the pipeline and would have the disadvantage that as the zinc is consumed the core of the anode remains and is attached to the cathode.

Magnesium anodes would suffer a considerable reduction in their effective driving voltage and the enhanced change in potential associated with the cathode would bestow cathodic control upon such an installation, that is, the anode would act as though it possessed a very low driving voltage. This would mean that low resistance shapes of anodes would be needed and ribbon or rod anodes have been developed for these conditions. The extra voltage needed to drive the current onto the cathode will have no effect upon an impressed current scheme as the overall driving voltage of the circuit will be large, say 24 V, compared with it.

Thus for large bare or poorly coated pipes a low resistance magnesium anode system could be used where the conditions are no worse than those described above, while zinc anodes, except in ribbon form, would be impractical. The use of zinc anodes would be possible in the event of the potential changes in the soil being less as are those shown in the lower graph of Fig. 110, as would occur were the pipe well coated. In each case the method of engineering the installation would be similar to that used on the short small diameter pipes and coated or wrapped pipes are similarly protected by these methods, Figs 112, 113 and 114 show the attachment of a typical anode.

The driving potential that the sacrificial anodes have relative to the pipe will be reduced by what ever ohmic potential change occurs in the pipe metal between the most distant points and the point of attachment of the anode lead. If this reaches the maximum value of 50 mV stipulated for short pipes then the driving voltage would be reduced by this amount. Sacrificial anodes are often electrically connected together by a busbar wire and this wire is then connected to the pipe. As up to 20 or 30 anodes may be so connected there will be some voltage drop in this anode lead and this potential will reduce the anode driving voltage, particularly of those anodes most remote from the point of connection to the structure.

FIGURES 112 and 113 - Attachment of anode cable to pipeline using thermite welding - mold in position on coated pipe with anode hole behind. (Photo reprinted with permission from Henry Gove, Esq., Romford, Essex, England.)

FIGURE 114 - Repairs of coating at connection point of anode cable with anode in background. (Photo reprinted with permission from Henry Gove, Esq., Romford, Essex, England.)

Impressed Current Protection. In the protection of this type of pipe with impressed current, the current used will be larger than that required for small pipes. The groundbed or anode group will, therefore, have a larger capacity and will probably encroach upon the design limitations such as maximum current densities of the anodes used. As the surface area of the anodes will increase more rapidly than their conductance through the electrolyte increases, then there will be a tendency for the anodes to require a higher driving voltage at these larger currents. This will raise the circuit voltages to about 20 to 30 V at which point it will become economical to increase the anode sizes beyond the electrochemical requirement in order to maintain a low circuit voltage.

The groundbed will be surrounded by a series of equipotential shells similar to those surrounding the pipe and at considerable distances, usually greater than the groundbed's maximum dimension, the equipotentials will be hemispherical. The voltage of the equipotentials that would exist were the groundbed feeding current to an infinite shell would be $I\rho/2\pi r$ in homogeneous soil where r is the distance from the center of the groundbed to the equipotential. Thus under these conditions the potential between two points, 100 ft and 100 yds from the groundbed, would be $I\rho/2\pi$ (1/100 − 1/300) which at 10 amp in a soil of 3,000 ohm cm, i.e. 100 ohm ft, it would be 1 V. Now if the groundbed is placed so that the pipe approaches within

100 feet of it and the pipe's most distant point is more than 100 yards away, when a potential difference might be expected between the soil at these points and this will be reflected in the pipe to soil potential. The effect of the groundbed can be greatly reduced when it has to be close to the line by burying it deeply in the ground. Now the equipotential shells will, in general, be spherical and the groundbed will have approximately half the influence of a surface anode the same distance from the pipe.

The current flow will be modified by the proximity of the pipe and the current flowing in the soil will be diminished as it gets further from the groundbed by the amount that is flowing into the pipe. The net effect of this will be approximately to halve the potential drop calculated from the assumption of hemispherical equipotentials. If the protection criterion adopted is that relative to a close electrode then the effect of the change in soil potential upon the potential measurements used to determine the degree of protection will be further reduced as the change in potential of the soil due to the groundbed proximity will include a large ohmic component associated with the soil in the pipe's vicinity.

The effect of the groundbed's closeness will be exaggerated by a non-ohmic interface resistance at the pipe surface that shows an apparent decrease in resistance at high current densities. The combined effect of the closeness of the groundbed and this interface condition is well known and is generally termed chaneling. It can be avoided by removing the groundbed to a place further from the pipe; this will involve extra d c cabling costs and must be balanced against the cost of the extra current required to achieve complete protection to the pipe when some over-protection is occurring near to the groundbed.

So far only homogeneous ground has been considered and if the ground is inhomogeneous some of the effects described will be greatly changed. In an area where there is a thin layer of low resistivity soil covering a high resistivity bedrock then the voltage gradients in the soil will be increased beyond those expected for a homogeneous medium and the reverse conditions will produce the opposite effect. As variations of a hundred to one in the top soil to rock resistivity are possible, large differences in the behavior of cathodic protection installations are found.

Particularly advantageous use can be made of a low resistivity water-bearing strata below an arid or desert country if the groundbed can be constructed at this level when an enhanced spread of protection is achieved. Where the low lying rock is lime stone or chalk the anode can dissolve the rock by its electrochemical reaction. In some limestones, caverns have formed around the anode which then becomes suspended in the water and may break up or its weight may break the cable or the cable insulation. The pipe trench may act as a land drain and dry out the ground around the higher parts of the pipe. The soil in the immediate vicinity of the pipe may also be washed away by this effect and the pipe can lose contact with the soil.

The soil potentials surrounding a pipeline can be changed considerably by certain pipe laying techniques. When the pipe is laid in a rock trench it is usually surrounded by a layer of soft, often low resistivity, soil; this may mean that a higher current density will be required at the pipe surface to achieve protection while the high resistivity rock will cause a very large potential gradient around the pipe trench. Such pipes cannot usually be protected by sacrificial anodes except when they have an excellent protective coating such as a tape or an extruded pvc sheath unless the anode, usually as a ribbon, is also located in the trench. The groundbed will be located in the area of lowest resistivity and this may be in the form of a pocket which extends to include part of the pipeline. Under these conditions additional chaneling will occur.

Similar conditions occur in permafrost areas. Here the pipe thaws an area around itself and this acts as a low resistivity tunnel through the otherwise frozen ground. It is difficult to find a pocket into which an impressed current scheme could be placed and in the absence of this the technique of ribbon anodes, described earlier, can be used. These anodes, however, have considerable limitations and care must be taken, both in their engineering and design.

The problems associated with the permafrost, the thaw bubble, the size of the pipe and the current distribution can be investigated by model techniques in the laboratory. The longitudinal resistance of the anode is of prime importance particularly towards the end of its life. Theoretically the anode should corrode first at the point nearest to its connection to the pipe, but this may be varied where the anode moves in and out of the thawed zone. Additional complications with salt concentrations mentioned earlier can add another disturbing dimension to this technique of engineering.

The spread of protection along a bare or poorly coated line will be controlled by the potential variations in the soil and when an impressed current system is employed the groundbed position will influence the soil potentials and control this spread, the maximum being obtained when the groundbed is remote from the pipe. Many buried pipelines will be well wrapped or coated and these may be protected by the techniques described for small pipelines which, by the electrical definition given for small pipes, they will be. From the mathematics of the engineering of the cathodic protection for large pipelines the potential gradient in the soil near the pipe will be the most important factor, particularly in high resistivity soils.

Modern coating techniques allow coatings of very high resistance to be used on pipelines and under these conditions the coating resistance controls the protection parameters. If, for large pipes, the product of the current required per foot multiplied by the soil resistivity is less than 1 amp ohm cm per ft then the coating resistance will take control.

Long Pipelines

The majority of buried pipes form long pipelines for the transportation of fuel or other products over large distances and the protection of these pipelines is achieved by two methods. Firstly, they may be treated as a series of short lengths of pipe and each section may receive protection from either sacrificial anodes or a small impressed current system as were described for short pipes. Secondly, and this is usually the preferred method, the pipeline is protected by a single or a small number of installations and these will use the impressed current technique. The current that is required to achieve protection will flow along the pipe metal and the potential drop caused by this will be of significance in the protection.

Consider a section of the pipe as in Fig. 115 which shows a pipe with a coating that displays an ohmic conductance. The pipe metal has a resistance of r ohm per unit length, is electrically continuous and the coating has a conductance between its inner and outer surfaces of g mhos per unit length.

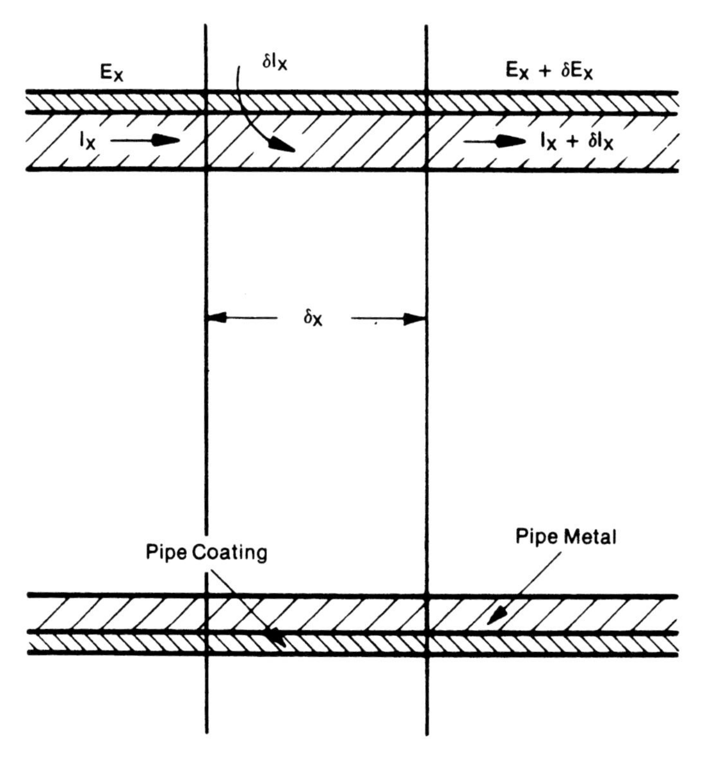

FIGURE 115 - Longitudnal section through buried pipe.

Cathodic protection is applied to the pipeline from a single installation and there is a single point of connection of the negative lead from the power source called the drainage point. This causes a potential change across the coating of E_x volts at point x where E_x is the relative change between the pipe metal and the soil, that is the swing potential. The current flowing in the pipe metal at x is I_x and it is flowing towards the drainage point. The current causes a change in the pipe metal potential which increases the value of E_x by an amount δE_x which will be related to I_x by the equation

$$\delta E_x = I_x \cdot r \cdot \delta_x \tag{6.9}$$

and there will be a change in the current flowing in the pipe δI_x by virtue of the potential E_x developed across the coating. These will be related

$$\delta I_x = E_x g \, \delta_x \tag{6.10}$$

These two equations combine to give the differential equation

$$\frac{\partial^2 E_x}{\partial_x^2} = r \cdot g \cdot E_x \tag{6.11}$$

which has two solutions

$$E_x = Ce^{-ax} + De^{ax}$$

$$E_x = \text{A} \cosh ax + B \sinh ax$$

where a equals $\sqrt{rg}$ and to find the correct solution boundary conditions for the practical cases must be inserted. Two conditions may exist, either the line is infinitely long or it is of finite length. In the first case at $x = \infty$ the value of E_x will be zero and at the drainage point where $x = 0$, $E_x = E_A$. The first solution will fit these conditions when

$$E_x = E_A \, e^{-ax} \tag{6.12}$$

In the second case the pipeline will be finite having a length $2l$ and protection is just achieved at the ends of the line by a minimum swing E_m. Thus at $x = l$, $E_x = E_m$. Now because the criterion of protection is associated with the end of the pipeline it will be simpler to consider the pipe as though it extended from the end towards the drainage point. At the end of the pipe the current will be zero so that $dE/dx = 0$. These conditions lead to the solution

$$E_x = E_m \cosh a\,(l - x)$$

$$E_A = E_m \cosh al \tag{6.13}$$

where E_A is the potential at the drainage point.

The current flowing at any point in the pipe will be given by equation 6.9 and the drainage current will be

$$I_A = \frac{2a}{r} E_A e^{-ax} \tag{6.14}$$

in the case of an infinitely long line, while in a finite line the current at the drainage point will be:

$$I_A = \frac{2a}{r} E_m \sinh al \tag{6.15}$$

The value of this current is doubled because current will be flowing to the drainage point from the pipeline on either side of it.

It has been assumed that the coating conductance g and the pipe metal resistance r are constant. In practice the pipeline would have to possess macroscopic uniformity as in the conditions below.

1. The pipe conductor must be of continuous and uniform resistance. This is true of a welded uniform pipe and of a screw coupled pipe while rubber flexible couplings or other resistive couplings can be bonded over to give practically similar conditions. If the line changes section and consequently its resistance per unit length, then it must be treated as separate sections.

2. The conductance of the coating is uniform over distances small compared with the length to be protected. In the case of good coatings with low conductance this is entirely a function of the coating itself but in high resistivity soil and with poor coatings the conductance depends to some degree on the soil resistivity. With bare pipe the coating conductance is more dependent on the soil resistivity, and the value of it can be derived from an equation similar to those used to calculate groundbed resistances. This is usually of the form

$$\text{where } C \text{ per foot} = \frac{1}{R}$$

$$R = \rho \frac{1}{2\pi \cdot 30} \ln \frac{K}{td}$$

where t is the mean depth, d the diameter, ρ the soil resistivity, and K a constant depending on the physical disposition of the pipe.

3. The soil is of uniform resistivity. This condition is imposed only for bare and poorly coated pipelines as the conductance of the coating is dependent upon it. Where there are wide variations in resistivity and the pipe is bare, the sections can be treated separately and the problem calculated by a method described later.

It is assumed that two conditions may exist: (a) that the immediate, 'before polarization' potentials of the structure are considered, or (b) that polarization has been completed and all measurements are taken after polarization. The only conditions which are not usually met by practical pipelines are those of uniform coating and true ohmic conductance. In the absence of the latter the mathematics become more complex, but the spread of potential can still be calculated while the variation in coating can be treated as a change in the value of a and again the spread calculated.

There are few pipelines which are infinitely long and protected at only one point in their length and so installation will be considered as being on a finite section of a line. If two or more cathodic protection installations are employed on a pipeline then there will be a point somewhere between these where the swing potential E_X will be a minimum and where there will be no net flow of current in the pipe metal. This point will act as the end of a finite section of line and its swing potential will be adjusted to the minimum required for protection.

The important parameter in the above equations, which are called spread equations, is a which has been defined as the square root of the coating conductance x the pipe metal resistance both per unit length of pipe. The units of these quantities will have dimensions resistance^{-1} $\times$ length^{-1} and resistance $\times$ length^{-1} so that their product will have the dimension of length^{-2} or a will have the dimension of length^{-1} so that when it is multiplied by the length protected the product will become dimensionless.

Pipeline Homogeneity. The values of g and r will vary with pipe diameter, the value of r with the pipe wall thickness and there will be great variations in the value of the conductance of each square foot between various pipe coatings, their thickness and method of application. It has been assumed so far that the coating conductance has been ohmic but this is not always true. The coatings described in Chapter 2 generally have a practically ohmic resistance and this will vary from 10^6 ohm sq ft to 10^3 ohm sq ft. Poor coatings will suffer heavy electro-osmosis under the influence of the potential E_x and this will cause a decrease in their resistance so that at the maximum value of E_x, that is E_A, the coating will be of lowest resistance. This will cause excessive current to flow onto the pipe near the drainage point as the change of potential of the pipe metal will be greatest there.

The pipeline may not be homogeneously coated but may be coated with a series of materials, each with a different value of g and the potential of the pipe will now be difficult to assess for even if the various values of g are known the mathematics becomes complicated. When the coatings are all of high quality and low conductance a reasonable approximation is possible and this will be sufficient for many practical cases.

If some of the coating materials are poor or are badly applied then the pipe at these points may require a sufficiently large current density for there to be a considerable potential gradient in the adjacent soil. As at some distance from the pipe the ground will be at equipotential and the sum of the potential drop across the coating and the potential drop through the soil must be equal for both the good and the bad coating. Because of the high resistance of the good quality coating, then the majority of this potential drop will occur across it while with the poor coating its resistance will be low and only part of the potential change will occur across it. This means that the potential of the pipe relative to a close electrode will depend upon the coating resistance; the good coating will cause the pipe potential to be more negative while a less negative potential will be measured on the poorly coated section. This close electrode potential reading is generally the accepted pipe to soil potential for cathodic protection criteria and under these conditions it will show wide variations.

The cathodic reaction at the pipe metal/soil interface will often cause a gradual change in the pipe potential by polarization. This occurs by the formation of a resistive film of calcium salts which are plated out from the ground and can reduce the current demand on a poorly coated surface by 60 to 80 per cent. Similar polarization is also found to occur on bare steel lines and greatly aids the spread of protection and reduces the amount of current required. Certain pipelines are attacked in the presence of bacteria, particularly the sulfate-reducing type; the result of which attack has the effect of depolarizing the pipe metal interface so that it displays zero resistance. Pipe potentials under these conditions are noble but the cathodic reaction corrects this condition; the interface slowly recovers its polarizing properties and eventually reacts as a normal, polarized bare pipe.

The second variable in the parameter a is the linear metallic resistance of the pipe which will be a function of the weight per linear foot and for a particular size of pipe will vary inversely as the wall thickness. The spread factor a will vary as the square root of this change in resistance and as the pipe wall thickness will vary little it will have a comparatively small effect upon the spread. Many pipes are laid with flexible couplings which general-

FIGURE 116 - Cable bonding across Viking Johnson using thermite welds. (Photo reprinted with permission from Henry Grove, Esq., Romford, Essex, England.)

ly are of high resistance and it is common practice to bond over these couplings so that electrical continuity is achieved. Where this type of coupling is used to join every length of pipe, the resistance of the bonds may be large compared with the pipe metal resistance and the use of heavier bonds to reduce the total resistance must be weighed against their cost and the increased spread of protection that may be achieved, Fig. 116.

Spread of Protection. So far it has seemed that the spread of protection depends only upon obtaining a large enough value of E_A to cause protection over the desired distance. With well coated lines it has been found that the coating becomes damaged at potentials in excess of a few volts, the voltage varying with the coating material, and a two volt swing, that is $E_A = 2$ V, is generally accepted as the maximum. This limits the spread that can be achieved with a particular spread factor and if E_m is fixed at 400 mV then al will be given by equation 6.13 as $al = \cosh^{-1} - 5 = 2.3$. Some coatings should not be subject even to this voltage stress and if E_A is limited to 1.5 V then $al = 2$ of if the limit that the coating can tolerate is 1 V than $al = 1.6$. Under such conditions, a reduction in a, which may be achieved by using a high resistance coating will be advantageous.

Potential limits are placed on high tensile strength steels used in modern long distance pipelines. The potential to copper sulfate is generally limited to -1.5 V, which is a swing of approximately 1 V. These pipes are generally extremely well coated and sensible distances between cathodic protection stations are achieved.

A further limit on the spread of protection is imposed by the increase in the mean current density required. If an even spread of protection were obtained then the total current required to protect a pipeline of length $2l$ would be $2gE_m l$ amp. While if the line is protected by a series of installations spaced at a distance $2l$ apart from the current required for the protection of a length $2l$ will be given by equation 6.15.

$$I_A = \frac{2a}{r} E_m \sinh al$$

The current in these two cases may be compared and the ratio will be

$$2a\, E_m \sinh al : 2rgE_m l$$

or

$$\sinh al : al$$

Now sinh al will be larger than al for all values of al greater than zero. The amount that sinh al is greater than al is infinitesimal for small values of al while when $al = 2.3$, that is when $E_A = 2$ V the usual maximum coating tolerance, then sinh $al = 4.9$ which is more than twice the value of al.

The variation of the spread with changes in a at various limits of E_A is shown in Fig 117 and the current required at various values of E_A is shown in Fig 118 which compares this with the current that would be required under conditions of uniform protection.

In the discussion of long pipes the groundbed has been considered to be remote from the pipeline but if this is not the case its proximity will have two effects. Firstly, it will enhance the voltage developed across the coating opposite the groundbed and this may lead to a swing of pipe potential beyond the maximum that can be tolerated so that the negative swing of the pipe will have to be decreased and less pipe can be protected. Secondly, the same mechanism that enhances the pipe to soil voltage will increase the amount of current entering the pipe there and this will tend to steepen the pipe potential curve in the same way that a non-ohmic coating might. Generally the currents used to protect well coated pipelines will be small and the groundbed influence on the pipe potential negligible.

The spread of protection has been calculated from a particular point but to ensure that the best engineering is employed this should be

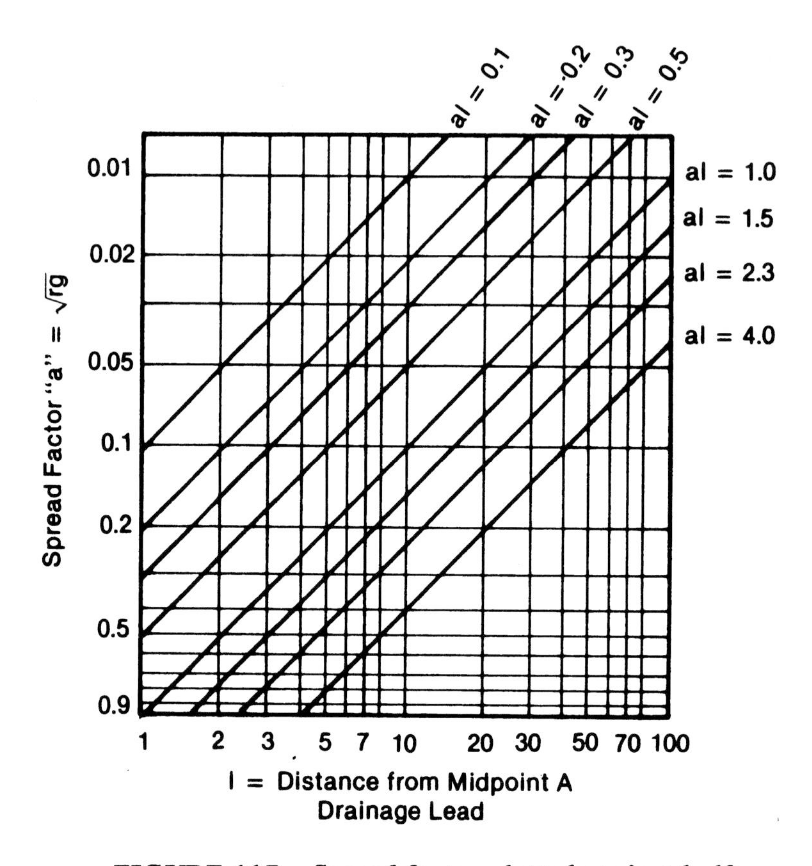

FIGURE 117 - Spread factor plotted against half distance between drainage points for various "al" limits.

calculated from a number of points along the pipe route. Often the availability of suitable power supplies will limit the choice to only a few points and the calculations can be made from these. Where there is a choice of location for the installation, particularly on a pipeline which is not uniform, the best location can be determined by considering the various sections of the pipe and calculating what current each section will require for protection. Now if these currents are multiplied by the resistance of the pipe between the section and the proposed drainage point these current × resistance moments should balance about the selected drainage point and these should be the minimum. Sections near to the drainage point should be weighted as they will receive a much higher current density by virtue of the high voltage that is developed across the coating close to this point.

Pipelines in Practice. In this chapter so far the various influences

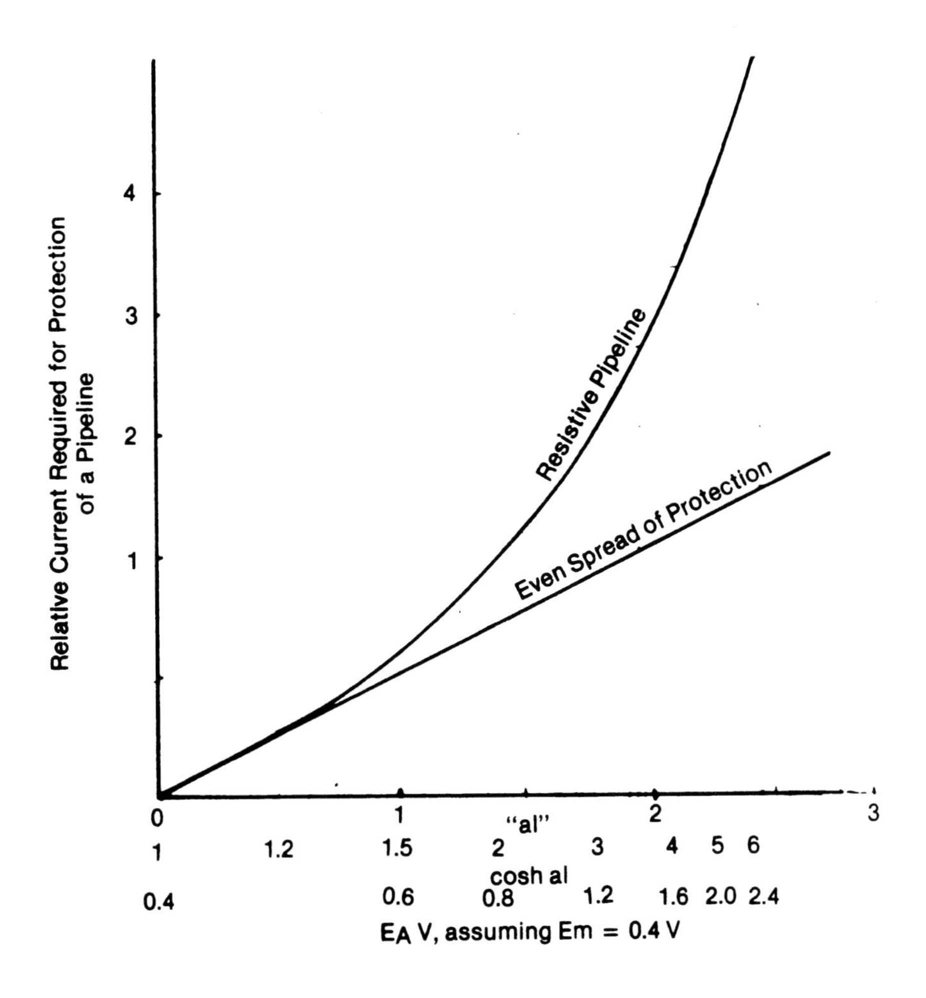

FIGURE 118 - Comparison of current required to protect resistive and non-resistive pipelines.

which might cause the pipe to soil potential to vary have been discussed. If protection has been achieved at one point on the pipe the factors that might cause a different value of potential to be observed at other points on the pipeline have been suggested and a number of pipes have been considered in detail where one of these factors has had the sole influence. The pipe properties that exert these influences may be summarized as variations in the pipe to soil potential caused by (1) the current density demand and soil resistivity changing along its length, (2) the depth of burial of the pipe relative to its diameter changing, (3) the proximity of the groundbed or anodes and (4) the ohmic potential build-up in the pipe metal caused by the cathodic currents.

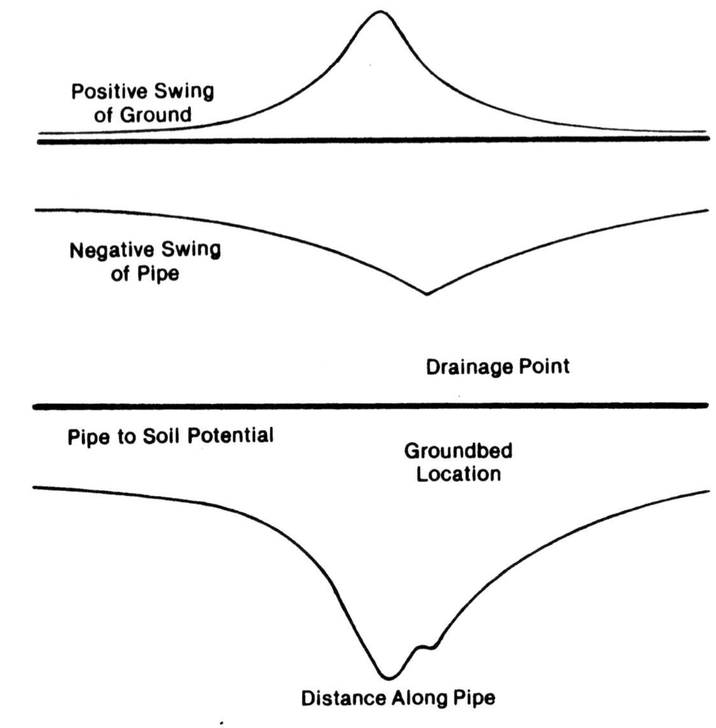

FIGURE 119 - Analysis of pipe to soil potential along route.

While these have been discussed separately and pipelines divided according to the influence these factors have upon the engineering of their cathodic protection most pipelines will suffer some variation in pipe to soil potentials due to each. The parameters that affect the pipe to soil potential can be divided into two groups, those that are properties of short sections of the pipe and the area surrounding it and are usually associated with the plane at right angles to the pipe route and those that affect long lengths of the pipe and are usually associated with the longitudinal properties of the pipe that is the spread of protection along the pipe line.

The former will be influenced by the first three variations listed above though the influence of the third will be small. The latter will be influenced by the last two variations and to a small extent by the first. The first three influences have been discussed in detail and the interaction of one upon the other is straight forward. The influence of the groundbed will be to provide by its proximity an assymetry to the potential field and these potentials will be added to those normally found in the soil surrounding the pipe. The ac-

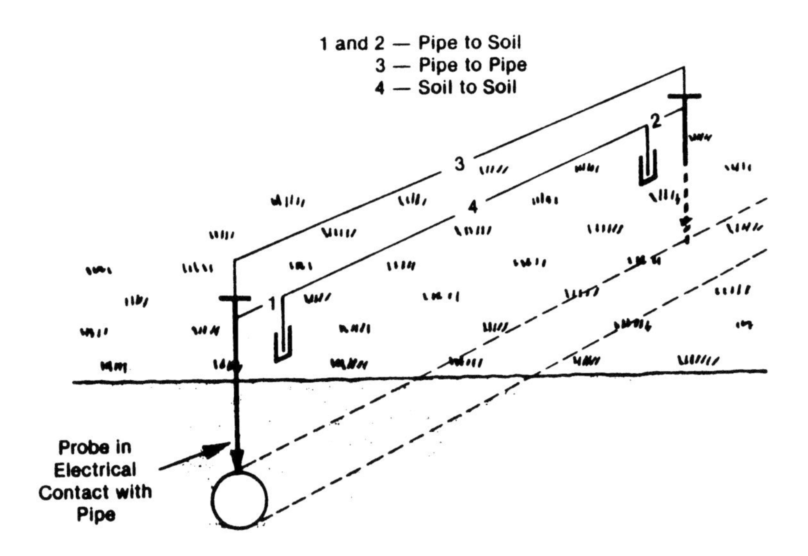

FIGURE 120 - Method of analyzing pipe potential.

cumulative effect of the longitudinal or long line parameters can best be illustrated by a plot of the pipe metal potentials and soil potentials relative to those before the cathodic protection was applied and these are plotted along the pipeline as in Fig. 119. The negative swing of the pipe metal will be caused by the cathodic currents flowing in it and the positive potential curve by the swing of the ground potential in the vicinity of the groundbed. The pipe to soil potential will be the difference of both of these.

With a well coated line the groundbed will influence so little of the protected length that its effect in causing a greater drainage current can be ignored but the effect upon the maximum tolerated voltage swing across the coating may be significant and will demand careful consideration particularly if the maximum spread of protection is desirable. At the other extreme, the protection of a bare or poorly wrapped line will be greatly influenced by the groundbed position while the ohmic drop in the pipe metal will be of secondary importance. The potential measured between the pipe and soil will be the result of both of these and it is useful to be able to distinguish between the two in the field. This may be accomplished by two means: either the pipe and soil potential may be measured to a remote earth, about 300 yards is usually sufficient, when the pipe curve and soil curve will emerge, or pipe to pipe and soil to soil potentials may be measured. This method is illustrated in Fig. 120 and from it curves similar to those in Fig. 119 are obtained. These figures may most easily be computed by automatic addition of the potentials which can simply be done using a computerized voltmeter.

Where a pipe is to be protected by a multiplicity of installations the engineering mathematics will be difficult unless the pipe is uniformly well coated. The protection of bare, poor or differently coated pipe is beyond relatively simple mathematics even if the whole range of parameters were known and so two field techniques are used to determine the spread of protection. In these a temporary source of current is used and the change in the pipe to soil potential is measured as it is applied to the pipe. The first method, called the current drainage survey, uses sufficient current to cause the pipe to display protective potentials and this may be operated from a single point or simultaneously from a multiplicity of points, the former being easier and the results to be expected from the latter may be deduced from it by using equations 6.12 and 6.13. The method is not flexible as a large amount of work is required to provide the full protection current. The second method employs a simple swing test technique where only a small current is applied to the pipe and the change in pipe to soil potential, though not enough to cause protection, is determined. The method, while quicker than the former, relies on a considerable extrapolation of the experimental data but alternative arrangements of the groundbed location and interval can be easily tried and where there is a choice of several sites for the installation this method will prove to be better.

The complete installation of a typical cathodic protection unit for a pipeline is shown in Fig 121. The pipeline passes through arable land and the protection is applied at one of a number of places that has electric power available. The groundbed, which consists of graphite or silicon iron anodes laid in coke breeze to form one long horizontal rod, is constructed in low lying wet ground of low resistivity. The anode and cathode cables are carried on poles set close to the hedgerows and the transformer rectifier is pole-mounted adjacent to the electric power supply.

Pipe Networks

Pipes are often laid in groups, either along the same route or connecting a series of closely spaced containers as in an oil refinery, and for the application of cathodic protection these pipes can be divided into three types, pipes in parallel, bifurcations and closely grouped networks.

Applications to the first type will differ from the normal pipe protection under two conditions, firstly if the pipes are sufficiently close together to cause some shielding or interference, and secondly, if the pipes are of different coating or linear resistance so that their spread factor a is considerably different and they are electrically bonded together. The former is illustrated in Fig 122, which shows a section of the pipe route with four pipes in it and the problem of spread of the protection may more easily be appreciated when the 'image' pipes are drawn in. If the pipes are poorly

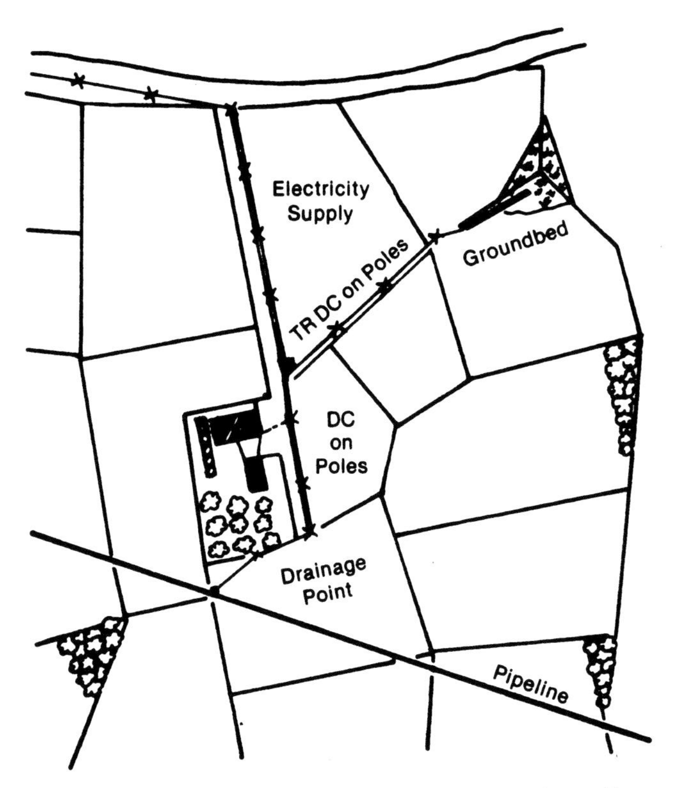

FIGURE 121 - Typical cathodic protection installation to protect a pipeline.

coated and the soil resistivity is high the outer pipes would normally receive more protection than the inner pair. Sacrificial anodes placed at A, B and C would give a reasonable spread of protection but they will not be very effective when the conditions are worst, that is under a high current demand in a high resistivity soil. If the pipes are protected by impressed current from a remote groundbed and are insulated from each other then the resistance of their cathode leads can be made such that the volts drop through these will compensate for that found in the earth so that less overprotection of the outer pair will result. If this method is employed the potential of the outer pipes should be measured relative to a half cell placed close to the pipe wall facing the inner pipes and their images.

If the resistances of the pipe metal and pipe coating are such that the spread factor a varies from pipe to pipe, then the pipes will be at different potentials some distance along their route, though they are connected together at the cathode drainage point. Each pipe, if they are insulated from each other, can have its drainage potential adjusted so that protection

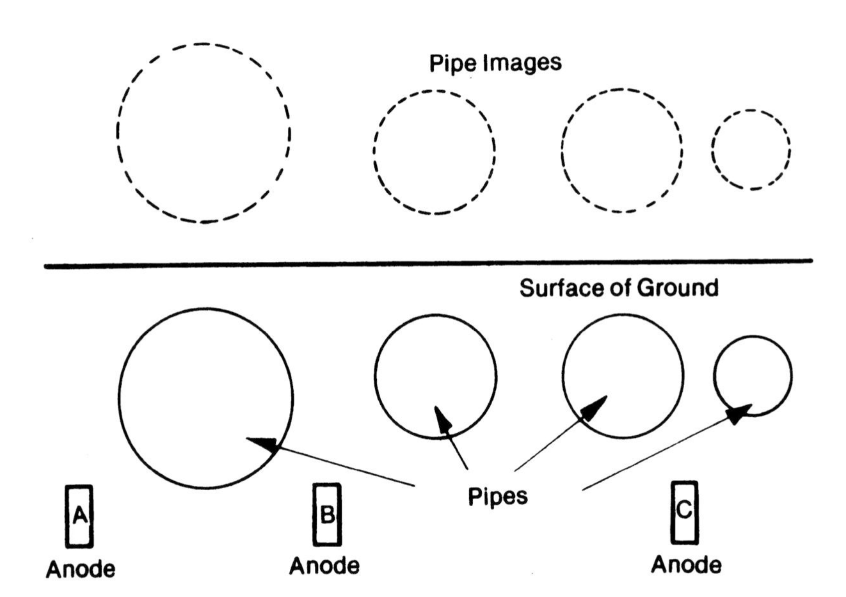

FIGURE 122 - Section through right of way with close parallel pipes.

is achieved throughout its length. If pipes whose spread factor varies considerably are joined electrically at a series of valves or other points, then the effect will be to achieve a mean spread where the resistances of the pipe metal and of the coating may be considered to be in parallel. The effect will be most marked where the inter-connections are frequent. Where they are not the pipe with the poorer spread will receive extra protection near to the junctions.

The second class of pipe networks are the bifurcations. If the installation site has still to be selected it is best to apply the protection from the point of bifurcation when the pipes may be treated as three separate ones each protected from one end. Otherwise the spread factor for each pipe can be calculated and at the bifurcation the conditions are that the two pipes *a* and *b* will have potentials E_a and E_b which will be equal and currents I_a and I_b which will add to make the current flowing in the third pipe to which the drainage lead is attached. Knowing these values and the spread factor for the single pipe, its drainage potential and the total spread can be calculated. The mathematics are not difficult and can be calculated simply on a programmed computer.

A model may easily be constructed to illustrate the spread along a pipe as in Fig. 123. The resistive strip represents the pipe metal, the electrolyte the coating conductance and the metal strip the soil or earth. By connecting the earth and resistive strips together in a series of models various bifurcations can be investigated and by moving the point of contact of the negative pole of the power source along the resistive strip the affect of various drainage points can be found.

The model itself is readily made, for example a thousand feet of 8 in. pipe has a resistance of 0.02 ohm and the coating might have a resistance of 10,000 ohm sq ft or 5 ohm. This could be represented by a foot of resistive strip lying in a dilute salt solution. A more elegant technique uses printed circuits designed to display various values of *a* and with various distance scales or an analog computer would give the same results. A simple digital computer program will give a complete analysis.

At refineries, chemical plant and crowded urban areas, pipes may be grouped together, crossing each other, and a number of these, often all, are inter-connected. The cathodic protection of such networks will be by anode control and two engineering methods may be employed. Firstly, the anode or groundbed may be made remote and the protection current drained from the structure by a series of cathode leads whose resistances can be controlled to achieve the desired protection. This method will prove to be effective and economical when there is some nearby low resistivity site, for example, at a refinery standing on the coast, the anode groundbed could be located in the sea. This system will not always work with 100 per cent success as some pipes may completely screen others; such effects will be local and additional protection by sacrificial anodes may be necessary.

The second method consists of distributing the protection as small units throughout the network. Sacrificial anodes are often used being buried according to the estimated current requirements and their expected output in the particular soil; this system needs an extensive initial survey, usually some additional anodes to augment the original design, and

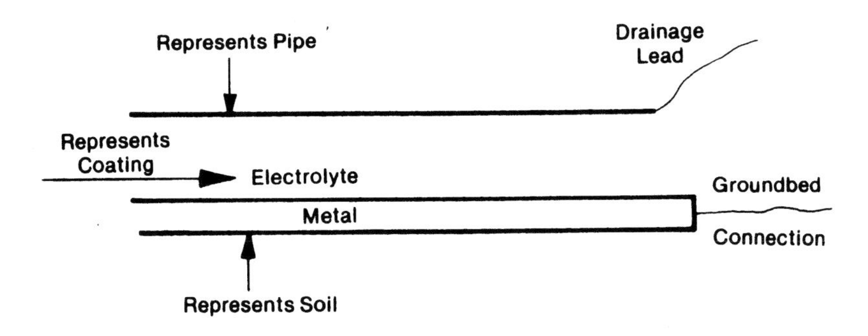

FIGURE 123 - Model of spread of protection along pipeline.

perhaps the introduction of resistance control to other anodes. A similar survey will need to be made at short intervals, probably every few months. Small impressed current systems can be employed and their larger current output calls for careful selection of the groundbed sites, although where there is little room and the bed rock is not of high resistivity the anodes can be installed at some depth. If this system is used it approaches the remote groundbed system just described. Periodic adjustment of the impressed current system is easily accomplished as the bad spots where protection is least easily achieved will be small in number, 5 or 6 to each unit, and inspection will be simple; permanent half cells can be buried at these points so that measurements take the minimum of time. A small transformer rectifier when enclosed is shown in Figure 124.

Automatic control can be used with such a system but the multiplicity of readings required means a considerable increase in the cost of the installation. Where the general level of protection varies, as opposed to changes in the detailed distribution, an automatic control from a single or one of a selected number of electrodes, can raise or lower the whole level of protection. Constant current control is used with anodes located in estuaries.

Rail and River Crossings. When a pipeline crosses a railway or a major road or passes close to some susceptible building, then the pipe is often enclosed in a sleeve. The usual technique is to place the continuous pipeline through the sleeve using a series of supporting insulators which carry the pipe and hold it some distances from the sleeve, usually a few inches, and electrically insulate it. The ends of the sleeve are sealed to the pipe, usually with a flexible synthetic rubber connector. The cathodic protection will now fail to reach the pipe inside the sleeve and there are a number of techniques used to ensure protection.

In some of these the void is filled with an inhibiting fluid often in the form of a gell which will protect the pipe. It will also protect the inside of the sleeve. In others, a conducting slurry is used and anodes are connected both to the pipe and separately to the inside of the sleeve. The cathodic protection achieved is local and zinc anodes are used because of their bactericidal effect.

The sleeve is usually vented and itself is cathodically protected by a local sacrificial anode system. This is because the sleeve is often driven beneath the roadway and its coating is damaged so that it takes a considerable current compared with its length. This would degrade the average resistance of the pipe coating if it were connected to the main pipeline reducing the spread of protection. Sleeves close to the groundbed can be protected and a resistive coupling would hold the sleeve potential at the normal and not the drainage point potential. A number of proprietary devices are available which combine some or all of these techniques and

FIGURE 124 - Pilla shell cast iron enclosure for urban transformer rectifier up to 2KVA. (Photo reprinted with permission from Hockway, Ltd., Croyden, Surrey, England.)

usually include potential measuring points, both for the pipe and the sleeve continuity and line current testing and sometimes electrodes are placed within the sleeve itself.

Relying on a complete seal and a dry void is not generally considered good practice as it is difficult to determine that the void is completely dry. A typical layout is shown in Fig. 125.

At major river crossings a separate section of pipe is usually installed

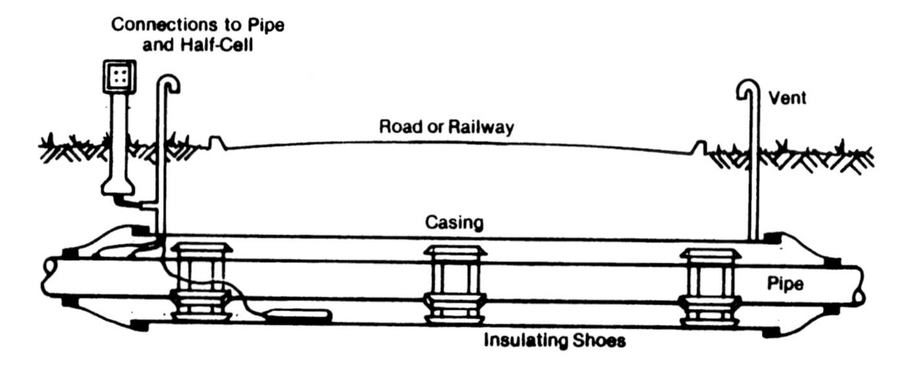

FIGURE 125 - Typical encased pipeline at road crossing.

and this is insulated from the main pipe by insulating couplings either flanged or monolithic, thus isolating the section of pipe from the main cathodic protection system. Under these circumstances either a separate system of cathodic protection can be applied to the crossing or the coupling can be bridged to provide cathodic protection. The pipe that crosses the river may have to be handled by different techniques and these, in some instances, lead to more damage than the bank pipe. Often the pipe will be encased in concrete, either as one or two blocks or as a continuous weight coating.

Where the pipe is isolated then the current in the main pipe will be disrupted and protection of the sections either side of the river will need to be undertaken separately. An alternative technique is to site the impressed current system close to the river crossing and to make separate cathode connections to both sides of the river and a third to the crossing pipe. Variations on this by bonding can also be used.

Pipe Insulating Joints. The river crossing is one example of where a pipeline is electrically sub-divided using insulating joints or couplings in the line. This is also done to avoid long line currents which can flow in the pipeline by virtue of geological or terrestial potential differences. These joints are made either at the flanged ends of a pipe, at a coupling on small diameter pipes or are prefabricated to form a short length of insulated pipe which can be welded in to the main pipeline. This latter technique is used in the majority of long high pressure lines while the flange insulation technique is used in short lines and in complex pipe layouts, as in a factory, refinery or power station. A typical insulating joint using flanges is shown in Fig. 126. This uses a face gasket and insulation of the bolts using sleeves and washers, Fig 127. It is usual to apply protection to the bolts themselves by leaving them in contact with one side of the coupling.

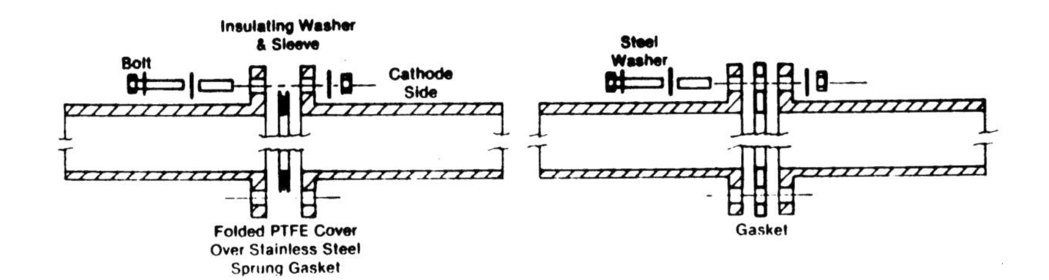

FIGURE 126 - Insulating joint at flange showing PTFE encased and solid gaskets.

FIGURE 127 - Gasket, sleeves and washers for insulating pipeline joint. (Photo reprinted with permission from P.I. Corrosion Engineers, Ltd., Hants, Alresford, England.)

The gasket itself must be carefully chosen and probably the best material is a sleeve of P.T.F.E. in which a sprung stainless steel gasket is housed. This will deform and retain its resilience should there be movement of the pipe. To sleeve the bolts usually means employing high tensile bolts so that the sleeve can fit over the bolt without enlarging the flange bolt hole. Where the bolt is to be protected then unless the pipes are at comparable levels of protection it is preferable that the bolt be protected at the expense of the pipe, that is, its connection is to the more negative part of the pipe, than the other way around.

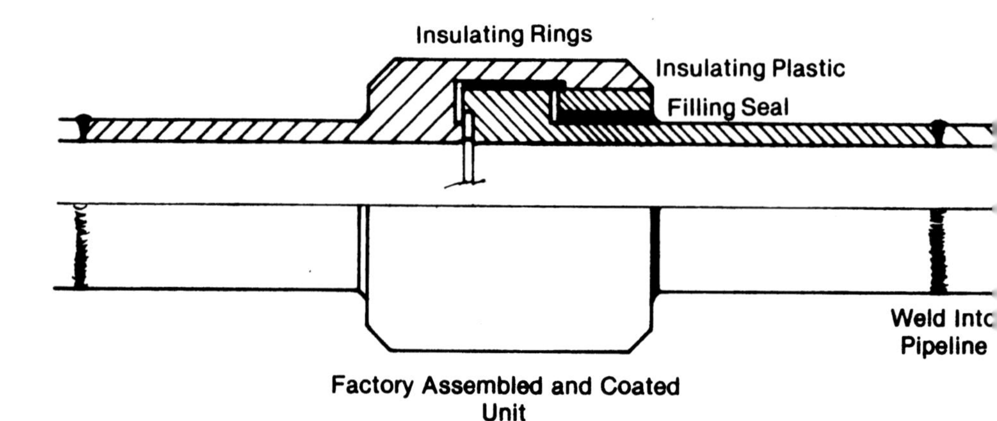

FIGURE 128 - Monolithic insulating pipe pup.

FIGURE 129 - Section of monolithic insulating joint in pipe. (Photo reprinted with permission from P.I. Corrosion Engineers, Ltd., Hants, Alresford, England.)

The alternative is to have a monolithic pipe joint which is made in a factory and these are usually constructed of screwed steel which has sealed gaskets within. The gasket material is a reinforced resin, (epoxy is most popular) and is set solidly at the works (Figs 128 and 129). The pipe will be at different electrical potentials either side of the joint and this will cause problems if the pipe is exposed to the soil outside or to a conducting liquid inside. It is usual to coat the cathode area exceptionally well and to not coat the anode area. However, as it is difficult to know in many cases which area will be which, it is usual to coat the pipe internally at least six diameters either side of the coupling.

The coupling is also going to be the point at which surge currents will discharge to earth, and to reduce this to a minimum it is standard practice to fit an electrolytic cell. The most popular type consists of two zinc electrodes, usually 60 in. × 2 in. × 2 in. which are held an inch or so apart with spacers, usually of soft wood, and the whole is placed in backfill of the type used for sacrificial anodes. The two anodes are connected one to either side of the joint (Fig 130). To reduce the high voltage surges that may occur in the line some form of arcing device is used. A simple type can be made by placing two aluminum plates close together, separated by a fine gap and sealed. Some layers of felted glass fiber will provide the gap and the whole can be encapsulated in resin glass fiber.

The nickel/potassium hydroxide cell can also be used and this provides an a c short circuit while blocking small d c potentials.

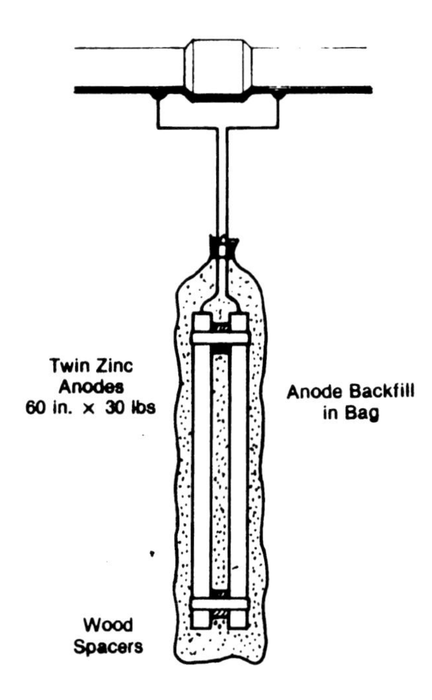

FIGURE 130 - Surge protection devices at insulating joint- soil electrolyte cell.

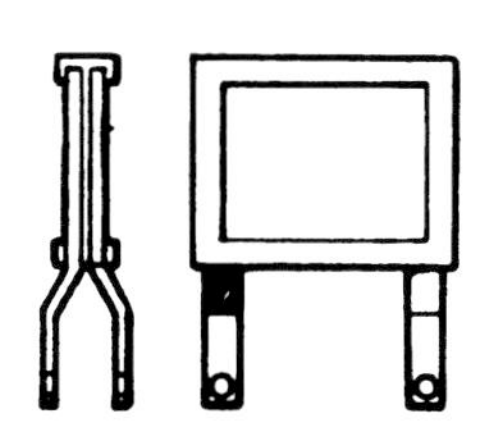

FIGURE 131 - Surge protection devices at insulating joint- air spaced aluminum condenser.

Occasionally insulating units are placed in pits or above ground. When they are it is sensible to cover the actual joint, particularly a flanged joint, with a resin glass fiber cover so that it cannot accidentally be short-circuited, say by a metal spanner; on breaking such a bond an incendive spark could easily occur. The problems of interference and other reasons for longitudinal separation of pipelines are given in the chapter of electrolysis and interference.

Modern long distance pipelines are protected over long distances and in many cases there is a limited number of points where power is available. Usually at the end of the pipeline, at a terminal or storage depot, power supplies are available. The pipe is isolated at this point so that cathodic protection can be applied separately and at different levels to the depot itself and to the pipeline. The pipeline potential at the impressed current station will probably have a potential of -1.5 V to copper sulfate to achieve the maximum spread. It would be uneconomical and, in fact, difficult to protect the remainder of the depot to this level. The isolating coupling or, indeed, in many cases two or three isolating couplings, are used to enable these different levels of protection to be used.

Vertical Pipes

Well Casings

Most well casings are constructed of steel tubes joined together to be in good electrical contact. The tubes are coated prior to installation though this coating cannot be relied upon to prevent their corrosion as it will be damaged during installation. At the top of the well one or two larger diameter pipes may be placed concentrically over the well tube and may enclose 10 per cent to 20 per cent of the total depth, the system being shown schematically in Fig. 132. Cathodic protection can be applied to the well casing tubes though only to the outer one where there are a multiplicity of concentric tubes. Because cathodic protection will cause a change in potential between these tubes, the space between them should be grouted with concrete or filled with clean inhibited sand if grouting is impossible.

It is difficult to measure the potential of the well casing though in some cases a half cell could be lowered so as to be outside the bottom of the casing and a potential measurement made relative to this half cell. The potential gradients in the tube metal may be measured by lowering a series of devices which will make metallic contact to the tubes down the well and from these measurements the current flowing in the tubes can be determined. Fig. 133 shows the type of potential variations that would be obtained by such a survey, the practical values varying with the geological strata and with the pipe resistance. A modification of the zero current ammeter similar to that used in pipe surveys could be employed to determine this current directly.

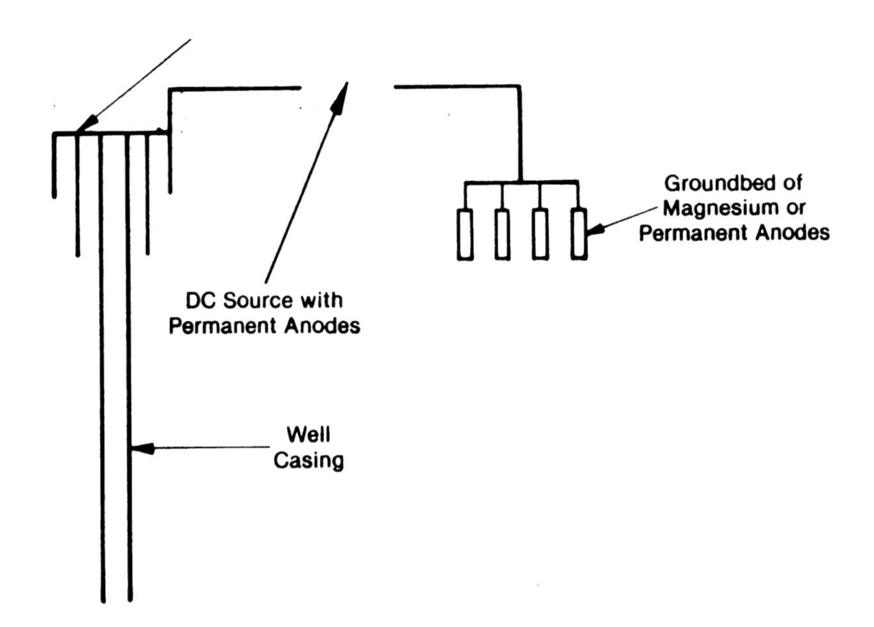

FIGURE 132 - Well casing with cathodic protection anodes.

The well casing will be protected when the metal potential curve indicates that the flow of current in the casing is increasing towards the top of the well as this means that the well must be receiving current and so is cathodic throughout its length. This is used as the criterion of protection.

Alternatively, a plot may be made of the potential against the log of current. There will be a break in the curve and it has been shown that where the curved section becomes a straight line as described earlier the casing will be adequately protected. This technique was developed by Haycock and has proved to be effective in a number of oil fields.

The spread of protection will be governed by the same parameters that affected the spread in a normal pipeline. The lengths of the casing will not be great, a spread of full protection of more than 10,000 ft. will rarely be needed and the current required for protection will usually be less than 15 amp, generally only 2 to 3 amp being sufficient. For isolated wells magnesium can be used, a typical anode installation may be a group of five or six high potential packaged anodes.

There will be an increase in potential at the top of the well, just as there is at the drainage point on the pipeline. In many cases this will detract from the driving potential of low voltage anodes so their use will not generally be possible.

Often the well is connected to a small pipeline or several wells may ex-

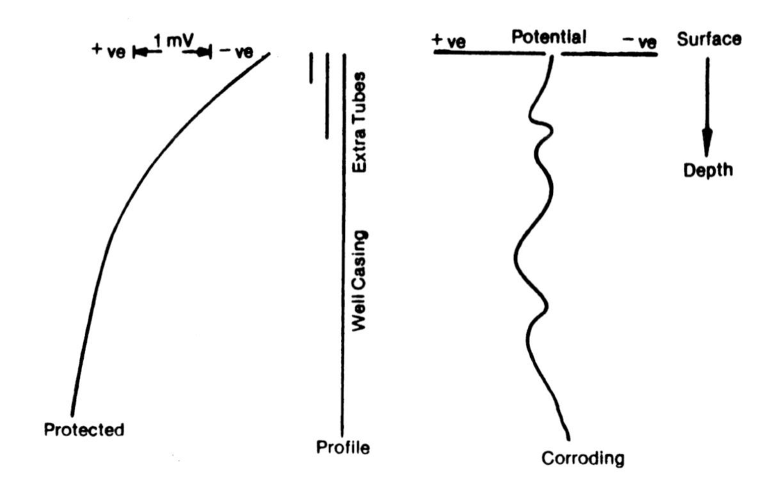

FIGURE 133 - Potentials along well casing.

ist in a group or small field, in which case, providing power is available, an impressed current installation would be preferable. The overprotection at the well head will mean that magnesium anodes, if used to protect a large group of inter-connecting well head piping, will be unattractive. Even with the use of impressed current it might well be that isolation of the wells will prove to be worthwhile. When a gathering line is connected to a well head then both may require an increased protection in order to achieve a spread. Where there are several wells there is unlikely to be any shielding of one well by another as their diameter is very small in comparison with their length.

Some wells have screens at their bottom end and these can be protected by the general system if the screen is in water or a conducting strata as it would be in an artesian well. The inside of the casing or pumps at the lower end will not receive protection by these techniques. Red water, which is often caused by the prolonged contact of the well and pumps, can be prevented by complete cathodic protection of the lower end of the well though this is not achieved from the outside.

Piles

Many buildings nowadays are constructed on steel piles driven some considerable depth into the ground. Often these are used solely to allow a concrete pile to be formed within them and their corrosion is of little consequence. When their corrosion is to be avoided the outside of the piles can be protected cathodically. The piles must be electrically continuous one with another or if not they must be bonded together. It is essential that the bonding is of sufficient cross section to carry the currents that are associated with the pile protection. In many cases more than 1 sq in. of steel will be required between any piles to achieve this and close to the cathode gathering points up to 3 sq ins. are usually necessary. It can be an advantage when the protection is applied from outside the building if the cathode connection can be taken from the innermost piles, when the potential drop in the inter-pile bonding will help to compensate for the potential changes in the electrolyte. Great care must be taken in this type of protection with the incoming services to the building and, as in the next paragraph, any electrical earths. It may be difficult to achieve sufficient spread of protection when the piles are close together, as they may be under a large building.

Two systems of cathodic protection are generally employed, either the building is surrounded by a ring of anodes, or anodes are placed vertically between the piles, the latter being generally installed by drilling or by driving the anodes in a similar manner to earth rod installations. Another technique has been suggested for laying these anodes using a high pressure water jet to create the anode hole. Where impressed current permanent anodes are installed beneath a building adequate gassing facilities must be made; sacrificial anodes, even if given an additional boost of d c power do not give off obnoxious gas though they may generate hydrogen.

The spread of protection will be similar to that found with pipelines and a high quality coating on the pile will be an advantage. To protect the inside of such piles grouting with concrete is advisable.

Smaller piles or footings are often used to support small steel structures such as electric power transmission pylons and these footings can readily be protected if they are electrically continuous, one with another. The use of galvanized steel for their construction reduces the current demand and protection can be obtained from small packaged zinc or magnesium anodes, the lower cost and longer life of zinc making it more acceptable for small currents. These feet are often encased in concrete and there can be rapid failure if there are breaks, cracks or voids in the concrete cover. When protected, it is not necessary to ensure that the concrete completely covers the steel foot and better earthing is obtained because of the low anode to earth resistance.

Electrical Earths

To provide adequate earthing facilities to an electrical system it is usual to bury large plates or rods in the ground to form an earth and to lower their resistance these are often surrounded by coke breeze or are constructed as thin rods and driven to great depths. In both cases the materials used are chosen for their corrosion resistance to prevent failure by open circuiting either a plate from its connecting strap or a long rod by its consumption close to the surface. The earths are usually formed from noble metals such as copper or cast iron surrounded by coke breeze. The earth and the incoming cable sheaths are electrically bonded together producing a large corrosion cell between the noble earth, which is the cathode, and the cable sheath. The electric cables may have galvanized wire armoring or lead or aluminum sheaths, all of which are anodic to the earth. The corrosion cell so formed causes rapid attack and premature failure of the cable sheath or armor near to the earth. Most cables nowadays have an outer extruded plastic sheath. While this greatly reduces the corrosion, it can concentrate the attack, either at breaks in the extruded sheath or at joints, or where the cable is earthed, say at a transformer, to the base of the transformer or the footings of outdoor switchgear.

There are two solutions to this problem. Either the earths have to be made effectively at the same potential or anodic to the cables or the earths must be constructed from metals which will not induce cable corrosion. The former may be accomplished by using a d c blocking device which has a low resistance to a c. This could be a d c bias voltage developed in a resistor inserted in the earth lead or could be a large condenser with a high d c resistance. The d c blocking device could consist of a resistor made of a series of parallel stainless steel rods inserted in the earth lead with a large current, say 100 amp, at low voltage approximately 1 V, output from a small transformer rectifier flowing through it. The voltage developed across the resistor can be varied, so that no corrosion current flowed between the earth and the cables; the circuit is shown in Fig 134.

A condenser device can be made by immersing pure nickel electrodes in a strong alkaline solution of potassium hydroxide. Small d c voltages of less than 1.7 V are blocked by the device though this level is reduced when there is mixed a c and d c to just over 1 V. a c and high voltage d c are offered extremely low resistance and a properly designed cell will take 100 kilo amps for a half second fault. The device has a slightly smaller d c fault capacity, Fig 135.

While these methods are suitable for earths already laid, new earths can be constructed of cathodically protected steel, and their current demand can be reduced by galvanizing the steel without loss of a c earth resistance. Such a scheme employing magnesium anodes has been in operation for many years in the United States and zinc anodes or impressed current could equally be used for this work. Coke breeze should be avoided in such an earthing system as it would not be compatible with the cathodic protection.

Where cathodic protection is applied to metallic earths, particularly copper rods, there is a danger that the polarization film built up on the earth metal at the general level of protection required for the associated steel work will cause the a c earth resistance to increase to a dangerous level.

It has been suggested that the earth could be constructed of anodes,

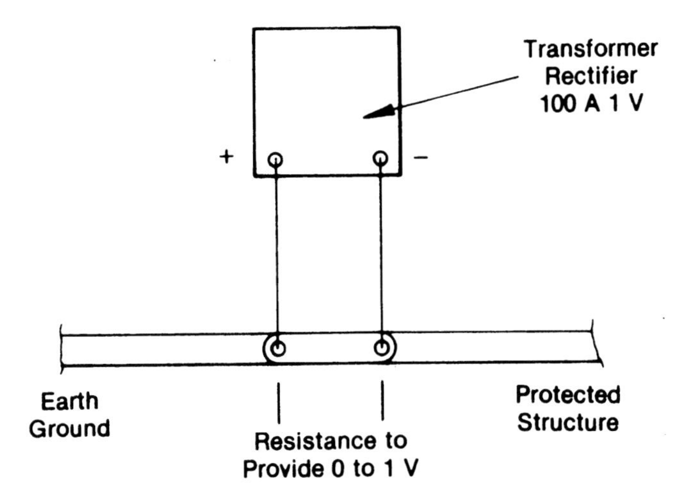

FIGURE 134 - DC blocking device using resistor generated voltage.

zinc being most suitable, so that the cables and any other metal work, such as transformer tank bottoms, would receive cathodic protection. As the zinc would be consumed the anode would need to be replaced from time to time, though with good design this could be at long intervals beyond 10 years. A mixed earth of anodes and galvanized steel could be used to provide some cathodic protection by causing the whole earth to have a sufficient driving potential to the protected cables. The use of pulsed techniques in protecting electrical earths may have advantages as the polarization film is extremely thin and would be of low resistance to alternating current.

Overhead Transmission Lines

The overhead electric transmission lines carrying a c across country may in many cases parallel or cross pipe routes. There are large a c potentials associated with these lines under operating conditions and these can be picked up by the pipeline. This may either be by magnetic induction where the pipeline runs parallel to the overhead cables, or the pipeline may pass through earth potential variations beneath the wires themselves. The voltage in the pipe may lie between 1 and 10 V and this can be detected by measurements to a distant half cell.

The first technique that is used to reduce this a c pick-up is the isolation of the line into smaller sections either where it passes beneath the cables or to reduce the length when it parallels the transmission towers. Corrosion on the line can be greatly reduced if this a c voltage can be earthed using low resistance zinc anodes in backfill, and it is possible to use

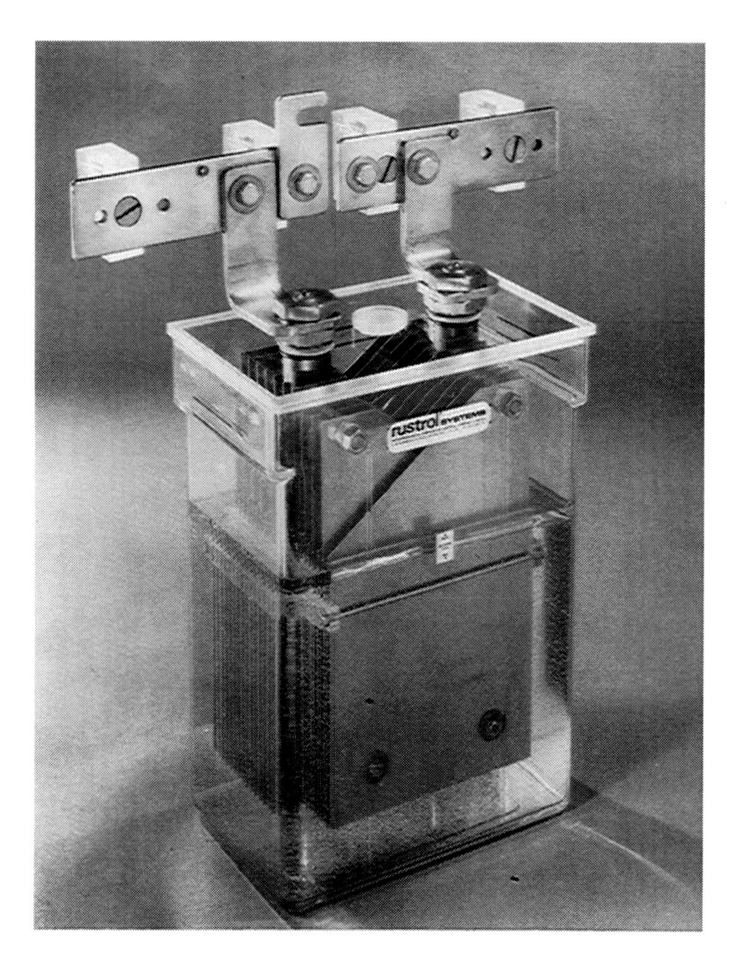

FIGURE 135 - Nickel plate/potassium hydroxide direct current isolating capacitors.

the a c power by rectifying it to provide cathodic protection. This use of zinc anode earthing and zinc anode coupling across insulating joints is the preferred method.

Alternating current itself does not cause serious corrosion but it can enhance soil corrosion perhaps because it is rectified randomly or because it reduces the polarization effects which inhibit normal corrosion. It is unlikely that a c current would affect the cathodic protection on a normal pipeline. While corrosion prevention is achieved by cathodic protection and this brings about a reduction in the a c current, there is a matter of safety to those people operating the pipeline and particularly those doing maintenance work or measurements on the line.

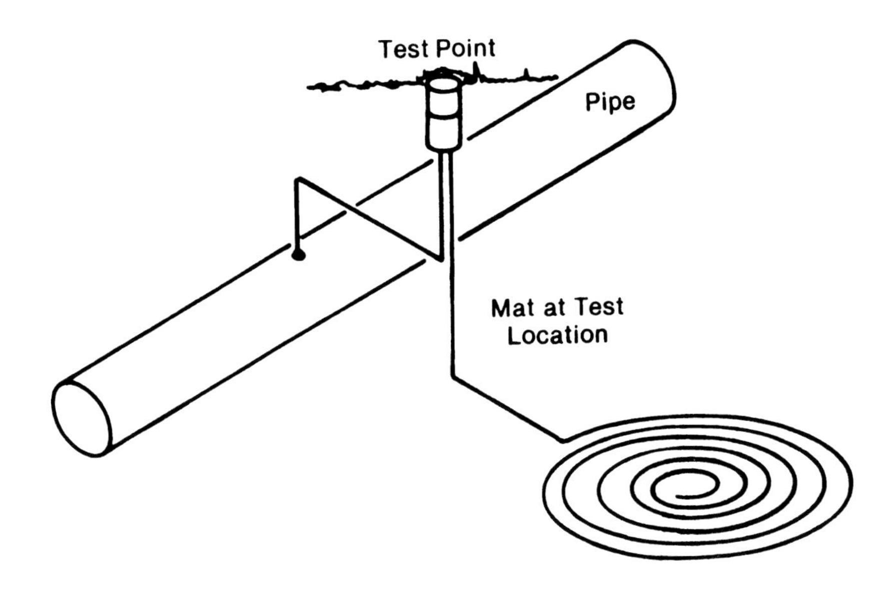

FIGURE 136 - Shaped anode mat at test point on pipeline subject to stray ac.

Fault currents are one of the considerable hazards on a pipeline and these are discussed in the chapter on interference in relation to the similar problem with H.V.D.C. In order to receive a shock an operator must be in good contact with the earth and the earth must be at a potential markedly different from his point of contact with the pipeline. One technique of reducing this effect is to ensure that the earth is at the a c potential of the pipe. If zinc electrodes are used to measure the potential of the line, or if zinc electrodes are used as the cathodic protection, then these and any test point should be located together so that the operator stands on ground which is at a similar a c potential to his pipe contact. Shaped anodes as in Fig. 136 can enhance this effect.

Where zinc anodes are used for grounding the pipeline, these may present a problem if the potential of the line exceeds the open circuit potential of the zinc. Where there is a utility right of way in which several services operate it is usual for none of these to have excessive cathodic potentials to avoid any interference of the systems. Where the zinc open circuit potential is exceeded magnesium can be used, and where the magnesium open circuit potential is exceeded, then the anode can still be a c coupled through a capacitor. One technique that can be used very simply is to connect the anode to the pipeline through a switch and metering arrangement which allows the anode to be used as a reference electrode for either a c or d c.

Under these circumstances the operator should pre-connect the a c ground to the pipeline before breaking the other contacts. The use of zinc ribbon as an earth mat at the point of contact is recommended.

While these precautions are necessary during the cathodic protection maintenance, there will be even more stringent precautions required during the installation of cathodic protection and also during the installation of the pipeline. These include problems that are associated with the massive earth, that is the groundbed, used in cathodic protection and techniques such as holiday detection which may be used while the pipe is above the ground. There are stringent codes of practice laid down on safety and the cathodic protection engineer has always to remember that he is probably dealing with the best electrical earths in the vicinity. The problem of electrical contact or close contact with the footings of towers and with the concentric neutral type cables is dealt with under HVDC transmission.

Pipe Materials

While the majority of large pipelines are made of steel or wrought iron there is a considerable mileage of small diameter service pipes and cables not made of steel. One class, the galvanized steel pipes, can be treated as poorly wrapped steel pipes whose surface will require about 1 mA per sq ft or less for protection. Much water pipe is made either from copper or lead and both might require protection which could be afforded by zinc anodes when the danger of overprotection is least. Lead is protected readily and small changes of potentials of 50 mV considerably extend its life. Copper may be similarly protected and as it is generally accepted practice to wrap these pipes with a petrolatum jelly impregnated tape the current requirements are low.

Aluminum is being used extensively as cable sheating and it is suggested as a pipe material for certain products. It can be protected readily and current densities of the order of 2 to 3 mA per sq ft are necessary. Because of the potential limitations on the protection it would be advantageous if the pipe were coated with a high resistant material and extruded P.V.C. has been used successfully. Cathodic protection can then be applied by zinc anodes which are intrinsically safe or by an impressed current system which has a control mechanism so that the maximum cathodic potential is not exceeded. Anodizing the aluminum would have a similar effect to a coating, reducing the current demand to about 1 μA per sq ft. The high conductivity of aluminum coupled with its low weight means that its linear resistance and hence its spread factor a will be low. Large swings of potential caused by groundbed proximity must be avoided.

One class of pipe type structures are the concentric neutral cables where the outer is made up of steel wires. The protection of these is similar to a pipeline but there are some problems. Firstly, the area of the metal in contact with the ground is much greater because approximately $\pi/2$ times the area is in direct contact with the soil. If there is not good bedding beneath the conductor then the area will be larger. Secondly, cathodic protection is unlikely to spread beyond the point where the outer wires are closest together. There will be deaeration behind this caused by the cathodic reaction and bacteria may grow. Zinc coating of the wires will provide a bactericide.

The linear resistance of the outer wire neutral will be that of the wires themselves. These wires will have a poor conductance compared with their earth contact surface and they will have a slightly longer length than the linear length of the cable because of their spiral. The spread of protection along these wires will be less than along a bare pipeline, but where the wire is galvanized or zinc coated, or where the core of the wire is copper not steel longer spreads of cathodic protection are achieved.

Storage Tanks

Buried Tanks

Many small capacity storage tanks are buried in the ground. These may be of several shapes, the most popular being a cylinder with convex dished ends whose diameter is approximately half its length. Most of these tanks are constructed of welded steel plates and are covered with a thin black bituminous coating to prevent atmospheric rusting. Cathodic protection can most easily be applied to a well coated tank when only a few tens of milliamps would be required and this could be supplied from a single sacrificial anode. The small current and the high coating resistance would eliminate any serious potential gradients in the soil surrounding the tank and protection might be assumed when the potential criterion relative to a half cell placed on the soil surface above the tank is reached.

When the tanks are buried as received, that is virtually bare, and then cathodically protected, the volts drop through most soils will be considerable. The lines of equipotential close to the tank will be similar to those for a large pipeline and they will change to hemispheres at some distance from the tank. By considering the image tank in the earth's plane it can be seen that the lowest swing of potential will be measured on top of the tank. This will be influenced by the groundbed position and the most positive metal potential will be recorded relative to a half cell moved to be further from the groundbed than the top center of the tank. Probably between 1 and 3 amp per tank would be required for protection and this could be derived from a number of sacrificial anodes or from a small impressed current system. The sacrificial anodes could be spaced about the tank while in the latter scheme a single graphite, silicon iron or magnetite anode surrounded by coke breeze and a small air cooled transformer could be used. Often such a buried tank is connected to filling, venting and emptying pipes; each of these will demand extra current and they must be checked for continuity with the main tank. Some of these pipes may, through electrical or other fortuitous bonds, be connected to a series of other services and if this is so either the anodes must be placed near to the tank to give anode positional control of the protection or an insulating coupling must be inserted into one or more of the pipes.

For safety reasons tanks that store petroleum or aviation fuel are often placed in a brick or concrete pit which is filled with clean sand. Ground water, particularly contaminated water, may rest in this pit and a cell may be formed between the lower and upper parts of the tank. Any cathodic protection system must have its anode placed below this water table and inside the pit. Probably only sacrificial anodes are suitable for this use as with impressed current anodes there may be a problem with gas blocking.

Buried storage tanks are often found in groups or farms and when they are well coated before installation little difficulty will be experienced with the tank protection. The spread will be good and a remote groundbed or an enclosing groundbed can be used, the current required being large enough for an impressed current scheme to be used economically. It is essential that the whole of the tanks and pipes are interconnected and electrically continuous; tests to prove this are necessary. If the site does not allow a remote groundbed then the anodes can either be installed at some depth in the ground or distributed amongst the tanks.

If a tank farm comprises bare tanks then the problem of protection is increased and the spread of protection will be controlled by the amount of electrolyte between the tanks and the location of the anodes. It will be necessary to distribute the anodes and also to allow for considerable overprotection of some parts of the tanks. The various configurations of anodes can be assessed by swing tests on site, by laboratory models or computer, though the latter are limited in their usefulness. Sacrificial anodes may be needed in such large numbers that they will be fairly evenly distributed while impressed current anodes, being fewer in number, must be placed in selected sites. The large currents flowing, some tens of amperes, may cause large potential drops between the tanks if the only interconnection is by small diameter pipes and so several distributed cathode leads may be necessary. A typical installation is shown in Fig. 137. Single cathode connections and single interconnections between tanks should be avoided as, in the case of a break, dangerously large currents may be interrupted and cause a hazardous spark.

Tank and Holder Bottoms

Many tanks, gas holders and other plants are built with a steel base which rests upon the ground. The base foundations are usually constructed

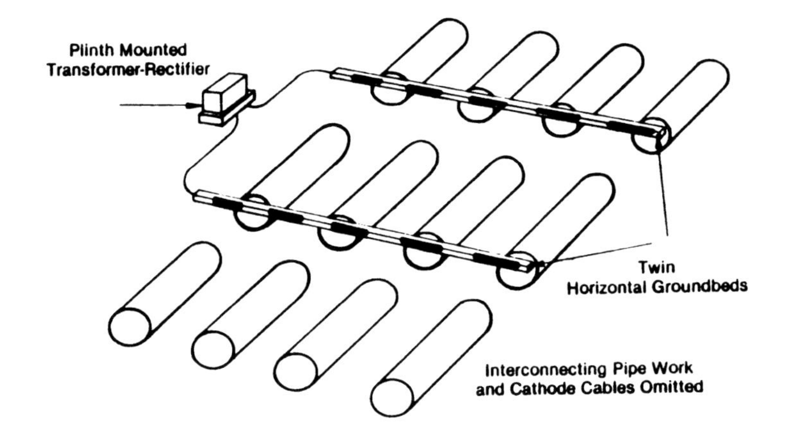

FIGURE 137 - Protection of a group of buried tanks.

of local rock and then covered with a layer of sand or soil to present a smooth uniform surface to the bottom plates. Corrosion of this lower metal surface occurs and where a high water table or other aggressive condition exists, this corrosion can be rapid. In groups, or farms, of tanks the electrical interconnection of the tanks by a pipe network often increases the corrosion problem by the formation of large cells with the whole of one tank bottom being anodic or cathodic to others. This corrosion can be considerably reduced by careful preparation of the tank base plinth and the electrolytic resistance will be greatly increased if a foundation of clean rock is overlaid with clean sand and then that with a further mat of sand-asphalt mixture.

To ensure even better protection the plates can be coated to within 1 ft of the plate edge, this bare area then preferably cleaned and primed at the site and a heavy layer of asphalt or a wide range enamel laid as a strip below its lower surface to cover the bare plate edges. Welding the plates will melt this strip and so complete the coating of the base; the choice of the right coating material and control of the welding are essential as coking of the coating material must be avoided. A superior coating on the tank bottom may be obtained if some of the tank plates are prefabricated and then coated so that only a few final welds are necessary. If this technique is adopted the center of the tank bottom should be fabricated as a single piece so that any bare metal near the welds will lie closer to the outer edge of the base.

Cathodic protection can be used to ensure complete protection against bottom plate corrosion and experience has shown that it is usually necessary. Ideally, the protection should be applied to a perfectly coated tank when there would be no problem of spread, but as indicated above, such a coating is difficult if not impossible to obtain and so the spread of protection must be considered. One of the main difficulties is that of measuring the potential at any point under the tank bottom, and there are two methods by which this may be done. With a new tank a tube of polythene, such as a domestic water pipe, may be laid from the outside of the tank to the center, and plugged at the buried end with a porous plug, the whole being used as a large salt bridge with several such pipes being laid under a large tank. This method is only possible with new construction or on complete tank reconditioning as the pipe would need to be protected by careful laying. With old tanks a small salt bridge could be inserted through the base and readings made from the inside of the tank when this is possible, a suitable bridge being described in a later chapter. This salt bridge would be accessible only when the tank was empty, but since the reading obtained then could be correlated to those outside the tank at its edge, further maintenance work could be carried out relative to these more readily obtained potentials. In using this method it is essential to determine whether the tank base has sprung clear of the ground when emptied as this would invalidate the readings.

Some estimate of the potential at the center of the tank bottom can be made from measurements at its edge, the resistivity of the ground and the current density required to achieve protection. The spread of protection will be controlled mainly by the electrolyte resistivity, somewhat influenced by the polarization curves of the metal interface.

First consider a tank bottom lying on a homogeneous soil and coated so that a small current density is required to achieve protection. Consider the particular case where the soil has a resistivity such that only a small potential variation occurs over the surface of the tank and if this is no more than 50 mV then the current density can be considered to be uniform over the whole of the lower surface. Fig. 138 shows the plan view and elevation of a tank that may be found in a refinery.

The current that will flow to the central area of radius x will be $\pi x^2 \sigma$ where σ is the current density required for protection. If this current is considered to flow through a hemispherical shell of electrolyte radius x, then the volts drop across this shell if its thickness is δx will be

$$\delta V = \rho \cdot \frac{\delta x}{2\pi x^2} \cdot \sigma \pi x^2 \tag{6.17}$$

where ρ is the resistivity. The total volts drop to the edge of the tank will be given by the integral

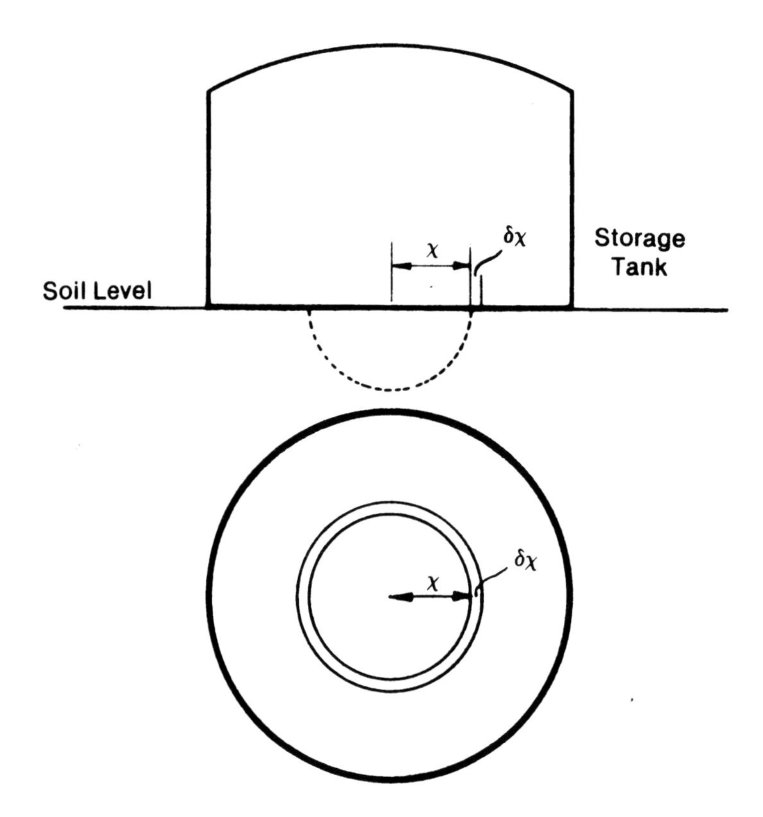

FIGURE 138 - Equipotential shell in soil underlying typical tank base.

$$V = \int_0^r \rho \frac{\sigma}{2} dx = \rho \frac{\sigma r}{2} \qquad (6.18)$$

Now if σ is in amp per sq cm, ρ in ohm cm, r in cm then V will be in volts. Say $r = 10^{-7}$ amp per sq cm, $\rho = 2{,}000$ ohm cm then

$$V/r = \frac{2{,}000 \times 1{,}000}{2 \times 10^7} \text{ mV per cm radius}$$

$$= 1.5 \text{ mV per ft diameter}$$

Now $\sigma = 10^{-7}$ amp per sq cm or 0.1 mA per sq ft and the product of $\sigma\rho$ will be 200 mA ohm cm per sq ft.

Thus a small tank of 40 ft diameter would have a potential change between the center and edge of 60 mV if laid on a homogeneous soil of 2,000 ohm cm if it required 0.1 mA per sq ft for protection. This means that to ensure a potential of the metal at the center of −0.95 V to a copper sulfate half cell the outer edge of the tank should be at a potential of −1.01 V. The values of σ and ρ can be readily found in practice, such a calculation made and where only small swings of potential are indicated by this method it can be used with reasonable accuracy. The total potential change between the center and the rim will be directly proportional to the tank diameter and on modern construction the potential factor should be below 1 mV per ft diameter.

It can be shown by the methods of classical electrostatics that the current distribution on the surface of a flat disc conductor is

$$\sigma \doteq \frac{I}{4\pi r(r^2 - a^2)^{1/2}} \quad (6.19)$$

where a is the distance from the center of the disc, I its total current and σ the current density. It can be shown also that the mean current density is $I/2\pi r^2$ while that at the center will be $I/4\pi r^2$ or half the mean value.

The metal disc will be considered to be at equipotential and for this case to be translated into a resistive model such as the tank bottom then there must be negligible polarization occurring between the tank bottom metal and the soil. If the tank bottom considered lies on a high resistivity soil and requires a current density for protection such that the resistive drop in the soil per foot is greater than the polarization at the interface, then it can be considered to be an equipotential surface, the above mathematical considerations will hold, and the current density at the center of the tank bottom will be half the mean current density over the whole tank.

This mean value will not be that found when the potential of the tank is swung to any particular value but must be estimated independently and then twice the minimum estimated current density must be applied to the tank. For example it might be considered that three mA per sq ft would adequately protect the steel base under uniform current conditions then in this case 6 mA per sq ft should be applied as a mean current density over the whole of the tank bottom so that 3 mA per sq ft arrived at the center. Thus an 80 ft diameter tank would require

$$\frac{40 \times 40 \times \pi \times 6}{1{,}000} \text{ amps} = 32 \text{ amps}$$

Equation 6.19 indicates that σ will be at its mean value when $(r^2 - a^2)^{1/2}$ is equal to $1/2\ r$ or $r^2 - a^2 = r^2/4$ that is $a = \sqrt{3/4} \cdot r$ or

$a = 0.866\ r$. This means that the mean current density will be found at a point 13 per cent of the radius from the edge. Even with 125 ft diameter tank this would only be 8 ft from the edge and it might be possible to auger a hole that distance. The current required to protect the tank at that point could then be determined and the current required to protect the whole tank would be twice that value. The above cases cover the coated tank bottom and the tank bottom on a high resistivity soil, the potential conditions in cases in between these depending upon the polarization of the interface. A calculation using both the described estimates will indicate the limits of the problem and suggest a suitable criterion for the tank protection.

Both the cases considered were concerned with soil whose resistivity was uniform particularly with depth. This will rarely be so in practice and a weighting factor must be used under heterogeneous conditions. If the tank rested on a high resistivity soil which covered a low resistivity strata the spread of protection would be improved, while if the tank rested on a soil which overlaid a high resistivity rock then a very poor spread of protection would result. It is suggested that a weighting factor can be obtained by measuring the surface resistivity and the resistivity when the four pins of the Wenner method are separated by the tank radius. If these resistivities are ρ_S at the surface and ρ_r at a separation equal to the radius then the spread can be said to have an efficiency of $2\ \rho_S/\rho_S\ +\ \rho_r$ and to use this factor, the voltage drop calculated in the first method or the current required in the second method is divided by the efficiency.

For example, an efficiency of 1.5 will reduce the change in potential between a measurement at the center and at the periphery by 1/3, while an inefficient system giving 50 per cent efficiency will double this potential difference. Similarly, the current required as estimated by the second method will be increased by a low efficiency system and decreased by a high efficiency system, though it should never be reduced as low as the mean current estimated for protection. This will fit the arguments a low efficiency will occur with a high resistivity base rock and a high efficiency with a conducting lower stratum. The efficiency will be unity when the soil is homogeneous.

No mention has so far been given of the anode location. Where these are placed so that they are at least one tank diameter divided by the efficiency away from the tank the spread will be reasonably uniform. The mean rim potential may be used with the first method and little or no variation in the current density pattern will be found in the second. If a plurarlity of anodes are used and these are equally distributed around the tank then uniform protection will result when their distance from the tank rim is less.

If n anodes are used these should not be placed closer than $1/n\ \times$ distance that one anode would be placed from the tank. Where anodes cannot be placed remotely and their distribution around the tank is difficult, they may be buried at some depth though this method is only practical where the lower strata are reasonably conductive. Care must be taken with impressed current anodes to ensure that adequate gas escape is provided so that gas cannot accumulate under the tank bottom.

While the above points indicate theoretical and empirical lines of reasoning they cannot be relied upon absolutely. The measurement of efficiency will vary from point to point and some mean value must be used. Further, this will usually indicate the nature of the ground around the tank and the conditions under the tank can only be found by inference, similarly the tank has been assumed to be uniformly coated but this may not be so. In each case the engineer must weigh these factors and design his cathodic protection accordingly.

Many tanks, particularly gas holders, are half buried in the soil and the problem of spread of protection with these is more difficult. The potential at the soil surface will be less than at the lower edge of the wall though this difference will not be great and it can be estimated by measuring the potential variation over the first few feet. The spread of protection over the bottom of the tank will be similar to those discussed though now ρ_S should be measured at a spacing equal to the tank depth and ρ_T at a pin spacing of the sum of this and the radius of the tank.

Groups of Surface Tanks

Most large tanks are built into groups or farms, where they are arranged in a regular pattern and are separated from each other so that fire or other hazards, such as settlement, are reduced. The tanks are usually served by a network of pipes bringing products or services, such as fire sprinklers, to them. Before cathodic protection is applied to a tank farm the electrical continuity of the inter-tank pipe work must be checked not only for resistance but also for its current carrying capacity. Where tanks are well spaced or the connecting pipework is small it will be necessary to measure the volts drop between the tanks at the expected protection current. If this is high, that is above 20 mV, then a distributed cathode lead system should be considered, and should a high resistance joint or coupling occur in the pipe work this should be bonded over. Each tank should have several connections to it, particularly if the tanks store inflammable fluid, as if a pipe is broken, as may be necessary for maintenance work, a spark might be initiated and the vapor ignited. For similar reasons it is usual to insulate the tank farm from the rest of the plant by inserting insulating couplings in the pipelines. It also stops excessive overprotection of the tank farm if long spreads are required on the incoming pipeline.

Having ascertained that the tanks are well connected together, then the resistivity of the ground should be measured including the change that occurs with depth and from this the efficiency of the spread and its effect on the current required or voltage criterion can be estimated. Cathodic protection can be applied to the tank farm by two methods using either a series of distributed anodes or using a remote groundbed. The former lends itself to the use of either sacrificial anodes or impressed current protection. The large potential required at the periphery will mean that only high driving voltage anodes can be used in comparatively low resistivity soils.

Tanks that have been carefully coated are idea for sacrificial anode protection. In this method the anodes are buried around the tank wall, there number being sufficient to give the required current. They are usually buried deeply so that the spread of protection from them is enhanced even when the anodes themselves are close to the tank wall. The anodes may be connected either directly to the tank wall or to a ring wire which connects all the anodes to one point on the tank. Regular anode inspection is necessary and the potential of the tank should be measured between and not opposite the anodes.

Impressed current installations employing distributed anodes usually give larger amounts of current from each anode group. The anodes are arranged to lie between the tanks and if these are separated by more than a distance, one diameter divided by the efficiency, then the anodes may be placed in groups so that each tank is protected from one groundbed only. At smaller tank separations the anodes must be distributed so that two or more groundbeds contribute to the protection of each tank. These schemes are illustrated in Fig. 139. The installation will be powered either by a multiplicity of small d c generators or by a few larger units. In the case of the latter some anode current control can be exercised by lead resistance variations though generally this will be small. The waste of power that may occur due to poor control of the individual tank potentials when using the large generator will be compensated for by the additional cost of the small generators which, if they have to be of flameproof or explosionproof construction, can be inordinately expensive. A balance in these costs must be reached and will be influenced by the user's power costs.

Where the tanks are widely spaced, as in the case where each tank is protected from a single groundbed, then the whole farm could receive protection from a remote or deep groundbed, and this should be placed to be at least the length of the tank farm away. Often the sea or a brackish estuary is close at hand when advantage can be taken of the cheaper methods that can be employed in these circumstances using marine anodes for the groundbed. Voltage drops in the d c cable can be kept to a minimum by using buried metal pipes as the cathode lead and insulating this with P V C or polythene tape.

Tanks are often earthed and equipped with lightning conductors. Earths of copper or scrap iron in coke breeze are to be avoided where possible and protected galvanized steel earths used instead. Polarization cells can be used to couple the tanks to the a c earth.

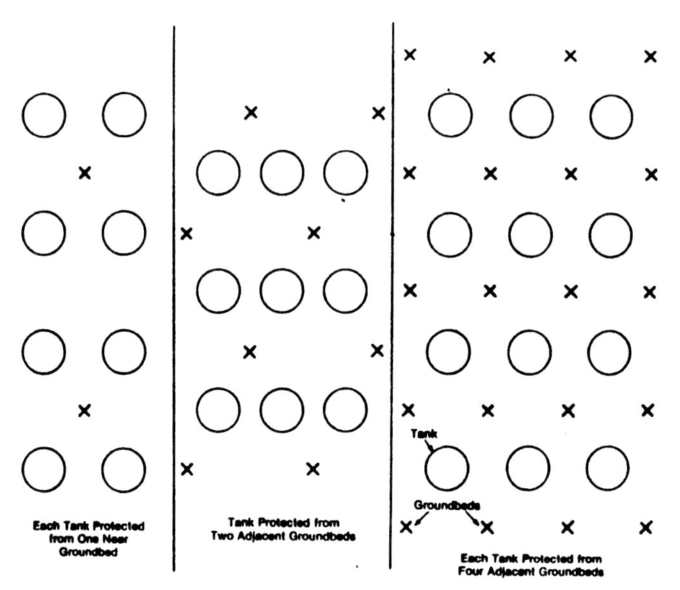

FIGURE 139 - Schematic plans of tank farms showing methods of protecting tanks from distributed anodes according to tank spacing.

Sheet Steel Piling

The last type of buried structure that is found in the ground is the large flat vertical 'wall like' structure. For example, the sheet steel piling used to strengthen excavations both temporarily, as on a building site, or permanently, as in a railway cutting. The higher cost of the cathodic protection of these steel piles has not led to its great application except where the corrosion of the steel would involve costs beyond its replacement, for example, a building may be using the steel as part of its foundations. Such sheet steel pilings may be used to support buildings on river banks or adjacent to embankments or slopes and where the river or estuary is saline it will be an advantage to apply protection from that side, but where the water is fresh then the groundbed is best located inland. The latter will be discussed in this chapter while cathodic protection in sea water will be the subject of Chapter 7.

Driven piling will be poorly coated as the method used to install it will remove much of the coating while where an embankment has been built up against the piling it should retain most of its coating. The advantage of a coating on the steel is twofold as it reduces the current demand and allows a spread of protection along the structure from a close groundbed. Generally, however, the piling will be bare, or nearly so, and the problem of protection under these conditions will now be analyzed. The usual piling is corrugated and this increases the actual area by about 50 per cent over that of the frontage of the piles. The lines of current flow close to the piling will be parallel and at right angles to its surface so that the volts drop per foot in the soil can be calculated simply from the current density required for protection and the soil resistivity. A typical structure of buried steel piles may require 2 mA per sq ft of frontal area and if these lie in a soil of 2,000 ohm cm resistivity then there will be a potential gradient in the soil of

$$\frac{2{,}000}{30} \times 2 \text{ mV per ft} = 133 \text{ mV per ft}$$

This potential gradient will be uniform for distances small compared with the extent of the piling and so sacrificial anodes will not be suitable except under conditions of very low resistivity or well-coated piles; even then only magnesium anodes can be used and there would be little spread from them. The best method would be to use a series of distributed groundbeds fed by impressed current. The current required to protect 100 yds of the piling described above, assuming it is 30 ft deep and corroding on both sides, would be

$$\frac{300 \times 30 \times 4}{1{,}000} = 36 \text{ amps}$$

If 100 amp cathodic protection units were used less than 300 yds could be protected from each of these and to achieve a low circuit resistance almost this length of anode would be required. Spread of protection would be by anode control so that it would be an advantage to alternate the groundbeds either side of the piling. The minimum potential swing would be found at the surface of the piling remote from the anodes so that control of the protection would be simple. The heavy current required for this protection would justify an initial boost of current when a lower subsequent current would be sufficient. The success of this method depends upon the soil composition but often a 50 per cent reduction in the initial current can be obtained.

Often the extension of the life of sheet steel piling will be sufficient, and here a potential swing of perhaps 100 mV will cause a 10 to 20-fold increase in the life at 20% to 25% of the current. Steel piling is interlocked but this does not provide an adequate current path for cathodic protection. It is usual to secure the steel piling by welding it so that it cannot subsequently move and this welding should be done with a view to providing an adequate current path for the cathodic protection. Alternatively, the piles can be interconnected by welding a continuous steel joist along its length.

Care must be taken with the protection of the anchor points and the stayrods as in the case of marine piling.

Steel in Concrete

Steel should be preserved in concrete by the alkaline nature of the cement used in it. However, the economical grades of strong concrete are usually too lean in cement for this protection to be achieved. Richer concretes, and many of these are of long standing, do provide proper protection to the enclosed steel and iron.

There are three conditions that can accelerate the corrosion of steel in concrete. Firstly, reinforced concrete may be in an electrolyte from which ions can permeate into the concrete mix. These ions reduce the effective alkalinity of the concrete by absorption in the cement matrix and this allows corrosion to take place. Secondly, a similar effect can be achieved by absorption of similar ions from sea spray, from an industrial atmosphere or by the use of de-icing salts and other chemicals on the surface of the concrete. Thirdly, the concrete will deteriorate in a normal atmosphere by the absorption of carbon dioxide into the cement matrix, again destroying its inhibiting properties. Concrete which is rich in cement and which overlies the steel reinforcing by at least 2 inches provides adequate protection to that steel and atmospheric absorption does not destroy the concrete certainly within the normal design span of the structure. Some concrete on setting is micro-cracked and many of these fine cracks aid the destruction of the cement filler and excessive cracking can reduce the effective life even of the best inhibiting concrete.

This corrosion attack can be accelerated by galvanic couples acting either within the concrete matrix but more often acting also from outside the structure itself.

The problems of cathodically protecting steel in concrete are twofold: firstly, the problem of protecting the steel in the concrete when this is the sole electrolyte surrounding the steel, that is the concrete itself is the corroding medium; and secondly, the problem of protecting steel in concrete when the whole is either immersed in an electrolyte such as sea water or the soil, or when it contains such an electrolyte as, for example, in a storage tank containing water. The corrosion of steel in concrete and the corrosion of reinforced concrete in the ground will be discussed in this chapter, while the problems associated with protecting reinforced concrete in sea water will be discussed in the next chapter, and the problem of protecting chemicals for water storage equipment which contains the corroding electrolyte will be discussed in a later chapter.

Reinforcing Rods—Rebar

Concrete is extremely strong in compression but not in tension and in the areas in a structure where tension is expected steel rods are positioned to add strength. These are shaped and sized to give the required load bearing properties and can be evenly distributed as in a fence post or the base for a roadway or they can be selectively placed as in a bridge beam where the major rods are shaped to lie in the areas of highest tensile stress. This change in location will at times bunch the reinforcing rods together and the main run of them may change from being close to the top of the beam at its ends and close to the lower surface at its center.

The reinforcing is held in place by iron wire and is often wired together to form a gigantic mass which is held by the mold or form work by spacers of wood, plastic or concrete. Alternatively, it is fixed at its end or temporarily held while the concrete is poured and the supports are successively removed as pouring proceeds. This may or may not occur on a practical site and the techniques of holding the reinforcing while strictly controlled by the designer may not be followed either because of practical difficulties or because of site conditions. However the reinforcing is fixed, it has to be done substantially because of the massive weight and high viscosity of the concrete; vibrators are used to ensure that no air is trapped in the concrete and that it is in close contact both with the form work and with the reinforcing bars. The reinforcing, therefore, must be very substantially held and the temporary securing devices are not always removed.

Much of the reinforcing may be in the form of pre-welded mat and many of the factory made custom bundles are welded together. Other appendages such as support brackets, pile tops, etc., may be welded to the reinforcing, either to locate them or the reinforcing bars within the matrix. These welds make the re-bar electrically continuous but they are usually not sufficient to provide the very low resistance current return path required for the cathodic protection of large areas of reinforcing.

The reinforcing has to grip and be gripped by the concrete to transfer its mechanical strength to it and to aid this it is often worked by indentations, ridges or by twists in rectangular rod. Rust and post-rust roughening are considered an advantage particularly as the millscale is easily removed by brushing as is the slag from welds. The re-bar is often left in the atmosphere to rust before use but occasionally this can cause accelerated corrosion when, for example, the bundles are held with stainless steel wire.

The corrosion of the re-bar can continue during the setting of the concrete and bi-metallic joints can cause voids to occur where there is gassing within the concrete structure. This is found to happen when some of the reinforcing rods are galvanized or when galvanized rods are held together with soft iron wire. Galvanizing can produce a smooth and almost greasy surface which is considered to have poor adhesion to the concrete: weathered galvanizing does not suffer in this way.

The coating of the re-bar will reduce the tendency for it to corrode but it also produces a smooth surface so that the bars have to be highly patterned. This means that the promontories will probably have thinner coating and present an increased risk of damage at site and during transportation. Inhibitors can be used in the concrete to reduce the rate of corrosion during setting but these have to be in sufficient bulk to treat the whole of the mix and have to be compatible with the concrete chemistry. In some applications, particularly where a lean economic mix is used, it is possible to pre-coat the re-bar with a richer concrete mixture to provide it with protection. This is extremely difficult on site except with the simplest geometry.

The second method of reinforcing is to pre-stress the re-bar. In this the reinforcing steel is held in tension during the curing process; after the concrete is fully cured it is released and the tension is transferred to the concrete, holding it in compression. The technique is favored for the construction of light beams as used in housing. Accelerators are included in the concrete mix and some of them, such as calcium chloride, can accelerate corrosion both during the curing and afterwards. Corrosion at this stage can be dangerous as stress failure can occur, particularly if the concrete is not evenly mixed and galvanic cells are established. Cathodic protection has been used during this curing process where it successfully protected the reinforcing wires. The mold itself was used as an anode and the rate of consumption of the mold was negligible. The reinforcing wires had to be made electrically continuous but this was not difficult as the tensioning plates were in extremely good contact with them.

The third general method of using steel in concrete is the post-tensioning method in which the steel is introduced after the concrete is set, usually into hollow tubes or channels, and is then tensioned so that the concrete is held in compression and this area between the steel and the concrete is usually grouted with either a cement mix or some other patent compound.

Many of the individual concrete items such as beams are cured away from their ultimate use and the reinforcing extends beyond the concrete unit and is used for fixing the unit into the structure on site. The concrete base next to the exposed steelwork can be contaminated and it is also possible for the exposed re-bar to corrode excessively, usually immediately where it emerges from the concrete. This is particularly true if water can rest on the surface where the re-bar emerges. In some post-tensioned concrete structures the tensioned steelwork remains exposed after the structure is completed.

Concrete and particularly concrete in which the steel is corroding has a comparatively low resistivity this being in the range from 3,000 ohm cms to 20,000 ohm cms. The lower resistivity concretes are those in which corrosion occurs most rapidly and usually catastrophically. The resistivity may be directional and where surface cracking occurs the resistivity at right angles to the surface will be lower than that parallel to the surface caused by the micro-cracks that extend inwards from the surface. In certain atmospheres, particularly salt-laden atmospheres, these cracks can become filled with salt and this reduces the resistivity along the crack direction.

The corrosion of the reinforcing will cause expansion and stress the concrete: close to the surface this causes extensive cracking and the brown corrosion stain so often seen. In many cases large sheets of the outer surface of concrete will break away from the first layer of reinforcing, exposing it.

Cathodic Protection

The reinforcing in concrete can be cathodically protected either using the concrete itself as the medium in which the cathodic protection is contained or where the concrete is in contact with the ground or water by placing the cathodic protection equipment in that electrolyte. The most difficult problems will be concerned with the spread of current and the cathodic protection within the concrete matrix itself.

The very nature of the majority of reinforced concrete will mean that the degradation of the concrete and the corrosion of the reinforcing will take place in the area nearest the outer surface of the concrete. Equally, this re-bar will require the most protection and from the discussions regarding the spread of cathodic protection it will be seen that cathodic protection beyond this initial layer of reinforcing will be difficult. In many cases the reinforcing will be in the form of rods and these will be spaced perhaps four to six diameters apart. Because of this the area of the steelwork that can be protected will be approximately that of the outer surface of the concrete. That is, if the area of the bars is taken, which will be $\pi \times$ diameter and the rods are spaced four diameters apart, then the area of the rods will be approximately 80% of the frontal area of the concrete. The additional rods of wires required to hold these rods in position will add the extra 20% required to equal the two. Because of this similarity in the areas the current densities are not defined as surface of rebar or of the concrete surface. Because there is a wide spectrum of results both through the nature of the concrete, the difficulty in obtaining even spread of protection, and knowledge of the exact location of the re-bar, the differentiation would in many cases be academic.

The current densities required to protect steel in concrete depend on the age of the concrete and on its alkalinity. The latter probably influences the polarization as much as it does the initial current. In general protection is achieved at about one half the current density required in the soil. The amount of alkalinity present inhibits the sulfate-reducing bacteria though this would not necessarily apply in the case of reinforcing that had been post tensioned and in which the composition surrounding the post tensioned cables or wires did not have a bacteria inhibitor or a bactericide present.

The problems of protecting the reinforcing, therefore, are considerable and in many cases rather than attempt to achieve full protection the concrete's life can considerably be extended by achieving a potential change of perhaps a third of that required for complete protection. There is also considerable difficulty in measuring potentials within the concrete and so other methods of determining the potential of the re-bar are used. The techniques originated by Pearson in eliminating the I R drop in the soil have been adapted for use in concrete as have the break-in-the-curve techniques particularly those used in deep well protection.

The use of zinc coated reinforcing provides a considerable degree of protection and it is possible at currents of the order of 10% of those used with bare steel to protect the re-bar adequately if the protection is applied when the concrete is comparatively new. Small potential changes will reduce the rate of corrosion to an acceptable level, about 1% to 3% of that normally expected and this is at about 10% of the current required to protect bare re-bar. It is unusual for the zinc coated reinforcing to be adequately interconnected within the body of the concrete as it is considered that zinc coated reinforcing is adequate without cathodic protection.

Epoxy coated re-bar can reduce the current requirement dramatically, but allowing for site damage between 3% and 5% of the current is needed to protect bare steel.

It is difficult to achieve protection using sacrificial anodes unless these can be placed in recesses which are filled with a conducting electrolyte. The sacrificial anodes cannot be embedded in the concrete itself as their corrosion is most likely to cause the failure of the concrete by spalling as happens with the corrosion of the steel re-bar. The use of a hollow, or tubular anode of zinc is possible when the anode itself would crush under the influence of its own corrosion product and this might find limited applications.

Similar problems occur with impressed current anodes and these have to be placed in a conducting backfill. One of the techniques that can be used is to place platinized titanium wire, either as individual anodes or continuously, into holes drilled in the concrete or into continuous slots. The anode can be backfilled with an alkali mixture which may consist of plaster with lime. The output from such an anode is limited to about 10 mA per linear foot of each 4'' individual anode, Fig 140. This means that a considerable reduction of corrosion in the concrete can be achieved by such an anode where one linear foot of anode will protect 10 square feet of reinforcing.

The spread of protection from the anode is difficult. If we assume that the concrete has resistivity of 3,000 ohm cms then a continuous anode would create a potential of 300 mV 8 inches from it and this would probably mean a halving of the area protected if the concrete was eight inches thick or each linear foot of anode would protect five square feet of rebar and continuous anodes should be placed not more than five feet apart.

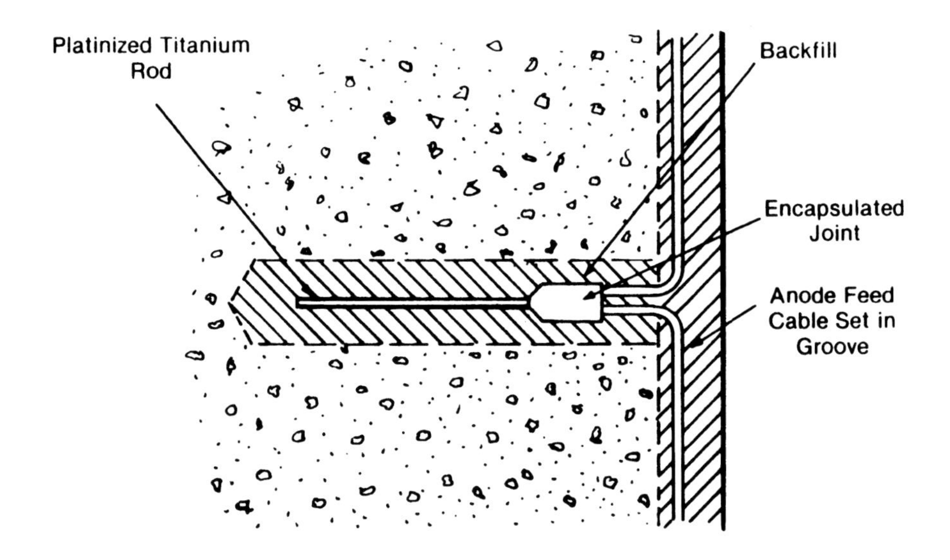

FIGURE 140 - Platinized titanium wire anodes set in concrete.

Hopefully, in many of the structures to which cathodic protection is applied, the concrete would be in good condition with its resistivity near 10,000 ohm cms. Under these conditions continuous anodes would be difficult to employ as it is doubtful that protection would spread over distances of five feet. Individual anodes, perhaps three inches long, could be employed and spaced at about 2½ feet. Such a system would be expensive but where buildings are already erected and suffering corrosion it may well be the only solution possible. Modern drilling and cutting techniques would allow anodes designed as miniatures of cathodic protection groundbeds to be used.

One of the most severe areas of attack is the corrosion of bridgedecks either by salt in sea spray, but more particularly by the de-icing salts that are applied in the winter. In this case it is possible to use a conducting asphalt on the bridge deck and to apply the cathodic protection from within the confines of the asphalt. For this purpose silicon iron anodes have been developed which are cast as flat plates slightly domed with a side connection into them. The conducting asphalt, which does not have the good hard-wearing properties of conventional asphalt, is laid about two to three inches thick with an overlay of road asphalt. The anodes are placed in the conducting asphalt and sufficient current is carried by this asphalt completely to protect the reinforcing wires within the deck. The anode is illustrated in Fig. 141 which shows a cross section of a highway structure.

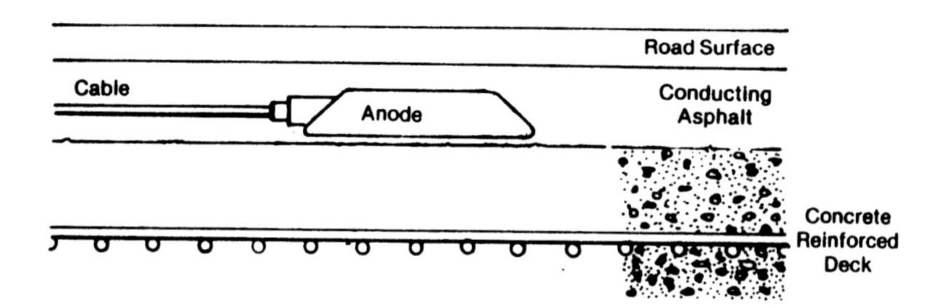

FIGURE 141 - Highway cast silicon anode in conducting asphalt.

A thinner layer of conducting paint can be used in place of the asphalt. This has the advantage that it can be applied to all surfaces including vertical surfaces and the underneath of bridge decks. The connection to the conducting paint can be made by anodes of platinized titanium either in the form of a continuous wire or as discrete anodes. Carbon fiber anodes are suggested as providing a dispersed connection to the paint film. As both the conducting asphalt and the paint film are two-dimensional, then it is essential in order to obtain a good spread of protection through them that their resistivity does not exceed 10 ohm per square feet.

Other techniques have been tried to mount the anodes and these include constructions which contain lakes of electrolyte mounted on the surface. Some of these, either in the form of a pond or as long strips, can be used where the upper surface of the corroding concrete is not used. Beams in bridges and other constructions can be treated in this manner and the anode holding cell can be kept moist by the techniques used in viniculture.

Reinforced Concrete in the Soil

When reinforced concrete is buried in the ground accelerated corrosion can occur. An enormous galvanic cell can be formed between the rebar in the concrete and any buried steel to which it is connected. The alkalinity of the concrete makes the rebar the cathode and rapid corrosion of the buried steel occurs. This has caused a large number of failures, for example on pipelines where the pipe metal is connected to the re-bar in a concrete valve pit. Occasionally the re-bar is connected for safety reasons to the electrical earthing which is made of copper and in this case it is the re-bar which fails as it provides protection to the electrical earth.

If the re-bar is continuous then when it is acting as a cathode it is a much larger cathode and when acting as an anode the corrosion tends to be more evenly distributed throughout the concrete structure.

Cathodic protection of the re-bar is easier and conventional protection techniques can be used if the re-bar is continuous. In large structures not only does it need to be continuous but the re-bar must be capable of carrying the protection current without excessive volts drop in relation to the 300 mV cathodic swing. One type of structure that has received a lot of attention in this respect is reinforced concrete pipe where spirally wound reinforcing was connected at the end of each pipe so that it could be cathodically protected. It was found that the spiral reinforcing bar had to be interconnected longitudinally to get sufficient spread of protection.

Where a structure lies partly in the ground and extends out of it then the cathodic protection applied within the ground does not spread through the exposed concrete to any appreciable height. As can be seen from the figures quoted for resistivity a spread of about five feet is the maximum that can be anticipated. This also is about the level to which the degradation of the concrete would extend from rising dampness and deicing salt spray on the structure.

When concrete is being laid, particularly in adverse weather, it is standard practice to cover it with some form of protection from frost or from very rapid drying. Occasionally, paper is used and is not fully removed before the next layer of concrete is poured. An electrolytic path exists through the wet paper to the reinforcing bar and this, not being surrounded by the dense alkalinity of the concrete, will act as an anode and will rapidly be corroded. The same can occur when gaps for other reasons are left in the concrete and it is essential that such expansion joints are filled with a suitable compound or with a proper seal.

Cathodic Protection in Sea Water

Chapter 7

Sea Water Properties

The largest bulk electrolyte is sea water which covers the majority of the earth's surface and, although it ranges from the Arctic to the Tropics, it has a comparatively uniform composition the major constituents being given in Table XI. The salt content of sea water is generally expressed as the salinity which is the total salt content in grams per kilogram of sea water, all the carbonates being expressed as oxides and all halides as chlorides. Because of the ease with which the halides may be estimated by titration with silver nitrate, sea water is often defined by its chlorinity and this is related to the salinity:-salinity = 1.8 × chlorinity. The specific gravity of sea water increases with increasing chlorinity and at 17.5 °C the specific gravity will be 1.0000 + 1.4 × chlorinity/1,000, that is a chlorinity of ten gives a specific gravity of 1.014. In the open sea the salinity varies between 32 and 36, the usual salt content being 3.4 per cent. Semi-enclosed seas in the tropics have higher salinities, the Mediterranean being 39 and the Red Sea 41, while salinities drop near the coast and particularly in seas, such as the Baltic, which are fed by a number of rivers and have only limited access to the oceans.

In cathodic protection the most important parameter is resistivity and this will vary with both the temperature and the salinity, as detailed in Chapter 3, sea water generally having a resistivity of between 16 ohm cm and 40 ohm cm. In equilibrium with air, sea water will contain dissolved oxygen and this will be present at a concentration of about 8 parts per million.

TABLE XI — The Abundance of Elements in Seawater (ppm)

Element	ppm
Chlorine	19,000
Sodium	10,500
Magnesium	1,270
Sulfur	880
Calcium	400
Potassium	380
Bromine	65
Carbon	28
Strontium	13
Boron	5

Excluding dissolved gases

Sea Water Corrosion

The rate of corrosion of clean bare steel specimens totally immersed in sea water varies very little with temperature or salinity and is reasonably constant throughout the world, being about 0.005 in. per year. The practical rate of corrosion attack on steel in sea water is locally much higher and can be attributed to a series of accelerated influences. Steel corrodes in sea water into ferrous ions and these are further oxidized to ferric ions with the formation of rust which covers the whole steel surface and limits the rate of corrosion. The corrosion of the steel by bacteria, which produce iron sulfide as a product, occurs underneath the rust but the attack is not rapid in the open sea.

On clean, unpainted steel samples the corrosion rate is small and the attack uniform while most steel structures suffer localized corrosion and pitting. Steel as manufactured is covered with a tightly adherent skin of mill scale, a conducting oxide which is highly cathodic to clean steel. Any break in this scale which reveals the bare steel beneath it produces a galvanic cell, and the low resistivity of the sea water causes rapid, deep pitting of the exposed metal. Mill scale does not produce the only bimetallic or galvanic couple that might occur on a steel surface as rivets, welds and other metals may cause equally large potential differences. While these corrosion currents may be as large as those generated by mill scale the anodic areas are not generally as concentrated and do not by themselves lead to heavy pitting. Painted steel surfaces are usually mildly cathodic to bare steel and holidays, damage or unpainted areas, such as those beneath the keel blocks, corrode as anodes. Often all three causes are present simultaneously and consequently more rapid corrosion occurs. The corrosion product in sea water tends to stifle the rate of attack and if this product is removed continuously the rate of corrosion increases; scouring by sand or shingle or high water velocities such as those found near to the propeller of a ship, causing this type of accelerated attack.

Corrosion cells are also established by variations in the electrolyte, usually by stratification with fresh water, or by oxygen concentration cells which may be associated with fouling or impingement.

Most metals in sea water foul with a variety of organisms which may be permanently or temporarily attached and many of these destroy the organic coating in their vicinity even if they do not cause other damage. Barnacles often cover large structures such as piers; these are believed to have a protective effect, while mussels are supposed to cause pitting beneath their place of attachment; no precise mechanism is known but the phenomenon is reportedly widespread. To kill this fouling, paints containing metals or metallic salts are used, copper and tin being popular. If such paints are applied directly to the bare hull their salts either by plating out the anti-fouling metal on bare patches of steel or by an electrical contact with the metal, can establish a bi-metallic cell with consequent accelerated corrosion. This attack is prevented by applying a coat of anti-corrosive paint to the steel and the anti-fouling paint over it.

Cathodic protection has no effect on fouling nor has it any appreciable effect on any anti-fouling paints. Cathodically protected steel does, however, provide an ideal surface for the operation of anti-fouling paints and their performance is usually as good as is found on the inert plastic or glass test panels used by the paint manufacturers. Metals that are not fouled in sea water, such as copper, because of their slow dissolution lose their anti-fouling effect when completely cathodically protected. This was realized by Sir Humphry Davy who suggested that complete protection of copper hull plates be avoided to maintain their anti-fouling property.

Sulfate reducing bacteria have been identified in most saline muds and when these are anaerobic the bacteria are active. Several cases of severe corrosion have occurred on ships that have rested on mud flats and rapid corrosion of the steel piles of some jetties is reported. Cathodic protection is successful against these bacteria, an extra 100 mV negative swing being considered necessary.

Cathodic Protection

The copper sulfate half cell used generally on land is not entirely suitable for potential measurements in sea water and, though it may by ingenious design, for a considerable time display the same potential, a very small contamination will cause current hysteresis. The silver/silver chloride half cell is compatible with sea water and this can be used in two forms; the element, a silver/silver chloride composition described later, being either immersed in sea water or in a solution of known chloride concentration, usually a saturated common salt solution. The former is most generally used in cathodic protection though its susceptibility to changes in the sea water environment makes the latter somewhat more reliable. Protection of steel is achieved when its potential is -0.80 V to the equilibrium silver chloride half cell, that is the former type, or -0.76 V to the saturated silver chloride half cell, both of these potentials corresponding to -0.85 V to the copper sulfate half cell. The potentials in this chapter will refer to the equilibrium silver chloride cell unless otherwise defined.

High purity zinc or the anode alloys can be used to provide reference potentials in sea water and this potential is not affected by variations in salinity or velocity. Zinc is usually used as the secondary electrode after comparison with a silver chloride half cell, or where an accuracy better than 5 mV is not required. The electrodes are large and cheap and considerable current can be drawn from them without polarization. This allows them to be used with simple voltmeters for potential measurements or to provide power to actuate automatic control equipment. A small anodic bleed is often used to keep the electrode surface active.

Automatic Control

Although sea water is relatively homogeneous, the current required to protect a metal structure in it may vary considerably. This can be occasioned by temperature changes in process equipment, by velocity as on a ship's hull, or by mechanical damage to a coating. In some cases these changes will occur rapidly as, for example, when a ship changes speed or enters estuary water. Usually the best method of ensuring complete protection is to design the installation so that the structure receives adequate current under the worst conditions, and automatically to control the output from the cathode potential. Details of the automatic control are discussed both in the chapter on Instrumentation and Control and relative to the particular application.

Steel

If a bare steel plate is immersed in sea water and made the cathode of a cell it will be polarized so that there is a logarithmic relationship between the cell current and the potential change of the steel. The steel potential change will follow the law

$$V = a + b \log \sigma$$

where V is the potential, σ the current density and a and b constants. In still sea water a bare steel panel may require from 4 to 7 mA per sq ft to protect

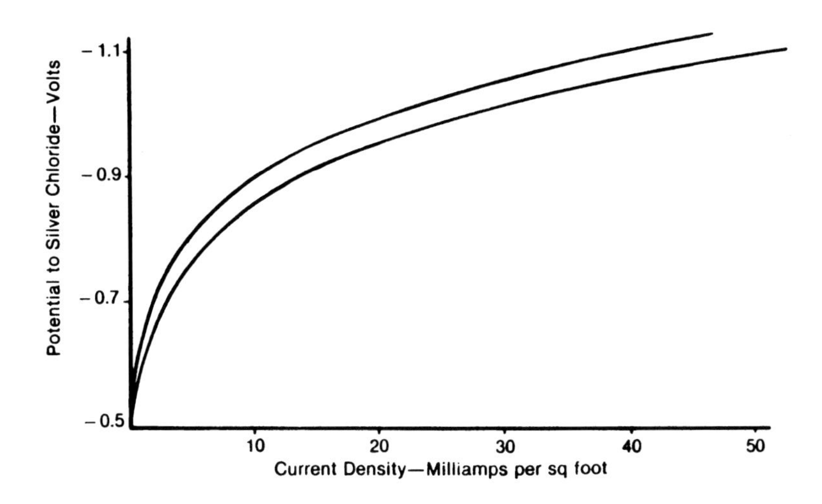

FIGURE 142 - Polarization of bare steel in sea-water.

it and the constants of the polarization equation, called Tafels' equation, are such that the current density has to be doubled for each increase of 100 mV in the polarization. The curves in Fig. 142 show this relationship for two particular cases where the steel required 5 mA per sq ft and 6½ mA per sq ft for protection. The cathodic reaction in sea water has two effects. It raises the pH value in the catholyte, that is it makes it alkaline, and it causes an increase in the metallic ion concentration close to the cathode. These two effects cause the precipitation of a 'chalk' which forms on the metal surface and consists mainly of calcium carbonate and magnesium hydroxide, the ratio of these depending upon the current density that is applied to the cathode. At low current densities the film is hard, almost completely calcareous, insoluble and usually tightly adherent to the metal while at high current densities the film has a larger content of magnesium whose salts and hydroxide are soluble and the 'chalk' film is spongy and soft. At current densities above 100 mA per sq ft this film is too soft to be used as a practical coating.

Several processes have been suggested for making these films hard and adherent and some of the most successful reported are made of successive layers of the different types of chalk. By using pulse techniques it is possible to lay down a very thin and very tenacious film of cathode chalk. This has the appearance of a bloom on the metal rather as on old furniture. It is suggested this is because the pulse does not extend over a sufficient period for bubbles of hydrogen to be formed and the film salts are laid down as a solid monolithic layer. For practical purposes the position can be summarized by

TABLE XII — Current Density Used in Ship Hull Cathodic Protection Design

Typical Coatings New Hulls	Self Polishing	Coal Tar Epoxy	Aluminum Bitumen	Conventional		
Deep Water to Deep Water	0.75	1.5	2	3	5	Design Current Density
Principally Deep Water	1.25	2.5	3	5	6	mA/sq. ft.
Occasional Scour	3	4	5	6	7	Typical Coatings
Ice Damage Frequent Scour	6	6	8	10	10	Retrofits
		Self Polishing	Coal Tar Epoxy	Aluminum Bitumen	Conventional	

the results given in Table XII which describes the current density and time products to give a certain reduction in the current density required for continued protection.

The current densities referred to so far have been those required with bare steel. The effect of paint will be to increase the current density at holidays to that required to achieve protection. For example, if only one mA per sq ft is required for protection where bare steel needed 5 mA per sq ft then the paint is 80 per cent efficient and the current densities in Table XII can be reduced to 1/5th. Alternatively the whole of the current density criteria can be translated into metal potentials, for example a current density of 20 mA per sq ft on to bare steel might produce a potential of -1.00 V and this would be so whether the current density were applied to a bare steel plate or to a series of pinholes in an otherwise perfect paint system.

At potentials more negative than -0.95 V, that is a potential swing of about -0.45 V, hydrogen is formed sufficiently rapidly at the cathode to be visible and bubbles may form. In some restricted places this is a disadvantage and the metal potential must be kept more positive than -0.90 V to prevent pockets of gas accumulating. Some of the hydrogen is absorbed by the cathode metal and travels through it although this is only a small effect and hydrogen embrittlement in normal hull steel has never been observed at the potentials used in cathodic protection. Where steel is well rusted the cathodic reaction, or the cathodically generated hydrogen, may reduce the ferric red rust to ferrous or black corrosion, this being the same reaction that occurs when the hydrogen evolved from zinc and hydrochloric acid are used to reduce ferric salts to ferrous salts. The result of this reduction is a general loosening of the rust scale and it may be washed away by water movement.

The reduced ferrous layer can be confused with the black iron sulfide layer that is the innermost corrosion product where bacteria are active and is sometimes called 'black rot.' Descaling of the protected object often occurs and the cathodic protection techniques can be extended to give this generally. The process is partly caused by the liberation of hydrogen from the metal, it having been absorbed at the bare metal areas and traveled along imperfections to reappear in its molecular form under the scale, and partly to the loosening, by reduction, of the rust and lifting this scale from the metal by the hydrogen gassing. This process is usually operated at between 300 mA per sq ft and 1 amp per sq ft and requires 8 to 10 ampere hours per sq ft for complete descaling; the current can be reversed finally to reduce the alkali on the descaled metal and to remove any polarization products. Careful control of the current density and the metal potential will allow good paint to remain on the surface and the cathodic reaction should not soften this.

The effect of the cathodic alkali on paints has been described in Chapter 2; in sea water this effect will be reduced by the renewal of the sea water next to the metal surface. This has two effects: it physically washes away the alkali formed, and the new sea water acts as a buffer preventing excess alkali formation. Most paint damage except that which occurs very close to the anodes is caused by bad surface preparation before the application of the paint, the underlying scale being loosened by the protection and removed taking the paint with it.

This theory suggests, and it is to be found so, that there will be less cathodic paint damage where the sea water is flowing past the metal with some considerable velocity than on static objects. The current required for protection will, however, be greater for steel in moving sea water than it will be in stationary water, and so the effect of the removal of the alkali is partly canceled out. Experiments on ships' hulls give an increase of protection current of between 50 per cent and 100 per cent for a ship underway over a stationary ship. Laboratory experiments usually give very much larger values than this which may be associated with the increase in velocity gradient immediately next to the surface in laboratory experiments. This should aid the transport of oxygen to the surface.

One effect of the chalk or calcareous polarization film is that it retains the concentration polarization products so that protective potentials are maintained despite breaks in the protective current; a well-chalked steel surface may remain at protective potentials for 48 hours. The protection of steel piles which are covered to different depths by the tide is an example of this, and it is found that the piles are protected to about half tide mark, above this there being a general decrease in the protection until none is evident near the high tide zone. On a practical jetty with piles extending only a short distance below the low tide level there will be a large increase in the area of steel to be protected at high tide and the average current density onto the steel's surface may drop. This will prevent the polarization of the temporarily immersed steel and the level to which protection is achieved will be lower. Jetty piles are often formed with spats of concrete or they are effectively painted down to half tide level and if either of these techniques is adopted, the whole structure will receive protection. Alternatively, extra anodes or the control electrodes can be placed in the upper tidal zone to ensure polarization.

Aluminum

The metal second in importance to steel in marine construction is aluminum, or rather, aluminum alloys, and because of their better corrosion resistance the aluminum magnesium alloys have been used rather than the stronger duralumin type. Aluminum displays a potential of about −0.6 V to −0.7 V to an equilibrium silver chloride half cell in sea water which is much less than would be predicted by its position in the electrochemical series, being normally in a 'passive' condition due to its protective oxide film. If the aluminum is amalgamated with mercury it displays its true, and much baser, potential and corrodes rapidly. Aluminum can be protected cathodically in sea water by a potential swing of 100 to 200 mV to a final potential of −0.80 V that is the same as the potential of protected steel.

The metal is attacked by dilute solutions of alkalies so that large potential swings are to be avoided in stagnant water. Aluminum of 99.5 per cent purity has, however, been successfully protected at a variety of potentials as negative as −1.2 V in a collection of sea and estuary waters without cathodic corrosion occurring. Zinc anodes have been used for some of these tests and even when the zinc has been made the much larger electrode, so that the aluminum is held at the zinc open circuit potential, no alkali corrosion has been observed in still water. The limit beyond which rapid cathodic corrosion occurs in moving sea water is greater still.

One effect of interest has been noted with some aluminum specimens that have a nobler metal attached to them: under cathodic protection, alkali may be generated at the noble metal and this may cause corrosion of the aluminum. In the tests in which this was observed the corrosion caused was cathodic, or alkali, in nature and in stagnant water extended upwards from the piece of noble metal.

A limited range of other aluminum alloys have been protected in sea water and though 100 mV swing has given them protection, the level of the onset of cathodic corrosion in different solutions has not been determined. Zinc anodes seem to lie within the safe range of the normal aluminum alloys used in sea water. Care has to be taken in the use of aluminum anodes to protect aluminum as a smear of the aluminum anode alloy could initiate corrosion of the aluminum hull.

Aluminum that is coated should be protected within the same potential limits as bare aluminum in static water and when there is any doubt pH measurements may be made on the metal surface, a pH value in the range 7 to 9 indicating protection. The corrosion products found in anodic pits on aluminum usually have a pH within the range 3 to 5 and pH values greater than 10 will indicate the possible onset of cathodic corrosion.

Anodizing as a coating has an extremely high resistance and aluminum alloys which have been anodized require only 2 μA per sq ft for protection; this means that a large spread of protection is readily achieved at very small current. The effect of large cathodic potential swings upon the anodizing are not known but the combined anodizing and cathodic protection of aluminum seems likely to provide cheap and ready protection in sea water.

Sacrificial Anodes

All three metals: magnesium, aluminum, and zinc, can be used to protect steel in sea water.

Magnesium acts as a high driving potential anode with about 0.7 V drive and the composition of the magnesium alloy used is not as critical as in other applications; the 6 per cent aluminum 3 per cent zinc with 0.1 per cent copper is suitable. This alloy is usually the cheapest of the magnesium alloys in most parts of the world. High purity magnesium, such as Dow's 'Galvomag' is not advantageous for general sea water anodes.

Two alloys of aluminum are being used: an aluminum/mercury alloy which acts as a low driving voltage anode, and an aluminum indium alloy which acts as a medium driving voltage anode having a drive of about 0.30 V to steel.

Zinc used in sea water must be free from iron and high purity zinc can be used when the iron content is less that 15 ppm. Aluminum, with either cadmium or silicon additions, kills the iron effect in zinc and produces an anode which is claimed to be better than high purity 'zero-iron' anodes.

Zinc acts as a low driving voltage anode having about 250 mV drive to polarized steel.

Impressed Current

Steel or aluminum can be used as consumable anodes in sea water and little or no polarization occurs to them. Both operate at higher efficiencies than in fresh water, steel being about 90 to 100 per cent efficient, and aluminum about 50 to 90 per cent. Neither is competitive economically with the permanent anodes and, except where a cheap scrap supply of steel is available; they find little use.

Of the permanent anode materials the lead alloys and the platinum film electrodes are the two most suitable. The platinum film anodes can be used universally except in some areas where there is sugar pollution from sewage, while the lead alloys can be used where a low current density is required and the anode operates in open sea water. Magnetite anodes, silicon iron with molybdenum and impregnated graphite can be used when freely suspended or carefully mounted. Metal oxide coated titanium anodes can be used both in open sea water and in saline muds.

Sources of d c current vary with the local power supplies, the most general source being the a c mains. Diesel or other heat engine generators can be used on ships or marine installations without spare electric power and solid state converters where high voltage d c power is available; solar power and wind generators have been used where little or no other power is available. Whatever the source of the cathodic protection current, the equipment providing it must be well designed to withstand the corrosive marine atmosphere that is inevitably found at the installation site.

Anode cables are best insulated with polythene, copolymers, or irradiated polythene and structural plastics near the anode must resist the attack of the anode gases.

Cathodic Protection of Inshore Structures

Cathodic protection in sea water can be divided into three groups of structures: first, those that are inshore, that is structures which are close to the shore and usually connected to it in some way; second, ships' hulls which are in active service; and third, offshore structures which are characterized as much by their size as their remoteness. Pipelines serving inshore installations and river crossings are treated separately from pipelines serving offshore installations.

Pipelines

Marine pipelines are cathodically protected in the same way as their land counterparts. Impressed current cathodic protection will usually operate from the land and this will mean that protection will be from one end in the case of a pipeline used for tanker loading or discharge, and at both ends for an estuary crossing or where the pipeline serves a fixed island terminal. The pipe will be coated and wrapped and usually protected by shielding or a weight coat of concrete or of some other material. It is essential that the shielding and weight coat should conduct the cathodic protection current so that the pipe receives protection underneath it. It is also possible that the pipeline will be treated with some anti-fouling composition where it emerges from the water.

Pipelines that travel some distance offshore need the best possible coating so that protection will spread from one end. If this is impossible then the protection may have to be augmented by anodes along the length of the pipeline. In this case the anodes can be attached to the pipeline directly, as for long distance offshore pipelines which are detailed later, or they can be laid on the sea bed close to the pipe.

The anodes that are used in the shoreline are constructed as normal pipeline groundbeds using an anode that will not be attacked by the nascent chlorine gas. Anodes that are mounted below the low tide line often rest on soft muds. These have to be constructed carefully so that they do not settle into the mud. Two techniques are used: one in which the anode is placed on a large sled and this has a considerable surface area so only settles slowly into the mud. The sled can be made of a variety of materials though care has to be taken that the anode reaction does not destroy it. Ships' hull anodes can be modified to sit on top of such a structure which may be made from a resin glass fiber trough which is filled with concrete. The alternative method is to make the anode light, making it either buoyant or semi-buoyant, so that it will not sink into the mud. The anode itself will be anchored by the cable feeding it. Two designs appear popular: one is an anode surrounded by a framework of hollow plastic pipes which are sealed and usually contain monocellular foam. In the other the anodes are attached to a plastic container which is hollow or if high pressures are involved filled with a non-crushable foam or low density hydrocarbon liquid (Figures 143, 144 and 145).

It is difficult to construct groundbeds that lie on the bottom of a river estuary or the seashore without running the risk that the anodes or the cable will be adversely affected by other marine operations in the area. Often a submarine pipeline will have restrictions on mooring and anchoring around it, but the anodes of necessity will need to be some distance away; local fishermen may cause problems at the anode or along the cable. While the cost of the anode itself may be small, the cable costs certainly will not and for this reason it is usual practice to place the anodes in the shoreline.

Where the pipeline serves a single buoy mooring or other technique in which the hose at the end of the pipeline is taken onboard the tanker, care

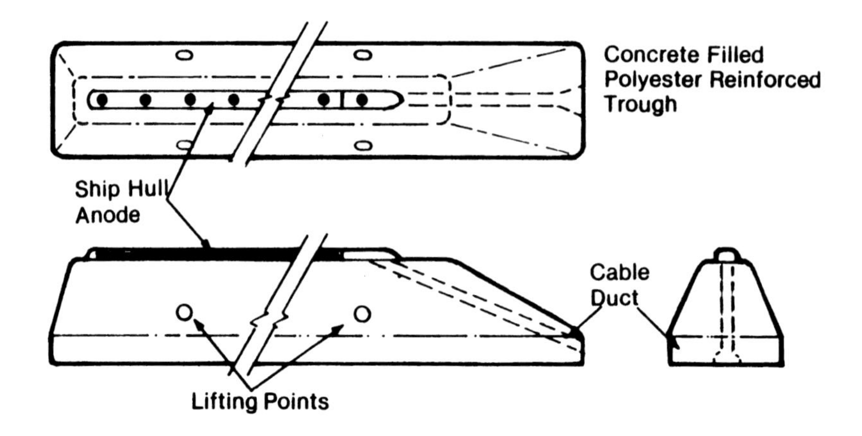

FIGURE 143 - Seabed impressed current sledge anode.

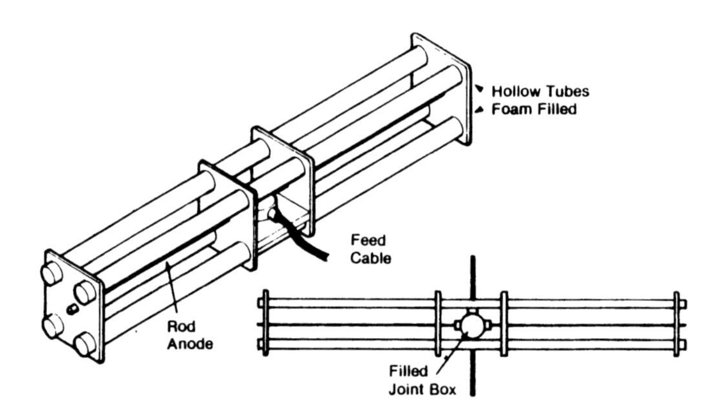

FIGURE 144 - Tubular supported anode.

has to be exercised that current does not drain through the vessel into the pipeline. There is no problem with this from the cathodic protection point of view, but there is a considerable hazard that on breaking any bond between the two an incendive spark may result. To overcome this it is quite usual the isolate the end of the pipeline, either by using an insulating flange or by the use of insulated lengths of hose. The hose itself is insulatedby a break in the reinforcing wire or there is sufficient resistance in the stainless steel wire to reduce the amount of current that would be drained from the vessel.

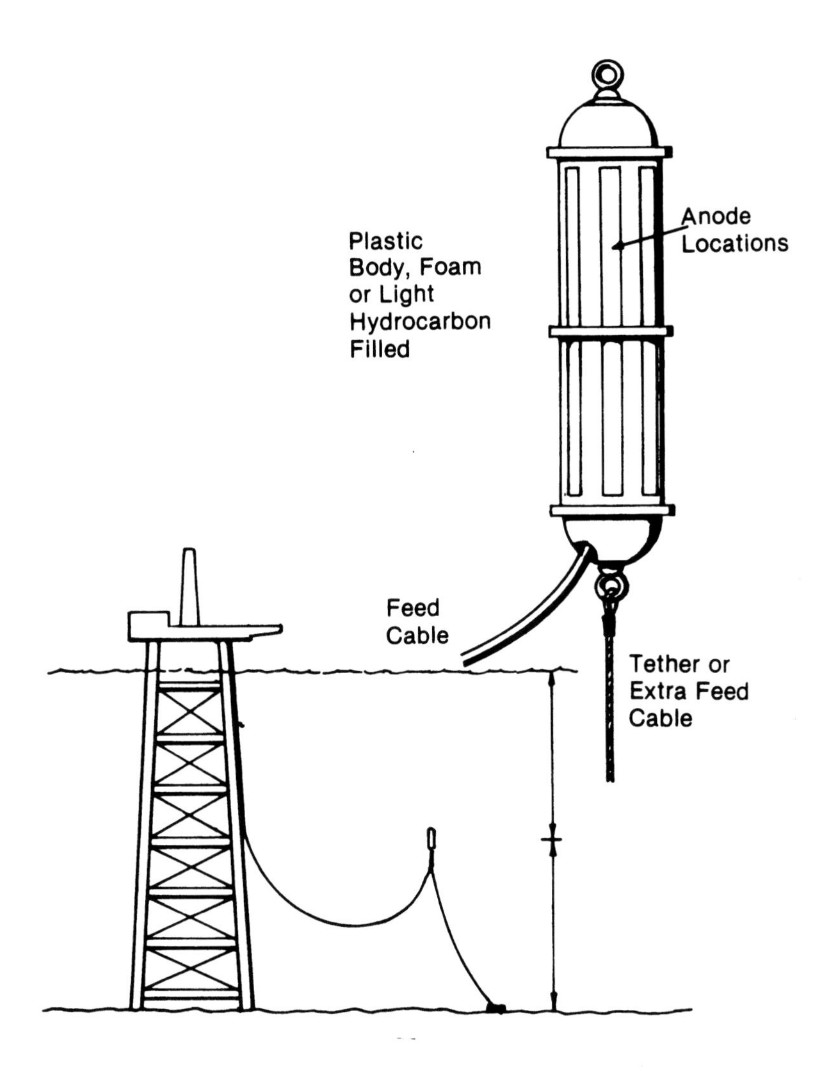

FIGURE 145 - Floating tethered anode.

When the pipeline ends in a dolphin or offshore terminal it can be protected from both ends. If the pipeline itself is to be electrically connected to the offshore structure then its protection must come principally from the land as it would be uneconomical to increase the protection of both the structure and the pipeline to cause a spread of protection shoreward.

Where the pipe terminates in a small structure such as a single point mooring buoy this will probably be protected by low voltage sacrificial anode and it is sensible to bond the pipeline to the mooring buoy. Both ends of the line will now be cathodically protected but it is essential to know whether the protection is spreading to the center of the line. One of the techniques described later must be used to give this information.

The same problem occurs with a river or estuary crossing where, again, the pipe will be cathodically protected from both ends. Where there is a single grouped installation of sacrificial anodes, these can be disconnected or the impressed current switched off at one end and the potential measured at the other. By repeating this process either end it is possible to calculate, by the spread equations given in Chapter 6, the adequacy of the protection at the center point. Where pipe is protected by a multiplicity of sacrificial anodes as, on the mooring buoy, it is not possible to do this. To measure the potential at the mid point of a pipeline that is either crossing an estuary or is connecting an offshore structure a number of small anode bracelets can be placed on the pipe similar to those used in long pipeline protection. The techniques developed for the measurement of the adequacy of protection for pipes that are protected by bracelet anodes can be used to indicate the level of protection in the vicinity of these small bracelet indicators. This is discussed more fully later.

Alternatively, conventional electrodes can be attached to the line before it is laid and the cables carried ashore from these points. An allowance must be made for the change in potential of the pipe metal itself and the better technique is to make a connection to the pipe at the point of attachment of the reference electrode. This method is expensive and is susceptible to damage.

Piling and Wharves

Sheet steel piling is used for wharf and sea defense construction and can be cathodically protected against corrosion both on the seaward side and the landward side. The piling itself is usually poorly protected by its coating, and during driving upwards of half of this coating may be removed. The piles are corrugated to give them greater strength, are usually interlocking and are driven sequentially. This means that the actual area of the metal is 50 per cent or more greater than the frontal area and that the piles need to be made electrically continuous.

The landward side of the piling will usually be lying in made-up ground which will be to wharf height, while the seaward side will penetrate the seabed, with a reasonable depth of tidal sea water in front of it. The corrosion of the sheet steel piling will occur both in the seabed mud, the soil and in the sea water, but most rapidly in the wind and water line and it will be essential to have a complete maintenance program, both for the wind and water line and for the buried and submerged areas. It will be possible to paint effectively down to about half tide level but below that the only effective technique will be cathodic protection.

The current density required on the land side will be about half that on the seaward side, and it is possible that protection applied to the seaward side of the piling will be sufficient to augment the protection of the landward side with the current flowing either beneath or around the piling. On the landward side there will usually be anchor piles with tie rods. These will require cathodic protection and their corrosion will be more critical than that of the piling in general as a localized attack on the anchor rods can cause piling failure by a lack of anchoring to the main piled wall. It is usual to take considerable care over the corrosion protection of these tie rods and they will be wrapped or coated most carefully. The piling itself, as in land applications, will need to be made continuous and it is advisable also to make the tie rod continuous with the piling. The protection of the anchors and tie rods behind the piling is usually carried out as a land based installation using some form of vertical anodes as deep well groundbed. Vertical anodes or other buried service pipes must be protected with the wharf structure.

On the seaward side, if the anodes are placed some distance from the piling they will provide a more even spread of protection, but this layout can be exceedingly damaging and ineffective if ships tie up at the wharf. To overcome this problem anodes can be placed in the recesses of the corrugation and a limited spread of current achieved. With impressed current systems one anode would need to be placed about every six corrugations, while with sacrificial anodes they would need to be placed at every other corrugation unless it is well coated. A typical impressed current anode system is shown in Figs 146 and 147 and sacrificial anodes are shown in Fig 148.

The impressed current anode, Fig 149, in order that it can be serviced, must be retrievable and the anode shown can be swung up for maintenance. To fix it more securely at its foot it is possible to use a magnet which, although perhaps not sufficient to hold the anode, will prevent its movement. The current off each of these anodes is, of necessity, small and there will be a complex network of cables feeding them. It is sometimes difficult to place these cables as the wharf may be fendered with either rubber or wooden fenders. If provisions for cathodic protection cables etc. can be made early on, the problems of protecting the wharf are much reduced. The transformer rectifiers for impressed current protection can be mounted in several places, in wharf sheds or godowns, incorporated into the lighting system and mounted either at the base of lattice lighting towers or on the smaller lighting standards or installed adjacent to same enclosure.

It is convenient if some form of reference electrodes can be permanently installed and wired back to the transfer rectifier; then either simple manual control can be operated or the system can be made automatic. Automatic control is useful where the dock or estuary water varies in salini-

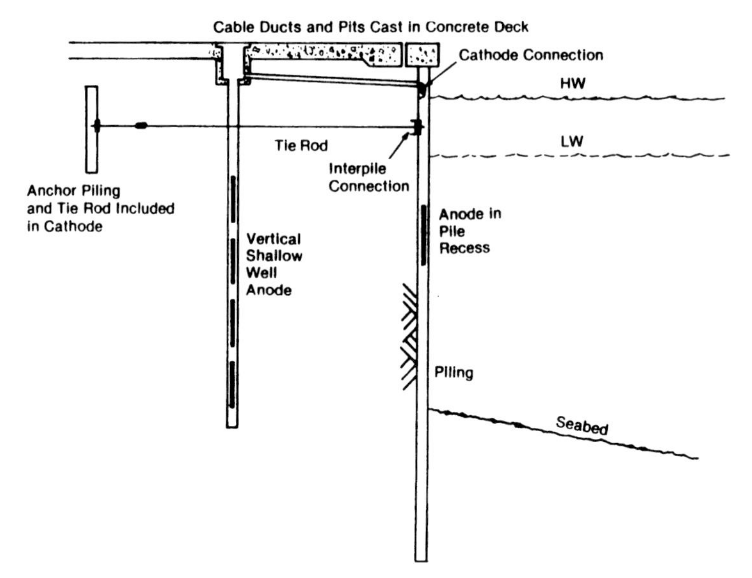

FIGURE 146 - Protection of sheet steel piled wharf showing anode location.

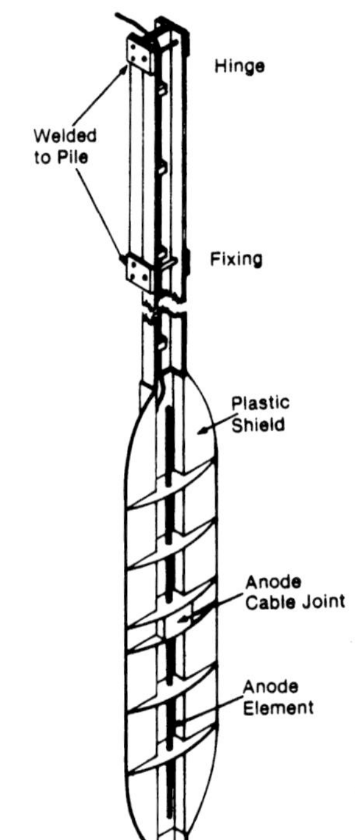

FIGURE 147 - Recoverable impressed current anode for pile recess mounting.

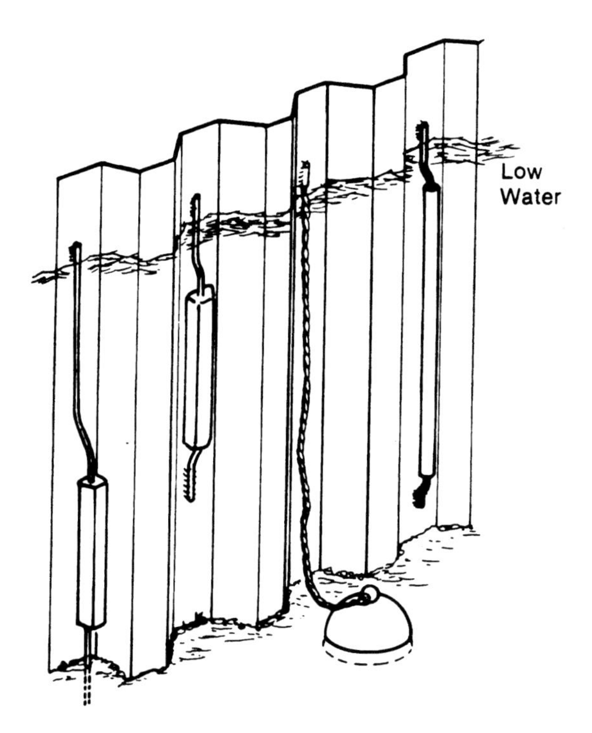

FIGURE 148 - Sacrificial anode systems to protect sheet steel piling.

ty, though it is essential that some limits are put on the system as it may be impossible at times to achieve full protection. This may be during a period of heavy rain which affects the water composition and flow in the local river, or at times of very low tide.

There are instances where impressed current protection has been applied to the steel piling on an estuary or inlet by placing anodes in mid stream and providing protection to both banks. Where these wharves are used by shipping considerable potential differences are found on the ship's hull caused by the current flowing through the sea or estuary water. The current enters the ship on the seaward, or anode, side and is often discharged from the ship to the wharf alongside. This, even in a short stay, can cause heavy pitting on the hull. Potential differences of as much as 3 Volts have been measured across a ship and the hull attacked with lines of pits that coincided with the spacing of the steel fender piles on the wharf. In one case a thin hulled vessel which regularly was moored between cathodic protection anodes and protected piling sank as a result of this concentrated attack.

FIGURE 149 - Retractable anode for location within corrugations of sheet-steel piled wharf. (Photo reprinted with permission from CORRINTEC/UK, Ltd., Winchester, Hants, England.)

It is not essential that the sheet steel piling be perfectly protected. If a change of potential of 100 mV can be achieved then its rate of corrosion will be reduced to 10 per cent and with a 200 mV swing only 5 per cent of the expected corrosion will occur.

There are a number of structures which are built using the same technique as wharves. These include footings for bridges, towers and conveyors. These usually do not have vessels tying up alongside them and so remote anodes can be used. Anodes mounted on the sea bed will give a good spread of protection or the anode can be stood off the structure. This latter practice is not recommended as there is always the risk that a small vessel under emergency conditions will come alongside.

Where the sheet steel piling is used in temporary works then protection by sacrificial anodes would be recommended. Small zinc or aluminum anodes could be attached before driving, though this may lead to difficulties in handling the piles. Simple techniques using sea bed anodes with a substantial cable are also possible.

Sheet steel structures are sometimes constructed as a series of circular cofferdams and these may be interconnected by a deck or roadway, each acting as a small independent island, or there can be further curved piling placed between them. In either case this gives bigger recesses in which to place the anodes, allowing larger impressed current outputs and a more even spread of protection. Again, it is essential that the piling is interconnected electrically, both pile to pile, and between the circular cofferdams.

It is usual for the cofferdam to be filled with dredged sand or gravel. This will have salt in it and natural rainfall will slowly leach it out unless there is fairly free access to the sea. The protection inside the cofferdam has to be designed with a view to it operating after much of the salt has disappeared and higher driving voltages are required.

Occasionally the steel wall has drainage holes in it and these may be always open or fitted with some form of flap which may require bonding across it. The opening should be carefully protected as corrosion at this point could mean an enlargement of the hole with a loss of fill in its vicinity and massive settlement of the wharf's apron.

This type of construction is often used in building dry docks. It is generally not necessary cathodically to protect the inside of a dry dock as it will usually be dry, being emptied except when ships are being taken in or out. The backfill behind the walls will be of made-up soil and this can be corrosive. There will also be many services in such an area which have to be carefully protected, both against corrosion and against interference or electrolysis. An added hazard in many docks is the d c equipment operating there: this may be cranes or other equipment, including electric railways and is referred to in the chapter on interference.

Dock gates, however, will always be wet on one side, they will be hinged, and it is necessary to bond across the hinges to avoid them being eaten away by electrolysis. The dock gate may be protected by either impressed current or by sacrificial anodes. Often the dock gate will be of hollow construction and fill with water. It will be difficult to place impressed current anodes within the dock gate structure because the anodes may exacerbate the corrosion above the water level by generating hypochlorite and by causing small pockets of chlorine gas to accumulate. Venting will not be sufficient to prevent degradation of the paint and corrosive attack. When sacrificial anodes are used venting of the spaces is necessary to avoid hydrogen gas pockets though the gas itself is not harmful to the coating.

Jetties

Jetties in this context are characterized by being built on individual piles. Although a number of other materials are used wood, concrete and cast iron, only the cathodic protection of steel piles will be considered. Some piles are tubular, resembling a pipe, others are made of rolled steel sections such as H-beams, and occasionally they are made of a hexagonal

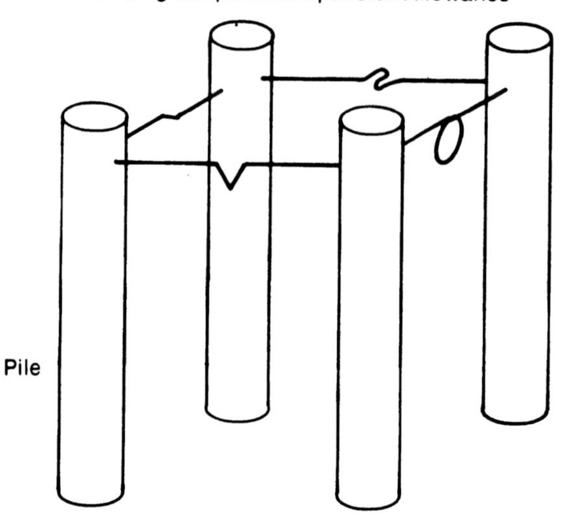

FIGURE 150 - Interpile bonding on steel pile jetty.

box which may be made from sheet steel piling sections welded together.

The piles are usually driven into the sea bed and support a deck. While many of them are driven vertically, at strong points piles may be driven at angles, called raker piles, in order to give reinforcement against the pressures of berthing. Many such structures have a berthing head with four dolphins and a roadway, usually with a carrier track to the shore.

In order that the jetty may be cathodically protected as a unit, it is essential that all the piles are in good electrical contact with each other. Steel piled jetties are sometimes capped with a concrete deck and the reinforcing rods in this cannot be assumed to give sufficient interconnection. If protection is considered at the time of construction, the reinforcing rods can be welded to the pile caps and additional rods built-in to provide continuity. Otherwise, connecting rods must be fitted and these should be welded between the piles. A minimum cross section of these is 1 sq in. and the rods should have some form of expansion loop or bend in them. Steel decks can be made continuous with the piles and require bonding across expansion gaps (Fig 150).

The piles of a jetty are usually driven under half their length into the sea bed, extend a similar amount to high tide, and then travel through a splash zone to the jetty deck. With concrete jetty decks the pile is often muffed with a reinforced concrete spat down to about low water mark: in

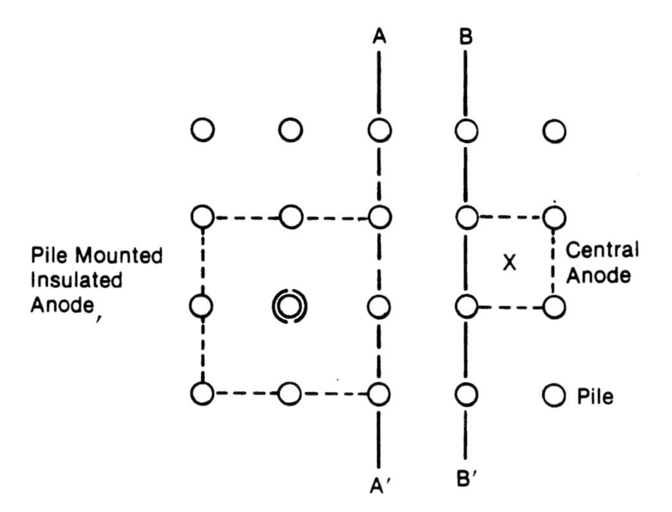

FIGURE 151 - Plan view of typical pile jetty with alternative anode locations.

some locations they are protected against ice by wooden slatting. The pile is usually coated and on being driven about 20 to 25 per cent of this coating is destroyed. Often the piles are welded on site so that they are lengthened and the top is cut off at cap level. Cathodic protection is effective to half tide level and maintenance can usually be carried out to the same point.

The piles of a jetty are arranged in some fixed geometric pattern with strong points and walkways. In the usual configuration the piles are in some square or rectilinear configuration as shown in Fig 151. The spread of protection to the piles will be influenced by their size, distribution, current requirement and by the resistivity of the water and of the sea bed. A relatively even spread of protection can be obtained by distributing the anodes throughout the structure so that there is an anode in every inter pile space and these anodes are placed so that there is a good vertical distribution of current. With a deep water jetty this may be by suspension of more than a single anode or by placing alternate anodes at different levels. The cathodic protection is usually installed after the completion of the jetty and a simple configuration using zinc or aluminum anodes is shown in Fig 152. A variation of this technique has been used in which the anode is cast over a very long insert one end of which is driven into the mud and its upper end is bent over and welded to the pile. This has been used on light roadways and is shown in Fig 153.

With impressed current protection fewer large anodes are used. It is not possible to protect the majority of pile structures from outside because

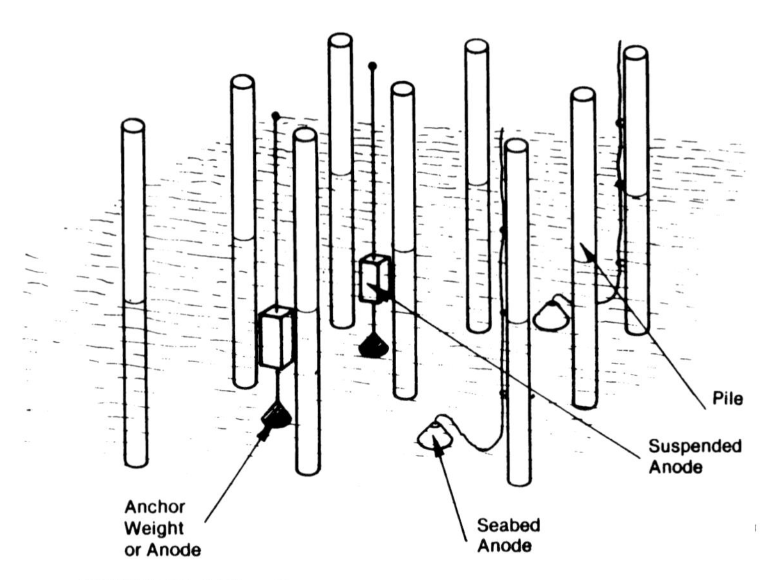

FIGURE 152 - Low voltage anodes protecting pile jetty.

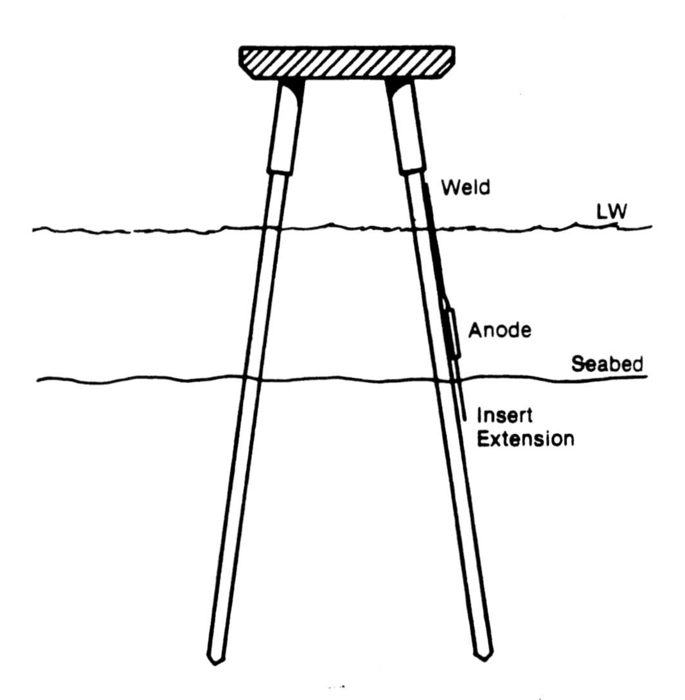

FIGURE 153 - Anode on trestle secured by seabed.

of the difficulty of spreading protection to the inner piles. In a typical jetty each pile will take 2 amps of protection current and so there may well be a 30 mV drop between two rows of piles if protected from outside. If the current has to flow to a third row of piles then this voltage drop is trebled and if the current flows to four such rows the voltage drop increases sixfold. This leads to excessive overprotection and a large increase in current which is uneconomical.

Because of the reduced cost of anodes and because they are now more rugged the tendency has been the place the anodes within the jetty and to protect the piles from within the structure. This has been encouraged by safety considerations, both to vessels that tie up alongside and to the vulnerability of the anode system when remotely located.

Anodes

Looking at a simple geometric arrangement as in Fig 151 an anode can be mounted to lie between four piles. This may be a suspended anode of lead/silver or a platinized titanium anode mounted off one of the piles. Two such anodes are shown in Fig 154. The protection is to the nearest four piles, and the next group of piles outside will be protected with more difficulty, probably each of these receiving half its current from the anode shown and half from adjacent anodes.

An alternative exploiting modern insulating materials is to mount the anode on a pile and insulate that pile in the vicinity of the anode so that it is not grossly overprotected. This is the basis of the anode shown in Figs 155 and 156 which will provide even protection on the same geometric pattern to the nearest nine piles or twice the number that are protected by the earlier arrangement. The system has the advantage that the anode is almost flush with the pile and is unlikely to be caught and damaged by ropes or by the passage of a small boat between the piles. Such an anode may have a capacity of 30 or 40 amps and provide protection to 15 to 20 piles. The anode itself can be of lead/silver alloy or platinized titanium, and both these have been used very successfully. The area behind the shield is protected, but not overprotected, by leaving a small gap between the anode shield and the pile for current to flow behind it. This system has been used successfully with many thousands of this type of anode and has superseded the other methods. A modified version fits H piles and both are shown in figure 157.

Both anodes may have a degree of flotation built into them so that they can readily be fitted onto the pile after completion of the jetty. This allows a diver to swim in with the anode and fix it very quickly. Arrangements with extended clamping mean that the pile can be pre-assembled at the surface and lowered so that no under-water work is required. The anodes can be fixed back-to-back to provide larger elements of protection.

The cable is carried up the pile in channel. This is much preferred to

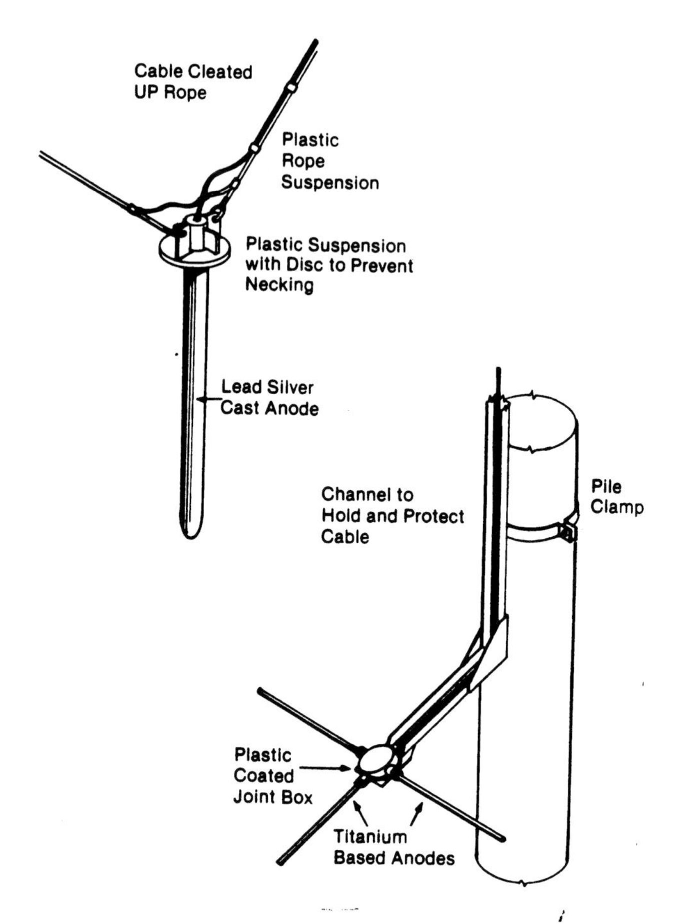

FIGURE 154 - Suspended lead silver and stand-off platinized titanium anodes for interpile locations.

pipe as the cable is equally well protected, indeed very often better protected as the channel is stronger, and cathodic protection extends to all surfaces of the channel, whereas water inside a pipe conduit or even condensation within it can cause its rapid failure. The anode cable is cleated inside the channel and two or more cables can be taken to the surface. It is usual to employ wire-armored cable and the wire armoring is anchored to the anode frame and so both are included in the cathode circuit. Although not recommended, if cable armoring is sufficiently secured it is difficult to damage the cable connector even if the anode is handled by it.

Cathode Leads

The cathode connection to the jetty is very important. A large, com-

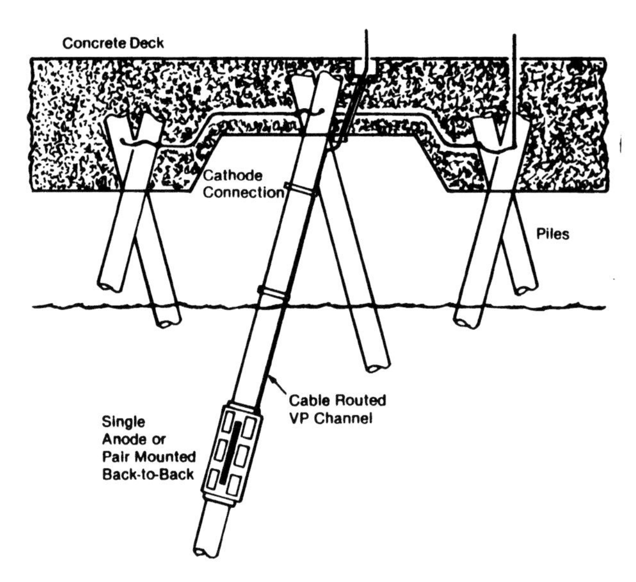

FIGURE 155 - Pile mounted Morganode and location on typical jetty.

FIGURE 156 - Morganode pile mounted anode.

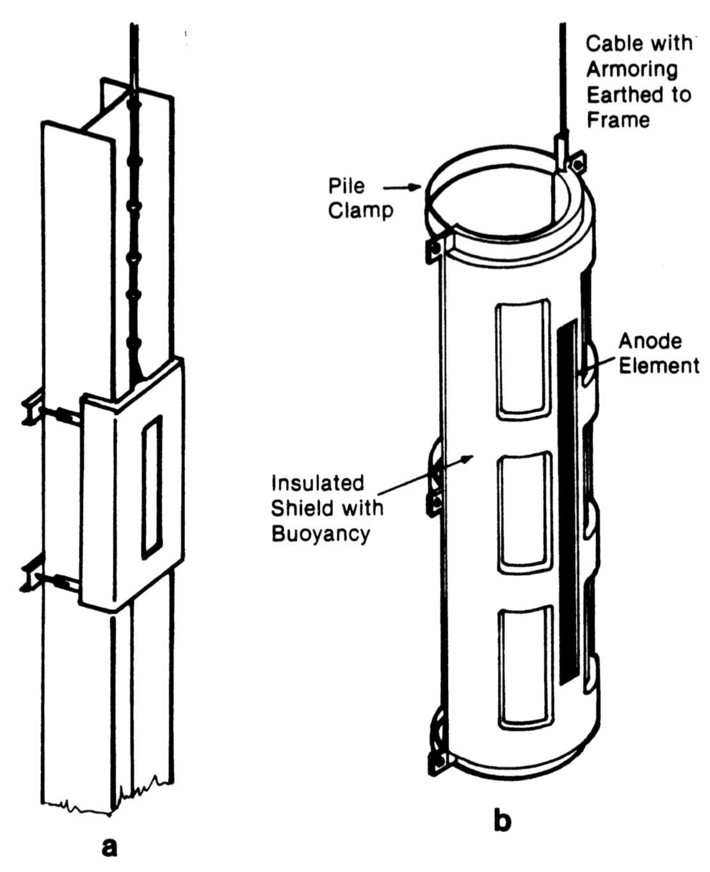

FIGURE 157 - H pile variation of Morganode.

plex jetty head may comprise a very heavy berthing head, sometimes with a separate sprung berthing beam, a roadway and pipe track, two inner mooring dolphins with walkways to them, and two outer dolphins which sometimes do not have walkways but increasingly are provided with them. The current that is given out by the anodes has to return to the transformer rectifiers through the jetty bonding system and the cathode leads. While a single transformer rectifier, say of 600 amp capacity, may be the cheapest technique of converting a c power into d c, an extensive network of cathode leads has to be used as the structure and its bonding will not be capable of carrying the current. If it is sufficiently massive to do so, the voltage drop within the structure itself will be of the order of hundreds of millivolts and this will be a variation on the cathode potential, that is the 300 mV swing, not on the overall driving voltage of the anode.

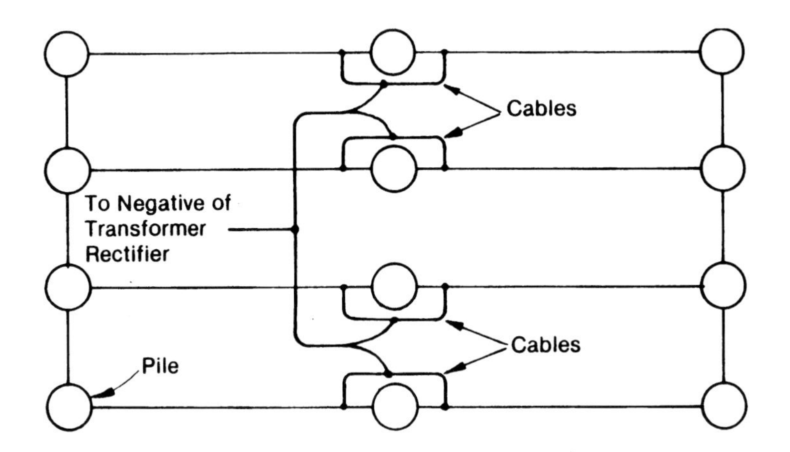

FIGURE 158 - Cathode lead distribution to reduce potential build-up in structure.

A distributed cathode lead system will be required in which the resistance or rather, the ohmic voltage drop of the cables are matched so that current can be picked off various points on the structure. This will be balanced so that each of the cathode drain points is at the same potential. There will be some degree of self-balancing in the system, rather as weight is distributed between railway carriage axles, Fig 158. However, on some of the more remote points on the structure, such as the distant dolphins, heavy cables may be required or resistive additions to the shorter cables may be necessary.

Electrical Engineering

With the advent of the larger pile-mounted anode it was found to be much more economical to install smaller transformer rectifiers each serving local anodes and each with a local single or split cathode connection. The anodes were rated at 20 amps per half shell and 40 amps per back-to-back shell so that power units that gave 40, 80 and 160 amps could be placed around the jetty to provide the protection. An a c distribution system is required but this is much cheaper than the d c system as it only has to carry a small current and volts drop in the line is not critical. The units can be balanced to give an even spread of protection throughout the cathodic structure and can be adjusted for small differences between the designed and the final structure; the cathodic protection arrangement usually being decided before the jetty was constructed.

Each transformer can be adjusted and the overall a c power can be controlled, either automatically or manually, as the jetty polarizes or as the

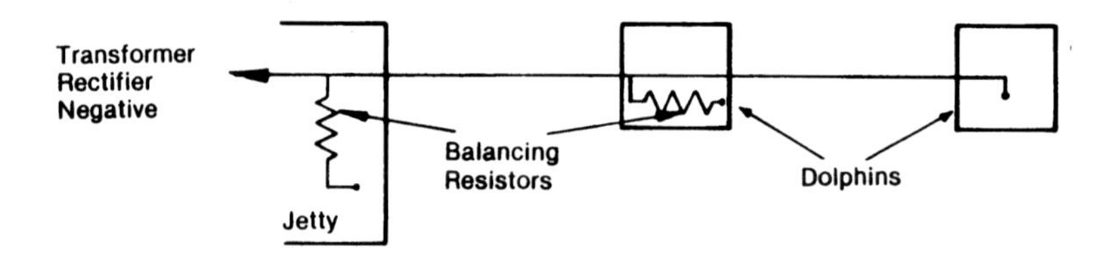

FIGURE 159 - Use of resistive connections to balance protection.

FIGURE 160 - Small distributed transformer rectifier with resin glass fiber lid on Jetty Approach Road. (Photo reprinted with permission from CORRINTEC/UK, Ltd., Winchester, Hants, England.)

current demand changes. This system eliminates the problems found with dolphins and with roadways. Fig. 159 shows two dolphins which are protected from a transformer rectifier at the end of the jetty head. Resistors can be placed so as to equate the cathode distribution or a single transformer rectifier could be placed on one of the dolphins with a much lighter cable to the outer dolphin and still achieve a better spread of protection. Similarly transformer rectifiers and anodes can be placed along the roadway, Fig 160.

Many jetties are used for fuel loading and unloading and on these

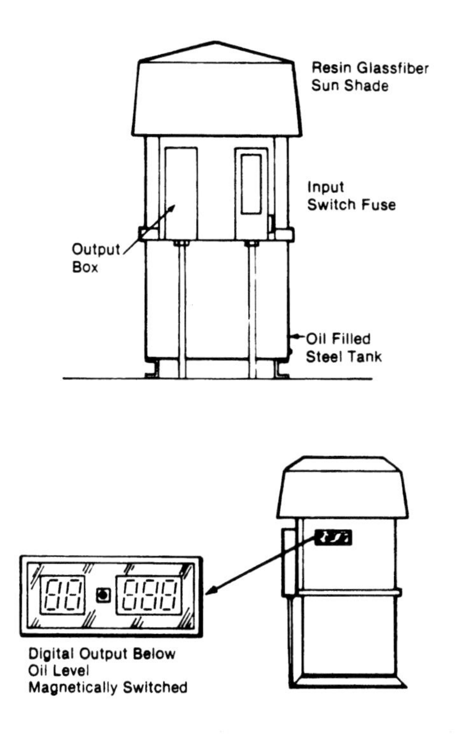

FIGURE 161 - Small multiple unit transformer rectifier.

there will be problems with the flameproof or explosionproof regulations. It is usual that part and only part of such a jetty is subject to the most stringent regulations and there is a hazardous area in which the equipment should be intrinsically safe rather than explosionproof. This involves placing the transformer rectifier and all the electrical connections beneath oil and making the connections into flameproof or explosionproof enclosures with an explosionproof switch. Such a transformer rectifier design is shown in Figs 161 and 162 and this has a steel tank which is well protected against the jetty environment, with a glassfiber lid which acts as a sun shield. Alternatively a clean-air purged enclosure can be used with conventional equipment.

On such a jetty the cables to the anodes and the cathodes leads must be wire armored and any joints in them made in explosionproof boxes which are usually filled with sealing compound on completion. The anode itself is under water and is considered to be intrinsically safe. The anode feed cable is wire armored, but because of the ease with which the anode may be handled by its cable during erection, the wire armoring has to be fully secured to the frame of the anode support. This also makes the armor part

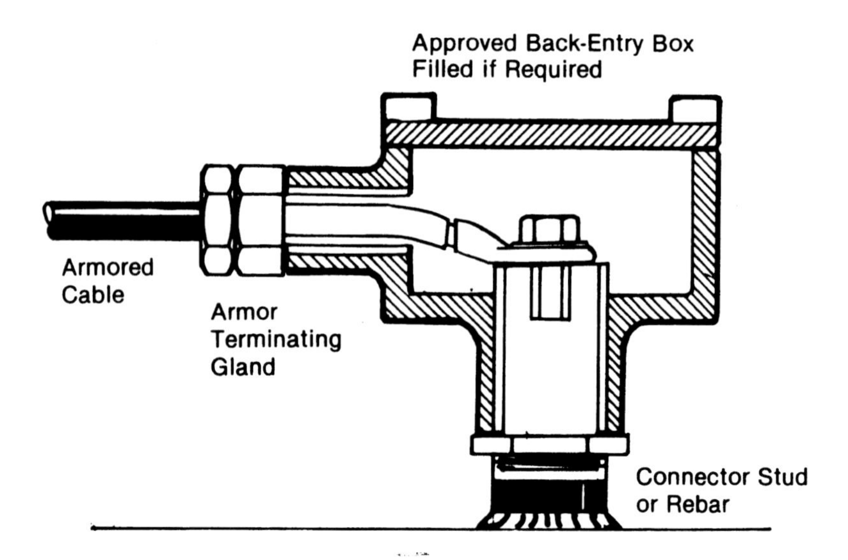

FIGURE 162 - Safe, flameproof or explosionproof cathode connection.

of the cathode circuit though it draws a negligible current as it will be p v c or other plastic, sheathed and also ensures that the anode framework, even if bolted on a well coated pile, is part of the cathode and receives protection as does the whole of the jetty.

The connection to the jetty itself must also be flameproof. The solution suggested by the author uses an explosionproof back entry box which is screwed onto a threaded stud welded to the structure, Fig 162. In a concrete deck the stud can be welded to an emergent rebar and located in a recess. The cables can be carried in ducting in a concrete deck or on trays or in conduit. Both can be built into the new structure and greatly aid in the cathodic protection installation.

It is necessary to check the potential of a structure such as a jetty on a regular basis, or to institute some method of automatic control, Fig 163. If the jetty is being surveyed with standard half cell equipment, then it is very useful to have prewired points to which the half cell, and the cathode lead, can be connected. Ideally in this prewiring jack points should be established at which half cells can be attached around the structure and the readings made at the transformer rectifier while this is being adjusted.

Permanent electrodes can be installed and these wired either to the individual transformer rectifiers, or if there is a master unit, then to that. Some simple variations make this even easier to operate by the introduction of mimic diagrams or by incorporating, say under the hinged lid of the transformer rectifier, a sealed plan of the jetty indicating the anode and electrode positions.

FIGURE 163 - Automatically controlled oil-immersed transformer-rectifier. (Photo reprinted with permission from CORRINTEC/UK, Ltd., Winchester, Hants, England.)

If automatic control is being used, either manual electrode selection or automatic selection of the lowest of the acceptable readings is made with the other potentials being displayed. The ability to select one electrode for control while displaying the others is useful. Where distributed transformer rectifiers are used it is possible to send controlled a c to the various units, or if there is ample a c electric power available, say at lighting circuit outlets, to send a d c signal to each unit which has its own slave power control within itself.

Bonding

Where vessels come alongside a jetty it is good practice electrically to bond the vessel to the jetty. A typical system including a flameproof switch suitable for a fuel jetty is shown in Fig. 164. This is to ensure that there is a very low difference in potential between the jetty and the ship though it is not the intention of such a bond that it should provide the ship with cathodic protection. This is sometimes called for and if such a system is to work it involves exceedingly large currents being drawn through this bond. For example, a 300,000 tdw tanker that is inadequately protected while at sea would, if it were to be protected on arrival at a jetty, need a current of a

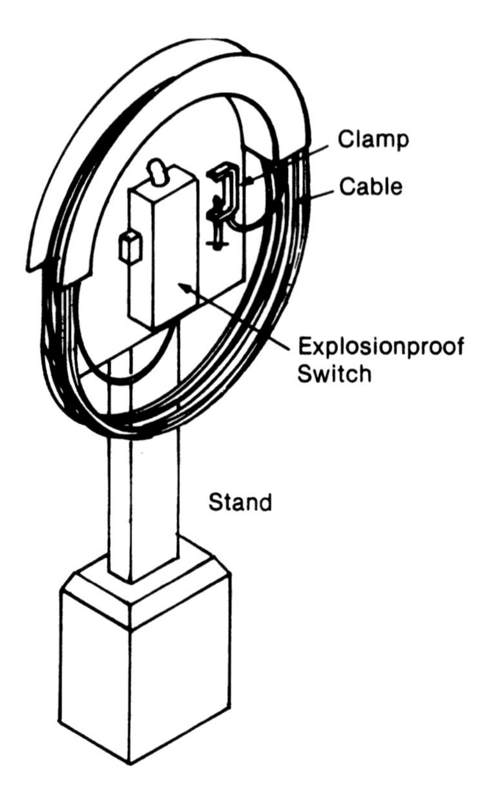

FIGURE 164 - Jetty to ship bonding system with safe switch.

thousand amps flowing through the bond cable. As the bond cable will not be the only connection, a break in any other chance connection would form an immense spark by the inductive effect of the current flowing. The bond connection may, however, carry a large current for other reasons and hence the need for a flameproof isolating switch which prevents sparking either on connection or disconnection of the vessel through the bond.

There will be a potential build-up in the water around each pile and around the jetty itself. At a current density of 4 mA per sq ft, the gradient immediately next to the pile will be 4 mV per ft. If current is flowing from a remote anode or, by some imbalance in the system, from the anodes in one area to the piles in another, then there will be a general potential disturbance in the water. This can be detected by surveying with a half cell from a small boat. Where the potential difference across the area where a ship will berth is more than 50 mV then the system should be adjusted to reduce this level or in the presence of a vessel the system switched down or off until it can be redesigned. A voltage variation of 50 mV in the absence of a vessel may be increased to more than 100 mV in the presence of a large ship which will effectively block most of the electrolytic path. The bonding of the ship to the jetty can be made much easier by the provision of permanent bonding cables.

The current density required for the protection of a jetty will depend to a very large extent on the coating applied to the pile. Bare steel being protected to -0.8 V to silver chloride, (+250 mV to zinc) will take 5 mA per sq ft. Piles that are site coated before driving will reduce this current density to an average 2 mA per sq ft. The designer has to allow for some coating damage beyond that caused in driving and construction and this may include scouring at the foot of the piles and damage to the coating by usage which may occasionally include spills. Where the water is stratified or where it flows rapidly there may be an inefficiency in current distribution or an increase in erosion which would leak to higher current densities. Particularly there are areas with high levels of suspended solids in the water, for example, the Cook Inlet in Alaska and the Severn and Humber Estuaries in the U K, where scour of the overall coating can be a problem.

Interconnection

Very often a jetty or wharf will be interconnected with services; on an oil dock by pipes, or on a loading jetty by a conveyor belt and by electrical machinery which will be earthed both to the jetty and through its electricity supply cables to electrical grounds or earths on shore. Fuel pipes and water pipes may well add to these interconnections.

It was comparatively easy, and certainly good practice, to isolate with insulating couplings the pipes that travel onto the jetty. With electrical equipment it is somewhat more difficult, but it is possible to avoid very heavy current drainage through the electrical earthing system. Where the interconnection is by cable the neutral of the cable system, usually the concentric wire armoring, will not be capable of carrying very large currents at the voltages which will arise from cathodic voltage variations. However, it can be wasteful of current and it can tend to polarize the earth network. It is possible that there will be some form of lightning arresters and care has to be taken that these are not made of copper in the submerged area or, if they are, that the engineer is aware of it. Cathodic protection isolation without affecting the efficacy of the a c earthing system can be maintained by using a polarization cell of nickel in caustic potash.

There is a particular problem with some jetty structures where piles contain transverse welds. The pile may appear to be perfect and adequately protected on cursory inspection but unless that survey includes the transverse welds the pile may fail catastrophically. There have been one or two such incidents on tubular and H-piled jetties where seemingly cathodically protected they have corroded by this mechanism.

Many of the piles will extend into the sea bed where it is impregnated with sulfate reducing bacteria. Under these conditions it is necessary to depress the potential of the steel by a further 100 mV. Generally this is accomplished, certainly on the inshore end, by lowering the anodes to be closer to the sea bed and causing overprotection in that area. The spread of protection through the mud will be enhanced by its lower current density requirement; usually less than half the requirement in the sea water area. Piles receive adequate protection when the pile at sea bed level is subject to the potentials associated with anaerobic bacteria. There have not been sufficient failures while employing this technique to warrant the excavation or removal of a significant number of piles to determine a better criterion. Two typical cathodic protection schemes for jetties are shown in Figs. 165 and 166.

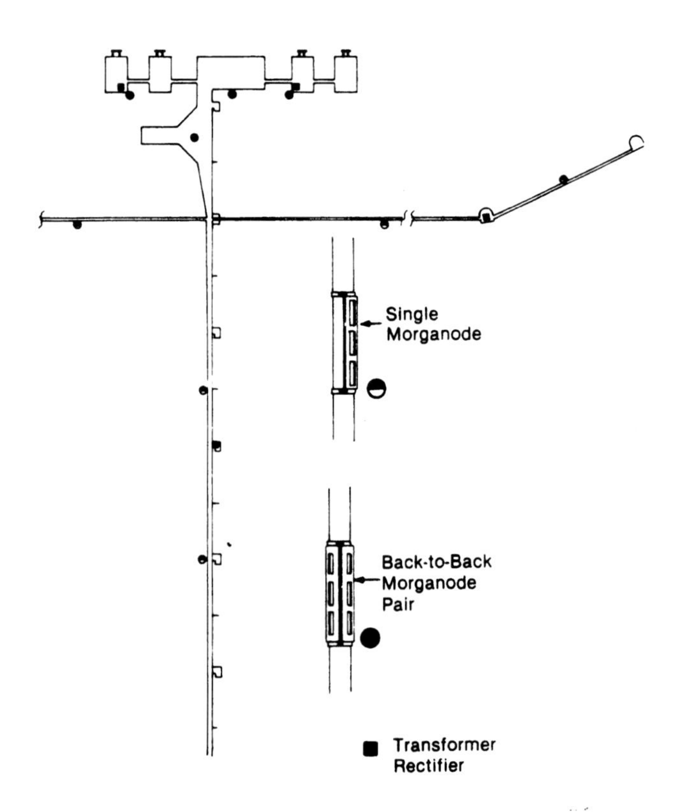

FIGURE 165 - Typical jetty layout using Morganodes.

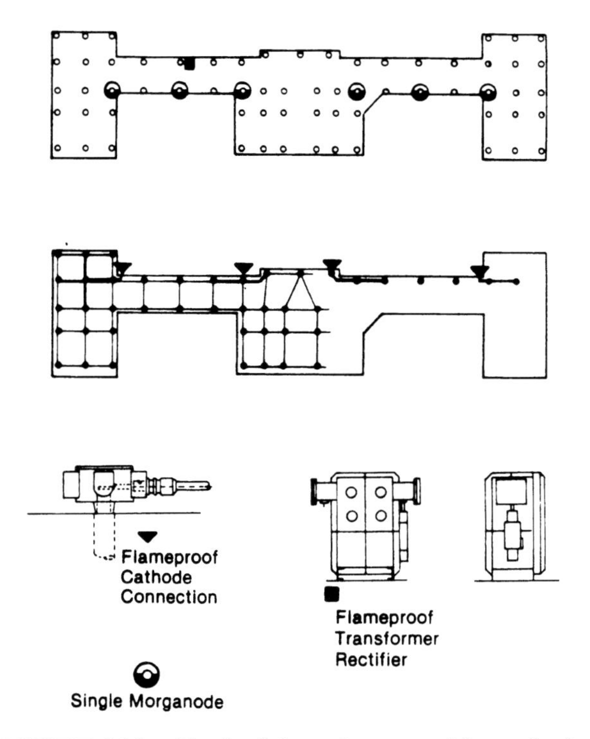

FIGURE 166 - Typical jetty layout with explosion proof equipment.

Cathodic Protection of Static Floating Structures

A whole series of static marine structures can be classified together by virtue of the fact that they float, the objects so far considered having rested on the sea bed or been built into it. The classification includes pontoons, floating docks, laid-up ships, ships fitting out, light ships, buoys, chains, nets and booms. They vary in size from a small buoy to a large floating dock or laid-up cargo or capital ship. The condition of their metal hull may equally vary from a well painted lightship to a badly rusted floating dock.

Despite these differences there are generally three ways of protecting them: close sacrificial anodes may be used, that is anodes mounted directly onto the body of the structure; sacrificial anodes may be remotely positioned, either suspended or laid on the sea bed; impressed current may be employed as the third method. Each structure will have its pecularities, many of which will be obvious, and the modification to the design equally apparent. For this reason the types of protection will be considered together with the methods employed for the particular structures.

Many features will influence the choice of the system. Where no power is available the choice of sacrificial anodes will be obvious, and where power is expensive a similar choice might be made. In protecting very small structures the capital cost of the d c generator might make sacrificial anodes most acceptable, while the lack of dry dock facilities or the cost of them may preclude the attachment of close anodes to large floating structures. The control of a great number of sacrificial anodes may be tedious and their uncontrolled consumption expensive, so that where electric power is available and reasonably cheap, the impressed current system will be preferred. The application of sacrificial anodes will be greater in number, the larger total current will probably be derived from the impressed current systems.

Directly Mounted Sacrificial Anodes

Small objects, such as buoys, will be best protected by the direct attachment of a small anode. The cheapness of zinc or aluminum and the added advantage of a low driving potential on a small structure will mean that these metals will generally be used. If the buoy is anchored by a chain or wire rope this will receive protection from the same anode. It is suggested that chain links must be electrically bonded together but accumulated experience has shown that the movement of the links which are below the water allows protection current to flow over a reasonable distance. Over long distances this type of protection might benefit from a higher driving voltage at the anode, say by the use of magnesium, or by the attachment of extra anodes along the length of the chain. This principle has been successfully extended to chain link nets of the type used for harbor defense and for similar nets and booms used to filter intake water, Fig 167.

Lightships can be protected by close anodes and these can be used to preserve the hull between refits and during the journey from dry dock to lightship station. The use of long life anodes is necessary though these could be augmented by anodes slung over the side as with laid-up ships. The mooring chains and anchors can be protected by anodes placed on special chain links built into the mooring chains.

The design of anodes for direct attachment to the structure is relatively simple. Streamlining is not important. The anode is stood 3 to 6 in. off the metal and the mounting may be by the steel strap used for the insert. The shape and weight of the anode will depend upon the life expected and the current output required. Plates of zincs 12 in. × 4 in. × 2 in. may be used to give a small current output, these weighing approximately 24 lbs. Larger anodes may be used and the design of these is a matter of selection by the engineer from the standard anode designs and the data given in Chapters 3 and 4. A smaller number of larger magnesium anodes than of zinc or aluminum can be used to protect large structures. Use of magnesium anodes will probably be attractive where there is a short life required for the protection. A typical anode and an illustration of an anode mounted on a buoy are shown in Fig 168.

Remote Sacrificial Anodes

The installation of anodes remote from the structure is recommended

FIGURE 167 - Zinc anodes attached to mooring chain holes. (Photo reprinted with permission from Federated Metals Corp., Somerville, NJ.)

FIGURE 168 - Buoy-mounted sacrificial anode.

where it is not easily docked. This may be either because the docking facilities are not readily available or that the cost of docking is to be avoided. The anodes can be placed on the sea bed or they can be suspended from the structure itself. A typical example of this is the protection of a laid-up ship.

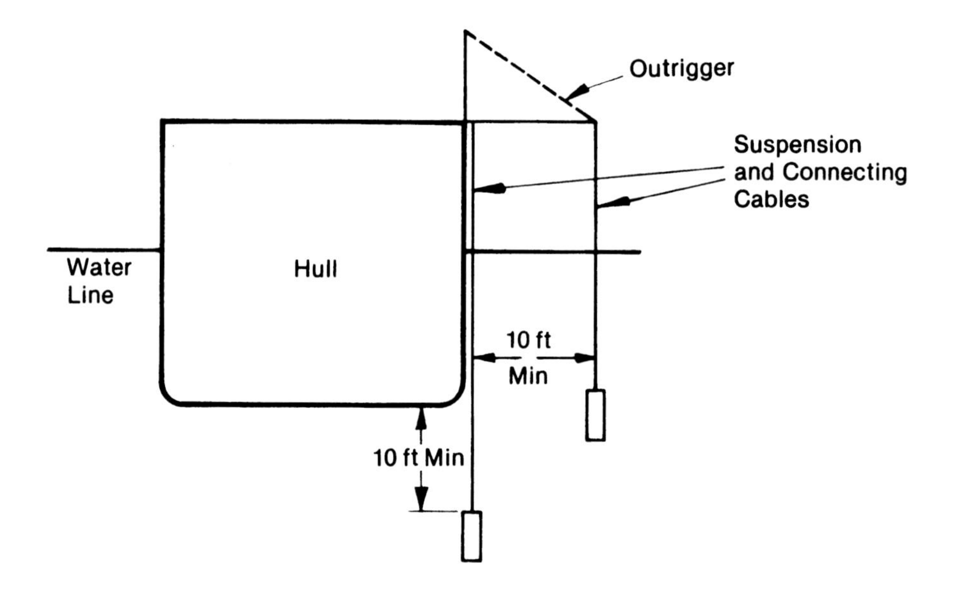

FIGURE 169 - Protection of laid-up ships.

In this, as shown in Fig 169, anodes are suspended either from the side of the ship so that they lie 10 ft or more below the ship, or by cables which are placed on an out rigger and achieve the same separation. If the ship is laid up for a short period, as during fitting out, magnesium anodes will be the cheapest as a smaller number of anodes can be used. For long term laying up anodes of aluminum or zinc would be used.

This system can also be used for small pleasure craft which may be laid up for periods during the winter or even between use by a weekend sailor. Here anodes of three to four pounds of zinc would suffice and they can be stored onboard while sailing.

Pontoons for jetties, floating docks and similar structures can be cathodically protected by this means. Fig. 170 shows cathodic protection of a small floating landing stage which could be achieved from a group of anodes placed on the sea bed. This leaves the structure clear of such encumbrances that may be caused by suspended anodes, and also means that there is no costly attachment. Where there is a flexible joint bonding across the bearing is necessary. Large floating dry docks are susceptible to the same type of protection and it will not be necessary to protect them while they are in the lowered position as this will only for the period when ships are moving in and out.

Impressed Current Protection

In the geometry of the remote sacrificial anode system the anode could

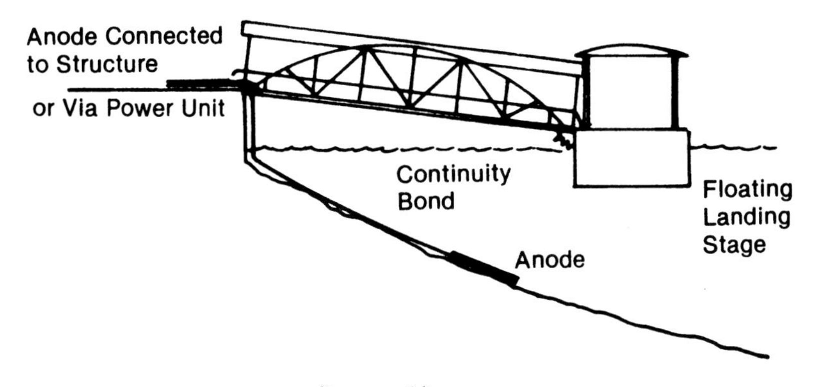

FIGURE 170 - Protection of pontoon jetty with remote anode.

be replaced with an impressed current system. Generally each impressed current anode will give out more current; in the arrangement in Fig 170 the anode could be an impressed current anode with a transformer rectifier at the connecting bridge. Such systems have been employed on pontoons used for loading ferries. The system used for laid-up ships could be similar and here considerable use is made of scrap steel anodes but the engineering of these is expensive, it being necessary to ensure that the steel was not preferentially consumed with early failure. A lead alloy anode for suspension was drawn in Fig 154 and an anode that could be used on the sea bed was shown in Fig 143. In the latter the anode is one designed originally for a ship's hull, and mounted on a sled. The sled is made of glass reinforced polyester with a steel framework to which the anode cable gland is attached and the wire armoring of the cable is connected.

Similar anodes could be constructed using platinized or metal oxide coated titanium. In both cases the anode would be of similar dimensions with a somewhat higher driving voltage for the coated titanium because of its increased back e m f and because it would be smaller in section, though of equivalent length. With a suspended platinized titanium anode it would be prudent to place some shield around the anode as lacking the weight of the lead alloy it will tend to move if its support cable is caught by the wind.

There is one big difference in the two techniques. In the case of the suspended impressed current anode the transformer rectifier is mounted onboard with the minimum of cabling. With the remote anode that is on the sea bed the cables have to be taken to it and this involves an area of wind and water line, say on a pile, or by carrying it through the shore line. The cost of the cable and the possibility of damage make this a less attractive proposition where other techniques can be used.

A number of ships can be protected when they are laid up from a single impressed current system which could also be used, for example, to protect the steelwork in an area such as a marina. With oxide coated and platinized titanium anodes hypochlorite is generated very efficiently and this could be used to act as a deterrent to the settling of fouling.

Some isolated structures can be protected by impressed current using generators from sunlight or from wind. In general the cost of sacrificial anodes is so low that these sources are not very attractive, added to which, with the wind generator there is always the possibility of a maintenance problem, and with the solar generator there must be some storage system to provide the current during hours of darkness.

The advent of wave power and some of the devices that are currently proposed for this will generate electricity. Some of this electric power could be used to provide protection either generally or to the devices themselves. One of these large devices manufactured from bare steel would, in fact, consume an appreciable percentage of its output in order to provide its own cathodic protection.

Protection in Fresh Water

Within fresh water the rate of corrosion generally will be lower; the exception to this is where there is some form of stratification and a differential salt concentration as may be found in a river mouth. The water will have a much higher resistivity and this will affect the engineering, both in the distribution of the cathodic protection current and in that some anodes will not be suitable. There will, to compensate this, be a lower current density required and in many cases, because of the lower rate of corrosion, complete protection to −800 mV to silver chloride or −850 mV to copper sulfate may not be necessary. Higher driving voltage anodes can be used for the impressed current system and operate at voltages normally associated with the cathodic protection of buried structures.

Cold water and water that has ice in it will present very similar problems. There will be an additional problem where the ice is formed massively, either where it is crushed under the jetty deck by ships coming alongside or where it becomes suspended from anodes or anode cables. Where ice damage is a problem consumable anodes, often or scrap steel, can be used.

Current densities can be greatly reduced by coating the steelwork and very often within the lack of tidal variation in many fresh water situations there will be less damage to the structure by suspended solids in tidal flow and by flotsam.

Chapter 7B

Ships' Hulls

The current density required for cathodic protection depends to a very large extent on the polarization of the metal surface where the electrolyte, as in the case of sea water, is of low resistivity. Paint or other coatings on the metal will give the interface a large resistance and the cathodic protection current will polarize breaks and imperfections in the coating.

The main depolarizing agent in sea water is oxygen and this is usually present to saturation. The amount of depolarization and the rate at which it occurs, and so the current density required for cathodic protection, will depend upon the rate of arrival of oxygen at the cathode. The movement of the steel relative to the sea water will be the principal influence upon this. Experiments at different velocities, mainly carried out under laboratory conditions in sea water pipes and on discs, indicate that there is a very large change in current density with velocity. This is not borne out in practice when a ship underway will require approximately twice the current that it requires when stationary.

Many of the experiments are carried out on small samples whether in pipes or on rotating discs and the velocity gradient off the surface is very large. For example, a velocity of 2 knots in a half inch diameter tube has a much higher velocity gradient than a ship's hull traveling through the water at 20 knots. There have been a number of determinations of the relationship between velocity and the current required by a ship's hull. These do not all agree and there is a great deal of difficulty in carrying out such an exercise on a moving ship as the distribution of current will change with speed for if anodes are placed on the hull of the ship an increase in current is accompanied by more overprotection and the efficiency of the distribution decreases. A mean of the results would be something like the curve in Fig. 171. It has been suggested that the curve should be flatter at high velocities while other workers suggest a second increase at higher velocities.

There are two approaches to the protection of a hull. Firstly, cathodic protection can stop corrosion so that the hull plates are not thinned, the welds not consumed and pitting is avoided. This will reduce the maintenance on the ship from these causes. Alternatively, and very much more importantly, cathodic protection can be used to maintain the hull's smoothness. The smoothness is a property of the initial paint film and the cathodic protection retains it by preventing the corrosion of the hull metal. In a normal sample of scale from an unprotected ship there is a sandwich of layers of paint and rust. Clearly the corrosion has taken place by some transport mechanism through the lower paint film and the rust has built up outside it.

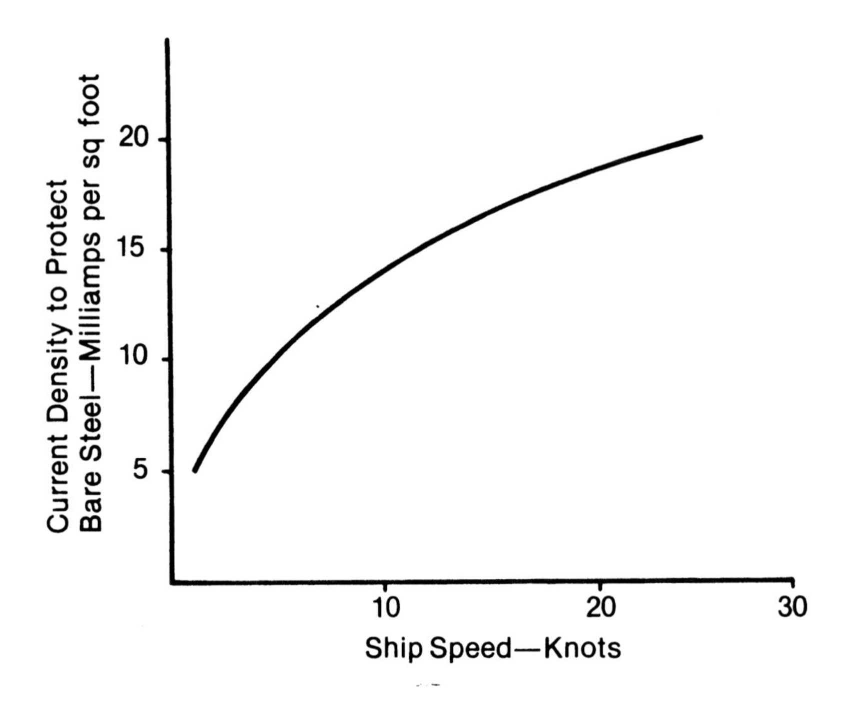

FIGURE 171 - Change in current density required for protection with speed.

This has led to a roughening of the hull, both of the actual metal surface and much more by the build-up of the corrosion-paint matrix. If such a hull is grit blasted the metal will appear pitted, generally with shallow craters, and on repainting there is a marked improvement in the performance of the ship. This improvement can be enhanced by the use of thick paint coatings and particularly those with self-polishing characteristics.

On most ships the two factors causing drag will be the shape drag and the surface drag. The shape or form drag will be a factor which relates the speed of the vessel to the ship's length while the surface drag will relate the drag to the under water surface area of the vessel. On a large tanker the major component will be the frictional drag of the hull if it is roughened by corrosion. The present cost of fuel indicates that to operate a 100,000 tdw tanker throughout its economic life with a smooth hull would save the owner $10m. This smooth hull, achieved by the initial application of high quality paint, can be maintained by the careful engineering of the cathodic protection. Indeed, as a spinoff from this protection the paint itself will have a much longer life and it is realistic to apply a top quality marine coating to the hull and to replace it not more than once during the 20 year life of the ship. Some touching up of damage may be necessary in dry dock and, of course, anti-fouling paint will need to be applied over it. This may be without dry docking the ship though underwater inspection would be necessary. The cost of avoiding a day's dry docking, both in fees and in delay, gives enormous savings to the ship owner.

The total current required to protect a ship will not only vary with its velocity but with the condition of its paint and its integrity. For example, on a tanker using a constant tension mooring system the ship's side will move vertically in relation to the berthing fenders. This will damage the paint and the vessel will have a large bare area at the center of the ship. Equally, a cruise liner may, in order to keep its schedules, enter a harbor where there is a minimum of water and the scour effect of its movement over and close to the sea bed will take large areas of paint off the bottom of the hull. Many merchant vessels will change their draft quite dramatically between being ladened and empty, although this is reduced by ballasting particularly with the advantage conferred on this type of operation by bulbous bows. A ship may also trade in different waters varying from the Red Sea to the St. Lawrence sea way. The design of cathodic protection to maintain the smoothness of the hull needs considerable engineering exellence.

Ships Over 2,000 tdw

This discussion will be concerned with ships which are generally above 2,000 tdw. Smaller ships will be considered separately.

Sacrificial Anodes

The use of sacrificial anodes on ships' hulls is the earliest form of cathodic protection. The technique was revived when magnesium anodes were attached to the hulls of ships in an effort to reduce the rate of corrosion. The anodes were successful though the limitations of magnesium became apparent. Firstly, it could not be installed directly but had to be controlled by some means. Secondly, the area immediately next to the anode was overprotected and either the anode output had to be restricted by encapsulating part of the anode or an area around the anode or a group of anodes had to be shielded with either a mat or an area of special paint. Thirdly, it was expensive. A number of workers set about developing a non-polarizing zinc anode or a more efficient aluminum anode. The reason for this was that the lower driving voltage of these two anode materials would not cause breakdown of the paint in the immediate vicinity of the anode, and that having a low driving voltage the anode would respond and give a degree of automatic regulation. When the hull demands a larger current because it was insufficiently polarized the anode would possess a larger driving voltage relative to the unprotected hull.

The development of anodes for the protection of ships' hulls proceeded along two different lines. The majority of the anodes were made for direct

FIGURE 172 - Streamlined zinc hull welded lug anode. (Impalloy, Ltd., Bloxwich, Walsall, England.)

welding to the hull having inserts cast into them that protruded and were attached to the hull, Fig 172. The anode corrodes unevenly, being attacked immediately next to the insert, and this is overcome either by coating the insert with a ceramic or by careful painting of the zinc-coated steel insert.

The alternative technique is to mount the anode over studs which had been welded either directly to the hull or onto a small doubler plate: this does not ensure a good electrical connection to the hull and a small welding tail is needed. The stud hole is filled and with some anodes by driving a close fitting short cylinder of anode material into the hole. The anodes are designed to protect a specific area of hull and the larger the hull the larger the number of anodes fitted, Fig 173. The specific area that can be protected by a single anode depends on the surface coating on the hull and on a modern vessel full protection would be achieved at considerably less than 2 mA per sq ft. An anode with an output of 1 amp could, therefore, be expected to protect an area of upwards of 1,000 sq ft. The stern area of the ship, because of the turbulence caused by the propeller, requires a higher current density so anodes tend to be grouped in this area. Because the anode protrudes they are usually arranged to lie close to the shaft line on a single screw vessel or beyond the blade tip area. Additional anodes are

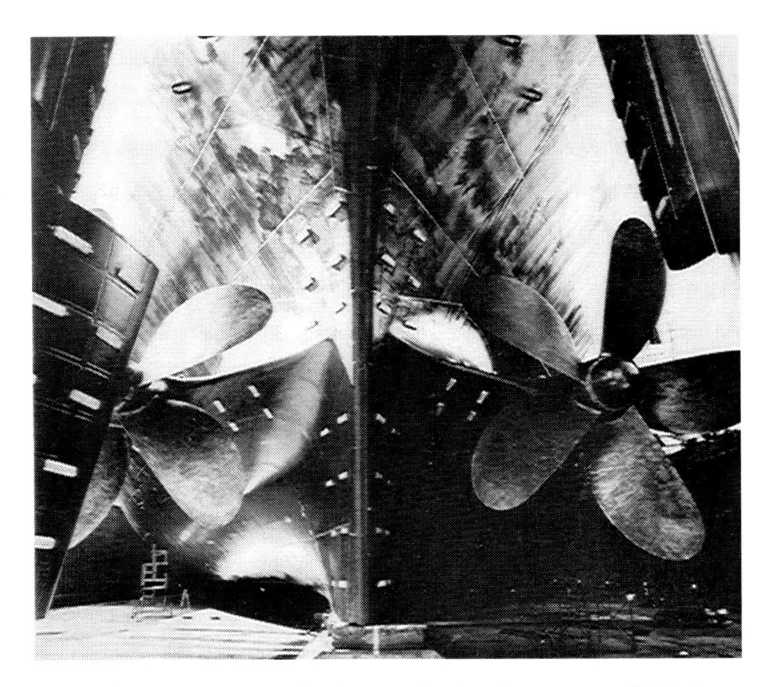

FIGURE 173 - Plethora of aluminum sacrificial anodes on stern of 300,000 TDW tanker. (Photo reprinted with permission from Federated Metals Corp., Somerville, NJ.)

placed on the skeg and on the rudder to provide protection independently to them.

Along the main center section of the hull the anodes are placed at bilge keel level and often on the bilge keel itself. Fig 174 shows a typical group of zinc or aluminum anodes and their positioning typically on the stern of a vessel. The number of anodes and their distribution around the ship have been calculated from the experience of the cathodic protection companies and there are standard arrangements for most of the proprietary anodes. Anodes cannot be placed in the forward part of the ship as they will be damaged by the anchor cable.

Anodes are usually fitted so that they will provide protection during the period between dry docking. This is usually two years and the anodes are replaced on that cycle. Some systems replace half the anodes at each docking so that there is a fuller consumption of the anodes. The cost of anode replacement is considerable as the anodes are replaced whether they are completely consumed or not. One of the major costs in replacing anodes is the staging that is required and anodes are arranged so that the majority of them can be replaced from a single level of staging which is moved along the ship. The time taken to replace the anodes may extend the docking period which is a considerable financial penalty.

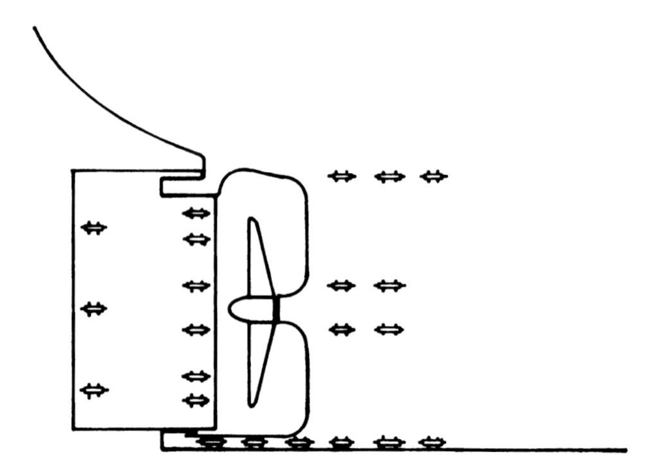

FIGURE 174 - Sacrificial anodes of zinc or aluminum on ship's hull.

It is difficult to predict the current that will be required during the two year period the ship is operating. To provide sufficient anodes to meet any degree of coating damage or deterioration may mean that the hull will be overprotected while a more cautious approach could result in loss of protection. The operations of the ship are equally difficult to predict, the wide variation in current demand means that even with low voltage anodes there will be considerable variations in the potential of the ship's hull. Where the hull coating is well preserved and suffers no damage the ship may always be protected but this will only be the case if the anode resistance is very low and the ship, for most of its life, is being overprotected. The anodes themselves will then form sufficient protruberances on the hull to introduce a measurable drag on the ship. It is suggested that this drag is negligible but the massive attachment required for the anodes on the hull belies this.

Sacrificial anodes can, therefore, be used to provide sufficient cathodic protection so that the welds are not consumed, the plates are not corroded, and even when much of the paint is removed, there will be only superficial attack in those areas. However, unless the owner is particularly fortunate it will not provide sufficient protection to preserve the smoothness of the hull paint.

Impressed Current Protection

The inadequacy of sacrificial anodes led to the development of systems of impressed current protection. These systems were developed contemporary with the major advances in ships' painting techniques and to a large extent the most successful systems were developed to work with the better paints, perhaps adding perfection to them.

The initial experiments with impressed current were carried out in the 1880's by Edison, whose beaker experiments had convinced him that a ship's hull made of iron or steel would corrode so rapidly that it would be uneconomic. He installed a generator in a ship and trailed from outriggers two anodes made of carbon. His engineering techniques were not well developed and the system, while providing protection, failed to last for any sensible time.

In the 1950's the system was redeveloped in two ways. Firstly, some streamlined anodes of platinum coated on silver were towed behind ships by the anode cable. There were a large number of teething troubles with this system, though it did illustrate that reasonable cathodic protection could be achieved from an anode trailing behind the ship. A more elegant solution was proposed by Rotterdam Lloyd Shipping Line, who trailed an aluminum wire behind their ships. This acted both as conductor and anode; as it was consumed more wire was wound out. The system provided good protection to the hull but it did not provide even protection (the bow and stern being overprotected) as had been expected. The anodes were trailed about 200 feet behind the ship and the distribution was not improved by extending this distance. If the anode was trailed closer to the ship then greater overprotection occurred at the stern. The anode position could be optimized but this would need some detailed supervision by the ship's crew.

The permanent anode was made of platinum and was about 4 ft long and 1/4 in. diameter. The main body of the metal rod was silver and was covered by 90 per cent platinum 10 per cent palladium alloy five thousandths of an inch thick. The silver core was used as it was felt that in the event of damage to the platinum coating silver chloride would form. This preceded the development of the platinized titanium anode which could be used now in its place with considerably more success. The anode was joined to a flexible cable, the joint streamlined and encapsulated in a hypochlorite resistant sheath. Because the velocity of the water past the anode when it was being towed, it was considered that the cable itself did not require this type of sheathing. Resistance of the anode was about 0.20 ohm and the anode trailed well behind the ship up to velocities of about 10 knots. The anode was easy to handle, there was little drag in the water, and its maximum output was 20 amps. The silver base was the limiting factor and the current could be increased considerably if the platinum was plated onto an anodizing metal. The anode failed, principally at the joint, where the rigid anode and flexible cable meet and on a trans-Atlantic voyage two or three anodes may be used.

The consumable anode was 1/4 in. diameter pure aluminum wire. It was uncoiled from a drum which was insulated from the hull and connected to the positive pole of a generator. The installations were made on vessels of

Liberty ship size and currents of 50 to 100 amps were required for protection. The wire was fed from the stern rail of the vessel and about 200 to 250 ft of the wire was run out, the last 30 to 40 ft of which trailed in the water. It had a low breaking strength and an eye fitted with a sharp V-notch was used to cut the wire should it foul the propeller. The part of the wire in the water has a low resistance through the electrolyte compared with the electrical resistance of its metal so that the total resistance of the anode in the water reaches a minimum value irrespective of the length submerged. This meant that whatever amount of anod was wound out it was always consumed within about 12 hours. One of the main problems with the system was that in even a moderate sea the amount of anode that is airborne varies and this changes the circuit resistance. The transformer rectifier or motor generator gave a varying output and it was difficult to control or to measure the cathodic protection potential and current.

Neither the wire nor the trailing anode could be used when the ship was in harbor without considerable rigging. The problems with both led to their demise but they had already shown what could be achieved by the use of properly controlled impressed current protection.

Hull Mounted Anodes

The operational disadvantages of the trailing anode are overcome by mounting the anodes on the ship's hull. Anodes must be designed to fit closely to the hull or to be flush with it. An anode a few sq in. in section will provide no measurable drag on the hull. The anode has to be insulated from the hull metal and be fed with current from inside the hull. Because of its close proximity to the hull metal the area immediately next to the anode has to be insulated otherwise an electrolyte short circuit would occur. This area is usually referred to as a shield and the area of the shield can be calculated for different anode outputs and shapes.

If the hull paint in the vicinity of the anode is capable of standing a potential of -1.8 V to silver chloride, that is a 1 V of overprotection, then the area of the shield can be calculated. There are two principal types of anodes: those that are linear and those that are point. A point anode lying on a flat hull will have equipotential shells around it that are hemispherical. The potential of any one of these relative to a remote point will be given by the formula

$$V = \rho \frac{I}{2\pi r}$$

assuming the hull to be planar; where r is the radius of the hemisphere in cm, ρ the resistivity in ohm cms of the sea water, and I the current. At low currents of, say, 5 amps, in sea water of 25 ohm cm resistivity an equipotential of 1 volt above a remote electrode will be eight inches, that is 20 cm, from the anode center. This will give a shield area of 1.4 sq ft, whereas if the anode output is doubled to 10 amps the area increases to 5.6 sq ft, that is as the square of the radius. The resistance of a disc on a semi infinite insulator is

$$R = \frac{\rho}{4r}$$

A line anode will have ellipsoidal equipotentials around it if it has a uniform output per unit length. The equipotential can be calculated in the same manner though now the mean of the two minor and the major radii of the semi ellipsoid will replace the radius of the shield in the point anode calculation. The output of the line anode can be increased by making it longer and this will increase the shield area almost in relation to the increase in the anode output, that is, the length by of the anode rather than the square of it. This can be seen by considering the original small anode of 5 amp output and placing a series of these side by side to achieve a larger output rather than increasing output from the single point. The line anode has advantages in respect of shielding when it is large and the changeover is at about 10 amps in sea water.

History

With this in mind some of the early work was concentrated on very long low output anodes by the British Navy, while the United States Navy employed an almost indestructable platinum point anode. The Royal Navy used steel anodes which were consumable and in simple wooden mountings proved the effectiveness of their approach.

In place of platinum a number of other anode materials were tested, including graphite and silicon iron. The graphite was manufactured in rectangular blocks and mounted in epoxy glass fiber holders which were carefully made to cushion the anode. As a variation of the point anode, shaft rod anodes of platinum were mounted off the hull in plastic and ceramic coated steel holders with insulating bushes. Fig 175 shows two of the graphite anode holders and the platinum rod assembly. The hull was shielded in the case of the permanent anodes by a neoprene sheet which was stuck onto the hull, while the steel anode was of such low current output that its holder provided sufficient shielding.

The Admiralty began to investigate lead alloys as anode material for use in sea water with this particular application in mind. Contemporary work by the Royal Canadian Navy used 2 per cent silver while the British Navy tended to 1 per cent. The anodes were made as replacements for the more conventional anodes in the form of flat sheets which were cast so that they could be held by machined resin/cloth holders. A solid single core wire

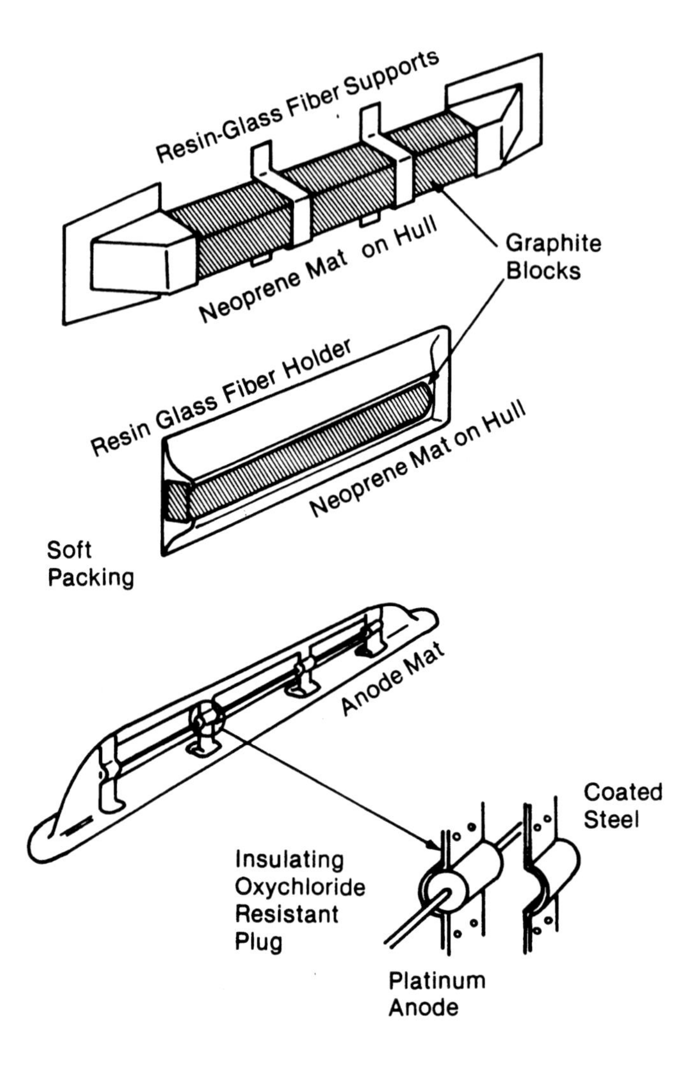

FIGURE 175 - Early hull impressed current anodes.

connection was taken through glanding at the back of the anode into the hull.

The author experimented with lead 1% silver 6% antimony and a longer anode using an extrusion of lead. Initially this was conceived as mounted in a plastic extrusion over studs onto the ship's side. The technique was abandoned in favor of a shaped extrusion of 2% silver lead which could be attached to a galvanized steel plate which was then fixed to the hull. The anode and insulators were in the form of a trapezoid approximately three inches wide and two inches high, the lead emitting surface forming the majority of the angled sides. This was attached to the galvanized steel plate over a neoprene sheet which was cemented to the plate under ideal conditions in the factory workshop. A cable was attached to the lead, the joint encapsulated and the cable taken to a hull penetration which lay within the plate area but was not critical to it.

The system produced excellent results, though the anode had a number of faults which detracted from its performance. The neoprene sheet became detached but the adhesive, which was an epoxy, was sufficient shielding in itself.

The combination of plate and insulator was replaced by an all polyester glass fiber anode of the same length—12 feet—and with the same shield dimensions—approximately 14 ft × 2 ft. This encapsulated anode was held to the hull by a number of studs which were welded to the hull and used to hold the resin glass fiber shield to the hull. Ribs in the shield were used to streamline these studs. The anode performed well, the shield was adequate, but mechanically the technique of attachment was poor.

At this point a new approach to the anode design produced the anode shown in Fig. 176. This anode is made of two components, the anode assembly itself, which is a lead/silver extrusion with 2 per cent silver, encapsulated in a resin glass fiber body with stud fixing points in it. The shield is a separate polyester glass fiber unit which is held by the anode body at its center and by a formed steel strip at its edge. The shield is cemented to the hull and an end faring allows the cable to be carried through the hull either immediately at the end of the anode or by carrying it along the hull in half round for a short distance. In areas where the stud welded strip is now allowed, the anode can be prefabricated onto a steel doubler plate and that welded to the hull. These anodes are entirely successful and it is found that they could be used over short periods while the vessel is underway at currents in excess of 40 amps without causing disruption to the paint film around the anode.

The assembly is sufficiently flexible to fit the curvature of the hull at almost any point and the technique of installation is simple so that the trades employed within the shipyard can undertake the installation, Fig 177.

In America the platinum anode was receiving similar development. This consisted of a disc of solid platinum connected to a stud at its center which penetrated the hull through a glanding arrangement. The disc is held in a lucite holder fixed by a circle of studs either directly to the hull or in a small recess. The anode was surrounded by a neoprene patch which was held at its edges by a steel strip. This was hardly more successful than it had been in Europe and was replaced by a mastic which is trowelled to the hull in place of the prefabricated shield. This is probably dictated by the area of the shielding and the difficulty that would be experienced in placing a large

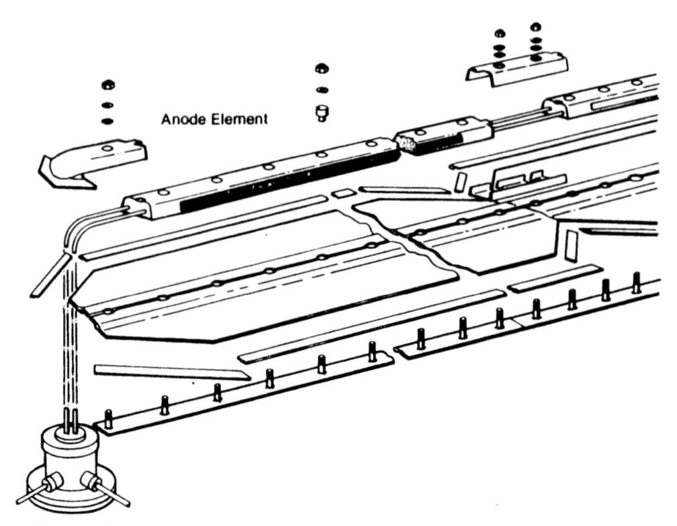

FIGURE 176 - Modern lead silver hull mounted imbedded current anode - exploded view.

FIGURE 177 - Stern showing anode and electrode on starboard side and light fouling on propeller where no dissolution of copper has occurred. (Photo reprinted with permission from CORRINTEC/UK, Ltd., Winchester, Hants, England.)

square or circular shield onto anything but a flat hull surface.

Other workers initiated developments which included lines of small discs, wires, ribbed plates, etc. The two distinct anode designs—the line and the disc, with the minor variations, from the principal basis of the majority of hull impressed current protection systems.

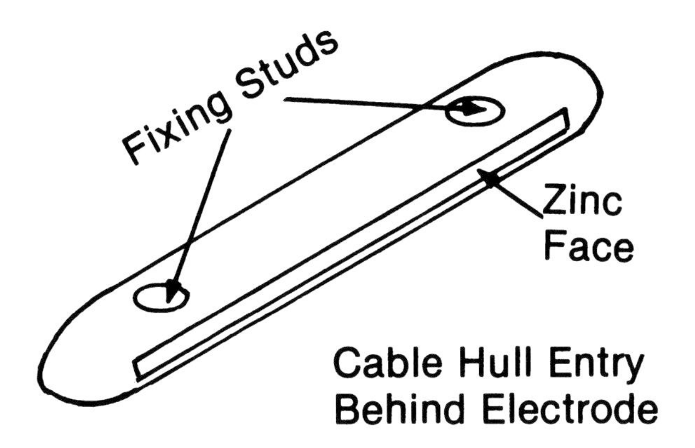

FIGURE 178 - Hull mounted zinc reference electrode.

Protection Current

With both anodes it is necessary to provide current and to control this power so that the hull is properly protected. To do this reference electrodes are needed to give a signal that can either be used to adjust the current manually or to actuate automatic control. Again, there are two developments. The author favors with his lead anodes, the use of zinc reference electrodes. These are very similar in design to a short length of anodes. The cable is taken from the center of the electrode through the hull and no shielding is necessary around the electrode. This is illustrated in Figs. 178 and 179. The alternative is to use a silver chloride electrode and to mount this onto or into the hull. There were attempts to use copper sulfate electrodes and to mount these in withdrawable holders in a short length of pipe. These were marginally successful but had to be changed frequently. Some silver chloride cells were used in the same manner but the better ones were developed to have a shape which was compatible with the hull and resembled the point anode. Silver chloride electrodes are covered with porous inert material and the silver chloride is in equilibrium with the sea water.

The choice of the electrode is dictated by its use. The zinc electrode was developed because it would provide a sensibly large signal, it was rugged and not easily damaged, but principally that the electrode provided a signal of +250 mV and that if this potential was diminished for any reason the cathodic protection was reduced as the controller saw it as an indication of overprotection. The silver chloride electrode has good stability but the potential between the hull and the electrode is 800 mV and any decrease in signal is indicated as underprotection and the controller tends to increase the cathodic protection current.

FIGURE 179 - Zinc electrode encapsulated in resin glassfiber insulating and streamline mount. (Photo reprinted with permission from CORRINTEC/UK, Ltd., Winchester, Hants, England.)

A 10% error in the silver chloride cell means an error of 80 mV in the protection whereas a similar percentage error with the zinc electrode is only a 25 mV change in protective potential. Taking this to a logical conclusion the author attempted to make electrodes using cadmium-zinc alloys but abandoned this as it appeared to have only minor advantages and considerable difficulties in development.

System Engineering

Having developed the reference electrodes, the next critical factor is the placing of the anodes and the electrodes on the hull itself. The first consideration is the size of the anodes and their location. It is an advantage if the anodes can be placed so that they will give the protection that is locally required around them. It is also a considerable advantage if standard anodes can be used and equally if a symmetrical system can be used on both sides of the ship. On these assumptions and calculating the current requirements of the then popular vessels, it appeared that an anode with an output of approximately 35 amps, which could be fitted to any area of the ship, would provide an ideal situation. Ships would then have systems which would have a 70 amp quantum and the largest tankers at that time afloat appeared to require 350 amps. The larger merchantmen were properly protected at either 210 or 280 amps. The anodes were arranged to lie two thirds to the after section of the ship and one third forward of amidships. This is to take care of the additional current required in the stern area and because there is an area on the bow where it is difficult to place anodes because of anchor cable damage. In tank ships anodes were not to be placed

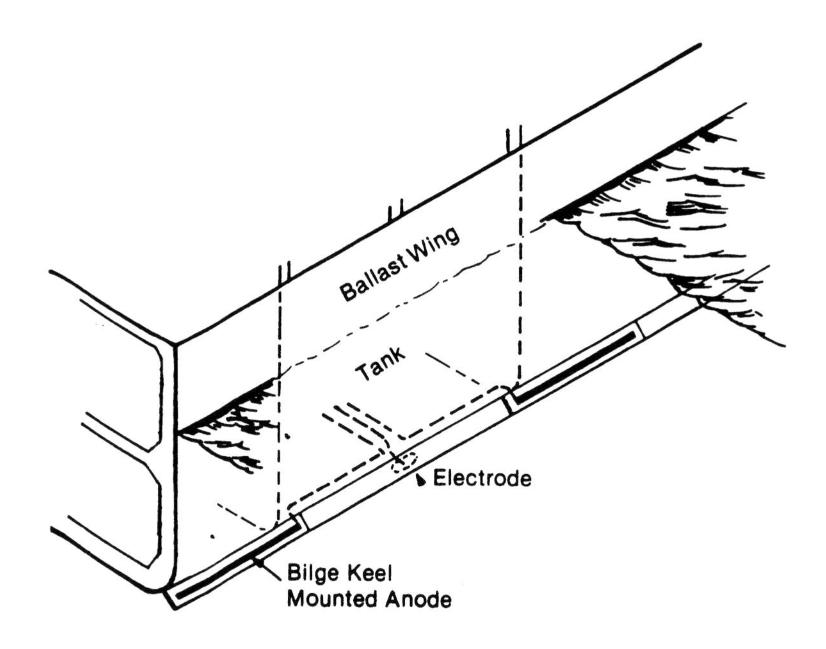

FIGURE 180 - Bilge keel mounted anodes.

so that their cables penetrated cargo tanks but they could penetrate ballast tanks.

In one of the first installations the anodes were placed on the bilge keel at a midship ballast tank and the controller placed in the center castle of the ship, the cable entering the hull in the engine compartment. The majority of the anodes were placed outside the engine room, Fig 180.

With the end-connected line anode the Classification Societies allowed the anode to be placed over the fuel bunkers with the cable penetration further aft directly into the engine room. The arrangement of some of the earlier systems is shown in figure 181. The cable penetration through the hull requires the use of a cofferdam so that there is double security at the point of hull penetration. Solid-filled cables, that is cable in which each individual strand is embedded in the insulator, are used; these can be sheared in the event of an accident and water will not penetrate along the conductor wire, Fig 182 and 183.

The platinum anode, partly because of its cost, but equally because of its capability, is designed to have a much higher current output. Fewer anodes are required and generally one or two of these are placed on the stern of the vessel only. In both systems there is the problem of placing the reference electrode. One of the earlier attempts at patenting a ship's hull system described the anodes as being located close to the stern of the vessel while the reference electrode was placed in the bow. By this technique it

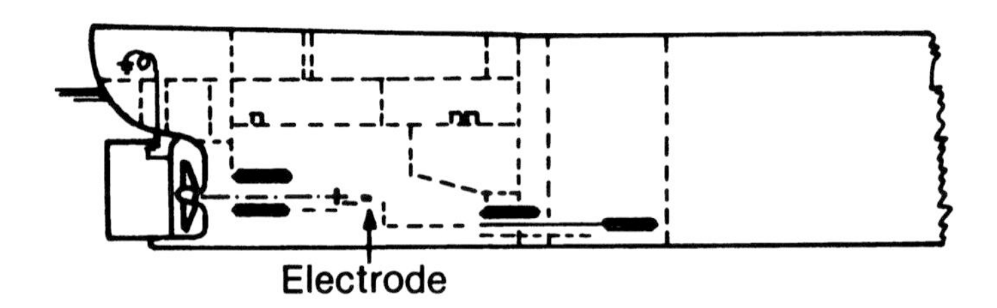

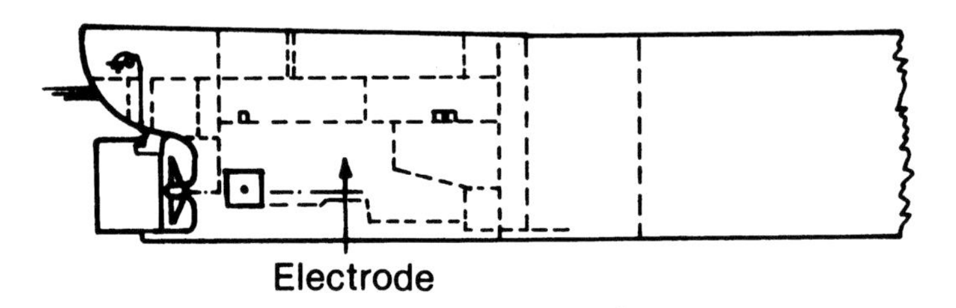

FIGURE 181 - Lead silver line anodes and platinum point anodes on hulls.

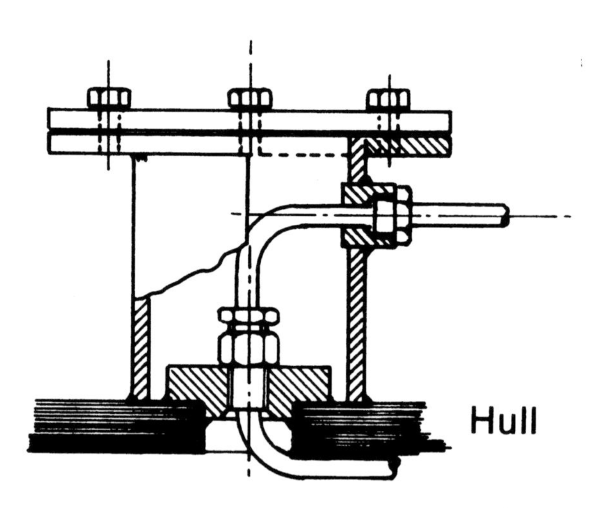

FIGURE 182 - Coffer dam for hull cable penetration.

was pretended that the hull would be properly protected with the reference electrode sited in the area of least adequate protection, that is, furthest from the anode. This is a fallacy as the bow of a ship receives additional protection by virtue of its shape just as the point of a lightning conductor receives most current.

a

b

FIGURE 183 - Hull penetration cofferdam (a) for anode cable and (b) anode recess with cofferdam for elliptical anode. (Photo reprinted with permission from CORRINTEC/UK, Ltd., Winchester, Hants, England.)

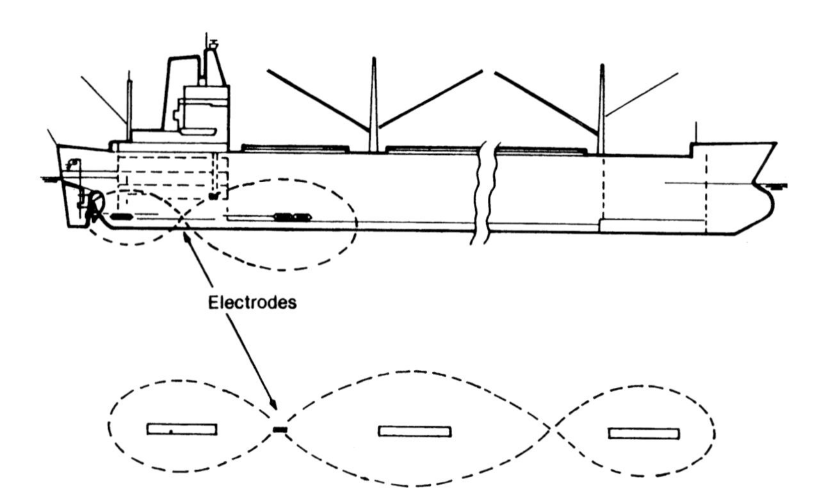

FIGURE 184 - Equipotentials around anodes showing reference electrode locations.

But the principal misconception lay in the use of this arrangement with automatic control. When the system was first switched on the controller was advised that the hull was inadequately protected and operated at maximum current while the ship slowly polarized. This period of polarization was of sufficient time for the excessive cathodic reaction around the anodes to remove the paint so that the current was mostly consumed in protecting the stripped area. The protection at the bow remained inadequate and the system was a complete failure.

The problem of the automatic control of cathodic protection is a difficult one and it appeared to the author that it might best be approached by causing the controller and the impressed current anode system to imitate an ideal sacrificial anode. This was done by placing the reference electrode at a point where it would receive a small amount of overprotection, or driving voltage, when the whole hull was adequately protected. If the output of the anode is now controlled so that the equipotential shell acts as a mythical sacrificial anode with a driving potential of say 30 mV and a very low electrical resistance; the control would then take care of itself as in the low driving voltage anode curves in Chapter 2.

Where more than a single anode is involved then a control point on their equipotential line would create a chain of the pseudo-sacrificial anodes, Fig 184. This system works extremely well, there is minimum overprotection in the vicinity of the anode and the system rapidly and accurately responds to changes in the cathode condition, whether this is caused by changes in ships' speed, in paint damage or in general paint deterioration. Using a low input impedance into the zinc reference electrode, the system automatically corrects for operation in high resistivity estuary water by reducing the signal from the electrode and thereby changing the potential criterion for protection and the nominal driving voltage of the mythical anode.

Surveys taken around vessels operating with this system are shown in Fig. 185. In one of these the ship is protected from a stern-only system but protection is achieved at the bow, Fig 186, and in the other the bow system when switched on and off indicates the revised distribution of current.

As ships became larger the number of anodes required to give protection grew. With the platinum anode the predicted hull current demand appeared to have been more accurately forecast. It was a problem to extend the line anode indefinitely as this would have resulted in an anode that was too long to be handled, and also the principal advantage of the line anode, that it could be installed from one level of staging, would be lost as the

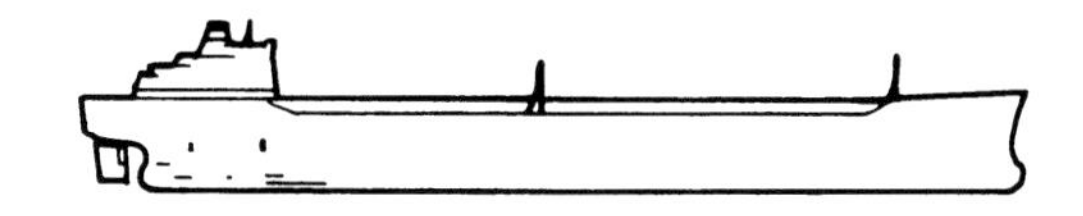

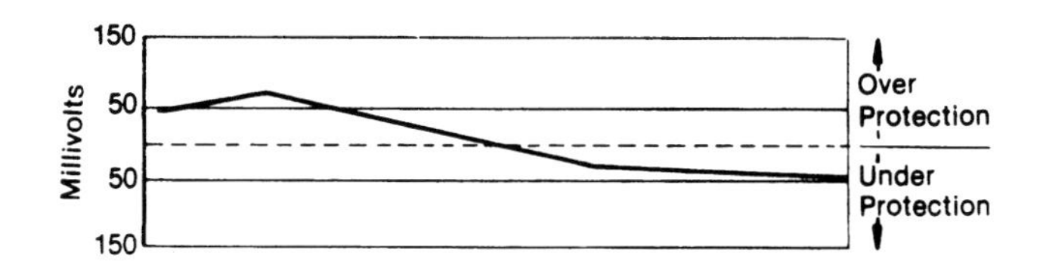

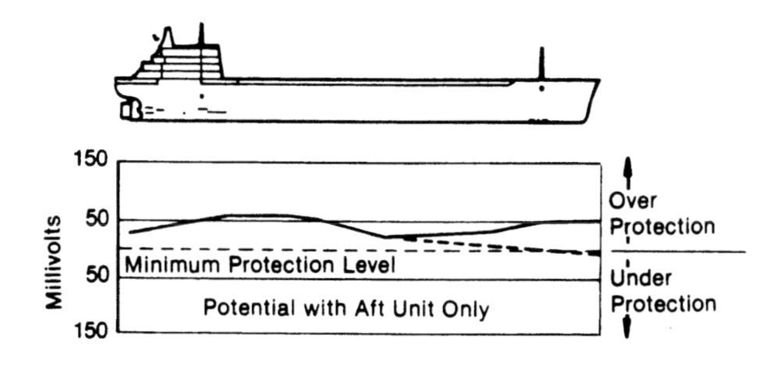

FIGURE 185 - Potentials around two tankers with impressed surrent protection.

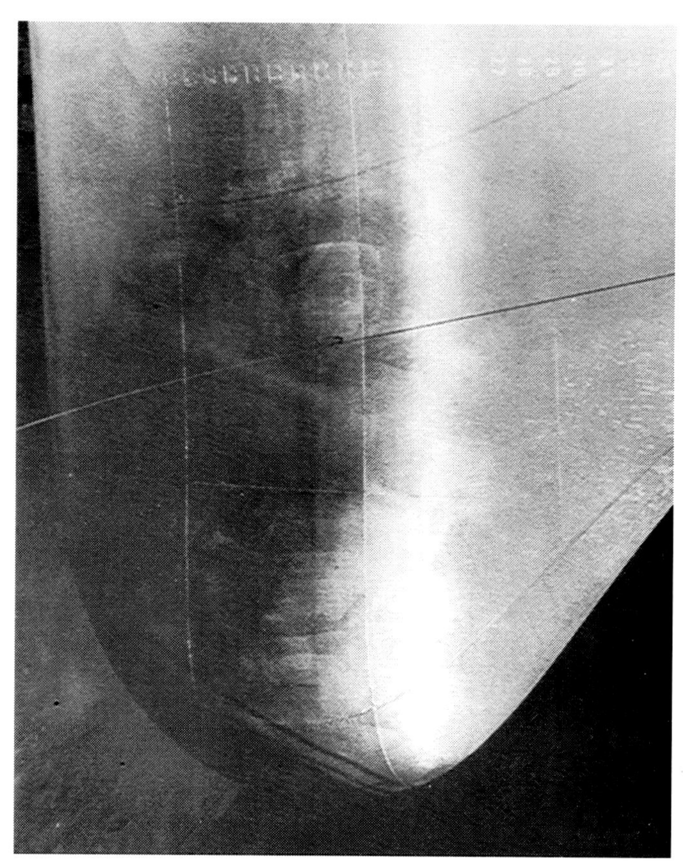

FIGURE 186 - Bow of super-tanker with all-aft impressed current cathodic protection system after 4 years. (Photo reprinted with permission from G.W. Moore, Esq.)

width of the center of the elliptical shield would become excessive. This was solved by making a segmented anode, that is in lengths that were interconnected by cable, so that the anode emitting surface is a series of lengths of lead alloy with non-emitting sections between them. The anode can be folded at the cable interconnections and it is detailed in Fig. 138. The shield now became a series of ellipses which is approximated to a parallel but slightly wider resin glass fiber plate, Fig 187.

The connection to the anode can now be made at one end or at any of the flexible joints in the anode. Almost invariably the connection is at one end and the Classification Societies have allowed the anodes to be placed outside the cargo tanks providing the cable was brought back and entered the hull at the engine room or other suitable space. The cable is carried along the hull protected by a half round pipe. This extends the scope for

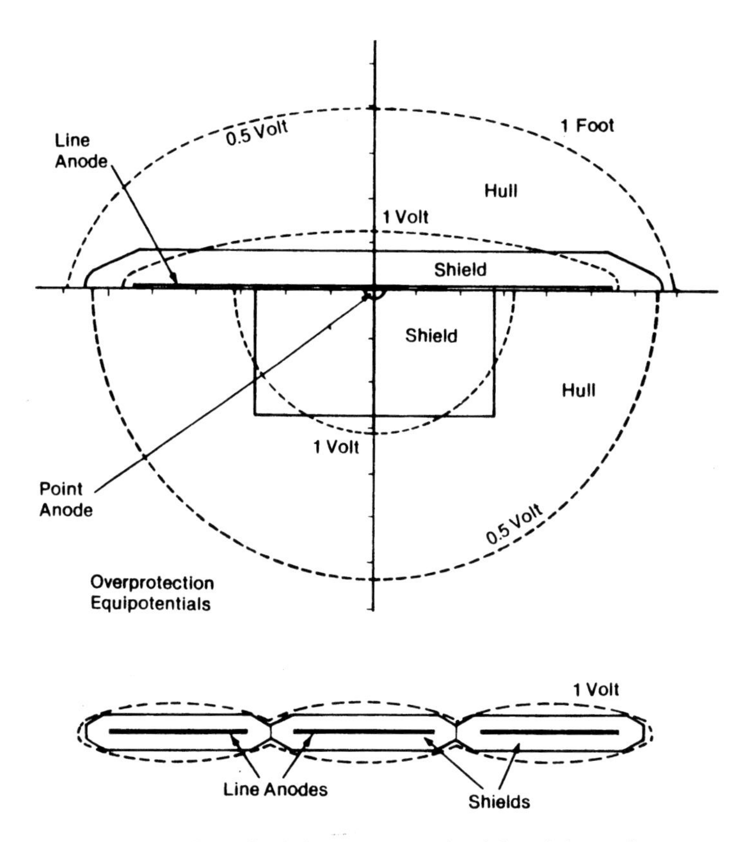

FIGURE 187 - Shield area to coincide with equipotential of 1V overprotection (a) point anode and line anode (b) chain of anodes.

protecting tankers as the engine room decreased relative to the ship's length as the size of the tankers increased, Fig 188.

This same technique had been used on the bow of tankers but it was found that anchor cable damage on a number of ships destroyed the anodes. To overcome this anodes are recessed into the hull. This means that the point anode has to be used as it is almost impossible to recess a line anode and retain the strength of the hull. Platinum coated anodes have more than adequate current output but in a circular configuration they require a high driving voltage compared with their minimum surface. In making hull recesses there are certain rules that have to be followed and the author devised an anode that fitted into an elliptical rather than a circular recess. The resistance of the same area of platinum anode is lower in the form of an ellipse than a disc, also the ellipse is easier to fit within the frames of the ship. The anode adopted is shown in Fig. 189. It is sur-

FIGURE 188 - Segmented chain of hull anodes extended along tanker hull from after engine room cable penetration. (Photo reprinted with permission from CORRINTEC/UK, Ltd., Winchester, Hants, England.)

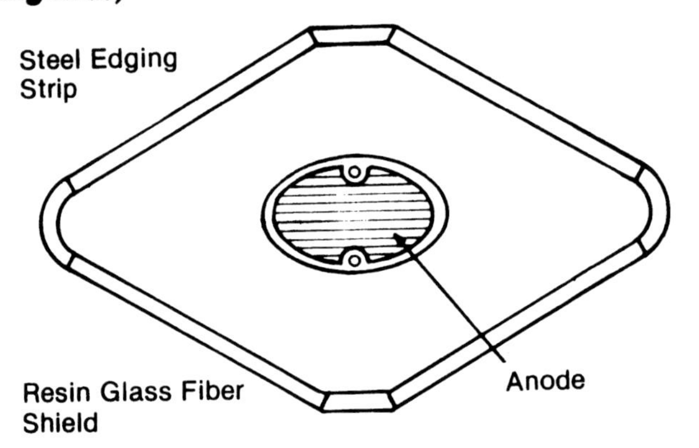

FIGURE 189 - Elliptical anode for recessing in bow area.

rounded by a shield formed into a near diamond to use the simple technique of holding the edge of the shield that had been developed for the line anodes, Fig 190. A similar recessed electrode was developed, though this, not having the same dimensional problems, was fitted into a circular holder.

Power Units

The first automatically controlled power units were transformer rectifiers operating with saturable reactors. The signal was amplified by a

FIGURE 190 - Recessed bow elliptical anode and polyester shield showing calcareous deposit around shield. (Photo reprinted with permission from CORRINTEC/UK, Ltd., Winchester, Hants, England.)

magnetic amplifier and fed into the saturable reactor to control the a c input to the transformer rectifier. The controllers operated extremely effectively and until they were installed on a ship that was well painted and its hull polarization time coincided with the response time of the reactor. This led to a slow oscillation of current which, although protecting the ship, was obviously unsatisfactory. Contemporarily silicon controlled rectifiers were developed and they appeared to give more rapid control which would follow the changes in the ship's demand without oscillating. The early controllers used an a c bridge with two silicon control rectifiers and two rectifiers to modify the a c input into the transformer. The system operates by switching on the a c for only part of its cycle. This produces an ultimate d c output which is a series of pulses at 120 cycles, the width of the pulse varying with the amount of current. The control is very wide ranging and as the current is being fed into a constant resistance and back emf then to achieve a 10:1 ratio in current, more than a 50:1 control ratio in power is needed. The controller operates very simply; it can use a signal from either one of the reference electrodes or from a combination of both. It indicates the potential and the output current. The control potential relative the zinc electrode can be adjusted and the system incorporates protection against excessive current being demanded by the ship. This system was used until the introduction of triacs reduced the cost of the bridge and as electronic techniques improved they were incorporated into the controller, Fig 191. The reduced size of the control equipment made it possible to build a small transformer rectifier and controller in a single case or to take the controller remote from the power unit with manual control both locally and at the

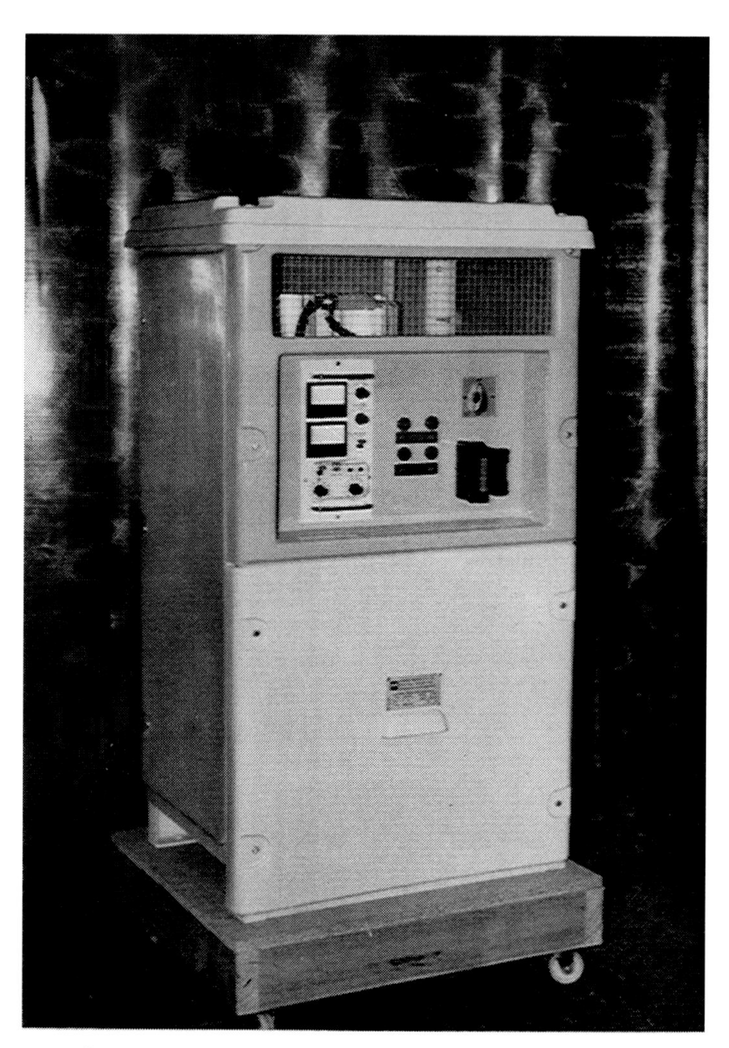

FIGURE 191 - Automatically controlled power unit for ship hull protection. (Photo reprinted with permission from CORRINTEC/UK, Ltd., Winchester, Hants, England.)

remote controller. It is possible from a small number of simple units to have a series of modes and types of control, for example a transformer rectifier with integral control, a transformer rectifier with remote control coupled with a slave control at the unit, and it is possible to operate a number of transformer rectifiers from a single controller.

This system of using a number of transformer rectifiers replaced the earlier technique of using one very large controlled power unit. On a super tanker this may mean a capacity of 2000 amps from which current was fed to anodes located around the stern of the ship. The cables to these anodes had to be balanced in resistance and this meant that very heavy cables had to be used for the more distant anodes. The new technique allowed the mounting of small units close to groups of anodes and to send control signals to them from a central controller. The mass production of these units produces economy and the units, being interchangeable, reduce the amount of spares carried on board and there is a considerable degree of redundancy in the system. The units, small and air cooled, are easily handled on a small trolley and taken on and off the ship for repair. Three more phase systems have been used with the larger power units but these are more difficult to operate, particularly at low power when any imbalance in the system can lead to only one or two phases operating. Where the controlled transformer rectifiers are connected to a local supply and controlled by a d c signal from the central controller then these can operate on different phases and a balance is maintained between them.

It was found that a number of ships' engineers were reporting units as defective or querying their efficiency when they failed to get the exact protection potential reading at the electrode. This was caused usually by changes in demand of the hull as, for example, when a tanker left a loading terminal or a cargo vessel left a port. The system current limited at maximum transformer rectifier output and for several hours failed to polarize the ship. In extreme cases it took several days for the ship's potential to reach the requisite setting, though it was within 20 or 30 mV of it within less than ten hours. In order to avoid this problem two steps are taken; the amplification of the control circuit is increased so that there is 100 per cent change in current over less than 10 mV, that is an almost imperceptible change in the millivolt meter reading, and secondly, the capacity of the transformer rectifiers is increased so that a more rapid polarization occurs. It was found that the anode which had previously been rated at a continuous 35 amps was capable of emitting 50% more current over a brief period of up to two hours without any detrimental effect on the hull paint; this raised the current output of the lead anode from 35 amps to a nominal 50 amps. There was no change in the protection achieved by this method but the operational advantages were considerable.

Vessels entering dry dock are found to be perfectly protected by the use of impressed current protection and it is possible in many cases to restore the original shine to the hull by rubbing it lightly. Ships that have been damaged retain the sharp edges on the metal when they return to dry dock. There is a very fine calcareous deposit on the bare areas of hull steel and this is remarkably more tenacious and smooth than is found in other cathodic protection applications. A similar calcareous deposit has been reported with pulsed protection. With an increase in the capacity of the units much shorter duration pulses are gated by the silicon controlled rectifiers and the pulses appear to be providing the very hard smooth cathode film which is reported to be rich in magnesium. This film matches the smoothness of the paint and a number of owners find that their vessels do not slow with age except on the fouling cycle. The paint is also preserved in a most remarkable way—owners are careful to achieve an immaculate paint finish initially and retain this as long as possible during the life of the ship.

Foremost amongst these are the naval users who have experimented with various coal tar-epoxy and other quality coatings which after more than 15 years' service have not been replaced. Some of the tanker operators who have a direct interest in paint formulation undertook similar experiments and vessels have already been scrapped that have not been painted more than once since their building. Self-polishing paints are being used in conjunction with cathodic protection most successfully. The combination of impressed current protection properly designed and adequately controlled with modern paints has given a new dimension to naval architecture. The cathodic protection system is flexible and can range from that used on a small, fast ferry with perhaps a capacity of 100 amps through a modern container vessel or liner at about 600 amps, to a super tanker or vlcc that may be equipped with a total current capacity in excess of 1,200 amps. Single systems can be used on the smaller ships or dual systems on larger vessels, particularly tankers where the bow will be separately protected and controlled from the stern. In many cases the second control system on a tanker may be pre-set rather than automatic although the cost effectiveness of this has virtually disappeared.

Design

To design a system of impressed current protection it is necessary to calculate the under water area of the hull. There are a number of formulae for doing this; the one favored by the author is

$$L\,(b + 2d)\,\frac{B_C + 1}{2}$$

where B_C is the block coefficient, L the length at the water line, b the breadth and d the draft of the vessel. The total current requirement can then be estimated from the paint and service of the ship. Table XII shows the current densities proposed for ships depending on their operation, age and paint.

From the total current the number of anode units can be determined and this is usually a symmetrical system, particularly on larger vessels. There is no fundamental reason on small vessels why the systems should be symmetrical but in terms of operating and reporting it is considerably easier to have symmetrical readings from both sides of the ship. The anodes are then dispersed about the ship with a concentration towards the stern; the latter half of the vessel takes approximately two thirds of the current. Anodes cannot be placed on the fore part of the ship unless they are recessed and it is difficult to place anodes in areas where there are special strength steels. Cable entry into certain tanks is prohibited as might welding be at subsequent dockings. With tank ships the anodes should be placed mainly on the stern and where cables can be carried to the anode along the hull these should extend as far forward as one third of the length of the ship.

Reference electrodes should be placed symmetrically so that they should each give the same signal. On large vessels duplication of the reference electrode is sensible as the whole system relies on the electrode. The electrode should be positioned to lie between anodes and where there asymmetrical arrangements of anodes, for example three anodes forward of the reference electrode and two anodes aft of it, then the reference electrode should be placed three fifths of the distance between the anode emitting centers. The emitting center of the anode is a weighted point rather as the center of gravity of a group of weights.

There are a number of commercial anodes available and the designer is able to select the system based on these anodes. Some systems have a standard unit anode whereas in other systems the anode output can be increased or decreased by a change in the basic dimension of the anode. The area to be shielded around the anode can be determined by the anode size itself, but where large outputs from single anodes are concerned, the nearness of the surface of the sea water will influence the size and shape of the shield.

Propellers

The hull of a ship really comprises the main body of the hull, the protuberances and appendages to that hull, such as the bilge keel, the sea water inlet and outlet, the propeller or propellers, and the rudder or rudders. The sea water inlet grids, bilge keel, etc., are protected with the hull and current flows into the scoop for some small distance. The rudder is protected by the hull system but it is necessary to bond across the rudder bearings as a considerable current otherwise will flow through them. A flexible cable is usually taken from the top of the rudder stock to the deckhead above. Wire brading once used for this was found to be electrically noisy.

The propeller, when it is turning, is electrically isolated from the hull by the oil film in the bearings. This does not happen immediately but is apparent after a few days' operation when the high points on the metal bearings have been reduced. A potential of 200 to 300 mV can be measured between the shaft and the protected hull. This is across the bearings and should there be any deterioration in the lubricating oil then severe electrolytic corrosion will occur. It is, therefore, necessary electrically to short-circuit the shaft to the hull, even if it was decided that the propeller did not require protection. With single screw ships another effect is noticed if the propeller shaft is not bonded, current is picked up by the tips of the propeller blade and discharged from the propeller boss onto the leading edge of the rudder. This does not happen with sacrificial anodes or with impressed current if sacrificial anodes are placed on the rudder.

To bond the moving shaft to the hull a number of materials have been used. The simplest is carbon or graphite brushes rubbing on the shaft, but these are inadequate and potentials of up to 100 mV can be detected even with a large brush system. A phosphor bronze brush, though a comparatively large one, if used with an electrical lubricant will reduce the potentials to less than 25 mV. The brush, however, has to be held with considerable pressure and not only are the brushes rapidly worn out, but the shaft is scarred and, eventually, worn away. A steel slip ring can be placed around the shaft but this is a considerable expense. Copper graphite brushes running either on the shaft or on a copper slip ring will reduce the potential to about 50 mV.

The best combination is silver graphite brushes running on a silver slip ring. One of the problems is that the slip ring is proud of the shaft and the oil film is centrifugally lifted from the shaft onto the slip ring itself. If silver graphite brushes of about 50 per cent silver are run on a silver plated track then this oil can increase the shaft to hull potential to 30 to 40 mV. If, however, massive silver is used, (though this can be perhaps 1/16th in. thick and embedded in copper) and the brushes contain more than 83 per cent silver, then the shaft potential is reduced to 10 mV and the system is not affected by small amounts of oil. One sq in. of brush is required for shafts up to 20 in. and beyond that 2 sq in. The brushes are usual 1 in. × 1/2 in. and are mounted in pairs in a balanced holder. The slip ring has to be split and twin brushes avoid loss of contact at the joints. This is not only the best but also the cheapest form of arrangement as the self lubrication of the brushes results in a low rate of wear and the small track that can be used is not excessively expensive as the material is easily worked into a slip ring, Figs 192 and 193.

The propeller on the ship will then be at almost the same electrical potential as the hull itself. Outside in the sea water, however, the single propeller is in an area that is shielded by the very shape of the ship and as it imposes a considerable current drain on the cathodic protection system, the potentials on the propeller itself are between 20 and 100 mV less than the protection to the hull. Because of the concentration of anodes in the stern area, there is probably 20 or 50 mV overprotection and the propeller receives almost the same protection as the hull. The anodes' location on the stern will influence the degree of overprotection in the vicinity of the pro-

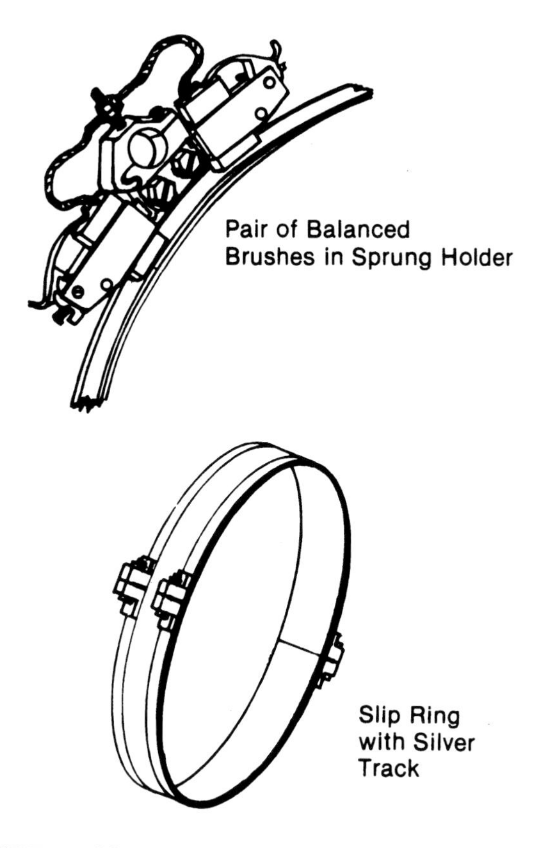

FIGURE 192 - Silver track slipring and silver graphite brushes.

peller. With twin screw vessels the problem is not so marked as the propellers stand proud of the hull and tend to receive more than adequate protection.

A number of propellers suffer from cavitation attack, while others suffer less severe impingement corrosion. This latter is completely suppressed by cathodic protection and is a form of differential oxygenation cell.

Propeller cavitation is a different phenomenon and appears to be electrical in its origin. The cavitation effect occurs when cavities collapse onto the propeller surface. Each of these cavities is at low pressure and is almost completely ionized. On collapse the charges that are present in the ionized gas are selectively collected by the cavity surface as it moves towards the propeller blade and this gives a resultant charge, either positive or negative, onto the blade which is then discharged as a conventional current. The order of magnitude of this current can be calculated if one assumes that

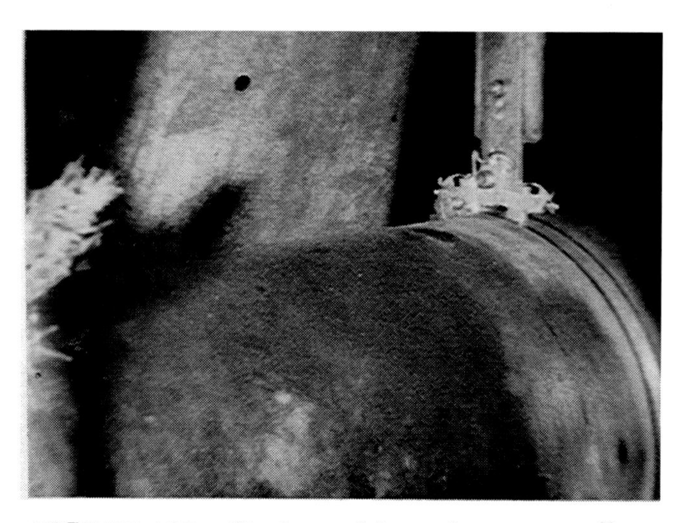

FIGURE 193 - Shaft earthing using copper silver tracked slipring with twin silver graphite brushes. (Photo reprinted with permission from CORRIN-TEC/UK, Ltd., Winchester, Hants, England.)

there is an inefficiency in total charge separation on the basis that if it is entirely random there will be some form of distribution in which the cathodic discharge will be beneficial and the anodic discharge will cause corrosion. This efficiency factor may be as low as 1 in 10^5. Consider a cavity 1 cc in capacity with one hundredth of an atmosphere pressure and if the cavity collapses 10 times a second, then on total charge separation the current flowing will be

$$\frac{2\ Ne}{22 \times 10^5} \times \frac{10}{100}$$

for singly charged ions. This current will flow onto an area of about 1 sq cm and assuming the 1 in 10^5 efficiency, a current density of 10 mA per sq ft will result; in corrosion terms a considerable current density. This will be the equivalent of an interference or electrolysis current and its magnitude will be given by some form of distribution as in Fig. 194. To the right of the curve the current will cause corrosion while to the left the reaction will be cathodic with no damage. If a potential gradient is applied to the surface through the collapsing cavity the effect will be to move the dividing line to the right and the area of the charge that will cause corrosion will be greatly reduced. This will be the effect of applying cathodic protection to the surface and if a sufficiently large potential gradient can be created there should be no perceivable cavitation attack on the propeller. Equally, the system should operate more effectively in fresh water than in sea water, where the

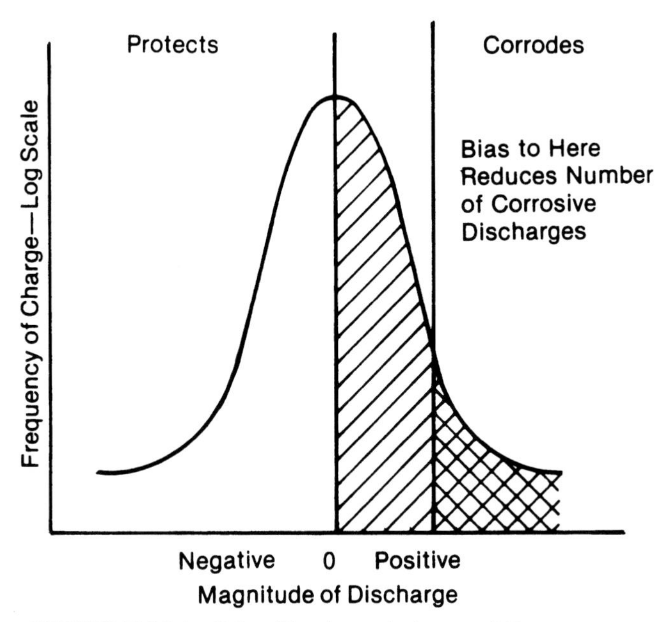

FIGURE 194 - Distribution of charge/discharge on cavitating surface.

potential gradients are steeper and it should work equally well with a steel propeller as with a bronze propeller.

To test this theory the Shell Tanker Group decided to fit a cathodic protection system based on potential gradient as designed by the author to a ship with a cast steel propeller. The system was operated and after six months there was no corrosion on the propeller though the operator had decided to coat the propeller as a second line of defense. At the six month docking, areas of the coating were removed and, unknown to the cathodic protection engineers, the propeller was deliberately damaged to ensure that cavitation would result. The propeller is still in service after 20 years and has outlasted the bronze propellers on the sister ships which are protected with sacrificial anodes, Fig 195. To achieve protection to this level the propeller shaft is negatively biased to provide a larger degree of protection. This biasing is done by the simple expedient of creating a potential across a low value resistor using a high current of approximately 100 amps so that when the propeller stops the system does not cause problems by passing large currents through the bearings, Fig 196. The system in this form has been fitted to a number of ships, mostly super tankers, where the tips of the propeller blades suffered very severe cavitation damage, Fig 197. The system has been entirely successful, though little use has been made of it on steel propellers.

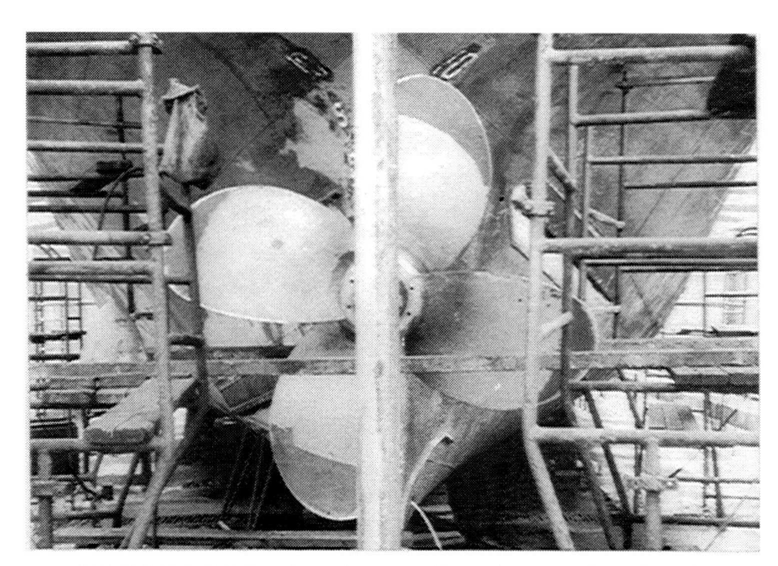

FIGURE 195 - Steel propeller after a decade of impressed current protection with bias applied to shaft. (Photo reprinted with permission from CORRINTEC/UK, Ltd., Winchester, Hants, England.)

Ducted and Shrouded Propellers

A number of ships are fitted with nozzles or shrouds on their propellers to increase the efficiency of the propeller and in the case of some shallow water vessels to act as a rudder. Protection into the nozzle is difficult and the propeller potential must not be depressed below the hull potential otherwise current may leave the inner face of the nozzle and flow onto the blades. Some overprotection at the entrance to the nozzle will provide complete protection, both to the propeller and to the nozzle itself. Where the nozzle is maneuverable it must be bonded as is a rudder, Fig 198.

Fast Small Vessels

The protection of vessels, such as naval patrol boats, light frigates, etc., poses some additional problems. The hull must be smooth and this is more critical than on larger ships, so that recessed anodes are used. A 5 amp circular anode is shown in Fig. 199 and this is surrounded by a shield of glass fiber. The anode itself is platinized titanium and the assembly is made for central fixing. Where a larger output is required the anodes are placed in pairs of threes into an elongated shield and there are two spacings available so that these can be positioned between frames, Fig 200.

Because of the low current demand secondary control, that is in the rectification circuit, is used.

There is an additional problem with this type of vessel. In operating at slow speeds it acts as a conventional ship, that is the stern requires more

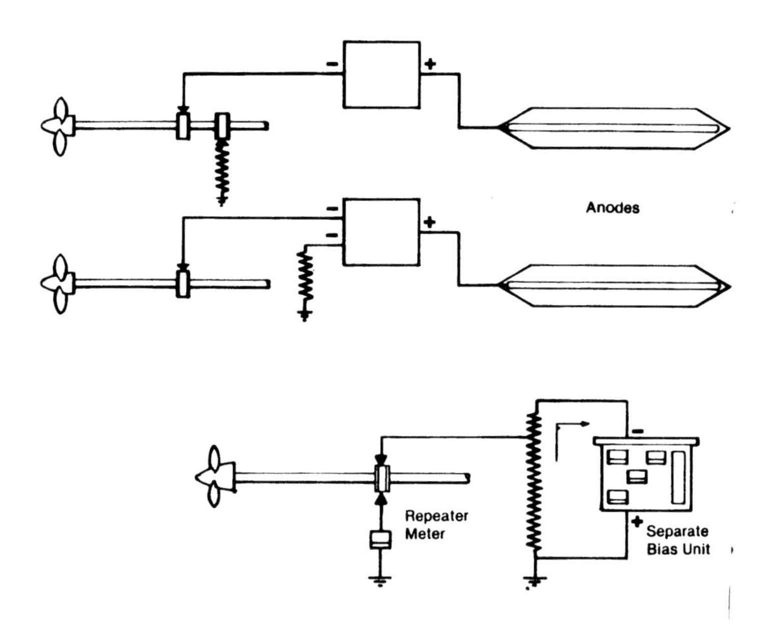

FIGURE 196 - Potential bias circuits for propellor protection.

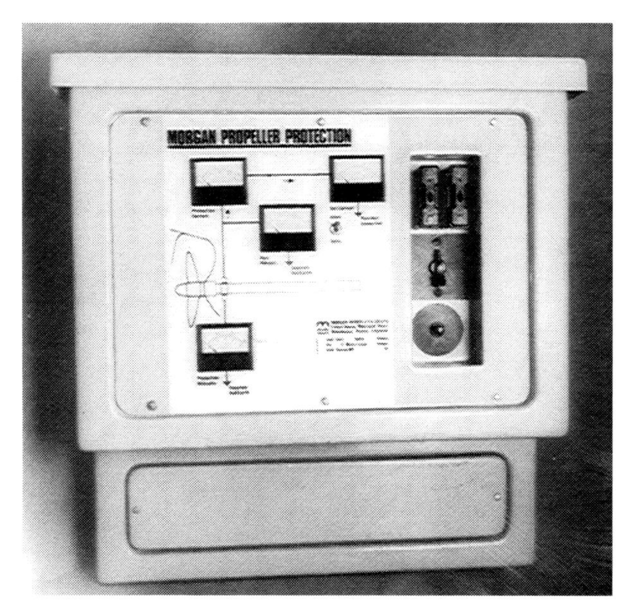

FIGURE 197 - Bias unit for propeller protection. (Photo reprinted with permission from CORRINTEC/UK, Ltd., Winchester, Hants, England.)

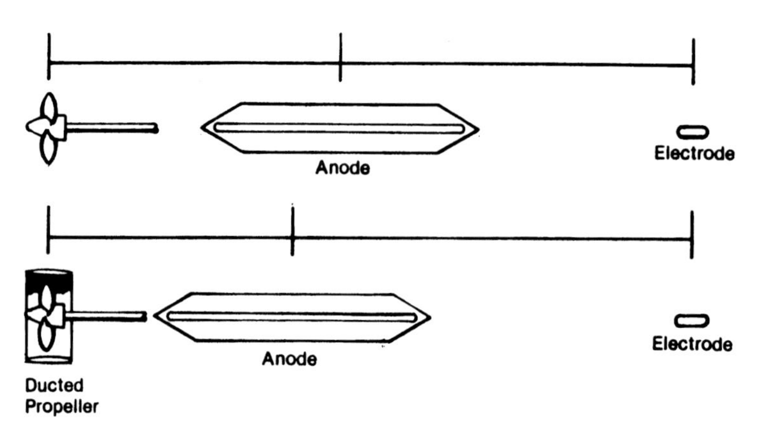

FIGURE 198 - Anode and electrode location with ducted propellers.

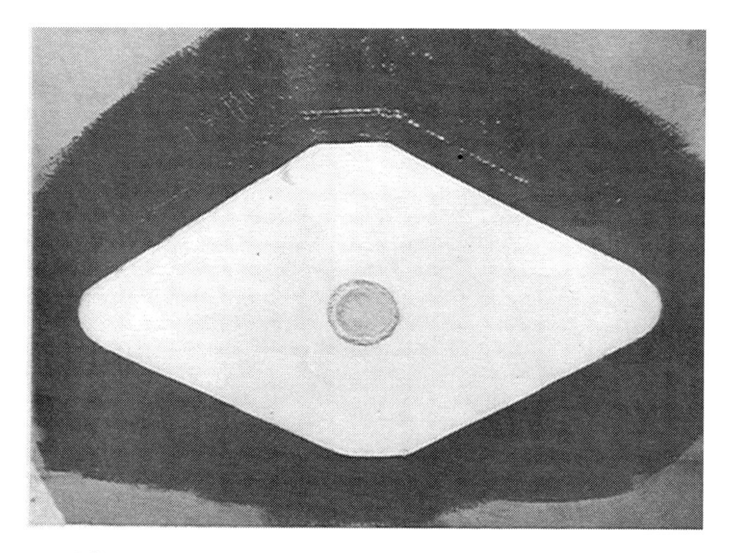

FIGURE 199 - Small disc flush platinized titanium anode with diamond shaped shield. (Photo reprinted with permission from CORRINTEC/UK, Ltd., Winchester, Hants, England.)

current, caused generally by the effect of the turbulent water around the propeller. As speed increases, however, many of the hulls start to semi-plane and the bow lifts out of the water with a massive increase in turbulence towards the stern. On a number of light frigates it was found that corrosion, even in the presence of cathodic protection, was occurring at the stern with attack on A and P brackets and on the rudder surface.

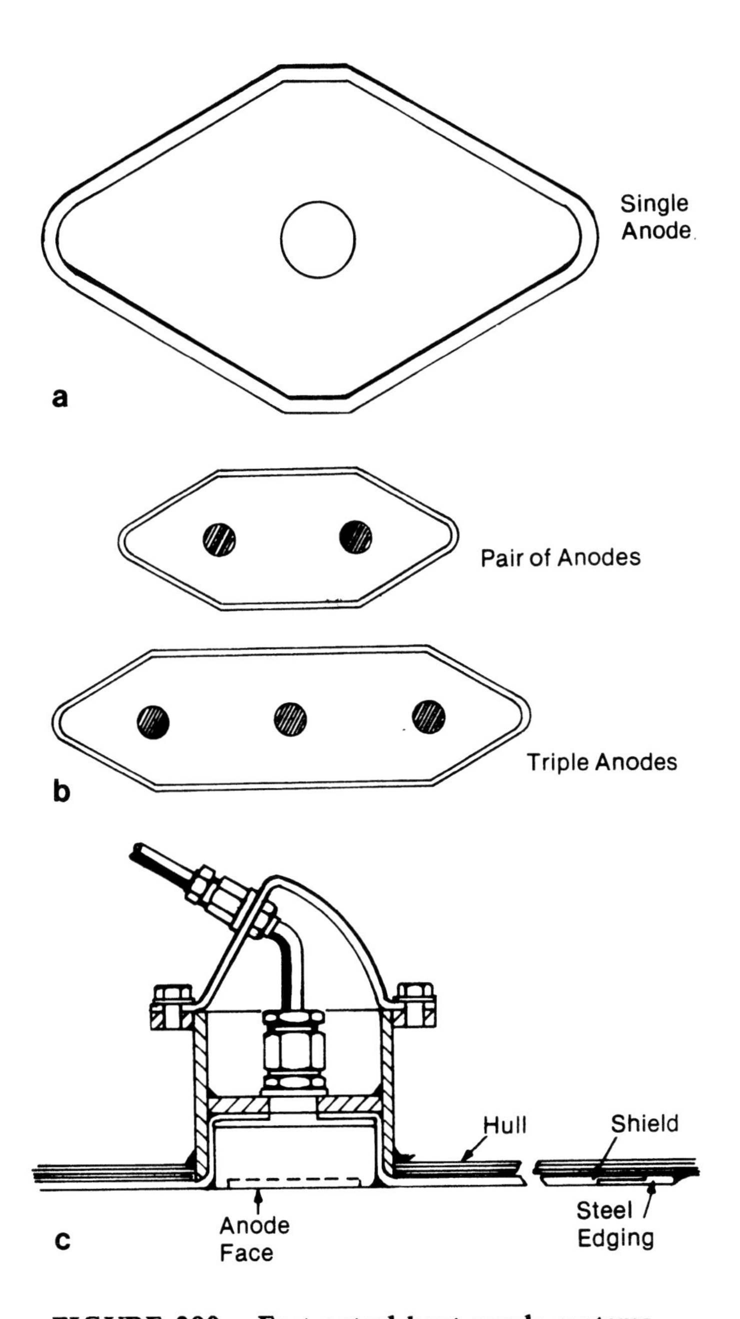

FIGURE 200 - Fast patrol boat anode systems.

A variable geometry system is used to overcome this problem using a dual control system which changes the distribution of current to the anodes as the overall output increases by having two separate sets of anodes operating from two controllers. These have different cut-on points and different amplifications giving a variation in distribution as the current increases. Complete protection has been achieved on vessels fitted with this type of control.

The protection of submarines can be engineered from the general information available. It is difficult to penetrate the pressure hull but the non-pressure hull can be fitted with anodes similar to those used in other vessels.

Small Pleasure and Work Craft

Small steel vessels can be protected by simpler techniques than those used in the protection of naval vessels. Vessels of more than 100 ft lbp have a permanent supply, either from their own generators or batteries or by being linked to the shore when laid up and have sufficient power to provide impressed current protection; the fast ships will benefit from not having the protuberances associated with sacrificial anodes. Simple anode systems can be employed with these and smaller anodes than the patrol boat type can be mounted almost flush with the hull. The use of trowelled mastics as opposed to shields at this size of anode is advisable.

For very small ships, particularly those that are made of glass fiber or wood with steel and other metal attachments, the use of a suspended anode when the vessel is not underway will probably give sufficient protection. A small impressed current system can be operated from the battery, particularly while the engine generator is operating, and this type of system can be used to prevent the corrosion of outboard motors. Solar panels and wind generators can be used to provide impressed current protection, particularly if they are used to recharge the ship's batteries.

Between these there are a variety of systems including hull mounted anodes, both of sacrificial and impressed current design. The low voltage anodes will have a number of advantages, not least amongst which is the need for no form of control. The owner who wishes to have control of his cathodic protection can use a system in which the anode also acts as a reference device. Other systems employing separate reference electrodes can be manually operated either using resistive control of sacrificial anodes or control of an impressed current power source.

The variety of options open and their attractiveness to the boat owner will, no doubt, decide the course that this type of cathodic protection will follow.

To the work boat, the small fishing vessel, whether of glass fiber with steel additions or of steel or aluminum construction, simple cathodic protection will be a boon. To the owner of a concrete vessel the interconnection of some of the hull fittings with the reinforcing will mean that a considerable amount of cathodic protection will be required. Other techniques can be employed but it will be difficult to isolate the metal fittings from the reinforcement on the majority of concrete hulls. The use of galvanized wire in the reinforcing may reduce the tendency for the reinforcing to act as a massive cathode. The problem is discussed for larger structures in the offshore section.

Some depth finding and other sonar equipment are deliberately made cathodic to the hull to protect the metal head; this causes corrosion of the hull, usually the weld around the unit. The cathodic protection to the hull has to be carefully designed to prevent failure of this weld which, mounted in the bottom of a steel boat, could lead to disastrous failure.

Vessels Operating in Fresh Water

Many of the structures that have so far been discussed are found in both sea and fresh water and vessels sometimes travel from one to the other. Inland waters usually have a resistivity in the order of 1,000 ohm cm as opposed to 25 ohm cm in the sea. The whole range of resistivity is found between the two in estuaries.

Although corrosion is not as bad in these waters generally, they do become stratified and polluted. It is sensible to apply cathodic protection to these vessels, and here the higher driving voltage of magnesium anodes can be an advantage. With impressed current systems there will be problems with the potential gradients across the coating close to the anode but in some cases, such as the corrosion of tug propellers and the corrosion of external heat exchangers on river craft, impressed current systems will probably provide the only solution. On other structures that are constructed in fresh water the low resistance shapes available with platinized or coated titanium can be used to advantage. Wire anodes, particularly those with copper cores, provide a low resistance extended anode which both aids in distribution and reduces the voltage of the operation. The limitation on the breakdown voltage of the titanium does not apply as the anodes will, in general, be fully platinized.

Care should be taken in the use of reference electrodes in fresh water and while zinc will generally operate its performance should be checked. Should the zinc fail to operate properly magnesium can be used as a reference electrode, though its accuracy is not as high as zinc, both because percentage errors are greater in absolute terms and the anode does not have such accurately fixed driving potential.

Operations in Ice

Ships such as Arctic supply vessels and ice breakers operate under very difficult conditions. The protection of these raises some interesting problems because the ice continually removes the paint from the water line of the ship. In many years' experience of protecting vessels that either operate full time or part time in ice, some interesting factors have emerged.

Firstly, the damage by the ice is much more limited than was first thought, and when the vessel is protected only about one third of the paint is moved by abrasion, the rest being lost by corrosion undercutting from these scratches. Secondly, the protection is not as difficult as it first seems and this appears to be associated with one or two factors. Firstly, it requires less current to polarize the hull in cold water or secondly; and perhaps more probably, very much smaller voltages are found between the anodes and cathodes on the hull so a lower swing of potential is required to suppress them.

In designing systems for ice breakers and vessels that operate in ice, a standard anode can be used if it has ice guards on it. A typical anode, with its associated ice guard, is shown in Fig. 201. These ice guards were designed in association with the naval architects of the Department of Transport in Canada and operated very successfully. Anode damage was not substantially greater than on the normal operating ships. The protection that was achieved was remarkable. The majority of the paint was preserved, weld decay and pitting were eliminated, and the anodes which were

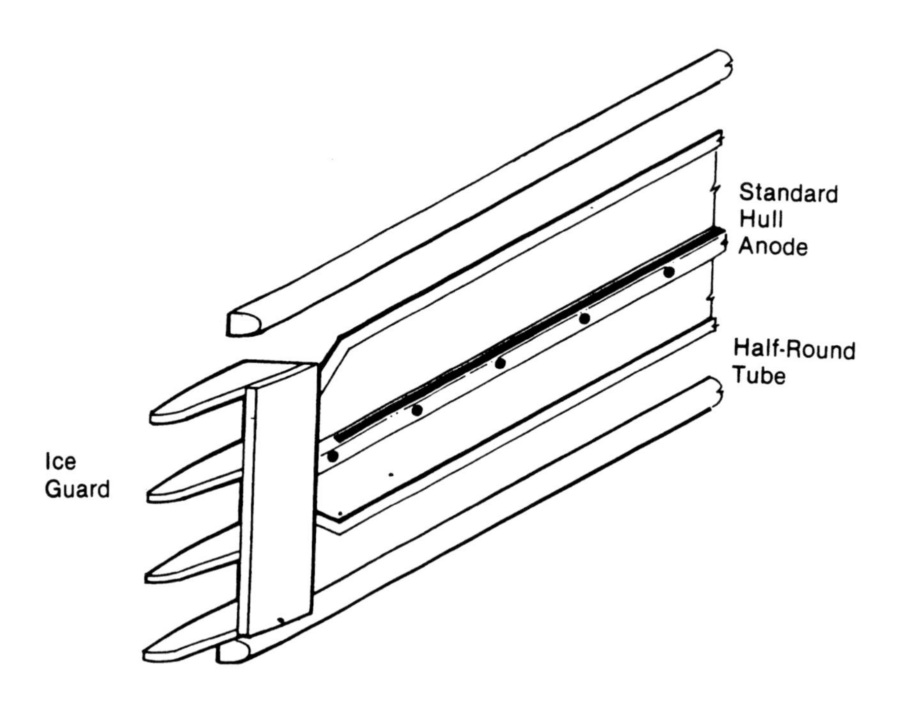

FIGURE 201 - Anode with ice guard.

lead/silver were not consumed at an appreciably greater rate. However, there was some concern that the lead peroxide film would be removed and if this occurs at regular intervals an anode consumption would be much greater because of the rapid initial loss of metal on peroxide formation.

On some Baltic ice breakers recessed anodes were fitted. These were of the elliptical type and were the standard anodes that had been used on the bows of tankers. It was feared that the shield would be removed by ice so the area was instead coated with an inert filled paint which had been developed in Finland. The anode was recessed to lie slightly below the surface and the edge of the anode recess was carefully rounded. After six years' operation the system has proved to give more than adequate protection and the owners are satisfied that they have eliminated the corrosion problems, particularly at welds, that they were suffering. The anode was platinized titanium and there was no evidence of the platinum being unusually consumed in the cold water.

Interconnecting Ships and Towing

When ships are interconnected as, for example, in lightening of one of the vessels, then the cathodic protection of one can have an effect on the other. It is possible for protection current to flow from one vessel to the other and for this to happen an equal current must flow through the interconnecting metal between the vessels. This can be via the reinforcing steel of the hoses or by the use of a metal gangway; marine aluminum gangways provide particularly good current carrying capacity. The problem is similar to that discussed in jetties and care must be taken that the main interconnecting steelwork is not broken close to points where there may be dangerous vapors. A tug is similarly connected to the ship it is towing but here the distances are too great for there to be any significant provision of protection current from one to the other. When an insulated strop is used at the end of the tow line, however, it is possible that the steel hawser can be affected by the potential field either around the ship or around the tug. In this case it is possible for current to discharge from the end of the tow rope and this will have a weakening effect upon it. On long tows the potential of the towing cable should be checked and it is sensible to interconnect this cable electrically to the tug if a plastic cable strop is used. Typical ship systems are shown in Fig 202.

Chapter 7c

Offshore Cathodic Protection

Offshore structures are characterized by three features: firstly, they are in sea or semi-sea water; secondly they are far from land or extended

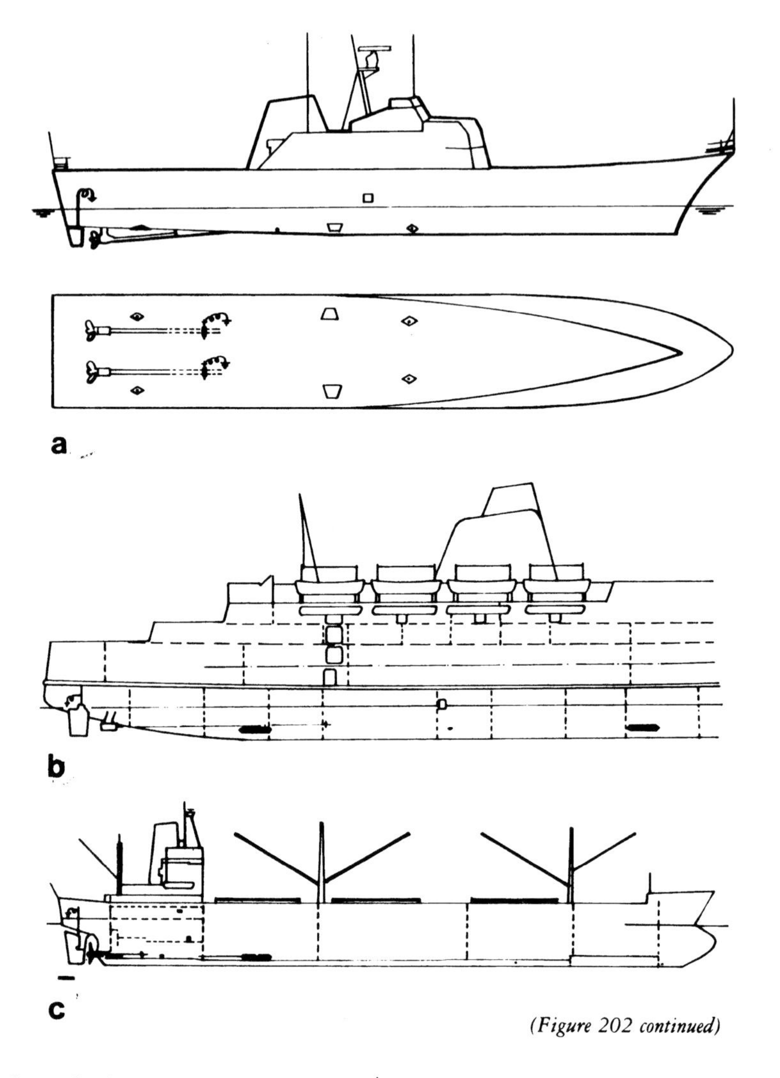

(Figure 202 continued)

from the land far enough to be remote; but most importantly they are large. The same cathodic protection techniques as are used in other structures apply but the scale factor, the sheer size of the offshore structures, is their most important feature.

The various structures offshore can be classified as

(1) Platforms
(2) Semi-submersibles
(3) Wells, Pipelines and Subsea Installations.

The first classification can be gravity platforms which are multitubular or have a single or small number of tubes, tethered platforms and vertical

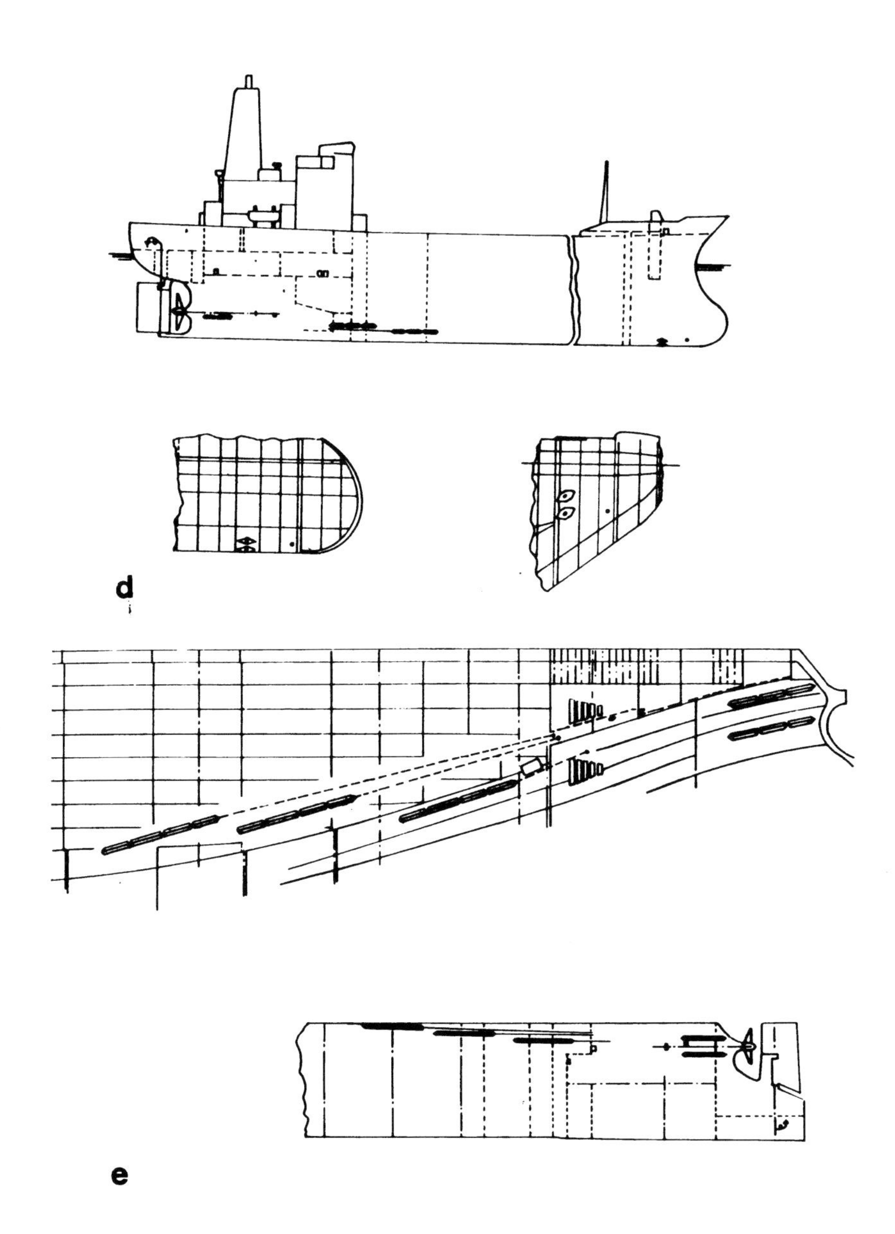

FIGURE 202 - Typical ship hull systems.

buoys which are floating but are held in position, guyed structures which extend from the sea bed and are held vertically by guy ropes and the tension leg platforms which are restrained by tensioned vertical supports. The latter overlap with semi-submersible structures, the second class, which are sometimes tethered but may operate by dynamic positioning. Some of these are based on ships' hulls and are used as storage or transfer vessels. The third category—wells, pipelines and subsea installations—are often associated with the other two categories of structures and otherwise are self-explanatory. In addition to steel structures a number of structures are made of concrete and these can be gravity hollow leg structures with either internal or external guide tubes. They can be piled into the sea bed or may just rest on the sea bed. Added to this steel and concrete may be used in the same platform or in adjacent platforms which are interconnected by pipelines or in some cases by walkways. Around these structures a number of working divers and submersibles operate.

Semi-Submersibles

Ships that are used offshore are protected by the same techniques as standard hulls. The storage tankers, the drilling vessels and the supply ships all can be fitted with conventional cathodic protection. Where a ship is laid up as a storage vessel care has to be taken that the reference electrodes do not foul, (the impressed current anodes will not) as this can give faulty readings. Where the ship is operating as a transfer vessel some care must be taken when other vessels come alongside. Bonding is essential but ideally both vessels should be at the same protected potential; this can be arranged automatically where the ships have impressed current protection using information retrieved through the bonding system.

The protection of semi-submersibles is best undertaken by treating them as a pair of hulls. The lack of dry docking facilities for the majority of them means that the best system will be an impressed current one using long life anodes. Such a system was fitted more than a decade ago to a semi-submersible and in order to extend the life of the anode the conventional lead/silver hull anode was clad with platinized titanium strips. The lead/silver acted to distribute the current and should the platinized titanium have failed for any reason, would have been an anode of last resort. The protection of the tubular legs which rise from the submersed hull and the cross bracing members can be achieved from the basic pontoons. Distributed power units will keep the anode cables reasonably short. Alternatively, jetty-pile Manganodes could be fitted around the inter connecting bracing members and these powered in a similar manner. When the semi-submersible is lightened these would be available for inspection, repair and replacement. This system provides cheap, effective protection.

The fitting of sacrificial anodes and their replacement is costly and the techniques that are used suggest a very low efficiency for the anodes, though this is explained by ascribing a requirement for a larger current density than they do under other circumstances.

Spar Buoys

Spar buoys, that is tall, vertical structures, are used for storage, for mooring and for flaring gas. These structures can be protected as extensions to the conventional buoy with hull type anodes attached to the body and additional anodes on the links of the anchor chain. The anchors themselves can be protected by sacrificial anodes which have to be prefixed before the anchor is laid. The anchor plate is often a massive drum which is sand or rubble filled and anodes inside this might better be of zinc which has a bactericidal effect; necessary because much of the sea bed mud contains sulfate reducing bacteria.

The body of the buoy itself can be protected by impressed current and this decision must be based on the size of the structure and the availability of power. The longer life that can be expected from impressed current systems is an advantage, as is the reduction of the dead weight of the anodes. Again, precautions have to be taken when off-loading the oil stored in the buoy, and either resistive hoses are used or bonding to the vessel must be complete. The techniques of mooring and access vary.

Offshore Platforms

There are a number of types of offshore platform. Perhaps the most common is the multitubular structure which is positioned on the sea bed and then anchored by piles. A typical North Sea structure may be standing in some 400 ft of water and will have four or six principal vertical or near vertical tubular members which will be cross-braced and strutted with smaller tubular sections. This will be prefabricated ashore and placed in position so that the top of the structure is some few feet above the wave tops. The jacket will then be secured to the sea bed with piles and capped with the working platform.

The main structural legs will be anchored to the sea bed by piles which will be driven through tubular guides and then grouted into position when driven; six or eight such piles may surround each of the corner legs. The platform depending on its use will have a bank of conductor guide tubes located within it through which the wells, averaging about 24, will be drilled. These guide tubes will be clustered either in a group or in a linear disposition. The structure will have to stand up to the rigors of the weather found in many of the offshore locations. It will be subject to massive tidal longitudinal forces and these will be increased by fouling on the structure and further increased if bulky cathodic protection is applied. The working load that the platform can support may be significantly reduced by this or, indeed, by the sheer weight of sacrificial anodes. The guide tubes often form a considerable barrier to the tidal flow and the water velocity between the tubes can be high. In some areas this is sufficient to cause an increase in current density in the zones of most rapid water movement and by the vibration of the guide tubes in the tidal flow.

The area between wind and water will be subject to wave action, more than on the normal inshore jetty whose location is selected usually with a view to some protection from the worst of the weather. The tide will rise and fall, though its range will be less than inshore. This presents a tremendous problem in the area of the wind and water line and has been solved by the welding of monel spats to the main legs and by designing the structures to have the minimum cross section traversing this zone. In the even more stringent conditions of areas such as the Cook Inlet the multitubular platform has been replaced by a monotubular or bitubular constructions.

Other techniques and spat materials may be used but are not presently common. The monel sheathing extends some 30 ft below low tide and this is generally considered to be as low as the average storm will penetrate. At the foot of the sheathing there will be a massive electrolytic cell.

The platforms will probably be interconnected by a series of pipelines, there will be risers within the platform, and there may be other tubes such as disposal tubes, sea water pump and discharge tubes. The platform will probably be surrounded by some discarded or accidentally dropped steel and there may drilling mud and spallings in the vicinity of the platforms.

A structure of this size will be subject to very heavy stress, particularly under storm conditions and this, coupled with the corrosive environment, will mean that cathodic protection is essential. It has been suggested that in the absence of cathodic protection the possibility of stress crack initiation during a 24 hour North Sea storm would be an unacceptable risk.

The crew onboard the platform will have a prime duty of obtaining the maximum oil or gas, or whatever their function is, and there will be a premium on accommodation and services so that although information on the state of the cathodic protection is essential, manned systems will not be attractive.

The steels that are used in the construction of the platform, and in particular in the high stress areas, will be of a type that will be susceptible to hydrogen embrittlement if excessive cathodic protection is applied. They will also be steels to which welding will not normally be possible and they will form a considerable percentage of the overall area of the structure.

The structure will be built in a yard which will have an industrial and/or marine atmosphere and in consequence it will corrode while being assembled. At the moment few structures are coated and those that are, have been coated most carefully but at great expense. Steelwork will roughen and lead to an increase of between 25 and 50 per cent in the actual area of the cathode as opposed to the superficial surface area of the steel:

this may help explain the differences in current density found between various structures. The structure may have incorporated into it storage tanks and/or flotation tanks. The latter should be removed after the structure is in position. The structure, though stationary, will be subject to tidal flows and these may be very high in certain operating areas. In estuaries the water may be stratified or contain considerable quantities of silt. In most offshore areas the water will be fully oxygenated and this will be consistent with depth. The steel members will generally vibrate and this movement will give a rapid replacement of oxygen to the cathode surface. This can increase the current density demanded beyond that of a normal stationary steel structure.

Cathodic Protection Design

The design of the cathodic protection for offshore structures has grown up from the oil industry and not from marine oriented corrosion engineers. In general the problem has been tackled by people used to designing protection schemes for pipelines, tank farms and oil refineries. There is, therefore, considerable duplication of the early work and the failures of marine cathodic protection. One of the main problems is the large current densities that are required and these can cause considerable voltage drops in the sea. A current density of 12 mA per sq ft can cause a potential gradient at the surface of 12 mV per ft and over the distances involved this can endow the cathodic protection with a considerable degree of cathodic control.

The structures are designed to operate for a period of at least 30 years and in many cases political considerations may require a longer life from the structure if it is decided to conserve the resources in the particular field. This poses a considerable problem to the cathodic protection engineer who is normally faced with design lives of the order of 10 to 15 years, it being assumed that after this period replacement of the anodes or major maintenance of the electrical equipment will be required.

The jacket, when first placed in position, will not have electric power available on it and this period may extend for six to 18 months. The initial cathodic protection of the structure must, therefore, be undertaken with sacrificial anodes. Following this period there is a choice of sacrificial anode or impressed current protection.

Sacrificial Anodes

The sacrificial anodes that are designed to have a 30 year life will, of necessity, give some degree of overprotection initially and as they are consumed the overprotection will reduce. The anodes themselves will need to be massive and the largest anodes that can sensibly be cast and handled will be about 1000 to 1200 lbs in weight and measure perhaps 10 ft × 10 in. × 12 in. They will generally be located clear of the structure by approximately

FIGURE 203 - Stand-off aluminum anodes for offshore. (Photo reprinted with permission from Impalloy, Ltd., Bloxwich, Walsall, England,)

1 ft and will be supported by an insert which will need to be strong enough to stand not only the stresses on the anode but also the pile driving shock, Fig 203. Where they are welded to the tubes it will often be necessary to use a doubling plate to reduce the stress.

Aluminum and zinc anodes give approximately equal performances and to ensure a long life the additional 50 mV drive of the Indium anodes will not be an advantage. The resistance of the anode will be less than that required for the 30 year life so that the anode output will be limited by the overprotection of the steelwork reducing the available driving voltage. Based on this system anodes would be distributed throughout the structure with additional anodes in the lower 10 per cent of the platform to provide protection to the piling and in the upper 25 per cent to cope with the depolarizing effect of storms. In general terms, about 40 per cent more anodes are located in these two areas, though where sulfate reducing bacteria are suspected a larger concentration of anodes at the base of the structure is recommended. The weight of anodes required can be calculated from the steel surface area and by taking a criterion of 15 mA per sq ft over a 30 year period. This will allow for overprotection in the initial 10 years, with the possibility of some lack of protection later. Monitoring of the system would be ideal and this is discussed later.

Should the anodes fail to provide adequate protection, then there are two or three options open to the engineer. Firstly, he can supplement or augment the anodes. To this end it would be useful if the anodes were made so that a further attachment could be made to them. This should not be necessary for many years when the original anodes have at least partly been consumed. Alternatively, additional anodes could be placed onto the structure itself, though this is an extremely expensive process which is quite unnecessary.

A form of bracelet segmented anode has been suggested for augmenting failed anode systems on tubular platforms. The system has little to recommend it as besides its excessive cost the anodes will corrode very rapidly, overprotecting the immediate steel, and with the possibility of a bacteria growth beneath the anode where this is deaerated by the presence of the anode.

The immediate contact with the platform members must be by a structural steel hoop as the anode itself will corrode away and lose contact with the metal. Point contact by some form of tightening hardened screw could lead to massive failure if the anode were to move when the hardened tips of the contact screws could act as a pipe cutter.

An alternative and much cheaper method of augmenting the cathodic protection is to use magnesium anodes and to place these remote from the structure on the sea bed. An array of anodes in a space frame is shown in Fig. 204 a and b, which was designed by the author for use in the North Sea. The assembly can be easily lowered to the sea bed and a single cable—in this case, a steel wire rope—is taken to the foot of the structure and attached there. Habitat welding made this a sound electrical connection or a pre-attached connection point could usefully be built into a platform should such a connection later be required. The inclusion of a prewired shunt for current measurement may be a useful addition, Fig. 204 b. Four of these units placed around a large North Sea structure are sufficient to provide it with almost immediate protection. In the case of a structure that requires its protection augmented it is a most cost effective arrangement.

Magnesium is designated an unsuitable anode for most offshore ap-

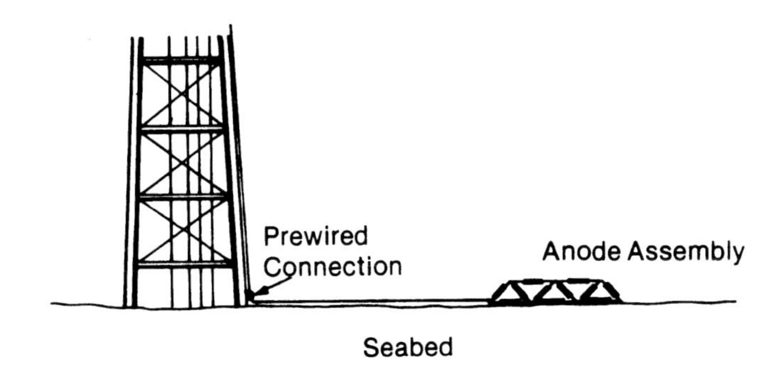

(Figure 204 continued)

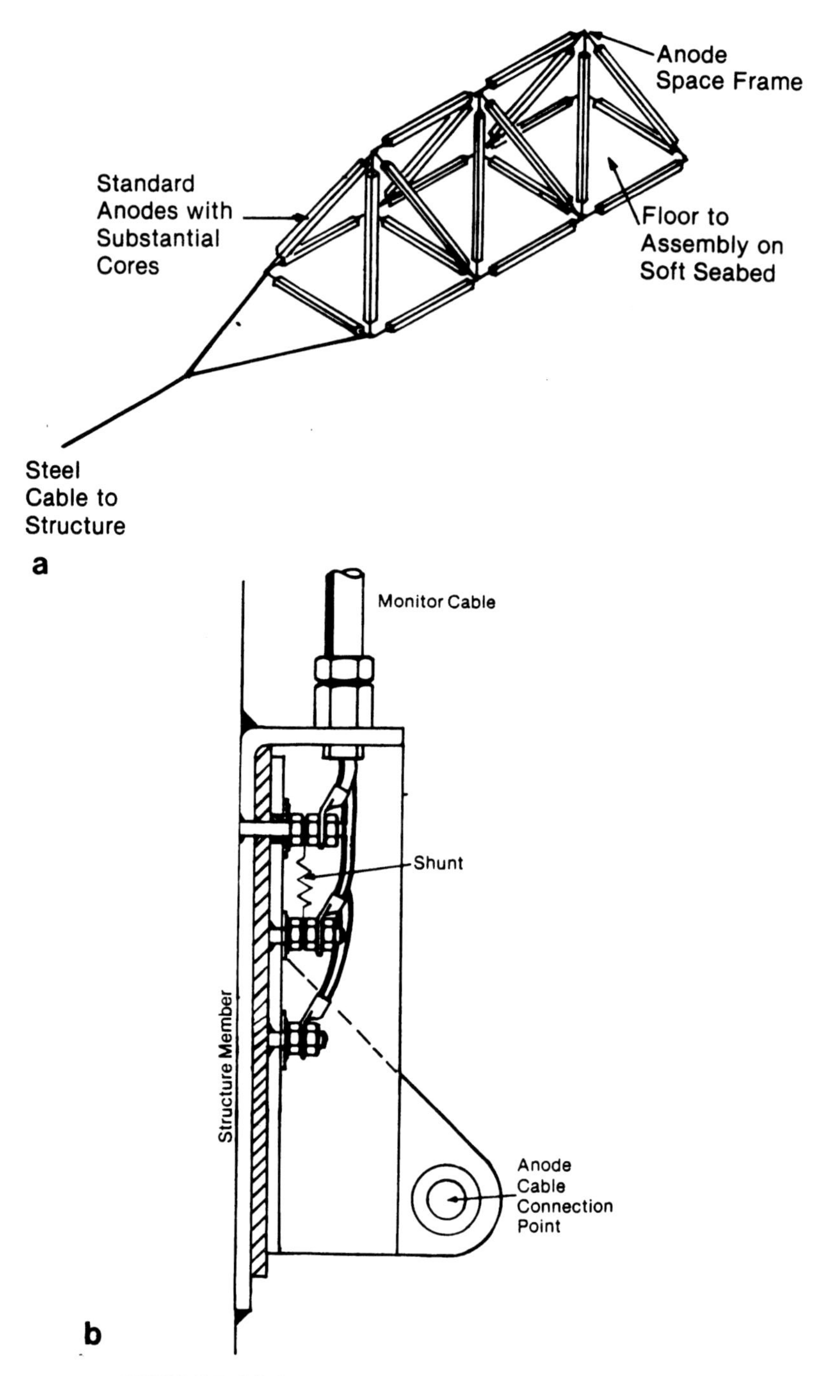

FIGURE 204 - Anodes in space frame arrangement and layout on seabed.

plications as it is considered to corrode too rapidly. This does not take into account the size of the installation and the fact that the surface of the magnesium in relation to its mass is reduced; a 10 to 15 year life can be expected from massive anodes. The simplicity of the installation would make replacement every five years very much cheaper than the use of multiplicity of low voltage anodes throughout the structure.

An augmenting system of impressed current could be retrofitted to the structure. This could be extremely simple and could be either by anodes placed on or above the sea bed when power cables to the anodes would be the principal problem or by anodes which were suspended or attached to a taut line within the structure. The latter would appear to be the better solution as there are no problems with stray currents or with damage to the anode cable. Simple systems have been devised that can form a string of anodes within the structure. One of these is illustrated in Fig. 205, and has

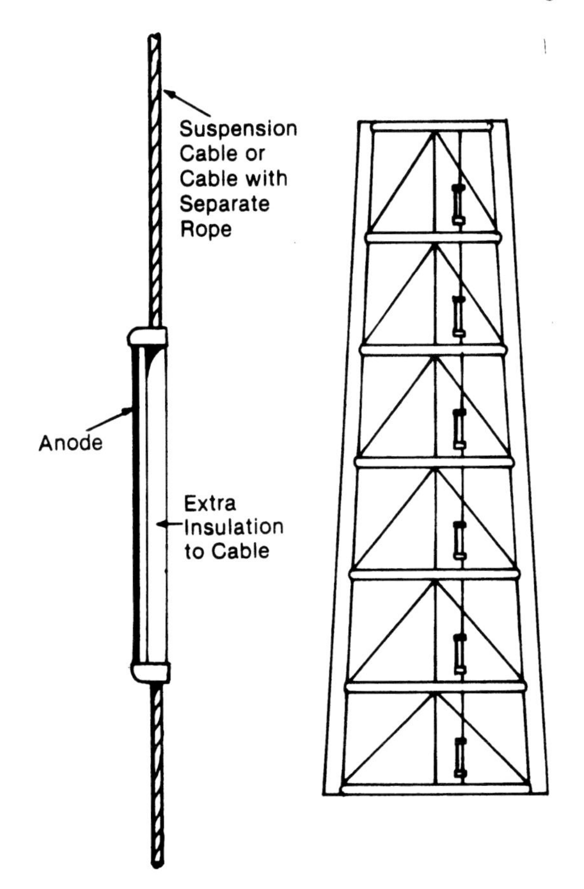

FIGURE 205 - Pre-wired connection to platform with built-in shunt.

been used on a number of platforms. Because there is already an existing sacrificial anode system the problems of distribution of the impressed current would be reduced and protection to the whole structure would be boosted.

The third technique which could be used is to polarize the structure from a temporary cathodic protection system. This could be done, for example, by the use of a small fishing boat with a generator onboard, holding itself off the structure by a steel cored cathode lead to the structure with impressed current anodes suspended from it. Polarization of a week or so, possibly at each of the major faces, should ensure a lower current requirement and full protection from an otherwise inadequate system. The economics of this type of operation are simple but would not be attractive where this type of operation was limited to a short period by weather.

There may on some occasions be problems not with the overall protection of the structure but merely confined to certain areas. One of the usual ones of these is protection of the guide tubes. In this case a string of impressed current anodes may well provide the answer as they will give a local boost to the protection.

Impressed Current

The design of an impressed current system for an offshore structure is difficult. The anodes that are used in the design will need to be held carefully and a multiplicity of cables will have to be brought up through the wind water line or within the tubular leg to the upper deck. In the design of the system scale is a most important factor. Some of the bigger failures of cathodic protection have completely ignored this making models of the platform and believing that this could be simply extrapolated to the large scale structure. In the majority of multitubular structures there will be a series of caverns within the structure and it has been usual practice to place one impressed current in the center of each of these. This means that the anode has to be held by some technique in the center of the section, Fig 206. This will then provide current to the enclosing members. Because of the size considerations it will be difficult to spread the current outside this first layer of steelwork. For example, if the current output from the anode is doubled the main effect will be to overprotect this first layer and only spread the protection to perhaps a further 25 per cent of the steel. This is the reason why massive increases in the impressed current have not resulted in a sympathetic increase in spread. It is much better to use one of the members that is between two such caverns and to apply the principles that were used in the jetty protection by attaching anodes to this member but isolating it from the overprotection that it would otherwise receive. This extends the protection to two, or in some cases possibly extends it to four such caverns, and, therefore, reduces the number of anodes without decreasing the evenness of the spread of protection.

FIGURE 206 - Rod platinized niobium anode mounted in twin resin-glassfiber insulated on pipe stand-off to protect offshore structures. (Photo reprinted with permission from Impalloy, Ltd., Bloxwich, Walsall, England.)

It is suggested that in impressed current systems on a platform there are advantages in using platinized niobium anodes, over platinized or coated titanium or lead silver. This seems to be a fallacy based on the increased capacity of the anode, whose only effect is to prejudice the longevity of the insulators next to the anode. Systems have been proposed that have replacable anode rods and that have duplicate wiring system that can be used on failure. There seems no point in sending a diver down to put an anode in place when his time will cost three or four times as much as the value of the anode and his life is unnecessarily at risk. It would be better to install twice the number of anodes even if they are to be used sequentially. It is suggested that one anode will cause interference on the other. This is a simple electrical problem which could readily be overcome and the potential variation on a parallel rod would be very low indeed. The use on one particular platform of duplicate wiring involved, in the original design, changeover links in flameproof boxes located in a hazardous area of the platform. From the number of anodes and the number replacement suggested they would only last approximately two thirds of the life of the structure and one of these links would need to be changed on the average every week at the peak of anode failure. To change over one link some 20 bolts had to be undone in an atmosphere which had to be made safe for working. Engineering this sort of impediment into the operation of the impressed current must be avoided.

A number of systems have been developed in which the anode is recoverable by the use of tubes which carry the cable and from the end of which the anode protrudes. In some of these the tubes carrying the anode were given a very gentle radius so that the anode would not foul on its descent. However, in attempting to recover the anode a long curve means that the anode cable is in contact over a large area and becomes almost impossible to move. This would result in the cable breaking and the anode being irrecoverable. Redesign of this type of system could lead to its better use. The tensioned wire anodes that are either held by a cable made strong by its armoring or an insert of by traveling over a tensioned rod, cable or rope, are ideal for augmenting sacrificial anode protection but seem to be difficult to justify as the principal cathodic protection system. Their construction is of necessity light and with currents of 10,000 amps on some structures, the engineering of such systems becomes difficult.

There are a number of designs for placing anodes on the sea bed. These as a prime system of cathodic protection where there is a multitubular platform cause tremendous difficulties which are described in the chapter on electrolysis.

Composite Systems

The basic reliability and simplicity of sacrificial anodes would make them ideal for platform cathodic protection were it not for their weight and bulk. Impressed current, on the other hand, becomes difficult when it is expected to provide complete protection and particularly complete spread of protection to the whole structure. There is a considerable case for systems which employ a mixture of the two using the distribution and simple properties of the sacrificial anode but reducing their overall bulk and weight by the application of impressed current. The system is not quite as simple as it sounds as the sacrificial anodes would be required to provide the initial period of protection, that is before the platform was placed on top of the jacket, and would also be required to recover and protect the structure should there be some failure of the impressed current system. This condition may be superfluous as the reliability of impressed current is equal to that of sacrificial anodes. The impressed current must, however, reduce the consumption of the remaining sacrificial anodes so that the lighter weight of anode is not consumed early, robbing the structure of the necessary spread of protection in the latter part of its life. This can be achieved by a number of techniques but anode shape and size will play a large part. Because there will be two conflicting demands—firstly, during the initial period before power is available and then for the continuous operation of the platform—either two different types of anodes or a shaped anode that will, through its life give two different types of protection is required. These latter anodes were developed for use in tanker ballast compartments and some return to those type of shapes of anodes may be what is required. Different anode metals can be used and if a rapid consumption of some anodes is required it is useful to build in automatic detachment of the anode insert. This can be

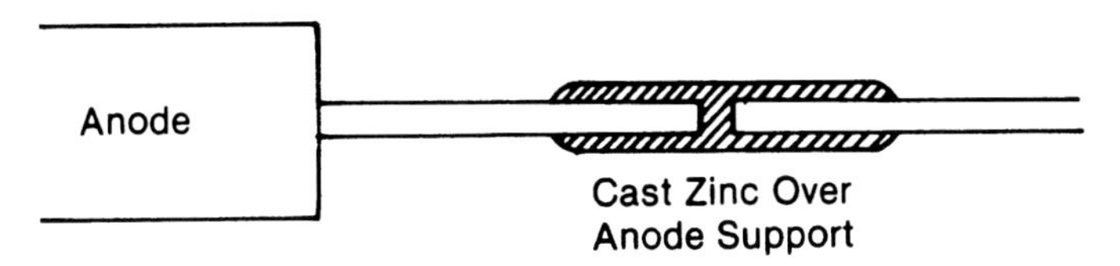

FIGURE 207 - Anode string for augmenting protection of platform.

done as in Fig. 207 by including in the magnesium anode insert a small zinc casting. When the magnesium is consumed it will corrode and release the majority of the insert.

Control of Platform Protection

Sacrificial anodes will operate without control but will need to be monitored. It would be useful to monitor both the anode output and the structure potential. While continuous monitoring will be useful measurements taken by divers or other means during period surveys may be sufficient. These should be done with considerable accuracy so that any work that has to be done on the structure can be predicted well in advance of the work period. Other essential work required for the operation of the structure must take precedence and this may mean where there is a short work window the indications of inadequacy or failure of the anodes must be made two or three years ahead of installation. Accurate monitoring will require potentials measured to within one or two millivolts and anode currents accurately determined. Impressed current protection will be monitored more simply, its current output and the potential of the structure will be available to the platform personnel. Automatic control will be required and this will operate from a small number of reference electrodes. As in the protection of ships' hulls these will need to be carefully placed so that the control electrodes give a sensible picture of the overall protection of the structure and not act as a mere monitor to one close anode. The information that is displayed will need to be immediately intelligible to the operating crew. Fig. 208 shows such a system in which the diagram indicates directly the adequacy or otherwise of the cathodic protection system. Beyond this are located the meters which indicate the readings of current voltage and potential and which allow further detailed exploration of other reference electrodes. This immediate and supplementary display of instrumentation will be necessary if it is to be read by the platform personnel. Alternatively, all these results can be fed into a small computer which will not only monitor but will correct and analyze the cathodic protection. This will indicate proper operation and will provide a prognosis for the system.

When this type of equipment is operated in conjunction with sacrificial anodes the location of the electrodes becomes even more critical. It is essen-

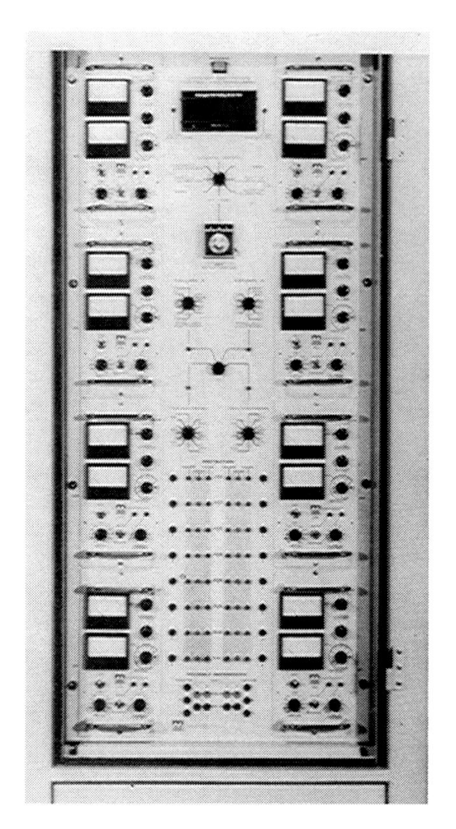

FIGURE 208 - Automatic control console for twin tower platform. (Photo reprinted with permission from CORRINTEC/UK, Ltd., Winchester, Hants, England.)

tial that sacrificial anodes are not either influencing the electrodes by themselves or in combination with the impressed current system indicating a better level of protection than is being achieved on the rest of the structure. This problem can be difficult, particularly if there is reason why an anode that was expected to have exhausted itself in the initial phase continues to operate and gives such false indications. The use of a computer to analyze and compare the results from different parts of the structure which should be given equal cathodic protection will be of great advantage.

These composite or mixed systems will probably be used increasingly and the advances in electronic computing and control will hasten this.

Single Tubular Structures

The problems with the multitubular gravity structure are simplified in the single or twin tubular structure of the type that is used in the Cook Inlet. However, when this type of structure is used it is because other operating conditions in the area make it necessary. These also affect the cathodic protection and, in the case of the Cook Inlet, there is a massive depolarization, indeed an almost wet sand blasting, of this structure. The anodes have now to be placed either on the structure, if they are sacrificial, or away from it if impressed current. The anodes on the structure itself, particularly close to the water line, can be damaged by ice while the impressed current anodes have to survive on a shifting sea bed. Anodes that float off the structure can be useful and some of these are employed on jack-up platforms which have four tubular legs to which anodes could not be attached. The problems associated with floating anodes are mainly ones of engineering; anchoring by a heavy cable is ideal in comparatively gentle sea conditions but off the rougher coasts huge anchors have to be used. There is a considerable problem with the use of remote anodes and both because of the massive overprotection to the outside of the structure and safety considerations with the operation vessels and divers in the area they are not recommended and should be prohibited. However, on this type of structure there is no alternative and anodes mounted on sea sleds or tethered to the bottom would seem to be the answer. The weight of lead anodes can be an advantage or anodes of platinized titanium or platinized niobium can be used. The problems with the detachment of the platinum film at low temperatures seem to be overcome and are certainly less on rods than they are on flat plates. The use of reference electrodes on such a structure is essential and recessed electrodes as used on the bow of tankers are to be recommended. These may need, however, to be further reinforced and to have some form of ice guard placed on them.

Tethered and Tension-Leg Platforms

In place of the gravity platform there are a number of proposals for tethered platforms, that is structures of much lighter construction which are secured at the sea bed, usually with some form of swivel joint or inbuilt flexibility in the tower, and the tower is guyed to make it stable. These structures are proposed for deep water applications and resemble a television transmission mast. The protection of the vertical structure is an extension of the protection of the gravity platform but the problem arises in the protection of the guy system. A similar problem arises with the tension platform in which a floating structure is held rigidly by vertically tensioned members which have a negative effect on the buoyancy of perhaps 10,000 tons. The platform, therefore, remains stable in all weathers, the tension being adjusted so that even in the most critical storms is not released.

These two types of structure are illustrated schematically in Fig 209. The tensioned guys or cables used in these structures are highly stressed and in the absence of cathodic protection would fail. Plastic ropes may be

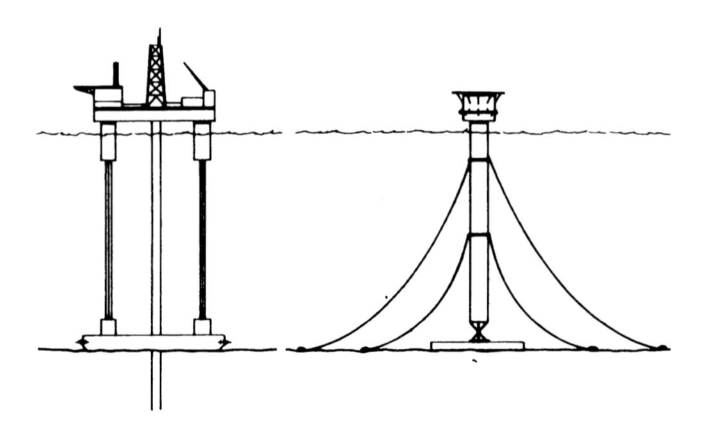

FIGURE 209 - Tension-leg and tethered platforms.

developed but at the moment it appears that only high tensile steel will serve. The length of these tethers and guys will vary and where it exceeds about 100 ft there will be considerable problems.

Wire Ropes

If wire ropes are used then these can be cathodically protected on the outer surface of the outer wires only. Inside this there will be a deaeration and there can be attack by sulfate reducing bacteria; there can also be crevice corrosion. The wire rope, therefore, has to be made with a mastic laid into it that is impervious to sea water, is not affected by the cathode alkali, nor is it consumed by anaerobic bacteria. The cathodic protection of the outer layer of the wire will be restricted because the surface area to diameter of each individual wire will be large. The area of the individual strand that is exposed will be πr and the cross section will be πr^2. This will be the equivalent surface to volume ratio of a pipeline with wall thickness equivalent to the radius of the outer strand not the diameter as would be expected. It will not be possible massively to overprotect the end of these wire ropes as this would mean overprotection of the structure or the anchor, which would be uneconomical, and invite the possibility of hydrogen embrittlement.

It will be necessary, therefore, to arrange for the guy to receive protection at a number of points along its length, or that it is constructed by some other technique which reduces its overall current requirement. The method by which the cathodic protection can be achieved along the length of the guy is limited. Sacrificial anodes could be attached but this would need to be done with care so that hard points do not appear in the wire. With impressed current systems it is the cathode connection that is important and it may be possible to make a cathode connection and to carry the cathode cables along the guy to the structure. The current can then be drained from the cable at three or four locations reducing the potential variation in the guy and providing protection throughout its length.

It may, alternatively, be possible to short-circuit the wires together at a series of points along the rope and to use the other wires as part of the overall conductor. Under these circumstances it may also be possible to have a core of a metal such as copper which has a superior volume conductivity and tolerate the loss of strength that would result. Techniques that would fuse or otherwise interconnect the wires, either at a single point or distributed over a short distance and would not be major stress raisers, could achieve this.

The guy can alternatively be coated. If the whole of the guy is sheathed then there may be problems with this bonding of the sheath to the wire rope. A better technique would be individually to coat the strands so that each of these is an insulated wire. Modern baked-on coatings could probably achieve this and then the spread of protection through the whole of the wire rope may be achieved from the ends.

In this respect it may be possible to enhance the protection close to the ends of the wire rope without overprotecting the whole structure. This could be done using the high resistance of the wire rope itself by cleaning off the section nearest to the platform and attaching a cathode lead to some point 50 ft along the rope. The potential then could be enhanced, say by 200 mV, and this would be dissipated in the uncoated section over the 50 ft while the spread in the coated section would be 10 to 20 times greater. This would be possible because the resistance of the coating film would be somewhere between 100 and 500 times greater than the bare metal and the spread factor would be improved by the square root of this figure.

The alternative for tension leg platforms is to use rod or tube as the tension member. For example, drill pipe could be used and this would be comparatively readily installed at the site. Providing a good connection is achieved at the joints, and that this is indeed a low resistance, then the drill pipe would be much more easily protected as it would have a much larger cross section compared with its surface area. The lengths of drill pipe could also be coated and, although some damage would result, protection over lengths of upwards of 1,000 ft would be achieved.

This technique would not require massive overprotection of the main platform but some overprotection would be necessary. Sacrificial anode protection of the floating structure would not be possible and an impressed current system with perhaps 100 to 150 mV overprotection would be used. At the sea bed the anchor system in both cases would be protected by sacrificial anodes and this being in more static water could be overprotected perhaps by 50 mV without excessive wastage of anodes.

Both the wire rope and the solid tethered members will vibrate very considerably in the normal sea conditions. This vibration will increase the rate of arrival of oxygen at the metal surface and a large increase in current density will be observed. This may be exceptionally as much as 100% increase. The use of dampening devices as on overhead transmission lines may be possible and some of the cathodic protection apparatus may be located within these.

Concrete Structures

A large number of gravity structures have been built of reinforced concrete. These usually are completed with massive tanks at the base and one, two or three hollow columns supporting the platform. There are two types of platform: those that have internal conductor tubes and those that have external conductor tubes.

The cathodic protection of the reinforcing in the concrete is comparatively straightforward if three rules are followed. Firstly, the reinforcing bar must be interconnected and must be so arranged that there is sufficient current path to interlink it all into the cathodic system. Secondly, the system must not have widely differing potentials relative to the water between the inside and the outside of the structure. Thirdly, wherever possible connection should not be made between the reinforcing and steel which is subject to free sea water. This has been dramatically illustrated where ladders, for example, that have been attached to the reinforcing and have corroded away in a matter of months.

The quality of the concrete will be an extremely important factor as will be the cover over the outermost layer of reinforcement. Good quality control in the concrete and a cover of 2 in. will reduce the corrosion to a minimum. In terms of cathodic protection the current density onto the reinforcing will vary between that required with good polarization about 2 mA per sq ft, to about a quarter of that value. It would be extremely difficult to judge the quality of the concrete on a structure and, indeed, some specifications of high strength concrete are too lean to form an effective barrier. In well made concrete current will not flow beyond the first set of reinforcing, whereas in badly made or lean concrete it may penetrate several inches further. The structure will be cracked, with a vast number of hairline cracks, and in some cases with considerably larger voids.

Many designers work on the principle that concrete requires about half the protection of the equivalent frontal area of steel. This protection can be achieved with sacrificial anodes which, of course, have to be attached to the reinforcing or, at least, interconnected to them, or by impressed current when the cathode leads, again, have to be carefully distributed. In many cases the concrete platform is surrounded by a steel skirt and attaching anodes to this, coupled with its interconnection to the reinforcing, give good protection.

Internal Conductor Tubes

Some of the structures have the conductor tubes placed within one of the columns. These, about 24 in number, extend downwards and are supported and guided by frames. These frames are interconnected with the reinforcing and cathodic protection is needed to prevent the corrosion of the guide tube frames and the conductor tubes themselves. This must be done without causing large variations of potential over the surface of the concrete where it is isolated from the internal steelwork or, if it interconnected with the reinforcing, protection must be provided to the reinforcing that is on the inside of the tower. Because the tower is enclosed it is likely that drilling mud will fall into the tower and it will be open to the sea water so that there will be a rise and fall with the tide. These openings, however, will not be great enough for wave action to be followed inside the tower. The guide tube frames could be protected by attaching sacrificial anodes to them treating each as a separate structure and providing enough current to protect it. To protect the conductor tubes it would be necessary to ensure these were all interconnected and that current could take a return path through this interconnection from the conductor tubes. This would be difficult to achieve at the limited driving voltage of the sacrificial anodes would probably mean that the lower end of the conductor tubes would not be properly protected. As the conductor tube would pass close to the guide tube then unless there was sufficient fortuitous interconnection there could be some electrolysis present. It is not possible to pre-attach anodes to the conductor tubes themselves as this would prevent them passing through the tube guides.

This problem does not arise on the massive steel structure as the steel itself provides the interconnection between the anodes and the conductor tubes and the conductor tubes have sufficient cross section that they are protected as a common cathode.

Impressed current can be used to protect such a system and this was carried out for one of the Condeep platforms in which anodes were suspended in the area between the guide tubes. Polyester glass fiber guides were placed on the frames during construction of the system, together with ropes which would be used to pull down the anodes, Fig 210. The anode assembly was made from platinized titanium rods which were connected at both ends to a cabling system. The structure was divided into four levels and the cables were taken to run through the first section of the anodes and the three cables to the lower section were carried past the anode through the hollow members of the anode cage. The system was supported on steel wire ropes so that it could be lowered into the tower from the limited space on the platform. Sacrificial anodes were placed on the lowest section where penetration into the mud would be restricted and these were made of zinc. Reference electrodes were placed at each level and cabling from these taken to the operating deck and from there to the control room. The disposition of the anode system was calculated using a model and the potential variations

FIGURE 210 - Resin glass fiber guide cones for string of anodes within offshore concrete tower.

to be found with different anode locations were determined from the model. The current density requirements were laid down by the Classification Society and the electrode positions were determined from the model.

The system has worked extremely well, though there has been some problem with the low current output that has resulted, the demand being very much less than specified, has caused heating problems with the transformer rectifiers. The system employs automatic control and even protection has resulted.

There is one considerable problem with this type of structure because the steel inside the tower is in contact with the wells outside the platform in the sea bed. These are at the same potential and the cathodic protection inside the tower is achieved by raising the sea water potential there rather than depressing the metal potential. The result is a potential gradient through the concrete. While the design was made to minimize this effect, nevertheless there is a problem as current may flow into reinforcing which is fortuitously close to the inner surface of the tower and interconnected to reinforcing which is close to the outer surface of the tower. On the outer surface corrosion could be accelerated by this process.

External Conductor Tubes

On other concrete platforms the conductor tubes have been placed external to the structure. Again, the cathodic protection can be applied in the same manner, but unless there is good interconnection between the various levels it will be difficult for anodes on the lower part of the structure to provide protection to the conductor tubes. The anodes will require separate cathode leads to the surface and in the case of impressed current there is sufficient driving voltage for current to be drained from each section separately. With sacrificial anodes these interconnections must be massive. A remote system of cathodic protection anodes could be used and it could be combined with local protection of the guide frames using sacrificial anodes and the conductor tube protected by impressed current. The small amount of current comparatively that would be used by this method would not cause hazardous potentials within the sea, but localized large potential variations could be detrimental to the concrete. Care would need to be taken in the design of such a system to avoid these variations.

The cathodic protection system protecting the conductor tubes would also protect the wells though the spread of protection to the wells might be limited.

Protection of Wells

The protection of sub-sea wells would be achieved by techniques similar to those used on land. Logging techniques to ensure the protection was being achieved could be used but this does not, at the moment, seem to be of great concern to the well operators.

Sub-Sea Completions

There is a growing use of sub-sea completions with a pipeline, whether it is a gathering line or an injection line, being connected to an assembly which is placed on the sea bed above the well. These systems can be adequately protected by sacrificial anodes. Anodes of the type used on the main structures are used, but care has to be taken where there open tubular structures that protection spreads inside these. Where there is doublt anodes should be placed inside them and in some cases ribbon anodes may be used. These areas are usually not sealed and are left as free flooding. A restriction on the amount of sea water entering this will increase the polarization and by de-aeration will reduce the cathodic protection current requirement. Where the structure is likely to be in anaerobic mud potentials appropriate to this should be sought and zinc anodes used.

Structure Painting

A small number of structures have been painted and the cathodic protection of these has been made much easier. Coating the whole structure is expensive and there may well be compromises in the techniques of coating which would allow the structure to be simpler to protect cathodically and yet not be exorbitently expensive in coating.

As much of a tubular structure is comprised of pipe-like tubes these could be coated using coat and wrap techniques of a conventional pipe mill and then built into the structure as coated components. There would be no need to coat the whole structure, though there is a considerable advantage in cleaning the steel in that it can be better inspected. The reduction of the current demand on the structure by half or even more would eliminate many of the problems found with the scale of the cathodic protection. This simple coating technique may well be more cost effective than complete coating.

There is also a problem at welds where there is a need to inspect these and also it is essential to achieve full protection at them. The weld, if not properly ground, presents a rippled surface which demands an increased current density. It would be useful at welds if an area either side of the weld, but not the weld itself, could be coated with an easily identifiable paint so that the diver on inspection is assisted in finding the weld and the cathodic protection, by virtue of the coating either side of the weld, is enhanced. Where the coating extends, for only 12 in., either side of a 4 in. gap at the weld there will be an approximately 50 per cent increase in the current density onto that area.

Submarine Pipelines

Most fuel is brought ashore from rigs and platforms by a pipeline which travels along the sea bed. This may either be on the surface, as are the majority, or trenched into the sea bed and backfilled. At the shore terminal the pipe is usually buried and often additionally mechanically protected for some distance out from the shore, this distance depending on the local conditions, often extending for a mile or more.

Pipelines that are laid over long distances are made from high tensile welded steel pipes which are coated, usually with a top pipeline specification coating, wrapped and then are overcoated with a weight coating. This usually consists of a concrete coating 2 in. or more thick into which is placed some form of reinforcing.

One of the techniques of laying which is most favored is to assemble long lengths of pipe, usually pre-welded lengths of mill pipe perhaps three or four pipes long and to weld them end to end on the barge, lowering them as completed into the sea. The joint that is made on the barge is cleaned, primed and coated after welding and sometimes weight coated with a prefabricated weight, sometimes just filled. Because of the laying technique the pipe traverses over rollers and is lowered via a boom which helps support the pipe. Anti-buckling devices are built-in at intervals. The cathodic protection of these pipelines can be accomplished by two techniques. When they are short they can be protected using the normal impressed current methods from the shore and from the platform. Usually, however, the pipes are too long for this technique to be used and the pipe is fitted with sacrificial anodes.

The usual form of the anode is a prefabricated bracelet which is either made of segments of anode attached to a pair of steel split rings, or the anode itself is cast as a hollow semi cylinder and the two halves are joined together. Two typical types of bracelet are shown in Figs. 211 and 212. The bracelet is usually made to conform with the weight coating on the line but occasionally the weight coating is insufficiently thick and then the anode is tapered, Fig 213. By doing this it will pass over the rollers and can be laid with the pipe. To ensure continuity it is usual to attach a jumper lead from the anode bracelet to the pipe. This is sometimes done by thermite welding and sometimes by other techniques, including welding to pads or lugs fabricated in the pipe shop.

The anode in operation is either resting with the pipe on the surface, when about one third of it will be covered with silt or sand that will build up on the sea bed unless it is exceptionally rocky, or it may be completely buried with the pipe. The product in the line is often hot and so the anode has to operate in warm or hot saline mud. Some aluminum anodes have been developed to work in this service while zinc anodes have tended not to react favorably under hot conditions. The aluminum indium alloys also have the additional 50 mV driving potential.

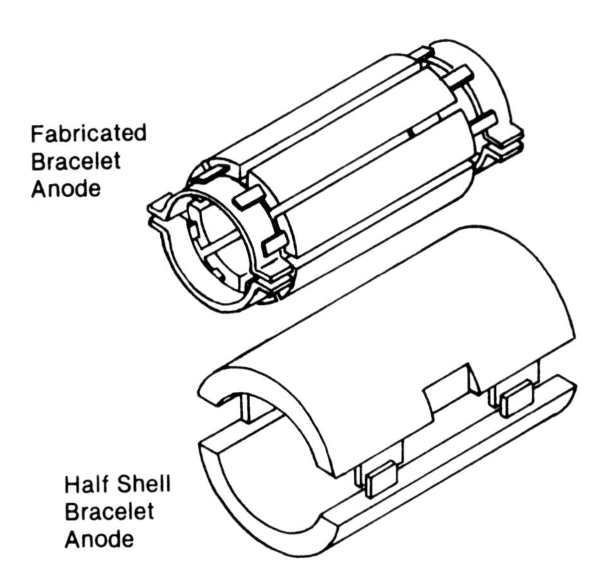

FIGURE 211 - Bracelet anodes for subsea pipelines.

FIGURE 212 - Zinc pipeline bracelet anode with two split semi-cylindrical sectors, the split allowing closer fitting to the pipe in the field. (Photo reprinted with permission from Impalloy, Ltd., Bloxwich, Walsall, England.)

Design of Protection

Where the pipeline comes ashore it is an advantage to use impressed current protection. The reason for this is twofold: Firstly, a very large length of the pipeline can be protected economically. This can extend some 10 miles out to sea on a well coated line. More importantly, as the pipeline approaches the shore it will be buried and gradually the area around the pipe will become leached of the salt by the fresh water bubble that extends under the sea. There will be a large increase in resistivity and the sacrificial anodes will probably be inadequate to protect the line. It is also the area where damage is most likely to occur to the pipe coating.

The current that is taken by the pipeline will be the same as the current

FIGURE 213 - Tapered bracelet anode for concrete weight-coated pipes. (Photo reprinted with permission from Impalloy, Ltd., Bloxwich, Walsall, England.)

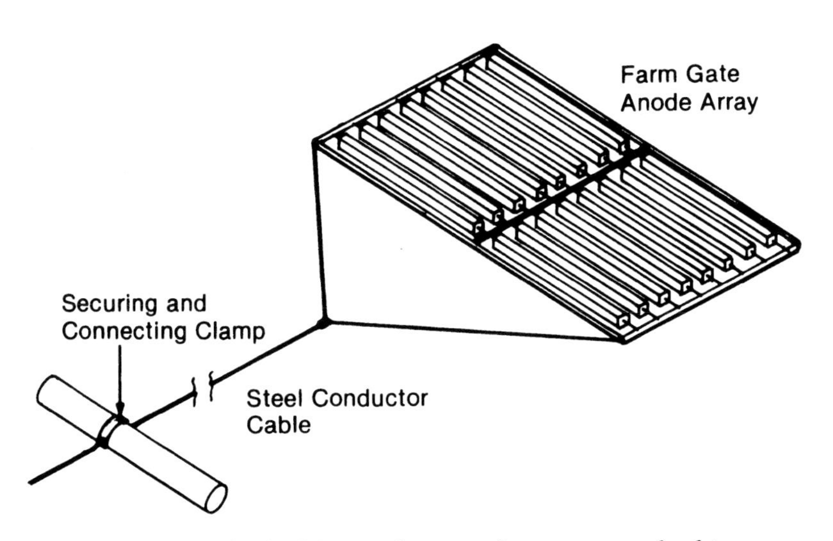

FIGURE 214 - Magnesium anode array attached to pipeline.

required for a land pipeline, with the exception that the difficult conditions under which the field joints are made may increase the current density at those points. An overall current density of about 8 to 10 microamps per sq ft will characterize the best coated lines and this may rise tenfold to average 100 microamps per sq ft. Where the pipe is damaged and the coating completely removed then the bare steel current densities will apply, that is about 3 mA per sq ft in sea water and about half that in mud.

The pipeline bracelets will be designed principally by the weight of anode required to provide protection. These may be placed at approximately 400 ft intervals and their size is based on the predicted anode consumption. In saline muds the anodes tend to operate with lower efficiency and where the pipe is partly buried then the exposed part of the anode will corrode very much more rapidly than the buried area. Taking these factors into consideration it is usual to over design a bracelet anode by approximately a factor of 1.5, which allows for the more rapid consumption should only part of the anode be exposed, and the stub wastage, that is the part of the anode that cannot be used towards the end of its life. Some designs require an allowance for the current that will be wasted towards the end of the anode's life where the supporting steelwork of the anode itself may become exposed.

In general this means that the anode will have a much lower resistance than that required to give the current output will tend to overprotect the pipeline and use its low voltage as a form of automatic regulation. Were the pipe to be severely damaged the overprotection of the bare area could lead to very rapid consumption of the anodes.

In a survey of bracelet anodes on pipelines it was found that close to platform structures the pipeline anoes invariably are helping to protect the structure. It is good practice to isolate the pipeline from the structure using an insulating coupling and this is usually placed above the water line. It is difficult to maintain complete isolation and some leakage of current between the structure and the pipeline inevitably occurs. The pipeline is similarly isolated on the shore end and this is particularly so if it is protected by impressed current as the extra swing of potential given to provide a good spread would otherwise be wasteful of current at the shore installation, comprising a tank farm and associated buried pipes.

There are some pipelines, such as risers, water pipes, etc., on the rig which are themselves protected by bracelet anodes. Here the pipe is treated as bare and the current densities of the platform are applied. Under these circumstances bracelets may be applied at intervals of 40 to 50 feet and where the pipe is hot at about half these spacings.

Some structures, particularly jack-up platforms, use pipe like legs together with a base plate or independent feet and a platform which is elevated. The vertical legs themselves cannot have anodes attached to them because of the lift and hold mechanism. It is usual to protect the base plate and feet of these platforms with sacrificial anodes and also to put anodes on the jack-up section so that it is protected when it is being moved in the lowered mode.

Either sufficient anodes are placed on the base to provide protection to the whole of the legs or, alternatively, a suspended anode can be used to provide impressed current protection. Lead anodes, often mounted from a small derrick, are used and although these are occasionally damaged the simplicity of the installation makes them the most economical method.

Anode Resistance

The resistances of the common anode shapes were described in Chapter 3 but it might be useful to indicate how these can be used in offshore structures. Firstly, a sacrificial anode that is placed on a flat surface which is painted will have twice the resistance of the anode itself plus its image in the hull, that is as though the hull were acting as a mirror; this will also apply when the anode is placed on a very large cylindrical member. If the anode has a length L width w and a thickness t then the anode will have a resistance as formula 3. that is

$$R = \frac{\rho}{2\pi L}(\ln \frac{4L}{\sqrt{2rw}} - 1)$$

If the anode is moved off the structure to stand, say, 1 ft from the surface the resistance will be twice that of two anodes two feet apart. This was derived in equation 3.15 and the additional factor in that equation will alter the resistance by about 10 per cent on a large anode. This will be an increase in the resistance of the anode over the resistance it would have if it were a long way from the structure. In approximate terms the anode placed immediately on the surface will have about 15 to 20 per cent more resistance than the anode which is 1 ft off the surface and this will have about 8 to 10 per cent more than the anode that is remote. The resistance of a bracelet anode can be derived as indicated in paragraph 3 by considering the two equivalent spheres and these will have diameters one equal to the outside diameter of the bracelet and the other equal to the length of the bracelet. The resistance of the bracelet were it a full cylindrical section will then be given by the formula

$$R = \frac{1}{2\pi} \cdot \frac{2}{(d_B + L_B)} \rho$$

but because the circular ends of the cylinder are covered with pipe the result will be an increase in resistance of 10 to 20 per cent. The formula then becomes

$$\cdot \frac{2 \cdot 3}{2\pi} \cdot \frac{1}{(d_B + L_B)} \rho$$

The resistance of impressed current anodes will depend on their shape. Whereas a hemisphere will have a resistance of $\rho/2\pi r$ the disc will have a resistance in which π is replaced by 2 giving the resistance of $\rho/4r$.

The resistance of a long extruded anode such as a lead anode can be calculated by taking the resistance of the whole of the appropriate semi-cylindrical section and increasing the resistance by about 20 to 25 per cent because only part of the surface is emitting current.

In many impressed current applications in sea water the back e m f of the anode material will be very important and will exceed most of the errors in resistance calculations. For other shapes the resistances are given in Chapter 3, which includes a list of the resistance of a variety of shapes by Dwight and the modifications to them in practice.

Current Requirements

There is a large range of current densities recommended for offshore structures. Firstly, there is a large variation in the definition of the current density. A number of people describe the initial current density that would be appropriate over an initial period of three to six months. This is quoted because it defines both the capacity of the anode and its electrical resistance. Alternatively, a mean current density may be given and this is then used in conjunction with the life required to define the weight of anode that is to be installed, usually in all-embracing figure which takes into account stub wastage, inefficiency in current distribution, etc., and relies on the limited variations between the most cost effective—usually the largest—anodes. In addition to these a minimum current density is often quoted and this is the final value that can be tolerated. It is employed in conjunction with the initial current density and often allowed for in the mean or average figure. The minimum current density takes into account the anticipated residual life of the structure and the need for full protection. Many anodes will reach this stage at the time when they have become much more spherical and there will be a smaller change in resistance with their consumption.

All three figures may be quoted.

In general, the mean values will be roughly 8 mA/sq ft in the general ocean environments, 6 mA/sq ft in the less arduous conditions, as are found, for example, in the Arabian Gulf and the Gulf of Mexico, while the more exposed oceans such as the northern North Sea will require 9 mA/sq ft. Initial values will be about 50% higher. There is one major exception where the seawater continuously depolarizes the surface and the principal area that this occurs is Cook Inlet when the current density remains at the high level of 40 mA/sq ft.

Most of the current densities are quoted at a temperature of 15° C, this being a warm, ambient temperature. In chapter 2 the formula for the increase in current density was given, that is

$$\sigma_T = \sigma_{15} \left(1 + \frac{T - 15}{55}\right)$$

Current densities are much lower in the mud, both in the light surface cover and in the denser seabed mud. These are usually quoted at 2 mA/sq ft, and again, because of the presence of sulfate reducing bacteria, this is a continuous value through the life of the structure.

In many areas storms are of sufficient intensity and the water is superoxygenated and this combination can cause depolarization. The increase in the current density required for protection usually reverts to the initial value and polarization will not be re-established for several weeks. During the latter part of the life of the cathodic protection system the anodes may not be capable of repolarizing the structure. This problem was discussed earlier in the chapter.

A number of the classification societies and a number of the oil companies have their own rules which must be followed on the structures that they classify or own.

Failure of Pipeline Bracelet Anodes

There have been a number of failures of bracelet anodes in the North Sea. These seem to have been in three categories. First, zinc anodes have failed through high temperature operation. This was to be expected and had been predicted many years before. The anodes disintegrate by intergranular corrosion.

Secondly, with half shell bracelets a number of them have lost contact with the pipe and the author believes this may be because when the pipe is pressure tested it expands, though only to a small extent, but this is transmitted entirely to the steel lugs which link the two halves. In some anode designs these are very short and are stretched beyond their elastic limit. The anode then is loose on the pipe and is free to roll. Some of the anodes appeared to have failed by this technique as far as can be gleaned from the submarine video pictures.

A similar form of failure could occur if the anode metal is consumed next to the pipe. This could occur by several mechanisms and a substantial coating on the inner surface of the anode or a redesign of the insert seems necessary.

The third form of failure is that the anodes have polarized. This seems to be a random failure and has been picked up on survey. The anodes that fail exhibit very low driving voltage whereas those that have become electrically disconnected exhibit the open circuit potential of the zinc.

When the anodes have failed it is almost impossible to replace bracelets on the line. One technique that was used on a failed North Sea line was to place mats of magnesium anodes either side of the line and connect them with a clamp and habitat welded to the line, a typical anode array is shown in Fig. 214.

The anodes are placed about 100 ft either side of the line and instead of an estimated twenty such arrays made of zinc anodes, only four, placed in two pairs, were required. This reduced the cost of the installation to less than one third and, as the installation was about fifty times the cost of the anodes, the difference in the efficiency of the current from the magnesium anodes compared with that from zinc or aluminum was trivial. The line polarized rapidly and, after six years, is still exhibiting full protection potentials.

The line could have been protected by impressed current. However, in the area where it came ashore there were problems with laying a ground bed in the sea or constructing it on land. The magnesium solution, although five times more expensive than the impressed current, was adopted because it could be done immediately, there being no way-leaves, electricity supply or legal problems to be considered.

Cathodic Protection of Structures Containing Electrolytes

Chapter 8

The protection discussed so far has been applied to structures buried in the ground or immersed in an electrolyte. The third division of cathodic protection engineering will now be discussed, namely the protection of structures containing electrolytes; for example, water storage tanks, water carrying pipes and ballasted ships' tanks. The variety of structures that can be encountered in this field would make the discussion of them all in detail tedious, so a number of these will be selected and their engineering considered as models for the variations found in practice. These will be broadly classified as tanks which contain fresh, usually drinking, water, tanks which hold sea water or are ballasted with it, pipes which carry electrolytes, including condensers and coolers in chemical plant.

Fresh Water Storage Tanks

The majority of such tanks will be used to hold water either to form a head or to provide storage. The water will be pumped into them, stored and then run out again either by gravity feed or pumping. These tanks will be made in a variety of metals—steel, cast iron, wrought iron, copper, galvanized steel and aluminum. Plastic, lead lined and wooden tanks are also used but these are not susceptible to corrosion. The cathodic protection techniques will vary with each metal but in general the criterion of sufficient protection will be that a certain potential or potential change has been obtained relative to a suitably located electrode. The tanks will be constructed in a variety of shapes and sizes and the effect upon the protection of variations in size must first be considered. Suppose a rectangular, open top tank as shown in Fig. 215 is protected by a centrally mounted anode, then if a similar tank of twice the dimensions is considered, its surface area will

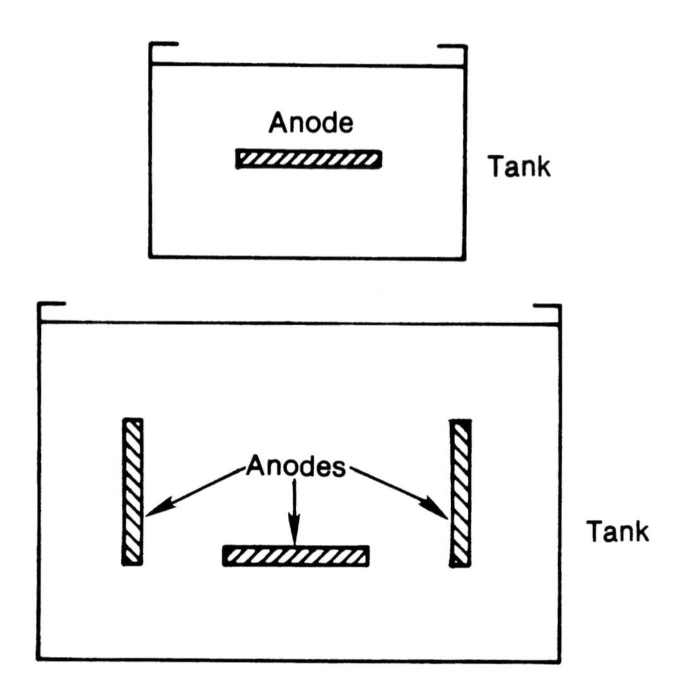

FIGURE 215 - Small tank (a) protected by a central anode and (b) larger tank with scaled up anode arrangement.

increase fourfold and four times the current will be required to protect it. If the anode is similarly scaled then its length will double and its resistance will halve so that the driving voltage will need to be twice that previously necessary. The surface area of the anode will increase fourfold and so the current density off the anode surface will be the same. The anode weight will increase by a factor of 8 so twice the life might be expected from a sacrificial or consumable anode.

If the installation has to be operated at the same driving voltage, as would be the case with sacrificial anodes and designed to have the same life as the original tank, then a modified arrangement, as in Fig. 215b, might be used. The difference in the voltage required to drive the cathodic protection and the other changes can be used to predict the type of engineering on a large scale tank from a model.

In a rectangular tank one of the most difficult regions to protect will be the corners and these will be of two kinds: those where three flat surfaces meet, and those where two surfaces meet. The corner will not have dimensions and if protection is achieved within 1 in. of its apex in a small tank and within 1/2 in. of the corner the metal potential is, say 20 mV more positive, then with the same metal and electrolyte these conditions will hold irrespective of the size of the tank. Generally protection of the corners will be difficult, particularly into the last inch or so of them. This will be exacerbated if the angles are more acute. The current demand into the corner of a tank may well be reduced because the electrolyte will not be disturbed and some deaeration will take place.

The argument of the difficulty of protecting into corners is often translated onto large structures, for example, the corners of an offshore platform. However, a corner is a corner, and these are usually radiused about 1 in. by the welding and no difficulty will be found in protecting under those circumstances.

Tanks that are coated or painted will be more easily protected both generally and in corners and where possible the corners of a tank should be painted or coated even when the main tank metal is not. Recesses in tanks and areas which lie beyond some pipe or other fixture are difficult to protect and either these sections must be coated or the protection asymetrically arranged to give additional current density, and hence polarization, to these zones.

Corrosion in fresh water will be limited by the high electrolyte resistivity and the formation of scale deposits. These conditions favor pitting corrosion and though the loss of metal is not great, penetration failure may occur within thin walled vessels. One of the most corrosive influences in fresh water is copper, which may be taken into solution from copper pipes or tanks and then deposited in its metallic state on the ferrous metal of the storage tank. This corrosion occurs when copper is present in quantities as small as one part per million. The small amount of copper which is plated on to the tank wall will act as a cathode and cause galvanic corrosion. The cell will have a potential drive of about 0.5 V and the high resistivity will limit the area of tank wall that acts as an anode and a deep pit will result, often with penetration of the tank wall.

Hot water tanks are more rapidly corroded than cold tanks. Other metals, if left in or introduced into, the tank in the form of chips, cuttings or turnings can cause equally rapid corrosion and cases of failure from screwdrivers, spanners, hammers, etc., being left in tanks are reported. Corrosion cells can be established by sediment or other non-metallic objects such as pieces of plastic or stones, which cause differential oxygen concentration and temperature variations; welds, oxide films and stress can also set up corrosion cells. Copper, tin plated steel, lead, zinc, brass, bronze, aluminum, solder, stainless steel, galvanizing, mild steel and cast iron are found in various combinations in some tanks and corrosion is accelerated by their presence. Often redesigning the tank to be of one metal, say galvanized steel, is the first step in corrosion engineering and one which the cathodic protection engineer should take.

Cold Water Tanks

Small header tanks and cisterns are now usually made of non-metallic plastic materials. Some household water storage tanks remain made of galvanized steel and their corrosion can be prevented by cathodic protection, a current density of 1 mA per sq ft will bring the metal potential to -1.1 V to a copper sulfate half cell. To protect the underlying steel body of the tank as opposed to the galvanizing, current densities of the order 0.1 to 0.2 mA per sq ft will suffice.

As copper bearing water is one of the main causes of corrosion in these tanks a method of protection has been suggested by Kenworthy in which the incoming water is passed over zinc plates to remove the copper. This adds considerably to the life of a galvanized or aluminum tank. Zinc anodes can be used in this protection. The small cisterns often have a ball valve and an anode can be attached to a modified silencer tube. Similarly, zinc is ideal for the protection of copper tanks and although this may not be serious the prevention of circulating water pick-up of minute quantities of copper will prevent corrosion elsewhere in the system.

Water storage tanks are often coated and several bituminous tasteless paints are available. Current densities are subsequently lower and a small zinc anode can be used. Where larger currents are required these are obtainable from magnesium anodes which are usually cast in the form of long rods with a wire or rod insert. In large tanks impressed current can be used and here the platinized or coated titanium anodes have an advantage in that they can readily be formed into low resistance wires; alternatively consumable anodes can be employed.

With either sacrificial or consumable anodes care has to be taken that no harmful or objectionable corrosion product is put into the water. Many anodes can be tasted, including the effect of impressed current anodes. Zinc can cause tasting and acts as a mild emetic. Magnesium and aluminum both produce oxides which are used as stomach powders. Many large buildings have storage tanks which are designed from segmented steel. These usually are better served with a small impressed current system. They may require up to 5 amps for protection at about 20 V so the cost of power and of the small air cooled transformer rectifier unit is low. Such a system is shown in Fig. 216. With permanent anodes some of the hypochlorite originally used in treating the water, which has been reduced to chloride, will be reconverted. Such a procedure might well be extended and would be effective against such organisms as those causing Legionnaire's disease.

In protecting water storage tanks great care has to be taken in the use of the copper sulfate electrode. Copper is a most dangerous contaminant as it readily plates out on the tank wall and most certainly on galvanized surfaces and fittings. This can be replaced by a zinc rod in a salt water solution or by the use of a silver chloride electrode. If a continuous reading is re-

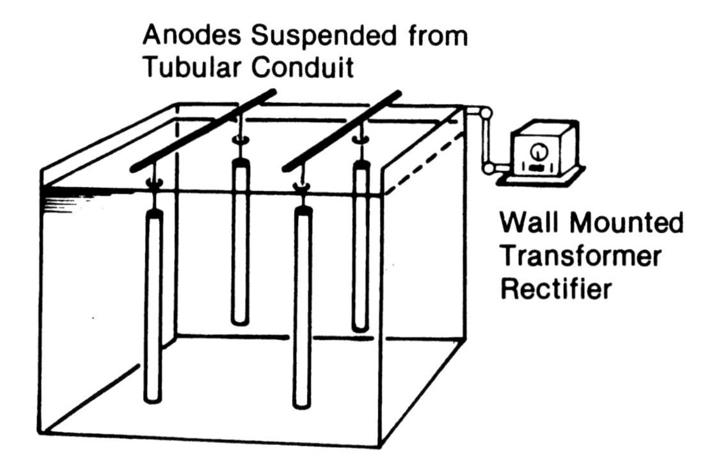

FIGURE 216 - Impressed current protection for small tank.

quired in the tank a zinc electrode may prove to be somewhat skittish in its behavior but a magnesium electrode, although not having the accuracy of the copper sulfate cell, will provide a reasonable indication of the level of protection.

Many areas of otherwise low domestic water pressure are served from a water tower. Some of these are rectangular tanks and their protection is as already described. A considerable number of the towers have a large central pipe which supports a spherical, ellipsoidal or cylindrical tank. Both the tank and the pipe hold water and both require protection. The top is easily protected with a group of impressed current or magnesium anodes. The riser pipe is protected by a chain of anodes and either magnesium sacrificial anodes or impressed current platinized titanium or consumable aluminum can be used. Fig. 217 shows a typical layout of such an installation.

Fig 218 shows a chain of silicon iron anodes in a concrete water tower protecting the steel and cast iron components which are bonded together.

Hot Water Tanks

Corrosion in hot water tanks is more rapid than in cold water tanks. The higher temperature causes a reduction in the resistivity to about half that of the cold electrolyte and stratification between layers of different temperatures will initiate large scale corrosion cells. The effects of dissolved salts and bimetallic couples will be increased at the higher temperature and reactions such as the dissolution of copper may be accelerated. Galvanized tanks suffer from a further effect: the zinc outer covering of the tank is normally anodic to the underlying steel and zinc/iron alloys, but at temperatures above about 140 °F, the polarity of these is reversed and zinc becomes

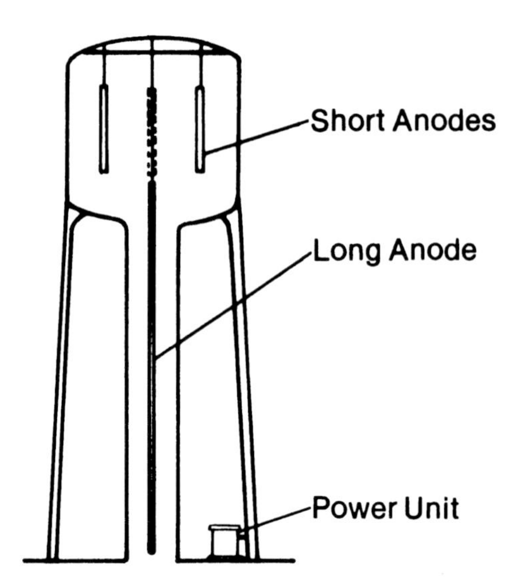

FIGURE 217 - Impressed current protection for steel water tower.

FIGURE 218 - Concrete water tower with string of silicon iron anodes protecting pipes, valves and reinforcing. (Photo reprinted with permission from Los Angeles Water Dept., Los Angeles, CA.)

cathodic to the underlying steel. The tendency to this inversion is decreased in the presence of chloride and sulfate ions.

The above effects have been associated with temperature alone but many hot water tanks are further endangered by having the heating element installed in them. Often these are made of copper, tinned copper or cupro-nickel alloy and they may be in electrical contact with the main tank so forming a large bimetallic cell. Even if the heating element is insulated from the main tank, copper may be dissolved from it, and plated on to the tank wall. The effect may be negligible upon the corrosion of the copper but one part per million of copper in a hot water system can remove the galvanizing from all the hot water pipes within a few months.

Galvanized hot water tanks of the domestic variety can be successfully protected by magnesium anodes; these must deliver sufficient current to produce 2 to 3 mA per sq ft current density on the tank wall. It is reported that galvanized hot water tanks either corrode within a few years or, if they last longer than this, remain sound for 15 to 20 years. This is supported by experiments on the corrosion of galvanized steel in hard water where the scale formed has a preservative effect. If this scale formation is prevented either because the water is soft or because some local corrosion cell prohibits it, the tank will fail rapidly. It is suggested that if a tank is cathodically protected for a year or so to form a deposit of chalk upon its surface, then it will last for a further 15 to 20 years even though the anode is no longer giving current.

Domestic tanks are of two kinds; either they are rectangular and of 25 to 40 gallons capacity for the average household, or they are cylindrical and somewhat larger. The rectangular type can be protected by an anode mounted on to the inspection plate or bolted to the tapped holes in the tank which take the bolts for securing the plate, and an anode about 18 in. long and suitable for mounting in several tanks is shown in Fig 219. The anode support is galvanized or aluminized mild steel and is slotted to fit various bolthole separations. The anode is fixed and the holding bolts are tightened against a lock washer which ensures good electrical contact between the tank wall and the anode. At the two holes where the bolts are in place the

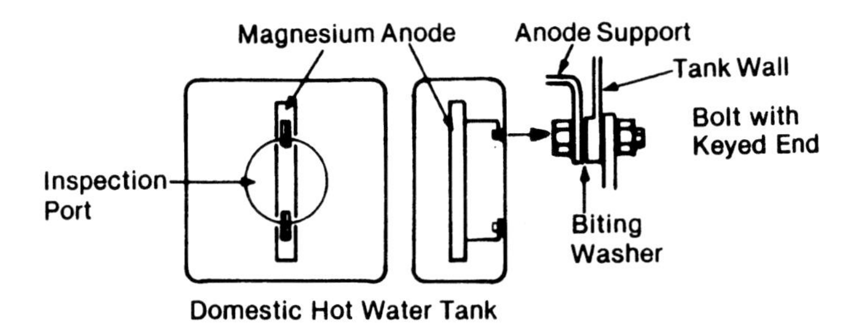

FIGURE 219 - Magnesium anode for domestic hot water tank.

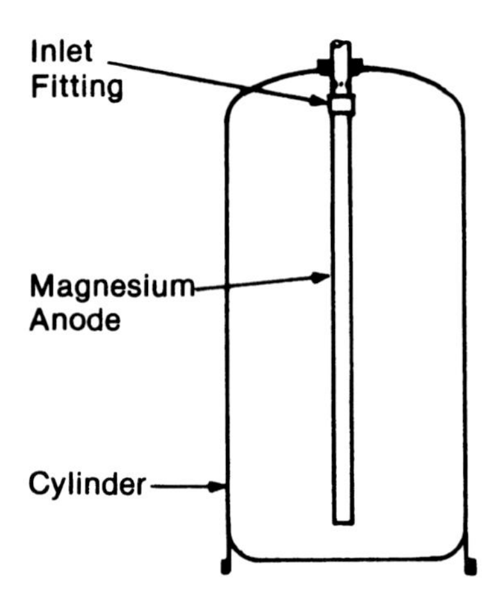

FIGURE 220 - Rod magnesium anode protecting hot water cylinder.

cover is held by nuts, the bolt having a slotted end so that it can be held by a screw-driver while the nut is tightened.

Larger cylindrical tanks are best protected by a long magnesium rod which can have a threaded insert or can be attached directly to a pipe fitting, often the inlet pipe being modified to hold the anode. Fig 220 shows such an anode fitted in the tank and similar anodes can be used for rectangular tanks in place of the one shown in Fig 219.

Copper tanks are often used in soft water areas and generally these do not corrode rapidly, but in certain waters the copper is dissolved and its presence in the water may cause corrosion elsewhere in the system. Zinc anodes can be used to suppress this and can be fitted in a similar manner; the anode should be designed so that protection is continuous throughout the tank life and a controlling resistor can be used. Similar protection can be given to the sheaths of immersion heaters. These are particularly prone to failure when fitted in copper tanks and a zinc anode can be incorporated in the immersion heater unit.

Industrial hot water tanks are usually made of steel and are not often galvanized; these are used for a variety of purposes and their temperatures vary from 80 °F to 200 °F. The current density required for their protection is about 3 to 4 mA per sq ft on the steel, though this varies with the water, the temperature and any bimetallic couples in the tank. The corrosion of these tanks can be caused by copper-bearing water, bimetallic cells or differential aeration caused by sediments. Platinized titanium anodes are preferred but graphite, magnetite and high silicon iron anodes can be used for cathodic protection in temperatures up to 120 °F if the impregnant for the graphite is carefully chosen. Molybdenum-rich silicon iron is reported to be satisfactory in boiling water. Consumable anodes may be used, though care must be taken that their corrosion product does not in any way contaminate drinking or processed water and that stratification does not cause premature failure. Magnesium anodes can be used though they are very rapidly consumed at high temperatures. High operating voltages are used with impressed current and platinized titanium anodes should be fully platinized.

Boiler feed water is often stored at high temperatures in steel tanks. The water is pure, the salt having been removed to prevent scaling, so its resistivity will be high, usually about 40,000 ohm cm, and pitting corrosion will result. The incoming water may be pumped as a hot steam-water mixture and the vibration may be sufficient to occasion exfoliation of the corrosion scale. Cathodic protection can be applied to these tanks and Fig 221 shows a system of fully platinized titanium wire used in such an installation. The tank may well be lagged and access to it limited. This will dictate the shape and layout of the wire anodes. Consumable anodes can be used and it is suggested that the use of high purity aluminum will confer the benefits of water treatment on the plant.

Because of the high resistivity and the high temperature it will be difficult to judge the adequacy of the protection using normal half cells. Swing tests of small steel coupons or high temperature electrodes such as the silver/silver sulfate half cells can be used.

There are several industrial machines such as bottle washers, pasteurizers, etc., which hold hot water in storage for their process. These usually employ steel tanks and where water has to be used at a constant high temperature it is often heated in these. As most industrial plants have adequate steam, this is injected into the water to heat it through steam coils which are usually copper, or through copper or copper alloy nozzles. Corrosion in these tanks is rapid and cathodic protection has been applied to a variety of them with complete success.

Anode to cable connection is difficult inside the tank so many of the anodes are taken through a holder in the tank wall. One such type of entry is shown in detail in Fig 222.

In many of the machines the water is continuously recirculated so consumable anodes would not be suitable as there would be a buildup of corrosion product and this could leave smears, for example, on the glass bottles in a pasteurizing plant.

Insulators, both for anode support and for the cables, have to be carefully chosen and fluoropolythenes or irradiated polythene can be used as sheathing. Inside the tank anode holders and insulators have to be carefully chosen. Some plastics are suitable and polyester glass fiber and epoxy glass fiber can be used with care.

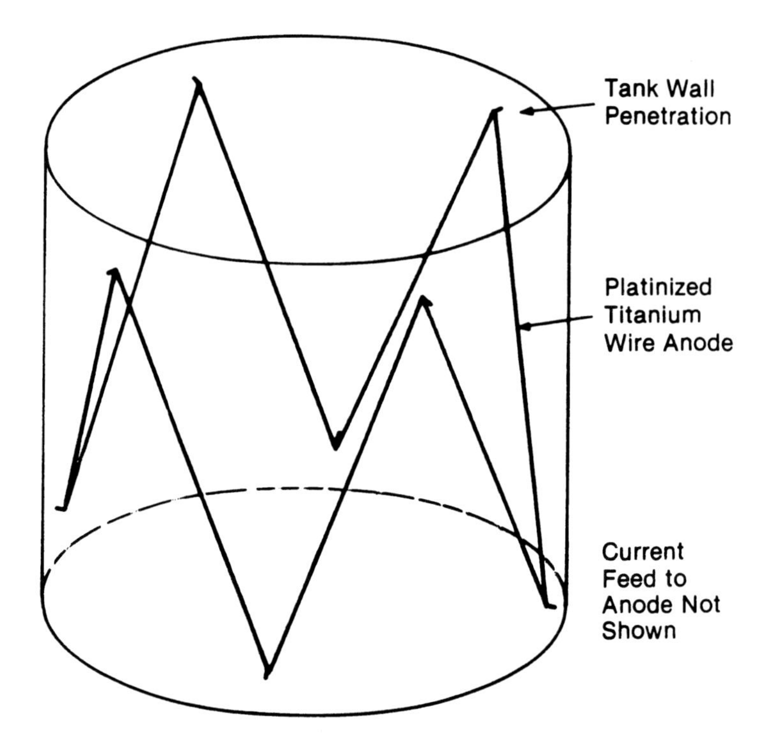

FIGURE 221 - Wire impressed current anode protecting cylindrical boiler feed tank.

Sea Water Storage Tanks

Sea water cooling systems and separators have large storage facilities often steel tanks similar to oil storage tanks. The crudely filtered sea water is circulated to prevent sedimentation and corrosion has been one of the major problems with these tanks. Coatings of heavy bituminous, coal tar and coal tar epoxy paints and concrete lining have been used with limited success. Cathodic protection is a much cheaper and more successful technique, though coating to below the wind and water line is essential.

Impressed Current Protection

Where the tank is open to the air impressed current protection can be used with permanent anodes. Where they are enclosed a steel roof will suffer corrosion. Reinforced plastic roofs can be used and these can have built into them the support system for the tank cathodic protection. Occasionally these allow sufficient light for algae to grow in the tank.

A large sea water tank will require 10 to 12 mA per sq ft for protection which may reduce to 3 to 4 mA per sq ft after initial polarization. This gives a current of about 300 amps, reducing to about 100 amps.

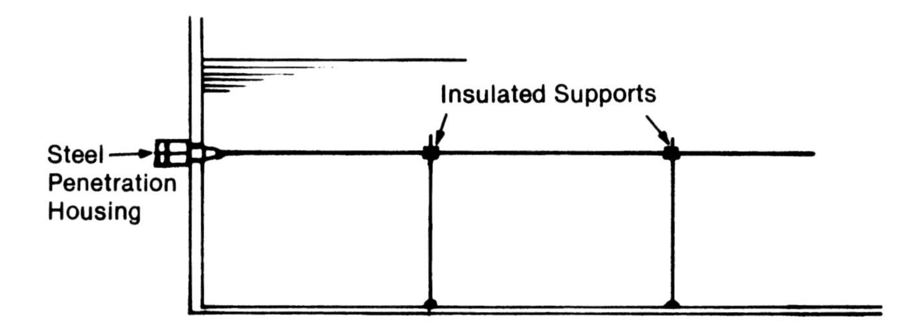

FIGURE 222 - Impressed current wire anode used to protect industrial tank.

Platinized titanium magnetite or lead alloy anodes are ideal for protection of such a tank and the suspension of anodes in the water would probably be the simplest and cheapest to engineer. The arrangement shown in Fig. 223 uses lead anodes which are an extrusion of 1 1/4 in. diameter and available as long anodes. The anode has a sleeve at its end into which the connection is made and sealed with a suitable resin. The anode can be suspended by its feed cable, the whole hanging from a support system. This could be achieved with galvanized steel rods or by the use of plastic rope.

Platinized titanium anodes would generally be very much lighter in weight and could be suspended from a rope framework. If the water is highly turbulent they would need to be restrained to reduce their movement and to prevent fatigue at the point where the anode feed cable enters the anode proper. A number of such arrangements have been developed commercially.

Sacrificial and Consumable Anodes

Where the tank is enclosed with a steel roof permanent anodes cannot be used unless there is a rapid replacement of the water causing complete absorption of the nascent chlorine as hypochlorite.

The application of sacrificial anodes is straightforward. The anodes of zinc or aluminum could be attached to the wall by standoff inserts or they could be suspended from brackets attached high on the walls. The center of the tank would receive protection from wall mounted anodes. The high driving voltage of magnesium would be of no advantage and their cost would be much greater. Consumable anodes could be used but unless there is a problem with mounting or with the space available in which the anodes can be placed they would not prove to be economical.

While permanent anodes may be obnoxious in closed tanks, consumable and sacrificial anodes can have a definite advantage where the water is stagnant and not in contact with the air. This condition may exist

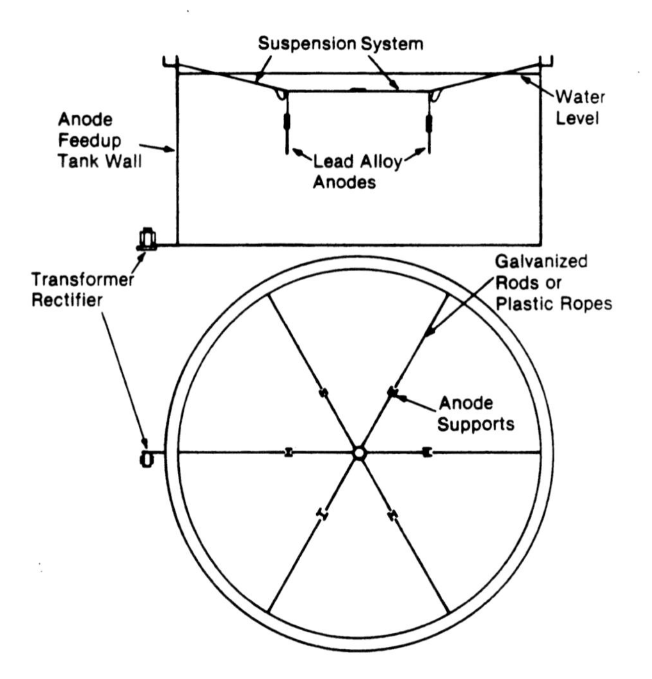

FIGURE 223 - Impressed current anode arrangement.

in an oil tank which is used for storage of separated oil. Cathodic protection without anode gas evolution will cause a gradual de-aeration of the water and a subsequent reduction in current requirements. Low driving voltage sacrificial anodes will make the most use of this reduction automatically, consumable or controllable sacrificial anodes should have their current output automatically controlled to take advantage of this.

Ballast Tanks

Tank Corrosion

Tankers, ore carriers and colliers usually carry cargo on their outward voyage and return in ballast. Some vessels are designed as oil/ore carriers and operate on a triangular route carrying cargo on two legs and ballast on the third. Each vessel is usually ballasted only in some of her tanks and these are selected so that either the same tanks are always in ballast or the tanks are ballasted in rotation. Most ships use sea water ballast, though this is not universal and many empty and clean their ballast tanks at sea so that their ballast water is then changed to clean sea water—the oil being separated and retained on board.

The majority of ballast tanks are partly coated, the top 5 feet being mandatory, with the application of cathodic protection below the normal ballast level. Many oil tankers have heating coils installed in their compartments to reduce the viscosity of crude oil and the tank is built to include the main structural members of the ship, various pipes, etc. These tanks corrode rapidly: sediment causes differential aeration cells as does oil-soaked rust or pockets of dry cargo. The heating coils are usually cast-iron or a non-ferrous alloy and these are generally cathodic to the steel. Even if the coils are insulated from the tank's walls, so eliminating the bimetallic effect, copper may be dissolved out of them and be deposited on the tank steel. Millscale often found on the plates and girders is highly cathodic to steel. Finally the stresses set up in working the ship, particularly in heavy seas, cause corrosion intrinsically by stress potentials and accelerate it by the removal of scale from the stressed areas. In the majority of carriers this results in a fairly uniform attack on the whole of the steelwork and very rapid corrosion on the hard spots of high stress, with pitting on some horizontal surfaces including the heater coils in crude tanks. Corrosion rates of 1/8 in. per year have been found on white oil vessels and often plates have to be replaced within 6 to 8 years after building.

There are four methods which could be used to counter this corrosion; these are:

1. The use of corrosion resistant materials for construction
2. Application of protective coatings to the steel to make it corrosion resistant
3. Cathodic protection during ballast periods
4. Treatment of the cargo and ballast water

Of these methods, cathodic protection has been shown to offer the most economic solution when coupled with a planned ballasting procedure and coating of the deck heads and sometimes other areas of the tank.

Cathodic Protection

The design of a cathodic protection installation in a ballast tank is affected by a great number of variables. In general, the current will be expected to descale the corrosion products already on the steel and then to maintain protection of the steel from the current applied during the ballast period. Such protection and its detailed design will depend upon the service that the ship operates, the cargo carried, the period of ballast, the ratio of this to the cargo period and the ballast water carried. Most shipowners will not be restricted to a particular cargo or exact schedule and the ballast tank conditions may vary greatly, while another owner may carry the same cargo on a well scheduled route. Because the tanks are enclosed and not well ventilated, the oxygen and chlorine evolution from impressed current anodes may cause rapid corrosion of the deck head if this is not completely under water. If the tanks are completely full then the chlorine will be dissolved in the water and at low concentrations will have no harmful effect upon the submerged steel. Carriers of light petroleum and other hazardous cargoes do not allow electric cables in their tanks even if the installation can be made electrically safe. These two factors mean that sacrificial anodes have been used almost exclusively for ballast tank protection. Magnesium was the most commonly used anode but the problems caused by sparks generated on impact have led to its total banning in ballast tanks of crude oil, cargo tanks and in the tanks adjacent to them.

Magnesium ribbon can be used to effect the descaling of tanks providing they are removed after use and they are only installed when the tank is in a gas free or safe condition. As cathodic protection is being applied to new building, fewer and fewer tanks are being descaled. However, as the process of descaling and subsequent polarization was developed for ballast tanks it is useful to review the early practice.

Rusty steel can be descaled in sea water at high current densities, the total amp hours per sq ft required for this will vary with the condition of the scale and the current density applied. At 300 mA per sq ft hard scale is removed in 24 to 36 hours, that is 8 to 12 ampere hours per sq ft, and at lower current densities a longer time and a larger charge of amp hours is necessary. Ballast tanks are not difficult to descale and 4 to 5 amp hours per sq ft. removes most of the scale at high current densities while at low current densities a the total charge density will increase.

If the effect in a tanker ship was carried over from one ballast time to another and a ship might require 5 to 6 ampere hours per sq ft for descaling in a single ballast period, and 8 to 9 amp hours per sq ft if the effect was cumulative over 2 or 3 ballasting periods. There are two reasons for this. Firstly, the descaling seems to depend on the overprotection rather than upon the current density so the process is more efficient at high current densities, and secondly, the tank metal is exposed to the air during the period between ballastings which depolarizes the metal surface and some charge, usually about 0.2 amp hours per sq ft is needed to reach a descaling potential. The current density required to protect the tank will depend upon the time the ship is in ballast and to a smaller extent upon the time between successive ballasting. The total charge per sq ft will not be a significant figure as a large amount of this may be consumed in obtaining and maintaining protection. If the tank potential is plotted against time then with a typical high potential anode system the results would be as in Fig 224 and the significant part of the curve is that above the protection level. Protection is reached after some 18 hours and if the tank were in ballast for three days, then the overprotection could be described by the area in millivolt days below the curve and above the protection level. In Fig 224 this area is 480 mV days. Under these conditions a firm calcareous

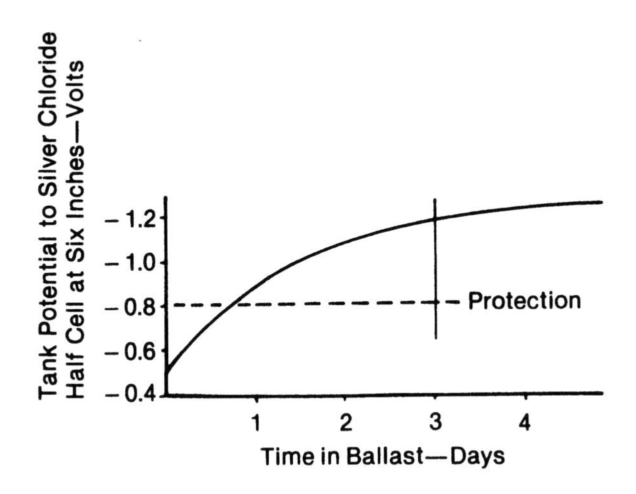

FIGURE 224 - Metal potential change with time in seawater ballast tank.

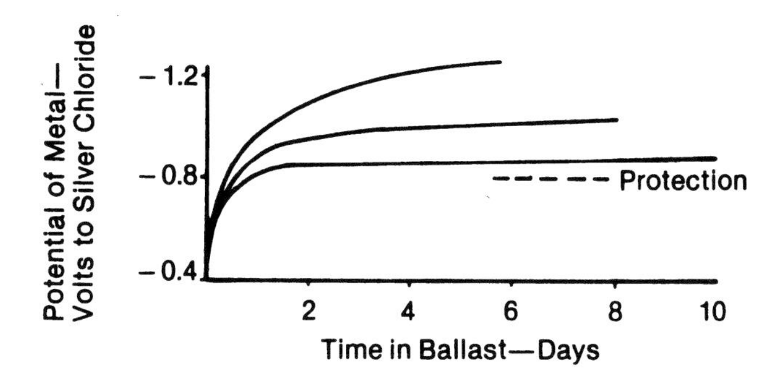

FIGURE 225 - Potential/time relationships employing different anode metals.

deposit should be formed on the metal and should prevent any corrosion occurring on the wet metal surface, in pits or beneath cargo when the ship is in ballast. If the ship had taken on ballast for seven days then overprotection would have continued longer and the current required to maintain this degree of overprotection would have been used unnecessarily. The curve in Fig 224 would be modified quite considerably by using different anode metals and engineering techniques. Two variations of the curve are shown in Fig 225 Curve (a) is the Fig 224 curve and this provides overprotection of 480 mV days in a three day ballast period. The system would be suitable for two or three day ballasting voyages. Curve (b) shows a system suitable for a longer period in ballast and a four day period gives 440 mV overpro-

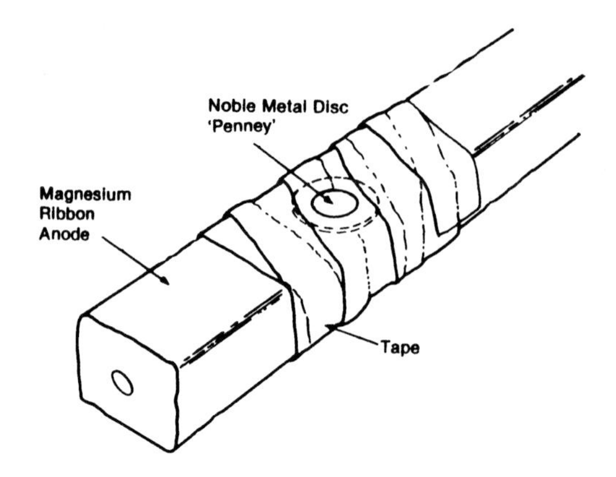

FIGURE 226 - Connection method for magnesium ribbon anode.

tection and a six day ballast 800 mV days. The curve (c) gives over 500 mV days in a 10 day ballast period. The structure potential of the tank is related to the current density and descaling anodes can be employed that will give all three types of curve. For descaling purposes alone ribbon anodes are employed. These can be made of magnesium which will give the high rate of descaling of curve (a) and aluminum and zinc which will generally give curves of type (b) and (c).

The ribbon anodes are attached in the tank, usually to structural members, using clamps. It is recommended that the wire core should be exposed. This is not necessary and an alternative technique was used by the author to great effect. A small metal disc, a 1 penny is the cheapest, is taped to the anode using P.V.C. tape. A small area in the center of the coin is then exposed by cutting through the tape and the pressure point of the clamp is applied to this. The anode beneath the tape is not consumed during the descaling process as sea water has only limited access to it up the tape tube and cannot penetrate the tape to disc seal. This is illustrated in Fig. 226.

The average anode ribbon will have an output in magnesium of about 1 amp per ft, in aluminum indium about 1/3 amp per ft and of zinc about 1/4 amp per ft. Magnesium anodes placed at a 16 ft pitch opposite a corrugated bulk head will have a capacity of about 100 amp hours per linear foot and the steel area affected by each foot of anode will be 20 sq ft, so that 5 amp hours of descaling capacity will be available. The aluminum indium anode will have to provide approximately 6 amp hours per sq ft and the zinc approximately 7 amp hours per sq ft. The anodes that are commercially available will require an individual calculation of their output, but the spacings will be approximately 12 ft centers for the aluminum indium and

FIGURE 227 - Zinc ribbon continuous anodes installed in bottom areas in cargo-ballast tank. (Photo reprinted with permission from Federated Metals Corp., Somerville, NJ.)

10 ft centers for the zinc. Ribbon anodes in a tank bottom are shown in Fig 227.

The protection of ballast tanks has a number of other important features. Firstly, the area of the tank wall will exceed the frontal area because of the frames and stiffening used, whether in the form of ribs, stringers or corrugation. Because of the high current densities used there will be a considerable volts drop on to the metal surface. This will require a very low resistance anode system and the anodes will generally be long, thin cylinders mounted close to the steelwork they are to protect. The major structural frames will have anodes placed either on or between them. There will be rapid polarization which was described in the properties of low driving voltage anodes. It was standard practice to use booster anodes in tank ships. There are two types of these: firstly, anodes that are used to give a positional boost, that is to increase the current density to some part of the tank that might otherwise be underprotected, and there were booster anodes which were used to give an initial high current output and hence polarize the tank. These were usually consumed in the first one or two ballast voyages and played no further part in the cathodic protection. A number of anodes had various fins or were otherwise shaped and these were called boosters as a misnomer. Some of the designs, however, were arranged so that the anode output varied throughout the tank life and this has had an advantage in that the gradual polarization that was built up was carried forward to a small extent from one voyage to another. These anodes had a programmed rather than a boosted output.

Anode Installation

The success of cathodic protection in ballast tank protection led in 1962 to the Classification Societies amending their rules to permit up to 10 per cent reduction in scantlings of bulk head plating, frames stiffeners, deck heads and other steelwork, providing that the system of corrosion control that was fitted was approved by them. They subsequently allowed a 5 per cent reduction in the main longitudinal strength members. Cathodic protection is one of the approved corrosion control techniques when used either with a minimal deck head protection, usually the top on 5 feet, or with a wider application of coating. The anodes themselves are subject to various rules and limitations. These rules apply only where there is a specific program of ballasting and cathodic protection can only be used where there is sufficient ballast time to allow the polarization to continue through the cargo or empty periods. On this basis in ballast only tanks the coating has to extend to below the normal ballast level and anodes may be used below that with the minimum of the upper 5 feet being coated. In cargo oil ballast tanks the same rules apply, though there are more restrictions on the anodes and their construction, type and location in the tank. In nonballasted cargo oil tanks there is a requirement for complete coating, but it is often prudent to use anodes if the oil has a sludge which it deposits and which may contain water. In many tanks that are used for ballast only or in cargo ballast, the coating will be extended to a lower level than required to encompass horizontal surfaces that are near the ballast line and often to cover all horizontal surfaces where sediment may collect.

The current densities that are generally required for cathodic protection are: in cargo clean ballast tanks about 8 mA per sq ft, in white oil tanks and in ballast-only tanks this is increased by 25 per cent to 10 mA per sq ft. In wing tanks and fore and aft peak tanks a higher current density is required because of the complexity of the surface which leads to some overprotection, and because in these tanks there is generally more movement of the ballast water giving better aeration and consequent depolarization of the cathode surface. Where the tank is coated it is usual to make an allowance for the cathodic protection at breaks and holidays of 1 mA per sq ft for this area. In the double bottom area where ballast only is carried 8 mA per sq ft is usually applied. In dirty ballast tanks and other cargo tanks an allowance has to be made depending on the type of trade. In general, dirty ballast tanks will leave a residual oil film on the surface which will reduce the corrosion. The current flowing in the cathodic protection system in many cases will de-emulsify the dirty ballast and increase this coating effect. Where permanent cleaning using high velocity water is used then the current density should be increased by 25 per cent and where the ballast period is short, as already described, again there is a requirement for extra mean current density, this will increase by 20 per cent if the ballast period is below 5 days, and by 30 per cent if it is of 3 days or less. The ballasting and cargo pipes within the tank are usually separately protected as these can have flanged jointed or insulated joints in them which will prevent the cathodic protection spreading to them. It is good practice to bond over these joints and it is usual now to fit additional anodes to the pipes often by means of a strap and clamp which can also be used to bond over the joint.

Fresh Water Storage Tanks

The above does not apply to fresh water storage tanks which are usually coated and have supplemental cathodic protection; unless there is a classification restriction, magnesium anodes are used. These anodes are similar in type to those used on land based installations though greater care may be taken in their fixing as the ship moves and flexes.

Anode Engineering

The anodes will be long and comparatively thin so they will need to have a substantial insert. Generally, this will be a rod or other metal at least 1/2 in diameter, and should extend throughout the length of the anode.

The Classification Societies have developed rules on anode positions and locations and these are as follows. Firstly, magnesium anodes are not permitted in oil cargo tanks or tanks adjacent to oil cargo tanks.

Aluminum anodes are only permitted in cargo tanks where the potential energy of the anode does not exceed 200 ft lbs. There are also restrictions on the placing of aluminum anodes where they could themselves be struck and this would apply to points such as tank openings and butterworthing openings. The height of the anode for this purpose may vary with the different Classification Societies, who may permit their fixing above substantial horizontal members from which they cannot subsequently fall. Aluminum anodes that contain magnesium are also not permitted by some Societies. In placing zinc anodes there is a similar potential energy restriction but as this exceeds 3,500 ft lbs. there is no practical limit.

The anode and its support have to be carefully mounted so that the anode does not resonate nor is it damaged by the high pressure water jets used in cleaning. The steel insert has to be securely attached to the tank itself and this is recommended either by welding, when a continuous weld must be used, or by a clamping technique, when the clamp must be approved and the bolting of the clamp must have sufficient locking devices in it. The anode must be attached to a single member and not be subject to stresses and neither the anode nor an anode support must be welded to areas of high stress.

The location of the anodes will depend on the type of service. In clean ballast tanks the anodes will be fairly evenly distributed with greater numbers at the bottom of the tank. There will also be an increase in the number of anodes close to horizontal surfaces and an alternative technique of design is to consider horizontal surfaces as though they had a 50 per cent greater area and position the anodes accordingly but basing the overall design for the tank on the lower current density of the clean ballast tank, that is 8 mA per sq ft. The anodes should be positioned so they lie between the main frames to provide adequate protection within these areas. Additional anodes should be placed in any section which appears to be partly isolated from the main tank, Figs 228 and 229 show typical installations.

Both in cathodic protection and in descaling an amount of hydrogen gas will be given off. Particularly in descaling, where there may be other contemporary work being carried out on the ship, great care must be taken to prohibit smoking or hot work close to the vents which must be adequate and must extend well above deck level. General cathodic protection of

FIGURE 228 - Zinc anodes fixed to bulkhead. (Photo reprinted with permission from Federated metals Corp., Somerville, NJ.)

FIGURE 229 - Low voltage sacrificial anodes in bottom of seawater ballast tank of super tanker. (Photo reprinted with permission from Federated Metals Corp., Somerville, NJ.)

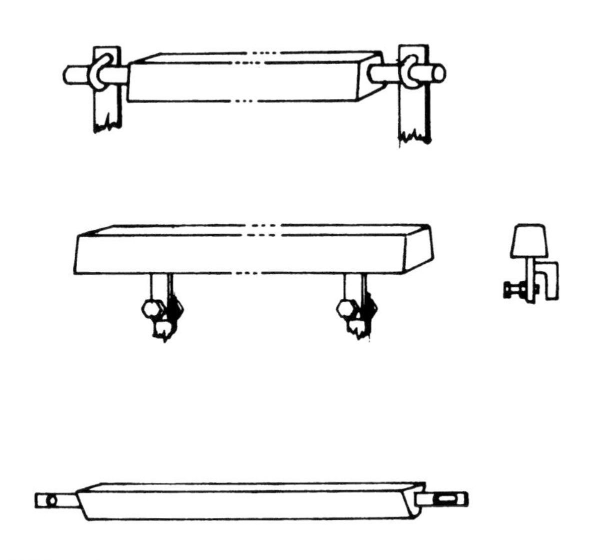

FIGURE 230 - Selection of anodes used in tank protection.

cargo and cargo ballast tanks will produce a lower volume of hydrogen which will vent rapidly and as these are anyhow in a permanent 'no smoking' area less danger will arise.

Many tank ships are nowadays being used to transport consumable cargoes. These may be in the form of grains, edible oils or wines. The calcareous film made by the anode will not adversely affect foods, though it is doubtful whether the health authorities would accept anodes with mercury in them remaining in the tank when an edible fluid was carried. Wines have a comparatively low resistivity and anodes can be rapidly consumed during the voyage. Wine, sadly, does not contain enough alcohol for it to be considered as a hazardous cargo and magnesium anodes could be used for protection. Anodes have been known to dissolve rather rapidly and claimed to impart to the wine a slight bluish haze. The addition of milk of magnesia to table wines might have some beneficial effect.

Tanks which have been cathodically protected or descaled will have a white calcareous deposit on their surface. This will contain a considerable amount of magnesium in the case of descaling, not necessarily from the anode but out of the sea water. If it is proposed to paint the tanks subsequent to their cathodic protection this will have to be removed. This can be done by various processes and in some cases where impressed current descaling is used a reversal of the current for the last 20 minutes will effectively remove it.

A selection of anodes used in tank cathodic protection together with their supports is shown in Fig. 230.

In calculating the anode resistance it is usual to use the formulae given in Chapter 3. The equivalent resistance of anode systems can be prepared by a simple technique which was developed by the author and G. W. Moore of Esso. In this the weight of anode material is used to provide the capacity of the system and its adequacy in delivering current is judged by adding together the length, width and thickness of the anode. This gives a very close correlation with the anode's resistance and the combination of the total length and total weight can be used as an effective criterion. Where tanks are ballasted with sea water it is usual to consider the sea water to have a resistivity of 25 ohm cm. It is bad practice to ballast with river or estuary water because these are often silted and may contain bacteria. However, ballasting in brackish water may be inevitable in such areas as the Baltic and other inland seas. A figure of resistivity should be obtained from the area and the anode design varied. Where the resistivity is less than 60 ohm cm the current density requirements will be the same as in sea water. Above that a lower current density will be required but this will not be sufficient to compensate for the higher anode resistance in these waters.

In making absolute resistivity determinations the resistance between the anode and a coated structure will be that for an anode close to the surface of a semi-infinite electrolyte. The majority of anodes in a tanker, however, will have their resistance reduced by the proximity of the cathode rather than increased by it. The formulae for estimating the anode resistivity are given in Chapter 3.

The anode will be slowly consumed and the overall ability of the anodes to provide protection should be calculated when the anode is partly consumed. Because there is a gradual reduction in the current requirement this can sensibly be taken as the ability to protect the tank fully when one half of the anode is consumed. Various anode shapes will change their resistance with consumption in different ways. The long thin anodes tend to be consumed more rapidly in their length which is the major factor, and less rapidly at the middle of the anode which has very little effect. If the final anode shape is an ellipsoid its resistance will be

$$R = \rho \frac{1}{4\pi} \cdot \frac{6}{x + y + z}$$

where x, y and z are the dimensions of the ellipsoid. Inevitably, x and y will be the same as the anode will be an ellipsoid of revolution about its major axis.

Anodes of other shapes can have their resistance deduced from the formulae in Chapter 3, and in descaling the anode resistance will be increased if very long lengths of descaling ribbon are used with too small a number of connections. As the anode is consumed its longitudinal resistance increases until only the very thin wire core is carrying the current.

Cargo Compatibility

Some of the cargoes that will be carried between ballast periods may affect the anode in that they form a blanket or coating on it which is not readily removed on the return to ballast conditions. If this is the case and a tar-like or waxy product adheres to the anode, then the cathodic protection will be ineffective. Most of the anodes that operate in sea water have a residual film on them which prevents the product adhering to the anode and it falls away under gravity as the tank is emptied. It is, however, useful to test this ability and anodes mounted vertically, as opposed to horizontally, will be found to shed this sort of covering more easily. Anodes that are close to washing machines are automatically cleaned. These anodes may come into operation ahead of the other anodes which are still losing product contamination and be more rapidly consumed. Anodes of more substantial cross section located in these areas would be advisable.

Pipes in Tanks

It was found that the ballast and cargo pipes in tanks corroded rapidly and there appeared to be two causes for this. Firstly, the pipes were not properly connected to the tank and, therefore, were not cathodically protected, but it also appeared that the potential gradients in many of the tanks were sufficient to cause interference or electrolysis on the pipes particularly close to flanges. Experiments were carried out and a simple technique of strapping using a tensioned connecting unit was evolved which the author applied successfully to a number of large tankers. Various proprietary variations of the technique are used and these almost inevitably involve some form of light clamp with anodes attached to it. The joint itself is bonded across and if the tank has a working atmosphere these can be by studs welded to the pipe surface.

The internal protection of these pipes is described later in the paragraph.

Cathodic Protection Life

It is usual to design the cathodic protection of a ballast tank so that it will adequately project between major inspections which are every four years. The period in ballast, of course, will be less than the four years and usually one to one-and-a-half years' anode life would be appropriate. Longer life systems can be designed; there are problems with an initial high current output causing anode wastage by overprotection, and economic penalties attached to the added investment in anodes with the weight of the anodes detracting from the cargo capacity.

Because the calculation of the life of the anodes has been based on certain ballasting assumptions it is necessary to inspect the system from time to time to ensure that there is both adequate protection and that the anodes will continue to give sufficient protection until their anticipated replacement. Visual inspection of the surfaces and of the anodes can be checked by using reference electrodes. The use of reference electrodes is not entirely accepted as any form of electric cable within the tank is frowned upon at any time. There are good reasons for this and though it might be argued that the possibilities are remote, they do, nevertheless, exist. When a tank is found to be inadequately protected or is likely to become so it is necessary to increase the cathodic protection. Where this is anticipated as, for example, where the owner is not entirely sure of his ballasting period, then either initial allowances are made for these variations or additional fixing points are built-in, alongside the anodes as fitted or at other places. Extra anodes can be fitted with cold work in the tank and some shipping companies and Classification Societies allow the use of rafts within the tank to fit additional anodes. This is not good practice and additional anodes in the accessible areas of the empty tank will usually increase the capacities sufficiently to prevent corrosion.

Heating Coils

Heating coils are usually insulated from the tank. Sometimes it is advisable to fit anodes to the heating coils as they do occasionally suffer pitting corrosion, this may be by electrolysis as in the case of the major pipes. Most coil manufactures have experience of this and from their observations the cause and solution to the corrosion can be achieved.

Pipelines Carrying Water

Pipelines were originally developed for carrying drinking water. Wooden pipes and lead pipes have been superseded by cast spun iron and steel. Plastic pipes are now used extensively and asbestos cement and concrete pipes are also used.

Large quantities of sea water are required for cooling in power stations, in refineries and in chemical plant. The sea water is fed through large diameter mains into the heat exchangers and process plant; steel pipes corrode rapidly. Among the techniques that have successfully been used to prevent their corrosion are coatings, linings and cathodic protection.

Cathodic Protection Design

The starting point in considering the engineering of such protection is the central cylindrical anode; the spread of protection is even and the circuit resistance is easily calculated by the method that was indicated in Chapter 3. The resistance between a concentric cylindrical anode 2 in. diameter and the wall of a 24 in. diameter pipe containing 1,000 ohm cm water will be given by the equation 8.1 which gives the resistance of 1 ft

$$R = \frac{\rho}{30 \times 2\pi} \ln \frac{D}{d} = \frac{1{,}000}{30 \times 2\pi} \ln 12 \text{ ohm/ft} \ldots$$

where D is the pipe diameter and d is the anode diameter. For the system discussed the resistance of 1 ft will be 13 ohm. Now the circumference of the pipe will be just over 6 ft and at a current density of 3 mA per sq ft, that might be required for protection, the pipe will receive 19 mA per linear ft from an anode driving voltage of 250 mV. This is not unreasonable and such protection could be achieved by the sacrificial or impressed current methods. The anode bulk is large, though it only occupies 1 per cent of the pipeline; smaller anodes could be used and only 400 mV drive would be required with a 1/2 in diameter anode. The water resistivity will affect the driving voltage directly and an increase in this will directly increase the voltage required.

If the pipe diameter is increased then the current required per foot run will increase at the same rate, while the resistance of each foot length will increase as the logarithm of the ratio of the new and old diameters; a reasonable driving voltage may still be required in large diameter pipes carrying high resistivity water. The pipe considered was bare but if it were coated the amount of current required to achieve protection would decrease, the circuit driving voltage would drop, and the coating would allow this system to be used in even larger pipes and at even higher resistivities. It is not easy to support the anode in the center of the pipe and will be more convenient to hold it somewhat off center. In this case, though the circuit resistance will decrease slightly, the current required to achieve protection all around the pipe wall will be larger.

The ease with which a pipe can be protected in fresh water, and the low driving voltage used, suggests that in the case of lower resistivity water and with coated pipe the central anode need not be continuous but could be made in short lengths each separated from the next. If protection, for example, can be achieved by using 5 ft long anodes separated by 5 ft, then the same protection could be obtained from 1 ft anodes again separated by 5 ft. In fact, the anode can be reduced in length, providing the inter-anode separation is not increased, until it retains only enough material to give the required life and a low enough driving voltage to be reasonably economical. The anode arrangement in the pipe might then be as in Fig. 231a. On a particular pipe the protection parameters will vary with the distance between the anodes, s in Figure 231a and this will have a maximum value beyond which the pipe will not be protected at the mid point between the anodes. In considering this arrangement the system can be divided by the lines aa' and that section considered separately. Because the system is symmetrical the arrangement is equally divided by the line bb' so that the small section length $s/2$ between aa' and bb' can be considered, Fig 231b. Consider the small length of pipe δx a distance x from bb', there will be a current of I_x flowing from the anode and this will cause a change in the potential between the pipe metal and the sea water. This swing of potential will be E_x and E_x will change in the small section δx by the ohmic resistance of this small disc of water multiplied by the current and this change will be δE_x where

$$\delta E_x = I_x \rho \frac{\delta x}{\pi r^2}$$

where ρ is the water resistivity and I_x the current. Similarly, I_x will change by an amount δI_x because of the current that will be caused to flow from the water into the pipe metal by the voltage change E_x and

$$\delta I_x = E_x \cdot g \cdot 2\pi r \cdot \delta x$$

where g is the coating conductance each unit area assumed for simplicity to be constant. These two equations lead to a differential equation whose solu-

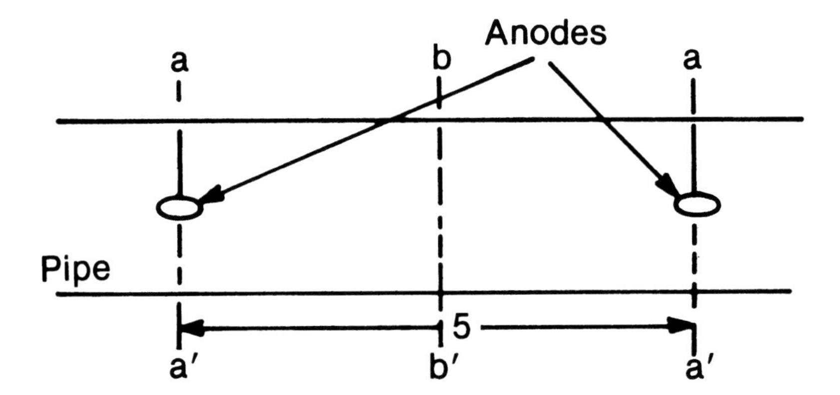

FIGURE 231(A) - Impressed current point anode arrangement in pipe.

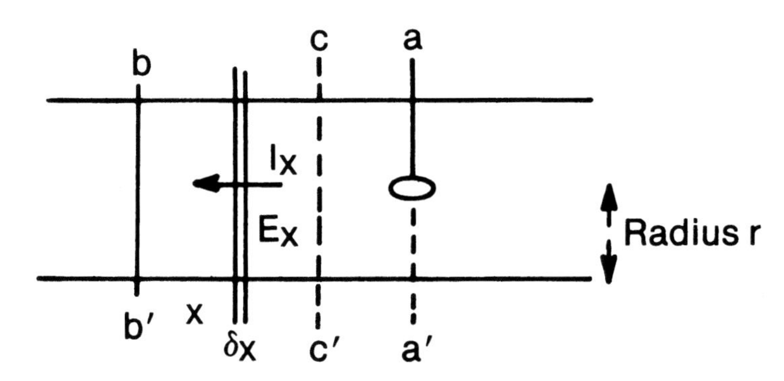

FIGURE 231(B) - Figure 231(A) with calculation parameters shown.

tion by mathematics similar to that used in Chapter 6 will be

$$E_x = E_b \cosh ax \tag{8.4}$$

where

$$a = \sqrt{\frac{\rho}{\pi r^2} \bullet g \bullet 2\pi r} = \sqrt{\frac{2\rho g}{r}} \ldots \tag{8.5}$$

and the current I_x will be

$$= \frac{dE_x}{dx} \bullet \frac{\pi r^2}{\rho} \tag{8.2}$$

$$I_x = E_b a \frac{\pi r^2}{\rho} \sinh ax \ldots \tag{8.6}$$

To achieve complete protection E_b should be the swing required for protection usually 300 mV, and for a particular value of ax, E_x will be a definite multiple of this. Now a will vary with the water resistivity, with the conductance of any coating, and with the radius of the pipe. If the ratio E_x to E_b is fixed, ax being a constant, the value of x will vary inversely as a. This means that x will vary inversely as the square root of the resistivity, inversely with the square root of the coating conductance and directly with the square root of the diameter.

The value E_x is the swing potential of the metal at some distance x from the point of minimum potential swing. If the anodes are separated by a distance $2x$ this is not the same value as the anode potential. The method of calculation assumes that the current is flowing through the water axially and the equipotentials are nearly perpendicular to the pipe center line. Around the anode these conditions will be very much disturbed and current will leave the anode radially, and it is reasonable to assume that the potential swing of the pipe metal varies as in equation (8.4) up to some point or line cc' and that beyond this the potential is influenced by the proximity of the anode. To estimate the maximum safe distance between the anodes it would then be reasonable to consider that this distance will be $2x$ as derived from equation (8.4) plus some distance of anode influence. The distance will vary with the various pipe parameters but it seems unlikely that the anode influence will be as highly dependent upon these. Assume for the sake of argument that the anode influence extends for one pipe radius either side of the anode. Then the distance between two anodes, where E_x is a fixed multiple of E_b and the pipes are identical, except in diameter, will be

$$s = kD^{1/2} + D$$

where k is a constant depending upon the pipe conditions and D the pipe diameter. The total extent of the anode influence has been guessed as one diameter for all pipes but this need not be so. However, it is likely to be small compared with $kD^{1/2}$ so that its exact value will not be of great significance. When the anode is off center from the pipe the extent of the anode influence will vary. With an anode that is only a small distance, say 1/4 diameter from the pipe wall, it may reasonably be assumed that the anode influence will fall to this distance either side of it.

In the protection of pipelines there are other criteria that may be considered to have greater importance and one of these is the rate of increase of the anode current with anode spacing. Second will be the ratio of the anode current to the current if the pipe received a uniform current density; the significance of the ratio will be coupled with the cost of the engineering.

The third criterion depends upon the coating material as there will be a maximum voltage swing before breakdown and this fixes a limit to E_X.

The pipe considered has been assumed to have an ohmic coating, that is the relation between the swing potential and the impressed current density was constant. Much of the worst corrosion occurs in bare steel pipe used to carry sea water and the relationship between the current density and pipe potential swing is not linear. The value of the swing of potential and the current flowing in the water can be calculated by considering small sections of a pipe of a particular diameter filled with sea water of a known resistivity. Each small length will present an ohmic resistance through the water and a resistance across the metal/electrolyte interface. The changes in E_X and I_X can be calculated for each small section and this has been done for a 48 in. pipe full of sea water in Fig 232. These calculations have ignored the anode influence and this can be assumed to be the same as in the case of the coated pipe when $kd^{1/2} >> D$.

At this stage the results of practical experience can be compared with the theory. From the curves of overprotection and current it would seem that in a bare 48 in. pipe the anodes may be separated by some 18 to 20 ft or 4 1/2 to 5 diameters. Some results had already been obtained from attempts to achieve protection by this method in bare 60 in. pipes at the British Petroleum Isle of Grain Refinery, where protection was achieved at anode separations of 20 ft but not at 30 ft anode spacings, and Carter and Crennell had suggested that 1 ft diameter sea water pipe could be protected at 6 to 8 diameters between anodes. The author and his colleagues have carried

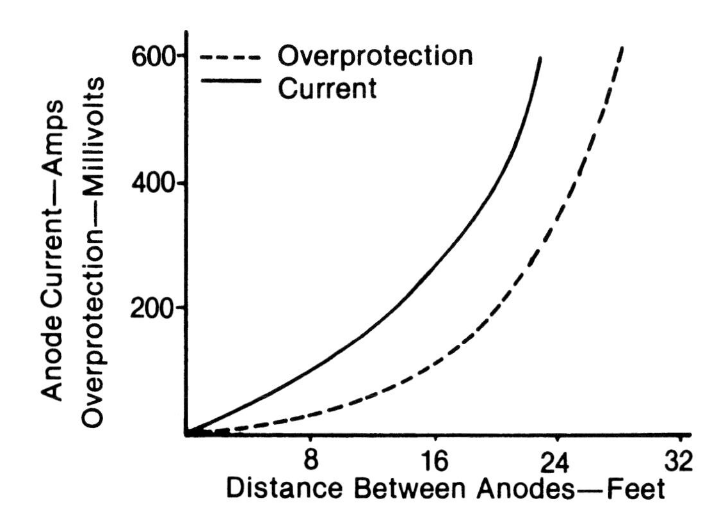

FIGURE 232 - Potential and anode current against anode spacing.

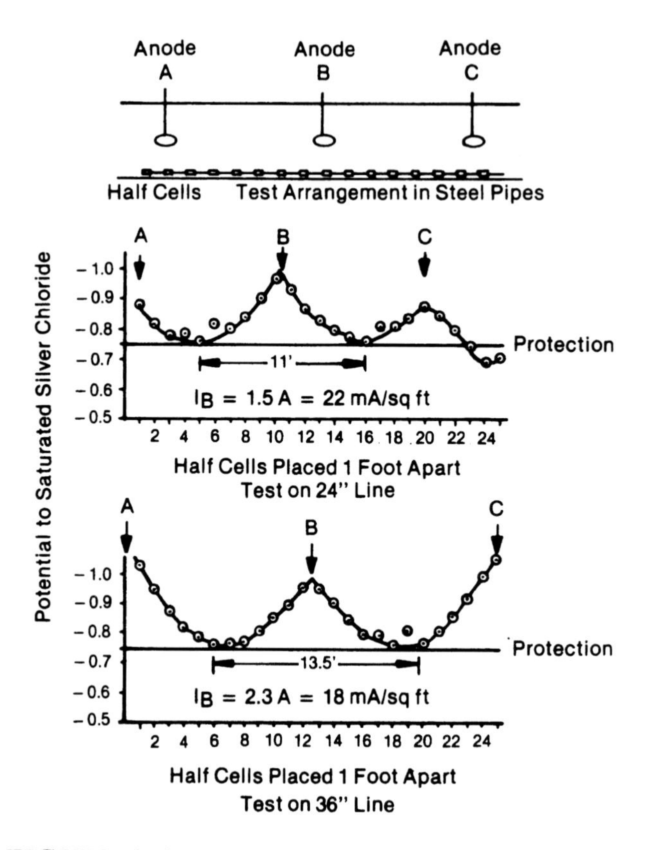

FIGURE 233 - Test arrangement and results on bare steel pipes carrying seawater.

out an extensive series of laboratory tests on small bare pipes of approximately 1 and 1 1/2 in. diameter and on some larger bare pipes in the Persian Gulf. In these latter tests three anodes were placed in a pipe and the maximum spread of protection from the center anode was measured. This spread could be varied by controlling the current into the two outer anodes, potentials were measured along the line at frequent intervals and curves similar to those of Fig. 232 were obtained. The test arrangements and results on a particular run of 24 in. and 36 in. pipe are given in Fig. 233. These were obtained by causing a fixed current to flow into the center anode *B* and then adjusting the outputs from anodes A and *C* until protection is just maintained between *A* and *B*, and *B* and *C*, and the distance between the two minima of the swings is measured.

The results of all these tests, that is the author's at 1 in., 1 1/2 in., 16 in., 24 in., and 36 in., and the other two reported results, are shown on

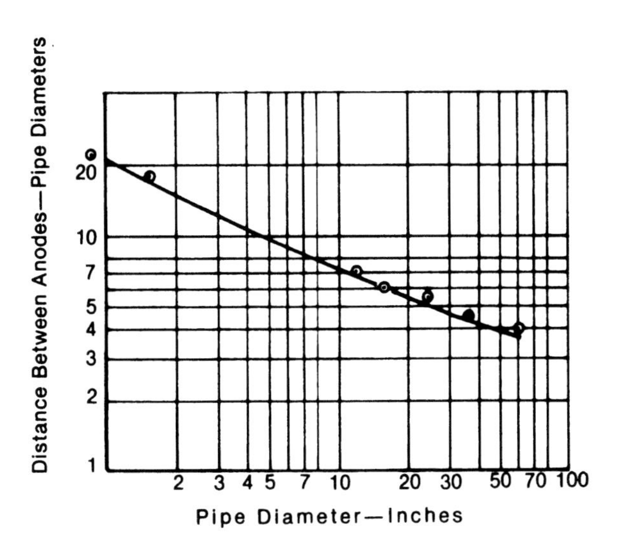

FIGURE 234 - Maximum spread between anodes in variety of pipe diameters.

Fig. 234, plotted as the numbers of diameters of the pipe between the anodes for various pipe diameters. The relationship given in equation 8.7 is also shown as a continuous line. The curve derived from equation 8.7 shows a reasonably good fit to the experimental results. The velocity of the water in the pipe will increase the current density required to achieve protection and decrease the spread. This effect is reduced on coated pipes and those that have been polarized so that the final spread of protection is almost independent of the water velocity. Polarization of the steel pipe can be improved by a reduction in the sea water velocity or by using the pulse techniques described in detail for ships' hulls. This latter technique has been used extensively in ships' cooling water systems with considerable success.

The spread of protection will, of course, be greatly increased by coating the pipe. The current required for protection of bare pipe seems to be of the order of 8 to 10 mA per sq ft and a coated pipe may only require 0.02 mA per sq ft to reach a protective potential so that in effect the coating resistance will have increased about 400 times. The spread of protection will increase and since in a bare 24 in. pipe there was a spread of 5.5 diameters between anodes, that is the part affected by *a* was 4.5 diameters, then if this pipe were coated as described above the distance between anodes would be 90 diameters or about 180 ft, whereas if there were a linear relationship this distance would have been 4,400 ft.

The coating resistance and the polarization of the steel could be determined using a continuous central anode and the water resistivity could be found. In the second series of experiments a small zinc electrode about 1/4 in. diameter and 3 in. long was attached to a flexible cable, rubber sheathed and insulated, of approximately the same diameter and this was introduced into the pipe through a sealable gland. The electrode itself had a small plastic shield so that it did not short to the side of the pipe and was carried downstream by the water flow. It was then retrieved by hauling measured lengths of the cable. A plot of the potential along the pipe was made giving results as in Fig. 233. This technique, though clumsy, was inexpensive and gave positive reproducible answers over 300 ft of line.

These methods are not suitable in practice and two designs of potential test points are available. The first uses a porous plug mounted in an insulated holder which goes through the wall of the pipe. The plug becomes saturated with water and so acts as a salt bridge which extends the electrolyte and the half cell can be placed against this plug to measure the potential inside the pipe.

The second method uses a small zinc rod which is mounted in an adaption of the anode rod holder.

If the pipe is of a peculiar shape and varies considerably with junctions and bends then it may be necessary to adjust the output of each anode individually and this may be done by a resistive control in each anode lead. If the pipe is a long straight run then the current required at a few random anodes can be used as a criterion for setting the other anodes, the whole series being controlled from a single power generator. The spread of protection achieved at bends and junctions in pipes is beyond the simple analysis used earlier. In practice the bend is taken as the equivalent of a piece of pipe the length of its outside measurement, while junctions can be analyzed by the techniques used for the protection of outside of pipelines, again the best position for the anode being at the bifurcation. Scour will occur more markedly at bends and at bifurcations. The additional risk of coating damage should be accommodated in the design.

Engineering

The engineering of the cathodic protection can be carried out by two different techniques. Firstly, where the pipeline has high resistivity water and is bare or poorly coated a continuous anode is used and this could be a wire of platinized or oxide coated titanium with a copper core. There will be some utility in skip plating the titanium. The ratio between plated and unplated area can be calculated by the techniques described. The wire can be mounted nearly centrally or it can be mounted in a continuous holder running close to the wall of the pipe. This latter technique might be preferred where inspection is required. Sacrificial anodes can be used in the same way and these can be placed along the wall of the pipe, though if the

FIGURE 235 - Twin zinc ribbon anodes installed to protect coated offshore line. (Photo reprinted with permission from Federated Metals Corp., Somerville, NJ.)

metal is bare a small shield around each anode will assist in the spread of protection. In fresh water magnesium anodes are used while in sea and brackish water aluminum or zinc anodes are used. The standard type of hull anode can be employed though it may need minor modification to comply with the internal curvature of the pipe.

In using a line anode for impressed current it has to be positioned to run along one point in the pipeline, whereas sacrificial anodes can be placed so that they successively lie at 9 o'clock and 6 o'clock or further divided to lie at, say, 4, 8 and 12 o'clock. If the pipe is occasionally left not full of water then it is useful to have anodes in the bottom of the pipe. In smaller pipes ribbon anodes can be used and these can be attached at about 20 ft intervals, the number depending on the diameter of the pipe, Fig 235. In high velocity water more frequent fixing is required to stop the anode moving. The weight of the zinc ribbon is such that it should preferably be placed towards the bottom of the pipe. This technique has been successfully used on ballast lines in ships where the advantages of deaeration in stagnant water are apparent. Similar systems have been fitted to sea water fire mains to ensure their integrity. In using any anode system in a pipe care has to be taken that particles of anode metal are not detached as these can, and almost inevitably will, fall in places where they may prevent valves closing properly, etc.

The second technique of engineering uses discrete anodes and relies on the spread of protection between these.

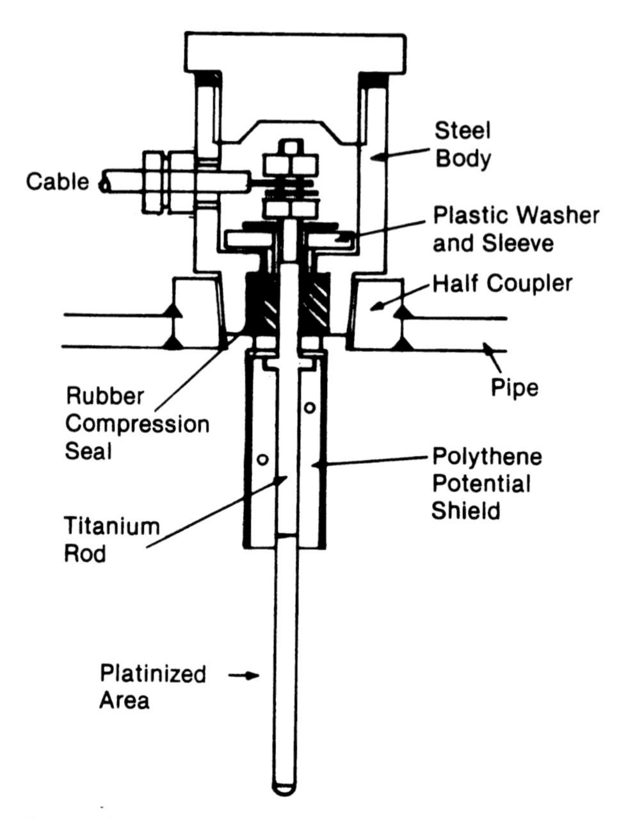

FIGURE 236 - Early platinized titanium cantilever, through-wall anode.

The early installations were carried out using lead silver anodes but these are large and heavy. A much superior anode is provided by the metal oxide or platinum film anodes on a substrate of titanium. The use of platinized titanium anodes in which part of the titanium rod is not platinized and is used as a support probably made its first and biggest impact in this field of engineering. The anodes were mounted by various techniques but it was found that hard mounting caused the anodes to snap off. Soft mounting of the anode, and an early design is shown in Fig. 236, were successful and the polyethylene cup was used as a shield so that the potential between the titanium and the surrounding electrolyte never exceeded 8 V. The shield contains an electrolyte which is nearer the anode potential than the opening of the shield as there is a small anodizing current that flows from the bare titanium. A much better design of anode has been developed and this is shown in Fig. 237. The rod is 3/8'' diameter and the unplatinized area is coated with a wound polyester glass fiber sheath. The

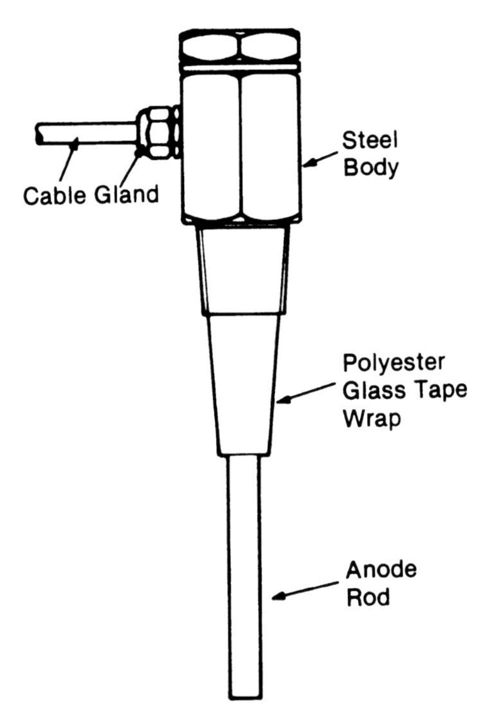

FIGURE 237 - Cantilever through-wall anode with polyester molding.

holder and the anode can readily be inserted into the pipeline. Some of these anodes have been wet tapped into lines, the boss being pre-welded. Where lines operate at low pressure this has been done by an expedient of drilling the pipe with a pneumatic drill and then placing the anode into the pipe through the drilled hole. There is some loss of water but it is much cheaper to wear waterproof clothing than to set up the elaborate system of pressure tapping the line.

The anode is inexpensive, both to manufacture and to fit. In many pipes it is considered prudent to place the anodes so that should one fail it is possible to operate backup protection from the adjacent anodes. It also is an advantage to place the anodes closer than the maximum spacing to reduce the average current density on the line. On big lines of 60 in. or more anodes can be placed on opposite sides of the line, a popular arrangement being at 10 o'clock and 2 o'clock.

Where there is a high water velocity and the incoming water may not be fully filtered the anode is angled into the stream so that debris, strands of seaweed, etc., are swept off the anode. Cases have been known where the flagging effect of seaweed has been sufficient to snap an anode. A smaller version of the anode was shown in Fig 222.

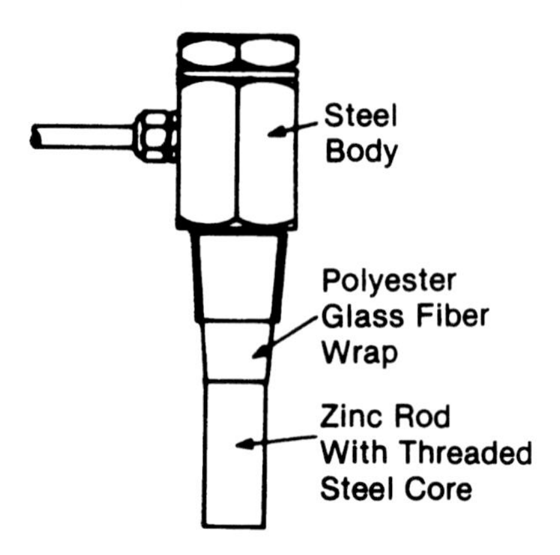

FIGURE 238 - Zinc electrode in through-wall steel fitting.

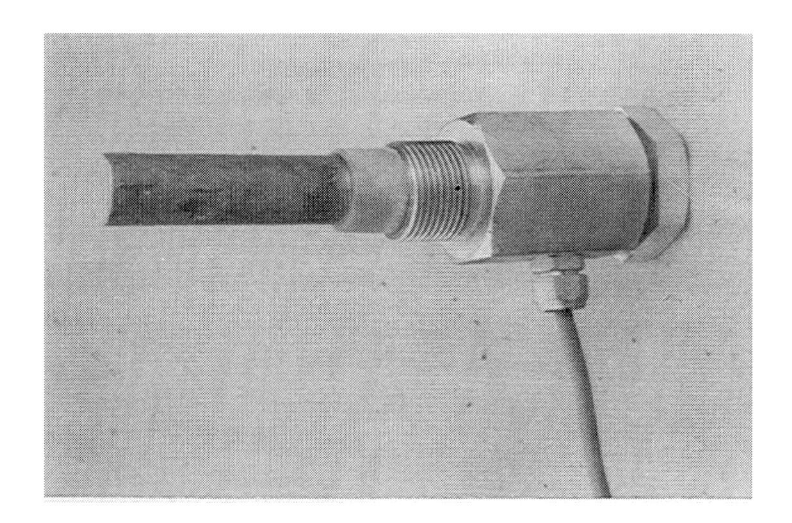

FIGURE 239 - Zinc electrode for the through-wall mounting complete with zinc coated steel holder. (Photo reprinted with permission from COR-RINTEC/UK, Winchester, Hants, England.)

A zinc electrode can be made to fit the same mounting and this is shown in Fig 238 and 239, and a silver chloride electrode can be manufactured complete with plastic protection to replace the zinc electrode.

Where there is considerable turbulence in the line the anode can be replaced with a flush or near flush disc anode and the type used to protect small ships can be employed in pipelines or a much smaller version in which the disc is attached to the holder can be used in pumps and other areas where there is a considerable restriction on the anodes.

Many sea water mains are buried in the ground or in equally inaccessible places and an installation from the outside of the pipe would be very costly. The inlet scoop on a VLCC is an example. The type of system described can be mounted inside the pipe or alternatively anodes based on stud fixing can be used. An anode similar to the ships' hull anodes can be used and this can be mounted complete with cable in a shape as shown in Fig. 240. The whole is held beneath an angle iron which is continuously welded into the pipe interior. The integral shield at the anode has proved to be highly effective. In large pipes two such strings of anodes have been employed and the installation from the inside of the pipe is simple and economical. The anode is designed for simple removal and an anode joint can be fabricated on site to the existing cable.

One example of large diameter pipes are the vertical risers and in these coaxial anodes can be freely suspended. Water tower riser pipes have already been described and other such structures are equally simply protected.

There are a few cases where the pipeline that is suffering corrosion has already been installed and there is very great difficulty in access. Under these circumstances it is possible to use the aluminum wire type anode that was trailed behind a ship and allow this to run in the pipe. By using a controlled rectifier it is possible to allow for some contact between the anode and the pipe wall. The technique is messy but can be used as a stop gap where installation otherwise would be nearly impossible.

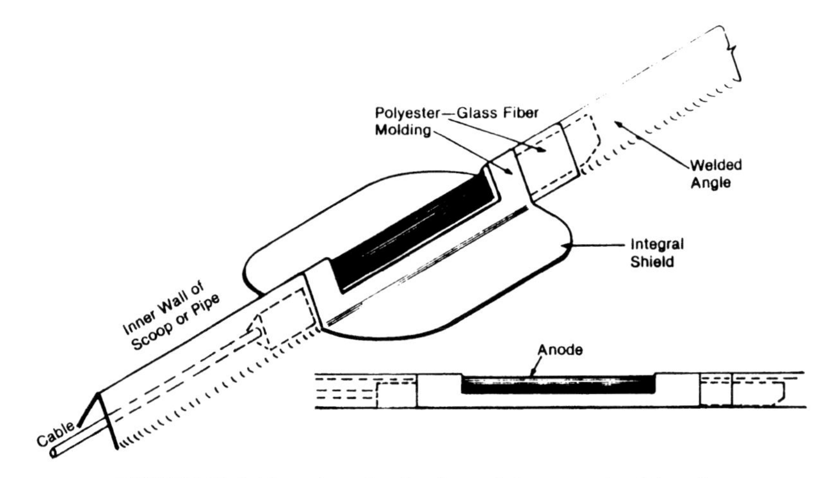

FIGURE 240 - Anode designed for use inside pipe or scoop.

Condensers and Heat Exchangers

The cooling water that has traveled through the pipes is used as the coolant in condensers and heat exchangers. In these the cooling water usually passes through a large number of tubes arranged so that several dozen in a group are in parallel and these are followed by other groups of tubes. The product that is cooled is passed over and around these tubes, the hot product reaching the outgoing water tubes first. Corrosion of these condensers can be rapid. They are made of a variety of metals, bronze tubes and tube plates with cast iron water boxes or aluminum bronze tubes with steel tube plates, etc. The use of these tube alloys has reduced tube corrosion considerably but has brought with it aggravated water box corrosion. Some condensers are made from steel and this is often galvanized, either before or after construction and all-aluminum heat exchangers and condensers particularly where the aluminum is compatible with the cooled fluid. Condensers and heat exchangers using titanium are now commonplace. This requires anodic not cathodic protection.

Cathodic protection can be applied to the water boxes of condensers and protection will spread some distance up the tubes. Where the water box is coated, or on small condensers, sacrificial anodes can be used. These are usually made of magnesium and bolted either directly or through a resistor, to the water box wall. The technique is similar to the hull mounting on ships and the anode is often surrounded by a small plastic mat. Zinc has been used more extensively and its small volume is an advantage. One use of zinc anodes is on aluminum dephlegmators which use sea water cooling. A typical small water box zinc anode is shown in Fig 241.

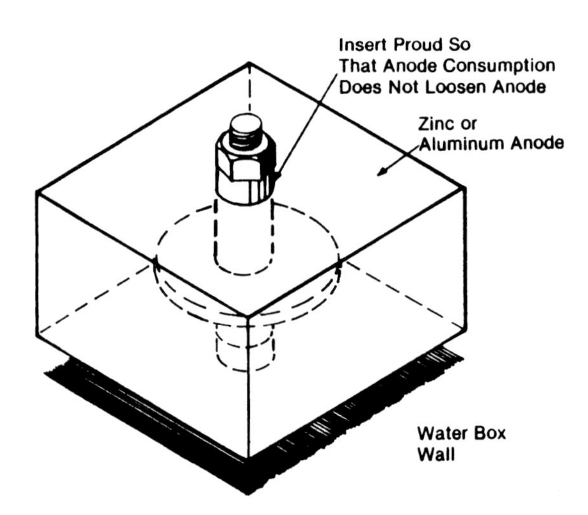

FIGURE 241 - Zinc anode for internal protection.

Initially there was some fear that the aluminum might suffer cathodic corrosion under impressed current techniques and that sacrificial anodes would not protect the tubes. The author suggested that the tubes should be anodized, and after this was done, the current density needed for protection was so low that complete spread, some 10 ft up 2 in. diameter tube, was achieved with no appreciable voltage drop from a single zinc anode in either end box.

Aluminum anodes have not been favorably received when they contained mercury, nor have mercury containing zinc anodes been used. Aluminum indium anodes would appear to not suffer from this defect but there is considerable caution in their use. This is similar to the problem with aluminum hulled vessels where it is feared that a smear of the mercury anodes might cause corrosion to be initiated along and through the smear into the cathode metal.

Most condensers will not be sufficiently protected by sacrificial anodes either because the large volume of metal required restricts the flow of the cooling water or because insufficient protection of the tubes occurs. Impressed current protection does not suffer from these objections and can be used in most cases. In sea, estuary or brackish water platinum surfaced anodes are preferable and can be fitted into the water box. They are small in size and their output can be controlled to give adequate protection. A typical anode has already been shown for pipeline protection, Fig 237, and these can be mounted easily into the water box and into the feed pipes.

Cathodic protection will be effective only à short distance, 8 diameters, along most tubes. Exceptions to this are galvanized steel tubes—where protection can be achieved some 15 to 20 ft up a 2 in. tube—and the anodized aluminum tubes mentioned earlier. The short spread of protection is, however, sufficient to protect against the majority of the corrosion as this occurs at the tube ends. This is caused by differential aeration, turbulence, impingement, stresses caused by expansion and galvanic couples. With some tube alloys care must be taken that the metal is not adversely affected by the cathodic protection.

Much of the most severe corrosion is caused by the closeness of the different metals in the tube plate and the water box. The galvanic cell formed when these are in contact has a high driving voltage and corrosion results in the corner of the water box which is the area most difficult to protect. This problem can be overcome by coating or painting the metal close to the junction when the cathodic protection will effectively suppress the corrosion. In most water boxes the spread of protection will be limited as the current density on to the tube plate will be high and so the tubes that end in the corner of the box will receive less protection than those at the center; any potential measurements made to judge the effectiveness of the protection should be made relative to the corners of the water boxes. The simple and effective anode shown for the protection of pipelines can be fitted around the

FIGURE 242 - Small air cooled automatically controlled transformer for condenser protection. (Photo reprinted with permission from CORRIN-TEC/UK, Ltd., Winchester, Hants, England.)

periphery of the water box to give a good spread of protection both to the edges of the water box and to the majority of the tube space. A number of attempts have been made to improve the distribution from the anodes. One popular technique was to use a wire which had a knitted polypropylene sleeve over it and to run this continuously around the water box. The technique never appealed to the author as any repair or replacement had to be made from inside the box and it always appeared to be susceptible to damage by people working inside. The spread of protection from it was particularly bad and the technique had been used with other anode materials without great success. The system chalked badly and this chalk tended to cover the anode proper. Automatic potential control reduces overprotection and small units, Fig. 242, are used for individual units or even single anodes. Other designs include hoop anodes but these seem particularly inappropriate as the spacing of the two seals through the water box wall has to be made to suit the particular anode, there being some difficulty in ensuring that the separation of the titanium ends is always the same. Electrically, there is an improvement if the hoop is widened and once this is done there is no reason why the anode should not be replaced by two straight rods and these further separated.

The corrosion further down aluminum brass tubes is not prevented by cathodic protection and when the whole of the cooling water system serving a large condenser is protected the water will be denuded of the iron that goes into solution from the rusting of the pipes and water boxes. On cathodically protected ships there is a decrease in the iron pickup from the hull. This led to earlier condenser tube failure as had been predicted by Brecken; in experiments with Esso, the author was able to replace some of the anodes with iron anodes and these were effective in reducing the down tube corrosion. At about the same time ferrous sulfate was being injected into the cooling water system to the same end. The ferrous sulfate, usually in the form of crystals from spent liquor out of pickling baths which contained sulfuric acid, was messy to handle, difficult to inject and formed a shock treatment. Much better success was achieved with a dissolving iron anode, and it appeared that there was a definite advantage in that the particles, which were macromolecular, from the anode had a tendency to move into the cathode film much more than the ferric chloride globules which were formed around the ferrous sulfate crystals. The effect was remarkable and appeared to be operating at an efficiency somewhere in the region of 100 times better than the ferrous sulfate. The results were more than marred by the distasteful techniques of handling ferrous sulfate and it had been reported occasionally that the crystals, complete with bag, were thrown into the cooling water culverts. In order to provide sufficient protection for a large tanker which would have a 10 ft diameter once-through condenser a current of about 5 amps was made to flow through a series of iron anodes from a small low voltage constant current power unit as Fig 243. These were made sugarloaf size and mounted on insulation into the sea water down stream of the condenser inlet. The anodes dissolved in about two years and their replacement was straightforward at condenser inspection. With condensers that had a return end additional anodes were placed there to boost the iron in the second pass. The system worked so successfully that it is now practice to not open condensers between dry dockings whereas previously large steam ship condensers were opened every voyage.

Protection can also be given to the associated screening equipment to the valves and to pumps in the cooling water system. Some of the screens operate as a moving band and in these although impressed current protection is practical the system has to be linked into the main cathode by a slip ring. This can be similar to a ship's hull slip ring though often the end of the shaft is available and as this turns at a much lower speed a plate and brush arrangement can be used on it. It is useful to reduce the current onto the band screen and to provide more effective protection inside the screened area by attaching sacrificial anodes. Small hull anodes or similar

FIGURE 243 - Small transformer rectifier for iron anode dissolution to extend 'protection' down aluminum bronze condenser tube. (Photo reprinted with permission from CORRINTEC/UK, Ltd., Winchester, Hants, England.)

can be employed for this purpose. Sea water culverts often have large gate valves and very often a lead-in channel. These can be protected using anodes which are mounted on the walls and these can be of the ship hull type, of the type that were placed in the corrugations of sheet steel piling or within perforated supporting plastic pipe. Similarly, the reference electrodes developed for other purposes can be used.

One type of condenser that is used extensively in the coking and steel industry is the vertical condenser which uses steel tubes and an open top box. Cathodic protection of these is difficult as they are bare steel. If the tubes are galvanized internally either individually before they are fitted or in smaller condensers after completion then there is a remarkable spread of current. This was discovered on some heat exchangers in a gas works and spreads of 20 ft along a 2 in. pipe are not uncommon. The process seems to be that the galvanizing provides a black e m f which is as effective as a coating at low voltage swings and has an almost negligible current drain. The spread occurs slowly, often takes a week to a month to develop and is a large improvement on that achieved on bare steel. The tubes will last very

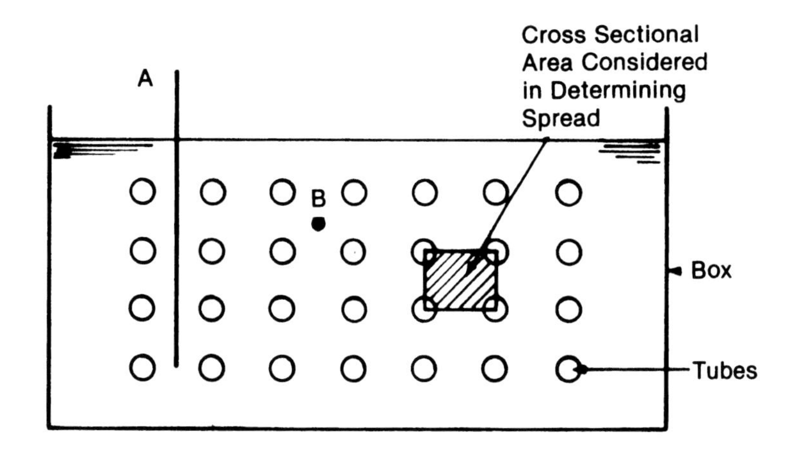

FIGURE 244 - Section of coil-in-box cooler showing anode arrangement.

much longer and this seems to be an economical method of achieving this type of protection without the loss of heat transfer expected with an organic coating.

Coil-in-Box Coolers

So far the cooling and corrosive water has been assumed to flow inside the tubes. Some coolers are arranged to have the product inside the tubes and the cooling water outside. Many of these coolers are used in the oil industry where the product is passed through a series of 4 in. to 6 in. steel pipes which are installed in a steel box filled with cooling water. The problem of protection is now somewhat different and, at any point, the section through the cooler would be as in Fig 244. The spread of protection up and down and along between the tubes can be analyzed in a similar manner to that used on the internal pipe protection. If the rectangular area bounded by the lines joining the centers of four tubes is considered, the area of metal will be πd per unit length where d is the pipe diameter and the electrolyte cross section or volume per unit length will be the area of the rectangle less $\pi d^2/4$. These are indicated in Fig. 244 and the current demand of the metal per unit length and the resistance of the electrolyte per unit length can be used to compute the spread of protection. In most cases the spread will be greater than that expected along a bare pipe whose diameter is equal to the tube separation. Protection can be achieved by placing the anodes in two positions; either vertical anodes as A or with some materials anodes as B can be laid parallel to the pipes. Most anode materials are more easily installed in the vertical position though their mounting and numbers make the job expensive. Lead alloy anodes, when they can be used, are able to take advantage of their flexibility and can be installed horizontally as *B*. The anode can then be taken out of the water at one end into a junction box and if this is done the anode should be taped from some inches below the water line.

Platinized wire anodes can similarly be engineered and both they and lead anodes can be snaked to provide better protection. The platinized titanium anode needs a copper core and care has to be taken in bending the wire to avoid making the titanium cover porous when copper will leach out of the core. Platinized titanium vertical anodes can be used and the type designed for pipeline protection can be installed from a channel running above the pipe structure. The use of anodes placed through the walls is difficult as the pipes may move in relation to the box by expansion whereas the other anode mountings can be attached to the frame which carries the pipes.

Some engineers use consumable steel anodes though the high current density required to protect the steel when the temperatures are high means that they need frequent replacement. Another effect in hot coolers will be the flow of the hot water to the top of the tank, the lower resistivity of this and the extra current required in its presence will make the vertical anodes corrode very rapidly near the water line. This can be overcome by augmenting the anodes with another set of short anodes near the water line, by using a tapered anode or by continuously lowering long anodes into the water.

One type of cooler is the returned tube cooler in which the pipes are hair pin in shape and used to carry the fluid to be cooled. In these cases it is possible to replace some of the tubes by sacrificial anodes and this has been effective in reducing the corrosion of the bundle. It is difficult to use impressed current anodes in these situations except immediately next to the cooler wall. The expense of engineering in what is often a comparatively small heat exchanger will mean that mitigation of the corrosion to perhaps double the life of the system will be the most that can be achieved. Occasionally it is possible to redesign the system so that cathodic protection may more easily be applied. This is increasingly being done, though in many cases only to the extent of providing bosses for the mounting of the anodes.

The platinized titanium group of anodes, while providing cathodic protection, also electrolyze the water forming hypochlorite. It is possible to augment this effect and provide both chlorination and cathodic protection. Designs using groups of plate anodes which have an overall cathodic protection effect as well as generating chlorine have been used but their success has been limited by the life of the plates. The field seems open for considerable development and it eliminates the use of chlorinators that provide high concentrations of hypochlorite which are always difficult and obnoxious to handle.

Chemical Plant

Some of the most costly corrosion occurs in chemical plant. The high cost of the plant, the possible loss of a valuable product and the high cost of shut down, all contribute to this. Cathodic protection can be used in most of the plant though the engineer must investigate the compatibility of his equipment with the chemicals being processed. Often sacrificial anodes give off objectional corrosion products and permanent anodes have to be used while in other cases the permanent anodes may generate chlorine or other gases which cannot be tolerated.

Cathodic protection can be used to prevent three types of steel corrosion which are found extensively in this class of plant; firstly it can counter the corrosion associated with welds and welding; secondly it can prevent stress corrosion cracking and even stop this after it has started; and thirdly it can prevent caustic cracking in alkaline baths.

Caustic cracking has occurred extensively in vessels used in the refining of alumina and cathodic protection has been successfully used to prevent this. Anodes of pipeline type were used and neither the platinized titanium anode nor the polyester glassfiber insulators were adversely affected. The failure, which was mainly at welds, stopped and the tanks have remained properly protected. A wider range of equipment has been protected in this manner by Champion.

Similar success is claimed with aluminum. A case of corrosion that the author investigated concerned an aluminum storage vessel of 99.5 per cent pure aluminum in the bottom of which an aqueous liquor settled. At an earlier stage in the process a catalyst was used and during the plant operation some of this escaped and settled in the storage vessel. Corrosion was rapid, pits of considerable depth were formed in under one year and failure was predicted within the next year. In some experiments with thin aluminum sheet lying in the aqueous liquor and covered with a few grains of the catalyst, the rapid corrosion was reproduced and the sheets were perforated within three weeks. Similar sheets, cathodically protected in the same laboratory tank, were in perfect condition after six months. This protection was achieved with zinc anodes which were acceptable to the operators.

Anodic Protection

There is much use of titanium in sea water and in other plant in coolers and steel is used in sulfuric acid plant. In both these cases the lowest corrosion occurs when the metal is held at an anodic potential and has a highly resistive film formed upon it. This is the basis of anodic protection proposed by Edeleanu.

The mathematics of the treatment are very similar to cathodic protection though, of course, the current flows in the opposite direction. The cathode is often made out of steel, very often of stainless steel, and sometimes of lead. The reference electrodes used in sea water applications with titanium are those that are used in cathodic protection while in sulfuric acid the silver sulfate of cell is used. A number of steels and stainless steels are susceptible to this treatment and extensive redesign of the equipment to accommodate the anodic protection has been made. The range of the potentials under which anodic protection is achieved is large but corrosion occurs at more positive and more negative potentials. Most of the electrolytes in which it is used are good conductors and the currents required are much lower than in cathodic protection.

The systems of automatic control are similar but there is usually a need for a very large polarizing current. This can be supplied automatically and is sometimes necessary only when the plant first starts up. The effects of this system are most dramatic where the heat exchangers of graphite have been replaced by metals such as titanium. Very often a short period of breakdown of the anodic protection will cause a rapid corrosion of the metal and complete backup systems are often installed.

Automatic Control

The automatic control of the protection of complete cooling systems and of the pipelines feeding them is now standard practice. The phase angle controlled thyristors that were used in ship hull protection are used extensively and there seems to be some advantage in the pulse technique.

Typically groups of anodes are controlled. A number of reference electrodes sense the potentials in various zones and can either be used for checking or can be used for control. Microchip technology has meant that it is easy to install processors which will select and test the reference electrodes, choosing the most suitable and controlling the current to the anodes to achieve the best protection.

Because of the multiplicity of small units the protection can be subdivided and often transistor control of d c is employed. It is possible to use individual controllers at each anode and combined anode reference electrode techniques have been developed. Methods using the anode itself as a reference electrode have been tried; some of these have not proved themselves in practice.

The whole system in, for example, a power station, can be controlled by a process controller which will indicate the adequacy of the protection, will correct faults and indicate them.

These systems are being extensively used and the reliability of some of them can be judged by the fact that in large tankers the cooling water system is now made of steel painted to hull specification instead of the more exotic cupronickel and aluminum brass alloys: these systems have performed over more than a decade without failure. Alternatively, the cost of

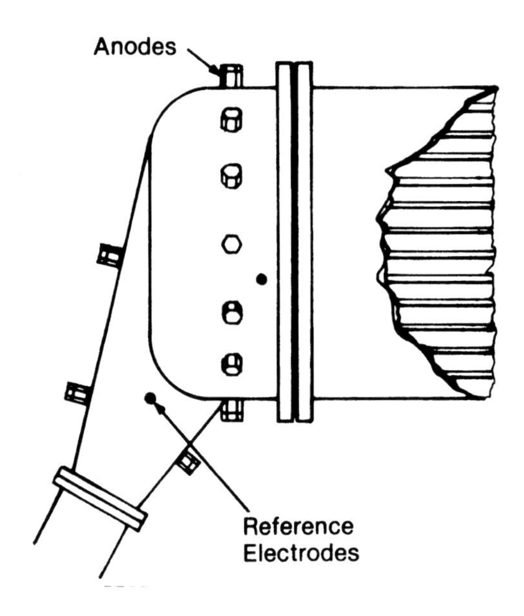

FIGURE 245 - Schematic condenser cathodic protection system.

failure can be judged where in some coastal power stations anode failure through inadequate engineering of the supports means that condenser tubes failed, that condensers have to be closed down or run at part efficiency with a loss of power generating capacity. Frequent breakdowns put some coastal power stations on low efficiency operation, believed to be in some cases as low as 50 per cent. Fig 245 is a schematic view of a condenser protection arrangement.

Stray Current Corrosion, Interference and Electrolysis

Chapter 9

Corrosion of a metal structure is the result of the flow of electric current from the metal into the electrolyte. This may be caused, as already discussed, by variations in the metal/electrolyte potential due to a variety of inhomogeneous conditions in either component. Similar corrosion can be brought about by variations in the electrical potential within either the metal itself or the electrolyte caused by a direct current flowing through it, and is called electrolysis, stray current corrosion or interference.

Although the words are used interchangeably it is suggested that the word 'interference' be used in terms of cathodic protection stray current corrosion, and the word 'electrolysis' used in other fields. The term 'stray current corrosion' being used generically.

There will be two mechanisms that will bring about stray current corrosion. Either the current flows in the structure causing it to have a variety of potentials or it flows in the electrolyte, developing in that potential gradients. A conventional corrosion cell of the type discussed earlier is shown in Fig 246a and is compared with a stray current corrosion cell, Fig 246b, in which the stray current flows in the structure and with the cell shown in Fig 246c in which the current flows through the electrolyte. The process shown in Fig 246b will be called the Structure Current Corrosion and that in Fig 246c will be called Earth Current Corrosion.

Stray Current Corrosion Cell

The induced corrosion current is considered an alternative current path for the stray current but this concept must not be carried too far for this alternative path will contain non linear components, as the metal/electrolyte interfaces, and such may be more easily appreciated by considering the potential drive in the corrosion circuit.

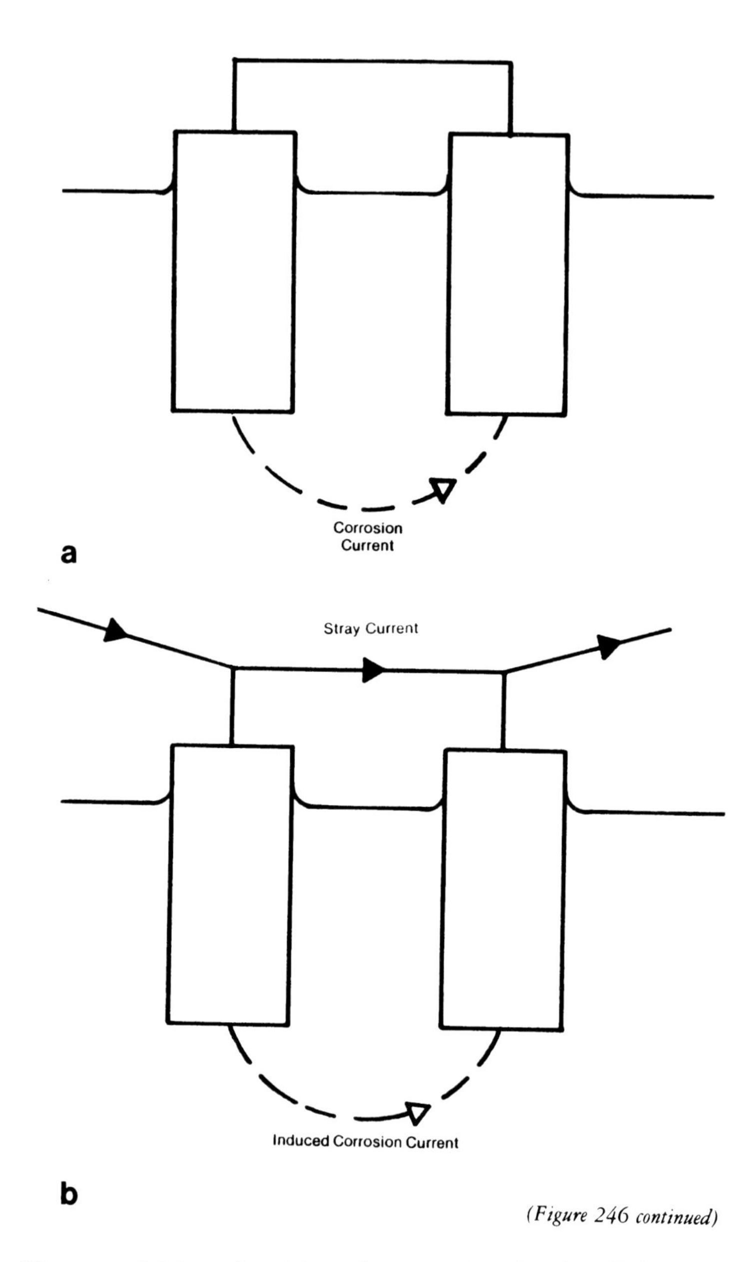

(Figure 246 continued)

The potential introduced into the corrosion circuit will depend upon the voltage drop in the structure or electrolyte. This will drive current around the corrosion circuit which will comprise the metal/electrolyte interface half cells and the resistance associated either with the structure or with the electrolyte.

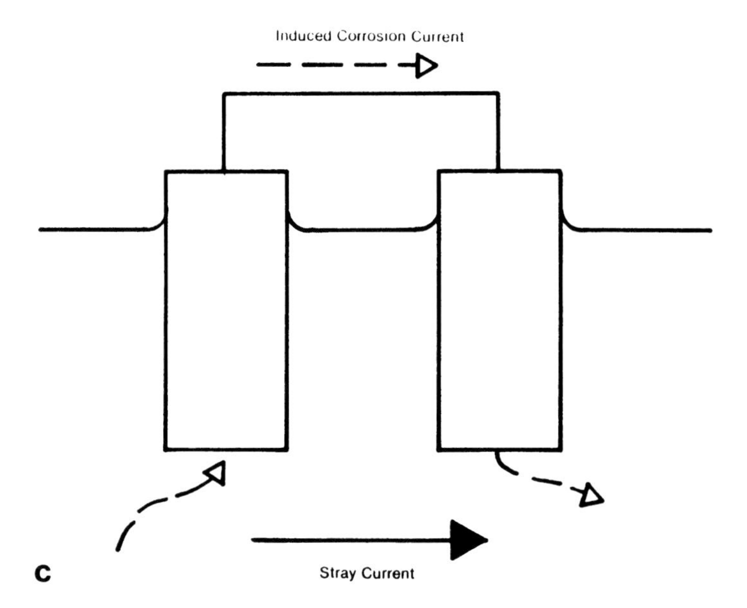

FIGURE 246 - Conventional corrosion and two corrosion mechanisms caused by stray current.

These circuits are shown in Fig 247, the structure current circuit in Fig 247a and the earth current circuit in Fig 247b. The driving voltage V_d in the figures will be badly regulated where the interface has a high resistance connection to the current path in the structure or the electrolyte. The corrosion current will be a function of the driving voltage, the circuit resistance and any voltage that exists at the interfaces.

At both eleectrodes, that is the anode and the cathode, the current flowing in the corrosion circuit will cause polarization. The anode polarization will be such that the anode metal potential will become more positive and the cathode polarization will make its potential more negative; that is the polarization at the anode and the cathode is the same as that found at these electrodes in a conventional corrosion cell. There is a difference between the open circuit potentials of the anode and the cathode in the conventional corrosion cell and the anode potential will be the more negative: polarization will reduce this difference, that is the anode potential E_a will be more positive than the anode open circuit potential E_a; these will both be more negative than the cathode potential E_c which will be more negative than the cathode open circuit potential E_c.

In stray current corrosion the open circuit potential of the anode and the cathode will be the same, assuming that no conventional corrosion is taking place, that is the anode potential E_a will, by virtue of its polariza-

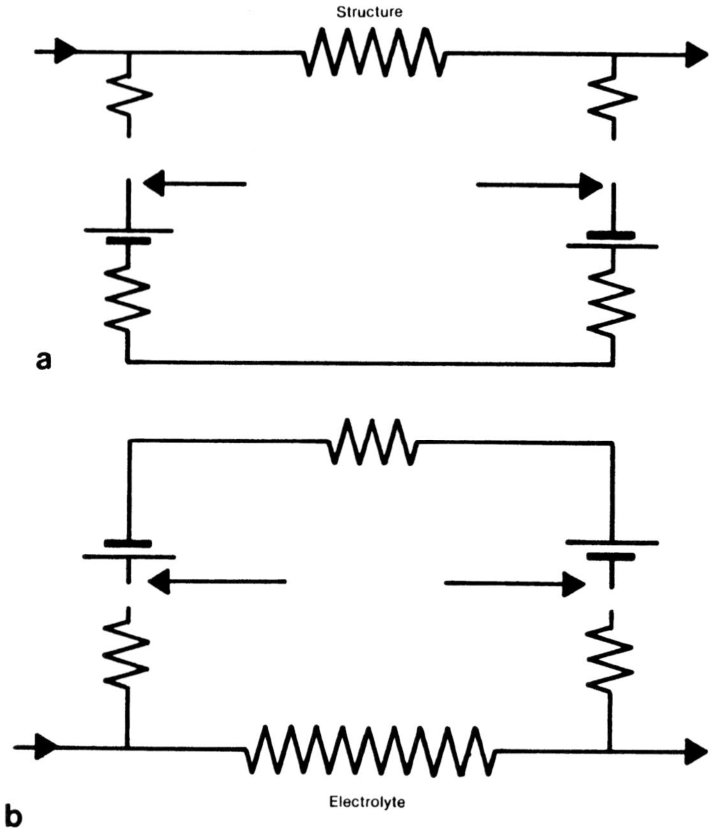

FIGURE 247 - Equivalent circuits of the two types of stray current corrosion (a) structure corrosion and (b) earth current corrosion.

tion, be more positive than its open circuit potential and similarly the cathode potential E_c will be more negative than its open circuit potential. Thus the anode potential will be more positive than the cathode potential in a stray current corrosion cell, this being the reverse of the polarity in the conventional corrosion cell. In a practical measurement of potential, the anodic areas on a structure suffering stray current corrosion will display more positive potentials to a close half cell than will the cathodic areas; these potentials will oppose the stray current driving voltage.

The corrosion current can now be defined by the driving voltage of the stray current V_d—less the polarization of the structure $\phi(i)$—and the circuit resistance R and will be given by the Ohm's law relationship

$$i = \frac{V_d - \phi(i)}{R} \tag{9.1}$$

Now the methods of reducing the stray current corrosion will be those that reduce the value of i without increasing the anodic current density.

Stray currents are generated from a variety of sources. Those found in structures are usually caused by earth return currents from some large d c system, for example, the return path from a light railway on a pleasure pier, the return of welding currents from a fabrication shop or the current flowing through the metal of a pipeline that is being cathodically protected internally. The large currents found in long structures as a result of electric or magnetic storms are usually of insufficient duration to cause any serious damage by this type of corrosion although they may cause damage by the large amount of energy they contain.

Earth currents are not nearly as rare as structure currents and result from the leakage of the returning current from large dc systems employing an earth, or not well insulated, return path. Earth currents are usually associated with tramways, electric railways and other areas where large amounts of direct current are used: two of these latter uses are electric welding and cathodic protection. With the former, except in shops and works, the duration of the effect is only a few days and serious corrosion can be prevented by taking simple precautions. The latter, however, relies upon earth currents for its function, and its ill effect, called interference corrosion, will be discussed in detail later. The large direct currents used in electrochemical processes, such as plating and anodizing, are invariably contained within a metal tank and cause no earth currents.

There is increasing use of high voltage d.c transmission in the electric supply industry. Some of the systems use an earth return path, though this is almost exclusively where this path is in sea water. More use is made of a two wire transmission system in which one wire is positive and the other is negative to earth. Out-of-balance in this system is corrected by earth currents, which can cause very large potential variations in the ground.

Stray Current Corrosion Prevention

The two types of stray current corrosion can be prevented by several methods. While each of these is, to some extent, effective, the best method will be to eliminate or reduce the stray current.

Equation 9.1, which defines the corrosion current, shows that it can be diminished by increasing the circuit resistance. One method of doing this is to increase the resistance of the electrolyte when structure currents are the source of trouble. The electrolytic resistance can be increased by surrounding the structure with a high resistivity environment such as clean dry sand. This will have a resistivity of several hundred thousand ohm cm and so will greatly increase the resistance of the electrolytic path. The structure resistance can often be increased by inserting insulating joints between various parts of the structure. In many long structures this must be done with care for a high resistance coupling in a steel pipeline might be short

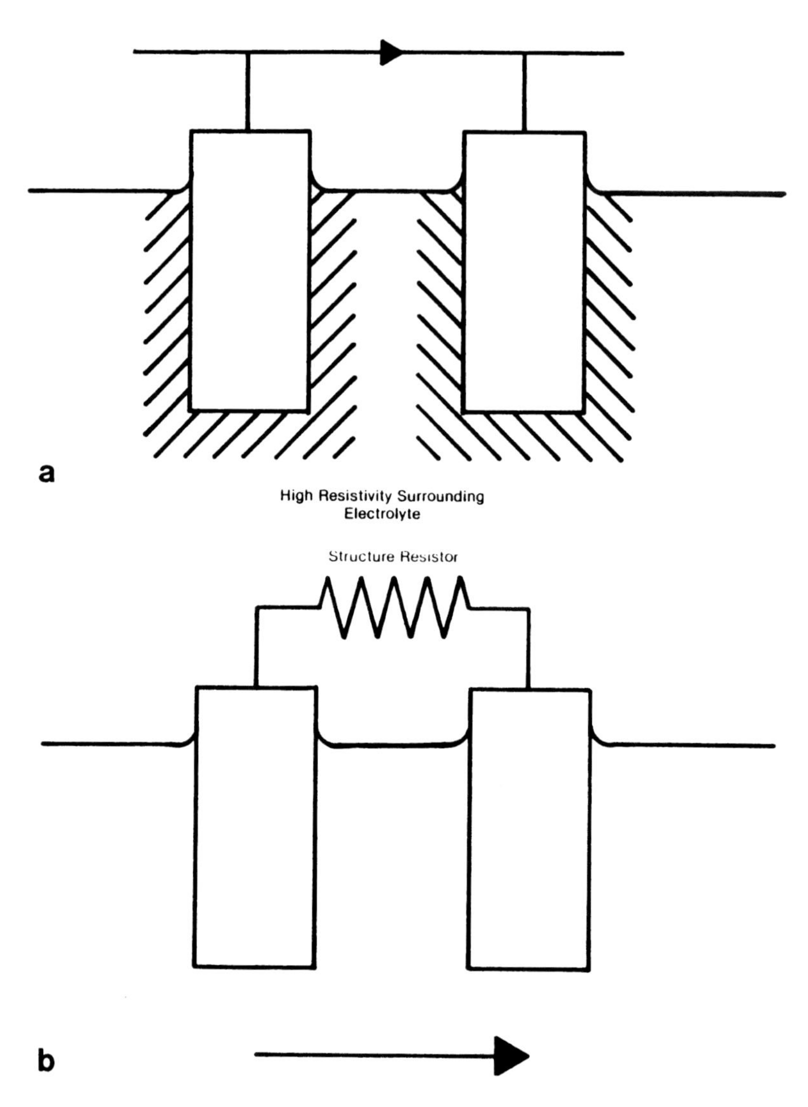

FIGURE 248 - Methods of reducing stray current corrosion by increasing circuit resistance.

circuited by the electrolyte inside pipe, the pipe then suffering structure current corrosion internally. This method, while reducing stray current corrosion, will not entirely eliminate it even if the joints in the structure are completely insulating as each short section of the structure will be subject to small amounts of this type of corrosion. These methods are illustrated in Fig 248a and 248b.

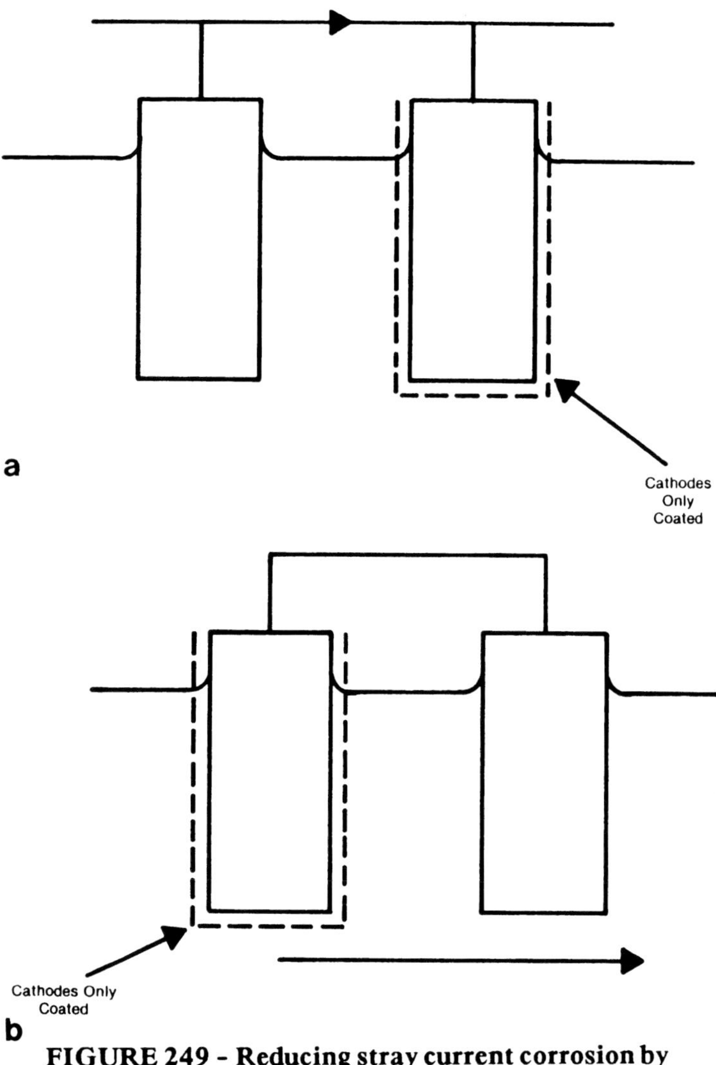

FIGURE 249 - Reducing stray current corrosion by structure coating.

The circuit resistance can equally be increased by painting or coating the metal/electrolyte interface with a highly resistive film, though where the structure is susceptible to failure by a pitting attack then only the cathodic area should be painted. This method can reduce the corrosion current enormously and a high resistance wrapping such as polythene tape might reduce the corrosion to 0.001 per cent of its original value. This method is shown in Fig 249. Often, as with a pipeline, it is difficult to determine at the time of construction the exact extent of the anodic and cathodic areas that earth currents may induce. In this case best practice would be to coat the whole structure and then to use some other method, such as cathodic protection, which would be made easier by the coating, to prevent the corrosion. Surrounding the structure with a thin layer of high resistivity electrolyte such as dry sand will have the same effect as coating, but because this disperses current it can also be applied to the anodes.

A reduction in the corrosion current could be made by increasing the polarization e m f; this will be difficult in many cases but the elimination of depolarizing agents, particularly bacteria, might be useful. One method of increasing the polarization of the cathode will be by coating it as described above when at holidays in the coating the current density, and hence the polarization, will increase. A method akin to increasing the polarization will be to introduce a back potential into the corrosion circuit. Where the structure is carrying the stray current this may be difficult as a back potential would have to be introduced into the earth, but where earth currents are the cause of the corrosion this back potential may be obtained by flowing current through the structure and if this is of small cross section, such as reinforcing bars in concrete, the method may considerably reduce the corrosion. The driving voltage associated with structure currents often results from a series of high resistance joints in the structure and heavy current bonding across these will reduce the driving voltage and hence the corrosion current.

Cathodic protection can be applied to structures suffering stray current corrosion. Either this can be used to give complete protection against all corrosion by causing the structure potential to be more negative than -0.85 V to a copper sulfate half cell or cathodic protection techniques can be used to eliminate the stray current corrosion without giving complete protection. In the former case the anodic area of the structure will require the larger current density and the spread of protection should be used to give the highest current density to that area. Where the major corrosive effect is the stray current corrosion and this alone is to be prevented, then cathodic protection can be applied to the anodic area only. The techniques of achieving this protection are similar to those already described and are illustrated in Fig 250a and 250b and the negative lead of the dc generator (or the connection to a sacrificial anode) is shown attached to the anodic area of the structure.

It is possible to achieve some degree of cathodic protection by using slightly different techniques from those commonly employed. In the case of structure current corrosion an additional electrode could be connected to the most positive end of the structure and if this is made of the same material as the structure, or a metal that is sacrificial to it, then current will leave the structure via this electrode rather than other parts of the structure. Similarly an electrode positioned in the most negative part of the electrolyte

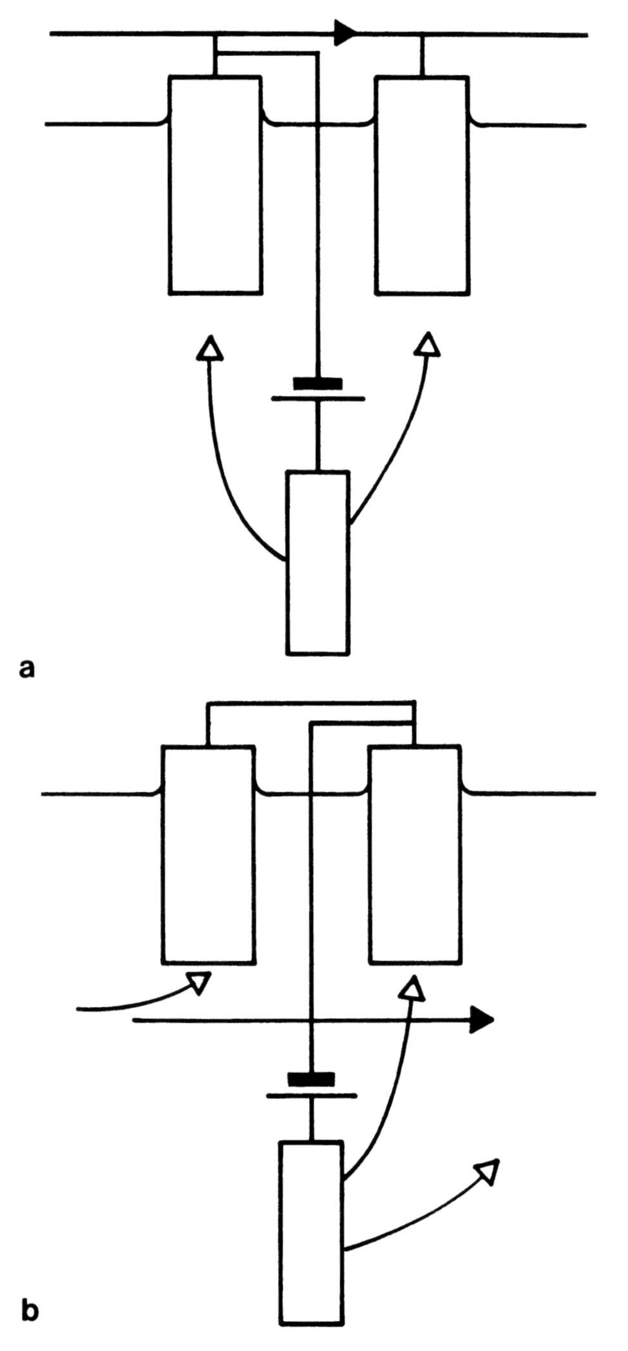

FIGURE 250 - Cathodic protection used to reduce stray current corrosion.

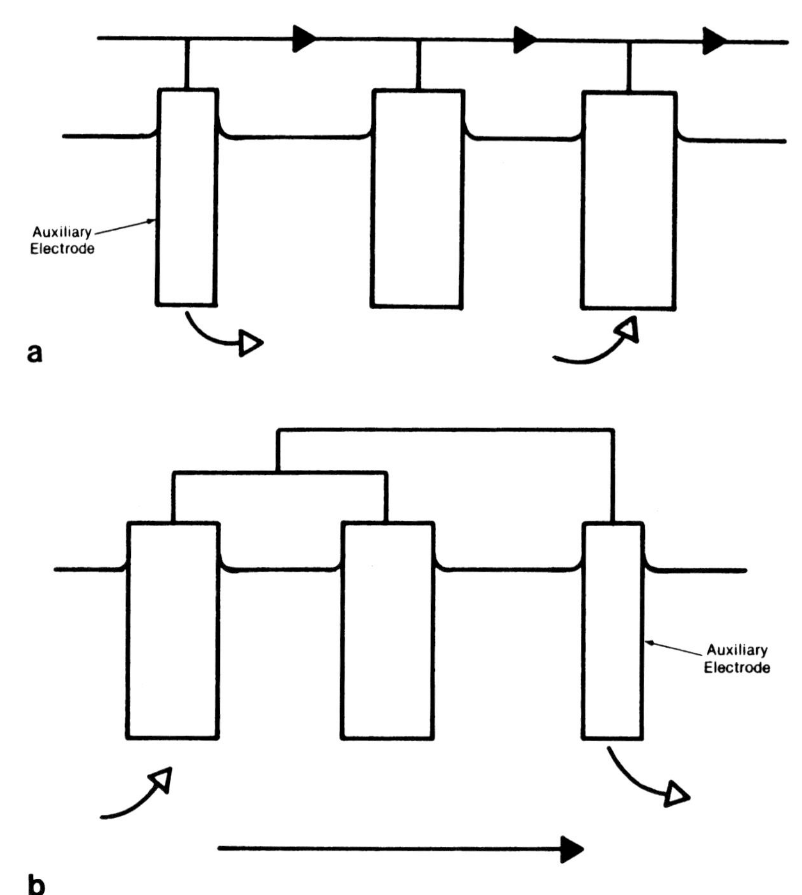

FIGURE 251 - Stray current corrosion reduced by using auxiliary electrodes.

would reduce the effect of earth currents upon the structure to which it was connected. In both cases the extra electrode is placed at the anodic end of the structure or remote from it where the most positive metal potential, relative to a half cell, is found. Fig 251a and 251b shows these methods. Where these techniques can be used then the cathodic protection drainage lead or anode could be similarly positioned and less power would be required to achieve protection.

Three additional methods of prevention are available with earth current corrosion where the sources of the stray current are accessible. The first of these consists of methods of reducing the driving voltage in the earth and the simplest way of doing this is to reduce the stray current by increasing the resistance of one or both of the d c electrodes.

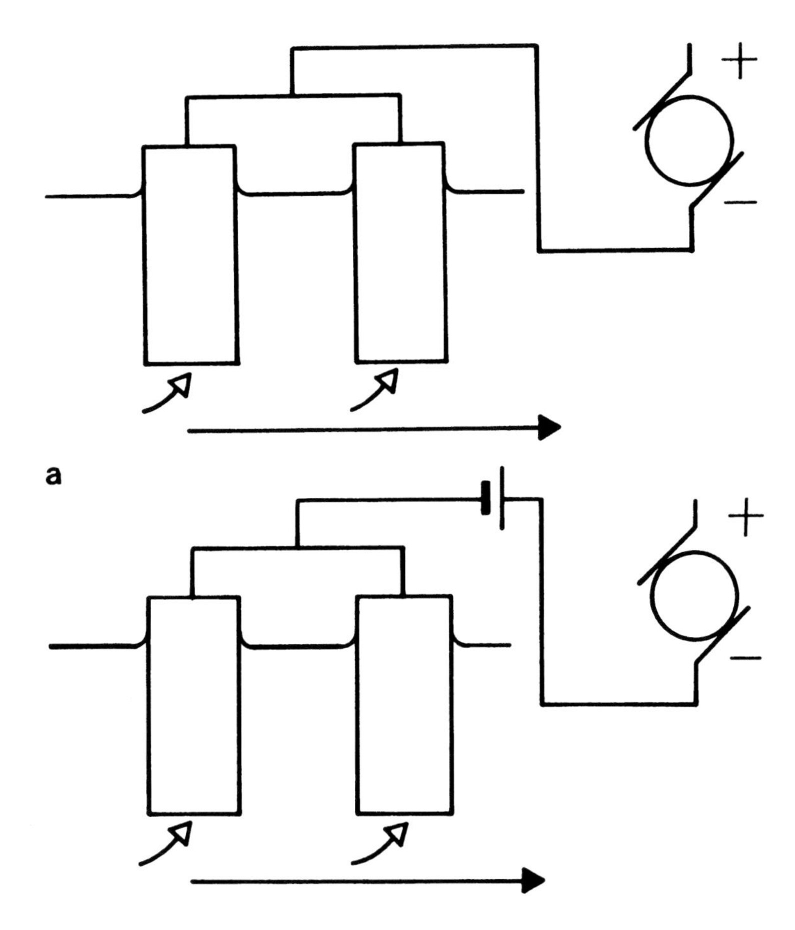

b

FIGURE 252 - Current drainage and forced drainage to negative pole of stray current source.

Another method of very limited application will consist of changing the resistance of various parts of the earth current path by introducing special electrolytes of high or low resistivity in some areas or by introducing sheets of insulating material into the electrolyte.

The third method, the earliest used of all these described and of most practical importance, is to connect the structure that is suffering stray current corrosion to the negative pole of the d c source which is creating the stray current. This connection must be of lower resistance than the alternative earth path so that current flows from the electrolyte on to the structure and thence through the electrical connection to the generator; this technique is known as *current drainage*. Where the conditions do not allow such a connection to be made readily then the structure can be connected to

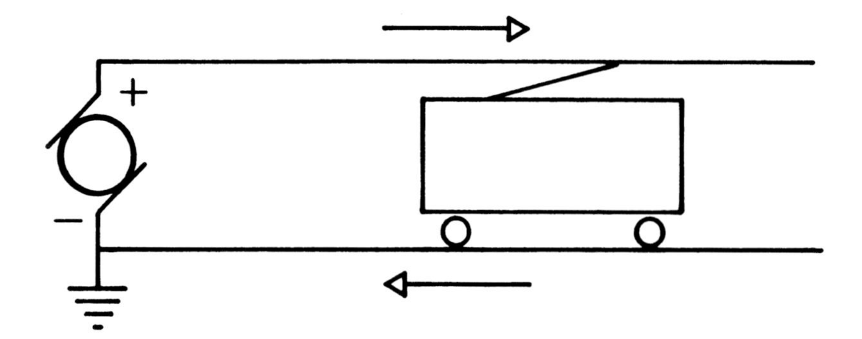

FIGURE 253 - Schematic tramway system.

the negative pole of the generator via a d c source which will increase the current flowing in the connecting cable; this being called *forced drainage*. The amount of current drained from the corroding structure can be controlled either by varying the resistance of the direct connection or by varying the d c voltage used in the boosted connection and so any degree of cathodic protection can be achieved by this means. These methods are illustrated in Fig 252a and 252b.

The limitations and cautions that apply to the application of low levels of cathodic protection must be observed in stray current corrosion. Bacteria may become more active under the influence of a small negative swing and polarization films which would otherwise protect a metal such as stainless steel may be destroyed.

Practical Stray Current Corrosion Problems

Tramways and Electric Railways

The general system upon which d c tramways and electric railways operate is shown in Fig 253. The station generator has its positive pole connected to the overhead wire or conductor rail and its negative pole connected to the return rail and to earth. Several vehicles will operate from each sub-station and one generator might serve two or three pairs of live conductor and return rails. The resistance of the conductor and return rails will cause the voltage to build up in the system and if each vehicle is considered to be consuming an equal amount of current then the potential developed in the return rail will be as in Fig 254. The total permissible potential change in this conductor is often specified by the Act which permits the tramway or railway construction and usually its maximum swing is less than 5 V. The shape of the rail potential V distance curve will vary with the number of vehicles, their load and their position on the track, as several systems may operate and be interconnected, these variations can be large.

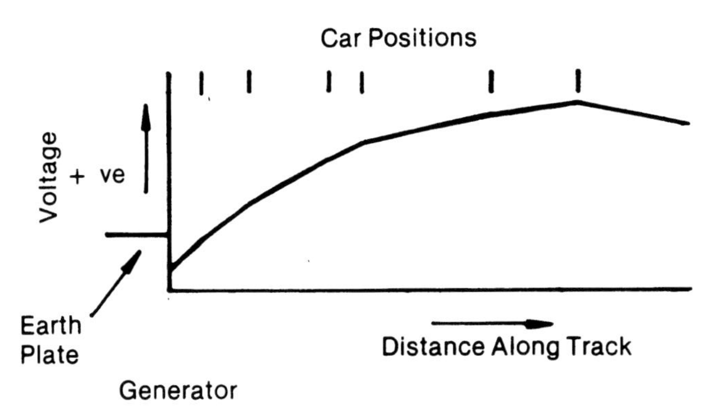

FIGURE 254 - Potential variation along the return rail of an electric railway.

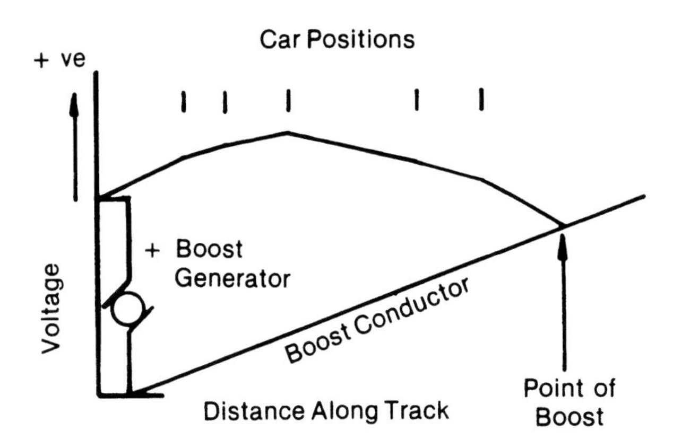

FIGURE 255 - Potential variation along the return rail of an electric railway with negative boost.

Some systems employ a device called *negative boost* which consists of connecting the negative pole of a boost generator to some point on the return rail near to the center of the system and connecting its positive pole to the negative pole of the station generator. Current is then drained from the return rail at the point of attachment of the boost and the total volts drop in the return rail is reduced. The potential of a typical return rail which uses negative boost is shown in Fig 255. The point of negative boost will act as a very poorly regulated negative potential and its voltage will fluctuate considerably with load. In addition to this some operators only use the boost generator when there is a heavy load on the system and this adds to the voltage variations that can exist.

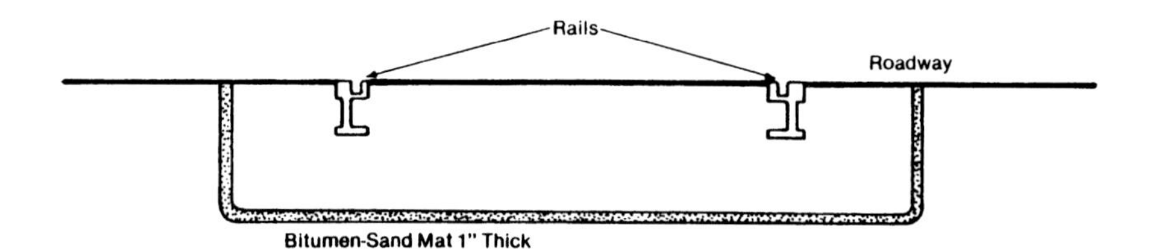

FIGURE 256 - Insulated return track.

The return rail will be seen to act as a structure carrying a large dc current and might be subjected to very heavy structure current corrosion. Also, because of the magnitude of the corrosion current, a tramway or railway using this system will create considerable earth current corrosion. The amount of this structure current corrosion and earth current corrosion will depend upon the voltages developed in the return rail and upon the resistance of this rail to earth. Many systems use low earth resistance rail tracks to reduce the resistance of the electrical return path while others insulate their return rails from the earth. This latter is not easily achieved and usually a bitumen/sand mat is used to surround the track. This is helped by the use of granite or similar blocks for paving the tracks and the method is illustrated in Fig 256. The technique is effective in reducing structure current corrosion as although the coating is on the anodic areas, these areas are surrounded by a current diffusing layer so that any attack caused by failure of the insulating mat will be distributed over a large area.

Railways and tramways that do not insulate their return rail systems will suffer structure current corrosion which may result in the corrosion of rail spikes, bolts or similar small items which can be expensive to replace. If this is the case, two courses are open to the railway operator, either the spike, or whatever part is corroded, can be insulated from the return rail by resin/fabric washers and bushes, or large pieces of scrap steel such as old rails can be buried close to the track and electrically connected to it. The current leaving that part of the rail system will then corrode the buried scrap rather than the railway components. Before undertaking either scheme it is advisable to check on the joint continuity as if these are offering a high resistance to the return of the current they will cause extra rail corrosion by the increased voltage drop across them and this condition is usually easily remedied. Though the last two methods of preventing structure current corrosion may be effective it is only the first and last methods that reduce the stray earth current that results from the structure current corrosion and so also reduces any associated earth current corrosion.

The magnitude of any earth current corrosion will depend upon the potential variations that exist in the ground around the affected structure. The return rail potential may be as shown in Fig 254 and 255 but the poten-

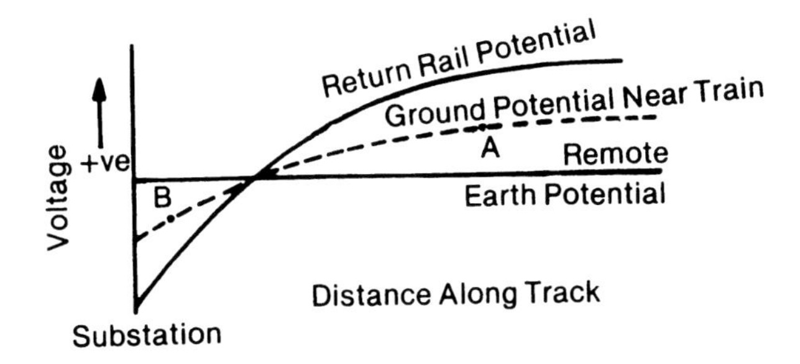

FIGURE 257 - Potential of return rail and ground in its close vicinity.

tial variations in the ground surrounding the track would be less pronounced. If the track potential is considered relative to an unaffected distant earth then the potential of some ground close to the track will lie between these two potentials. This is illustrated in Fig 257 where a return rail potential similar to that of Fig 254 is considered.

A pipeline, buried cable, or similar long structure may be laid so that it approaches the track at points *A* and *B* and runs through other areas between these points. The difference in the ground potentials at *A* and *B* will be the driving voltage and, as this may be as large as 1 volt, heavy earth current corrosion will result. The pipe in the vicinity of point *A* will be acting as a cathode and will be receiving current while the pipe near point *B* will be acting as an anode and will corrode. The rate of corrosion will depend upon the corrosion current that is flowing in the pipe and the rate of attack will depend upon the current density at the anode. This will be influenced by the length of pipe that is at the potential of *B*, that is whether the pipe runs parallel to the track for some distance or whether it crosses the track so that only a small length is in ground which is at the potential of *B*. This corrosion could be reduced by coating the cathodic areas of the pipe but these areas would be difficult to identify even in a static earth current corrosion system, and impossible in the system described above as the potentials in the earth and the cathodic and anodic areas are continuously changing.

Should the corroding pipe pass close to the generating station it would be possible to connect it to the negative pole of the generator. This would depress the pipe potential at *B*, and even more so at *A*, the whole pipe will act as a cathode and under some circumstances it is possible to give the pipeline complete cathodic protection by this method. Where the pipe does not approach the generating station, three courses of action are available to the engineer. It may be possible to connect the pipe to the negative boost cable which will be at a much greater negative potential than the track and this connection will usually be made via a resistive bond which can be varied to adjust the pipe potential to the most suitable value.

Secondly, the rail potential itself will be more negative than the pipe and the corrosion of the pipe may be prevented by bonding it to the rails, the point of bonding must be the most negative one available or the effect will be reduced. Where the potential of the rail varies greatly, it may be difficult to determine the best location for this connection and, indeed, this method may, because of this fact, be unsuitable. Where the return rail is generally at a highly negative potential relative to the pipe, then a plate of scrap steel or a sacrificial anode could be buried close to the track so that the anodic corrosion occurred on this rather than on the pipe. This method is useful where either the pipe owner or the railway operator does not want the systems to be metallically bonded together.

The third system, which can be used universally, consists of bonding the pipe to the negative rail via some electrical device which will maintain protection on the pipe. This might be a rectifier so that current flows only when the pipe is positive to the rail or a transformer rectifier which performs the same function but additionally causes current to flow from the pipe to the rail even when the pipe is slightly negative. The rectifier, or reverse current switch as it is frequently called, has been developed to meet this need and several workers have evolved switches which function under various conditions. Usually such switches are placed at several points and the switch is only 'closed' in the section that is connected to the most negative point on the return rail. As this point varies the switches open and close and ensure that the pipe is always protected. The transformer rectifier acts in a similar manner and provides a rectifying source of d c voltage which means that it can be used as a very low or negative resistance rectifying bond. The transformer rectifier would be connected in the circuit as shown in Fig 258 and current will only flow from the pipe to the rail.

Cathodic protection by almost conventional means could be used to prevent the corrosion of the pipe. Where this is done because any bond is undesirable, the groundbed anodes should be placed close to the most negative area of the soil.

The use of any technique which introduces current, either dc or ac, into a railway system must be carefully designed so that it does not affect the signaling and safety devices.

When any protective equipments have been installed it will be necessary to test them, both to judge the effectiveness of their operation, and to arrive at the correct setting of any adjustable apparatus. In many cases this will involve studying the potential of the pipe under a variety of working conditions. The half cell used in this determination is best placed close to the pipe metal surface, within 1 ft, so that effects remote from the pipe are not included. A recording high input impedance voltmeter should be used to determine the potential and the system adjusted so that the pipe

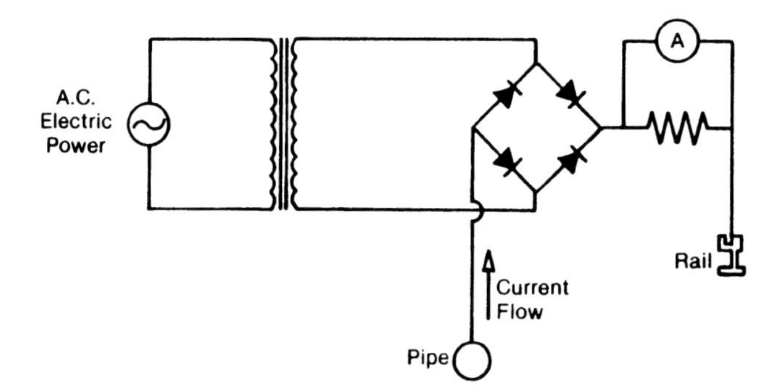

FIGURE 258 - Forced drainage from pipe to rail using transformer rectifier.

always displays a cathodic potential. If at first it is not possible, without using large currents, to eliminate all positive or anodic potential swings the system should be allowed to run for a month or two as polarization will often reduce any tendency to develop positive potentials. Potential readings should be taken at a variety of points and in urban areas the interference effect of the pipe upon other structures close to its path should be considered.

Where an electric railway runs on a metallic structure such as a pleasure pier or long bridge, the return rail currents flow in the metallic structure and this structure may suffer structure current corrosion. Piers built into the sea might be particularly prone to this type of corrosion as the electrolyte path would be of very low resistance.

The electric railway discussed employed a d c negative earth return. If a positive earth return had been employed a similar problem would have arisen. In this case the associated earth current corrosion would have occurred at points remote from the substation and any bonding would now be to these distance points which would have the most negative potential. As these points would move under various load conditions, the bonding systems would be difficult to operate and transformer rectifier bonding should be used. The use of resistive bonds to the negative conductor would be expensive because of the power they consume.

Welding

Electric arc welding is used in metal fabrications both for large structures such as ships and platforms and in shops where smaller components are fabricated. The currents used for this welding are large and often the earth return system is poor. Ships are usually welded while in a dock basin and the generators are sometimes placed on the quay side. The cables used to bond the ship are often inadequate and the welding generator may additionally be earthed to some metal work on the quay so that some of the current returns to it through the water. Alternatively, welding may be taking place on two ships from the same generator, earth return cables being taken to each ship; and when work is being done on only one of the ships some of the current returns to the generator through the water and the return cable attached to the other ship. The welding current returning to the generator under these conditions will flow through the dock water and if there should be any scrap steel on the dock floor this current may be concentrated at one or two points on the ship's hull where rapid corrosion would occur. Alternatively, the ship may have been recently painted except where she lay on keel blocks. The paint film resistance may be high and so the majority of the corrosion would take place on the bare patches. Generally welding will be of insufficient duration for the ampere hour charge to cause significant corrosion if it is distributed over the ship's hull; concentrated at a few points, however, it could cause serious damage by a pitting attack or by loss of weld metal. The best method of preventing this type of corrosion on ships is to place the welding generator onboard the ship. Similar problems occur with offshore structures particularly where the welding sets are located on a work barge alongside.

Direct Current Systems

Most countries are now without any low voltage d c distribution systems for public electricity supply, although many d c systems are still used for electric trains and similar motor applications. The d c is often obtained by local generation or by rectification of the a c mains and, within the system, earth returns may be used. Where this method is employed in urban areas much stray current corrosion will occur. The return system can be analyzed as was the tramway system and similar measures taken to reduce the corrosion. Often the d c installation can be modified to reduce the earth current effects to a negligible level. In some cases rectification is used to power mains electronic equipment. When this is the major load, as sometimes happens in the summer, d c currents can be detected flowing in the earth returns.

High Voltage Direct Current Transmission

Electrical energy can be more economically transported as high voltage direct current because there is an increase in the capacity of the conductor, particularly when it is near the breakdown of the insulation. Occasionally transmission across sea water or saline water estuaries is made with a single d c cable and an earth return through the water path. Massive electrodes are used, constructed in a manner similar to cathodic protection electrodes and often made of the same materials. There is, of course, a large potential build-up around the two electrodes but these are removed to isolated areas and ships are advised not to moor near them; also their magnetic compasses may go berserk in the area of the cable. The electrolysis effect of such a system is well known and it is carefully planned to avoid harmful effects. The d c power is then converted to a c for forward transmission.

A more popular and generally used technique is to transmit HVDC as positive and negative lines either side of earth and to use the earth electrodes only to carry any out-of-balance in the current in the two lines. This out-of-balance is kept to a minimum but it is almost impossible to ensure that at times large currents do not flow in it.

The problem then becomes one of an electrical disturbance with the earth electrode acting under different conditions as an anode or as a cathode depending on the degree and polarity of the out of balance. This complicates the techniques for eliminating electrolysis.

One solution to this may be to make two earth electrodes separated by some considerable distance and use rectifiers so that one is always an anode and the other always a cathode. It is much less of a problem dealing with an electrode which, while at times it may have no current flowing through it, as only as a cathode. This technique would make the electrolysis problem less complicated and would allow bonding to the cathode and particularly forced drainage to the cathode with automatic control to provide the solution. The anode area could be similarly treated, though this will only be acting as an anode when the other area has no current flowing through it. Discharge of the current on remote parts of any metallic structures might be difficult to control but at the least in the area of the grounding electrode these structures could be well protected.

A similar problem occurs where the direct current is associated with radio or cable transmissions and earth electrodes are buried deep in the ground to avoid electrolysis or because there is a deep water table with comparatively low resistivity.

In some cases the continuous use of the electrode as an anode will mean that a large concentration of anolyte will develop. This may destroy the rock structure, as in the case of limestones, with the formation of a cavern. In other cases there may be an accumulation of alkali in the region of the cathode and this may percolate through to the potable water system. In general these effects will be small but it is as well to recognize the cell that will be formed in the ground in this type of system.

Internal Cathodic Protection of Pipelines

The currents used to protect bare steel pipelines internally when they are carrying sea water are large. A typical large diameter pipe may require 2½ amp from anodes spaced every 20 ft along the pipe; the resistance of the pipe metal will be low, possibly 4×10^{-6} ohm per foot, and the voltage drop between anodes due to the cathodic current from one anode will be 2 $\times 10^{-4}$ V. A current of 100 amp may be supplied to 40 anodes, 20 either side of the transformer-rectifier, and the cathodic current will return to this same point along the pipe metal. The metal potential will change, by virtue of this current, by an amount 0.2 mV between each pair of anodes per 2½ amp cathode current following. Thus the total voltage drop in the pipe would be

$$V = \frac{n(n+1)}{2} \times 4V \quad \frac{20(21)}{2} \times 0.2\ mV = 42\ mV$$

over 400 ft of pipe and could cause heavy corrosion of the outside of the pipe. The remedy in the case of such a pipe is simple and consists of two steps, first dividing the cathodic connection into two parts so that only 20 mV potential change occurs and secondly using cathodic protection on the outside of the pipe. The same d c power source could be used for this providing sufficient capacity, perhaps 5 per cent more, is available.

Occasionally the cathode connection from the transformer rectifier is made at a point where an electrolyte is present that is not part of the cathodic protection circuit. This could be, for example, a connection to a frame of a ship that was partly immersed in bilge water. Around the negative connection there will be a build up in potential and if this is a thin metal structure significant potential changes can arise.

Interference From Cathodic Protection Installations

The currents that are made to flow through the corroding electrolyte to achieve the cathodic protection of one structure will cause potential changes which may bring about earth current corrosion of a neighboring body. Potential changes in the electrolyte will be associated with both the anode and the cathode and interference, as the resulting corrosion is called, can be classified by the source of these potential variations.

The electrolyte in the vicinity of the anode will be made positive and if a structure, for simplicity consider a pipeline, passes through this area then it will suffer some earth current corrosion as a result. Current will enter the pipe metal close to the anode or groundbed and will leave the pipe at places remote from it.

Close to the protected structure the ground will be given a negative potential swing and a pipeline that passes close to it will suffer corrosion by current entering the pipe at areas remote from the protected structure and leaving the pipeline at the point of closest approach. These two types of earth current corrosion may be called *anode* and *cathode interference* respectively.

Often a long structure such as a pipeline will pass close to the anode then, later along its route, pass close to the cathode; this will occur frequently in urban or industrial areas. Stray earth currents will then enter the pipeline close to the anode and leave it close to the cathode. The effects will augment each other and produce a combined type of earth current corrosion.

If the structure that passes close to the anode does not approach the cathode it may, nevertheless, pass close to another body that itself passes close to the cathode. In such a case the structures will have current flowing in them as the current picked up in the anode vicinity by the first body will be transferred to the second body and discharged close to the cathode. At the point of cross over or close approach of these two foreign structures, the current will flow from one to the other and so corrosion will be induced at this point, which may lie in a region of equipotential quite remote from the cathodic protection system.

Finally, it will be proper to consider now the corrosion that might occur inside vessels that carry or hold electrolytes when potential gradients exist in the container wall, such as the voltage drop across a resistive coupling in a cathodically protected water pipe. The five types of stray current interference are illustrated in Figure 259 a, b, c, d and e.

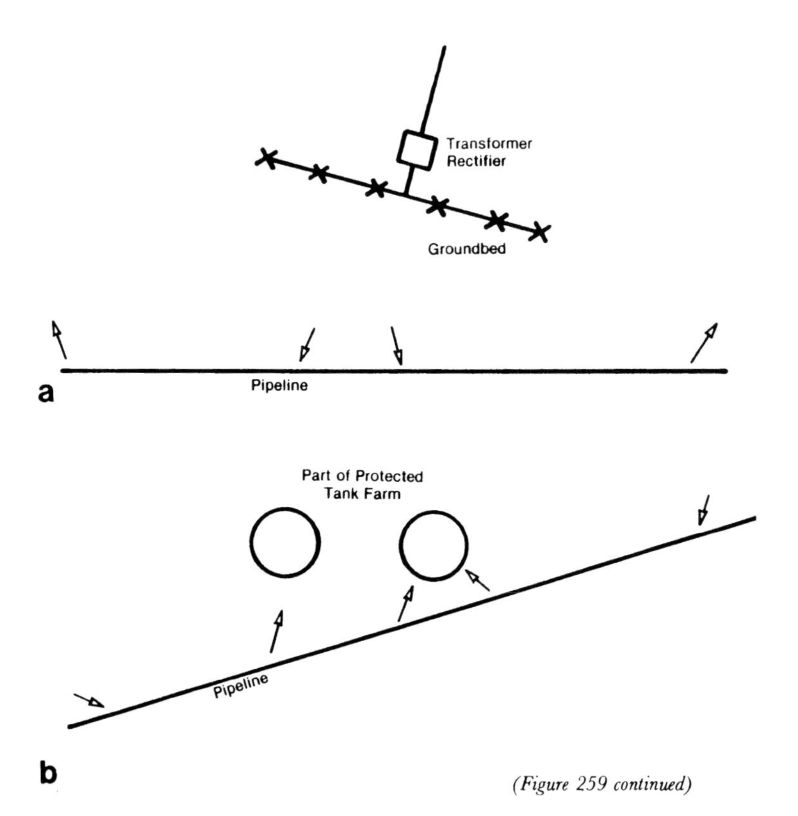

(Figure 259 continued)

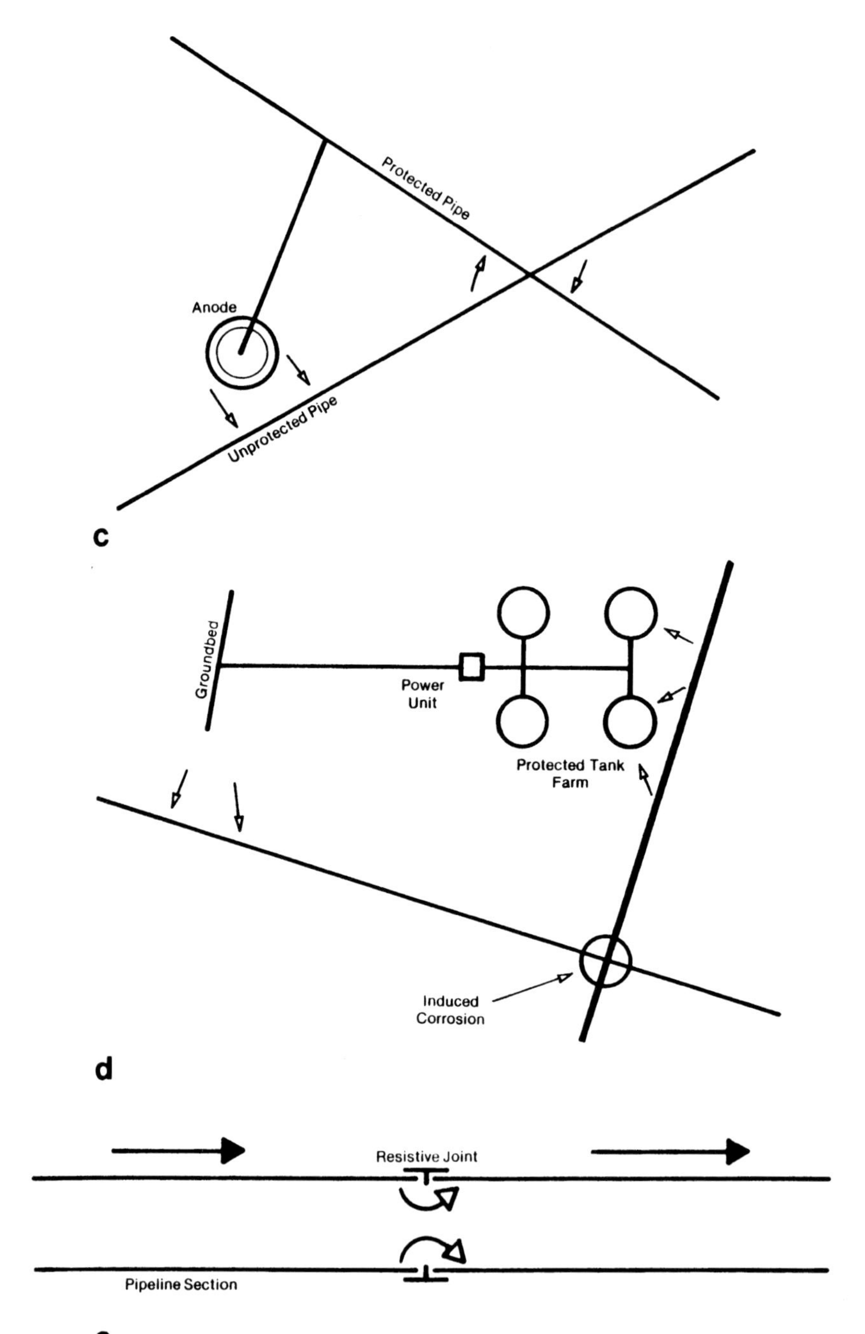

FIGURE 259 - Interference from cathodic protection installations (a) anode (b) cathode (c) combined (d) induced and (e) resistive joint.

In all cathodic protection interference the right approach to a solution is to design the protection system so that it causes the minimum of interference. It will not always be possible to avoid interference, and when it is not, the design that promises least trouble should be selected and the corrosion that might be caused by interference reduced by the methods described below.

Anode Interference

The ground close to the anode will have a positive potential whose value will depend upon the groundbed shape, the soil resistivity and the groundbed current output. To points some distance away, usually greater than its maximum dimension, the groundbed will act as though it were hemispherical and the potential in a homogeneous soil at a distance r cm will be

$$\frac{I}{2\pi r}\rho$$

where ρ is the resistivity in ohm cm and I the current in amperes. Using this formula it is possible to plot the potential along a section of the ground occupied by a foreign structure.

The maximum positive swing will be very close to the groundbed and so the potential in the vicinity of a nearby structure can be reduced by careful selection of the groundbed site. The positive swing can be reduced in soils where there is no marked increase in resistivity with depth by installing a deep groundbed. In general the gradients are half those found with a surface groundbed and twice the pipe distance to the groundbed is now the appropriate distance to take. The area of any foreign body (as distinct from the one being protected) closest to the anode will be receiving current and so will be acting as a cathode. With most metals this will not have any harmful effect, in fact, partial protection will result, and the current will leave the pipe by leaking away into the surrounding electrolyte some distance from the anode. In the simple case of anode interference this leakage will be over a large area, the anodic current density will be very low and the resulting corrosion slight.

This effect can be remedied by two methods. The area of the structure nearest to the groundbed can be coated so that the amount of current picked up is reduced and a lower current density will leave the anodic sections of the foreign structure. Alternatively, or additionally, sacrificial anodes can be attached to the remote areas of the structure which are acting as anodes. This method can be considered as providing an easier path for the leaking stray current to flow to earth, any output from the sacrificial anodes in excess of this providing some cathodic protection to the structure.

To test the effectiveness of this method it is best to measure line currents, if this is possible, in the structure to ensure that the effect is being limited to the area which is being treated.

Cathode Interference

Cathode interference is usually more serious as the stray current will be collected by the foreign structure over a large area some distance away but will be discharged, possibly at high current density, in the close vicinity of the protected structure. The voltage gradient in the soil will depend upon the cathode shape and for large bulky structures a formula similar to that for a groundbed will apply. Most interference will result from buried pipelines and the voltage gradient surrounding these will vary as the logarithm of the distance from the pipe as described in Chapter 6. In either case the foreign structure may only be within the area of negative potential for a comparatively short distance and corrosion of the foreign body might be severely concentrated.

The negative potential swing of the ground is greatly reduced around well coated structures and one of the best methods of minimizing the effects of cathode interference is to coat the protected structure.

If this is a pipeline then it may easily be coated and a P.V.C. or polythene adhesive tape will provide an exceptionally high resistance coating at a comparatively low cost. The length of pipe to be coated is difficult to determine exactly as it will depend upon local conditions, but it can be estimated by measuring the potential swing of the ground relative to a distant unaffected earth and, where this is considered insignificant, say at points 40 ft either side of the pipe, then 80 ft of the pipe should be coated or wrapped. Some potential effect on the ground may be found at great distances and the engineer will have to use his discretion in determining an insignificant potential change. This residual potential change will remain as the driving voltage in the revised interference corrosion cell. Where the cathode structure is large and bulky, coating may not be possible and while cathode coating reduces the amount of corrosion, it will not eliminate it.

Three methods are available to do this. The foreign structure may be bonded to the protected structure so that it too receives cathodic protection. This bond can be made resistive to limit the amount of protection that the foreign structure receives and to adjust such a bond the potential of the foreign structure should be measured relative to a half cell placed close to it at a point where most corrosion is to be expected. This method for two pipes crossing at right angles is shown in Fig 260, in which the soil is cut away to illustrate the method; an excavation is not made in practice. The resistance of the bond is adjusted so that the potential of a foreign pipe is no more positive when the cathodic protection is switched on and preferably it should become a little more negative in sympathy with the protection system. When this technique is adopted, coating the cathode will reduce the

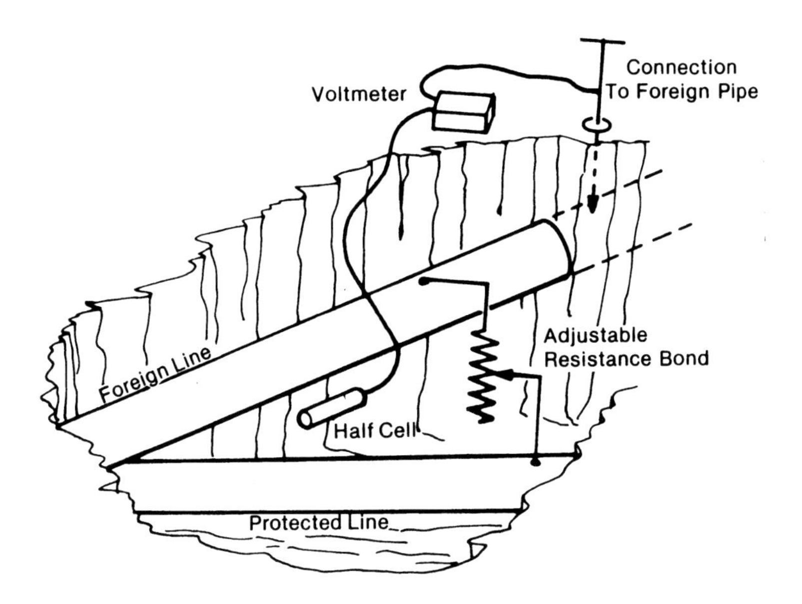

FIGURE 260 - Adjustment of resistive bond between protected and foreign structure.

amount of current required to achieve the correct potential condition on the foreign line.

The effect of the current drained on to the protected structure must be considered in its cathodic protection engineering. If the point of attachment of the bond is some distance from the drainage point then the resistive volts drop in the protected structure, that would accompany even the small current, would be large and techniques—either mathematical or by models—similar to those used in Chapter 6 should be employed to check this. A multicombination meter with facilities to make all the measurements is shown in Fig 261.

A second method of preventing cathode interference corrosion is to place a sheet of metal similar to the foreign structure between the two structures and electrically connected to the foreign body. The main anodic attack will occur on this extra metal and in many cases the structure itself will not suffer. In theory, this metal should be as close to the cathode and as far from the foreign structure as possible. The excessive current density that this will locally produce on a bare cathode may cause heavy polarization and so such a piece of metal is best located about half way between the bodies. The piece of metal will only provide an easier path for the interference current if the potential graident in the soil is in excess of the ohmic polarization and other potential changes that are caused by the current flowing from it.

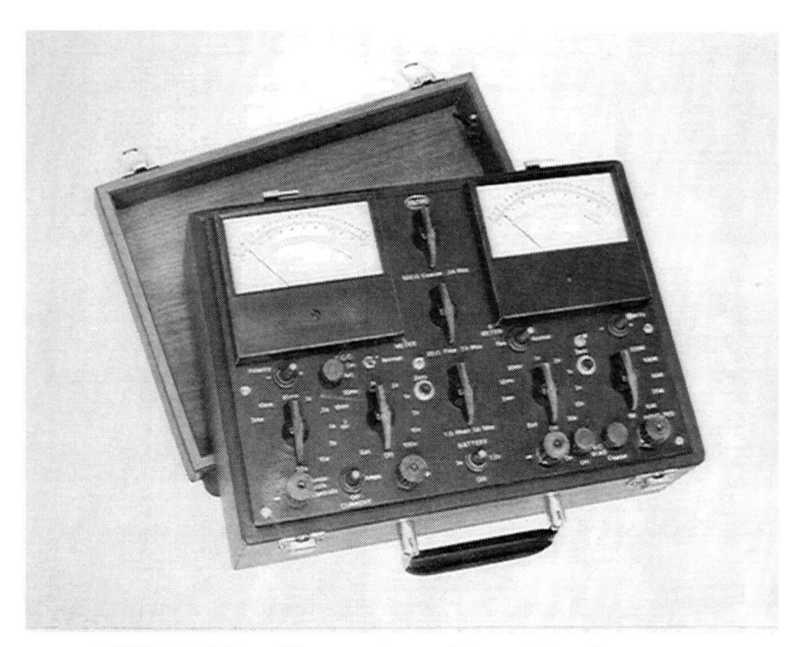

FIGURE 261 - Electronic multicombination meter using analogue meters. (Photo reprinted with permission from M.C. Miller, Inc., Upper Saddle River, NJ.)

Often this balance will be slender and as a third technique a sacrificial anode metal should be used in its stead. The driving voltage required is not high, indeed it can be reduced by using a resistive anode lead so that zinc is probably best suited for this purpose. The sacrificial anodes need to be placed near the cathode though not now necessarily exactly between the structures; this makes the engineering easier and a life of some 20 to 30 years can readily be obtained. Again, these methods are most easily used with a coated cathode. The last two methods of prevention have avoided any direct metallic link between the structures and when utilities are not anxious to interconnect their pipes and cables these two methods should be used.

One special type of cathode interference is worth analyzing separately. This occurs where a limited part of a structure is being protected and at some point an insulating joint is used to isolate that region. Probably the simplest example is a pipeline which is protected as far as an insulating coupling or flange. The pipe on the protected side of the flange is, therefore, the cathode while the pipe to the other side suffers cathode interference as shown in Fig 262. The part of the unprotected pipe closest to the insulating flange will corrode and this corrosion will be rapid as the two pieces of pipe will be very close to each other. This condition will occur with all structures and its presence can be detected by a potential survey close to

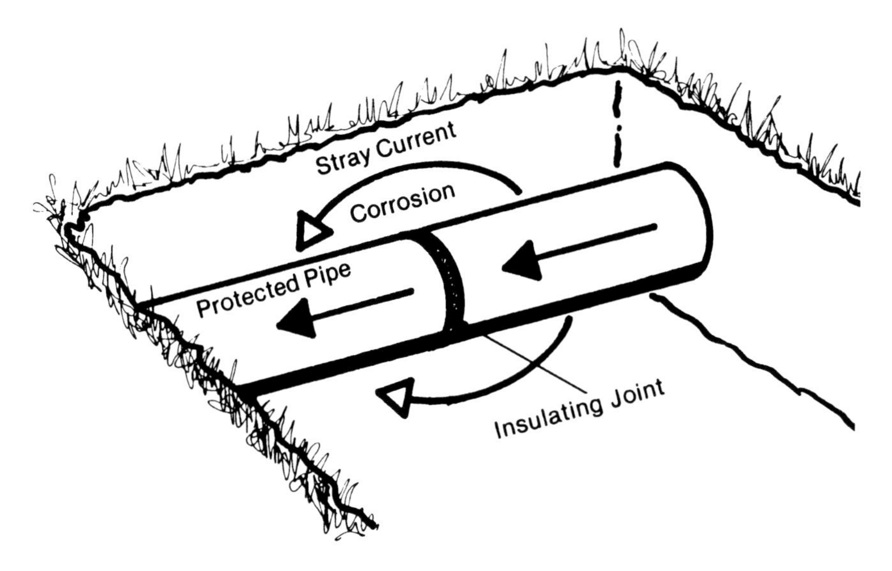

FIGURE 262 - Stray current corrosion and insulating flange at end of protected line.

the insulating joint. The end of the protected pipe will show a more negative potential near to the insulating joint than the rest of its length and the unprotected pipe potential will be most positive nearest to the flange.

The rate of corrosion due to the interference can be reduced by coating or wrapping the protected pipe close to the joint and reducing the coating resistance on the unprotected pipe. Again, a plastic tape might be most suitable for the former and the distance it should extend can again be judged by the potential swing of the soil around the unprotected pipe prior to the extra coating as suggested above. If the criterion of an insignificant potential change is taken as 10 mV then the driving voltage of the interference earth current corrosion should be reduced to this figure. Complete protection can be achieved using one of the other methods described above and usually a resistive bond or sacrificial anodes will be most easily engineered.

Many pipelines are made of short lengths of tube joined by flexible couplings which are usually resistive if not insulating and a section of such a pipe may be made electrically continuous by welding straps over the joints prior to receiving cathodic protection. If the end of the section consists of a series of couplings of doubtful resistance then either one coupling should be made completely insulated, and subsequently treated for interference, or several of the couplings made slightly resistive so that the cathodic protection is successively reduced. This latter process should be extended beyond the two or three couplings necessary to do this in case line movements or other effects cause some couplings to short circuit.

In the case of an insulating flange an insulating gasket is employed and the bolts are insulated. The bolts should connect to the cathode side or isolated completely. The cathode connection is preferred where voltages are low and the small amount of corrosion that will occur by virtue of the bolts' cathodic current pickup will be distributed over the flange by the insulating washers. The bolts themselves can be coated to reduce this effect.

Combined Interference

The corrosion suffered by a structure under combined interference will be greater than that suffered by a structure under either anode or cathode interference. The increase in the effect will depend upon the closeness with which the corroded structure approaches the anode and the cathode and the methods of preventing this corrosion will be the same as with the separate effects. However, as the pipe passes close to the anode it will probably also approach the cathode close to the drainage lead and so any potential build-up in the protected structure as a result of bonding will not be serious. If this is not the case, that is the foreign structure does not cross the cathode near to the drainage point, the interference effect may be greatly reduced by some current drainage off the foreign structure near to the power source.

Induced Interference

Induced corrosion caused by combined interference involving two structures is difficult to anticipate or detect. Where it is found the area can be treated by the methods suggested for cathode interference. If no maps of the structures surrounding a cathodic protection installation are available then some detailed testing will be necessary to ensure that no induced interference effects are occurring. Two methods are most useful, firstly the structure potentials can be measured as the cathode protection system is being switched on, though this type of survey is difficult and may be misleading if it is not performed in great detail. The second method consists of measuring line currents in foreign structures. This is only possible on extended 'pipelike' structures but fewer measurements give considerably more information. When any stray currents have been traced to extinction, further interference effects can be ruled out.

A particular case of induced corrosion occurs when a foreign structure that would normally be suffering combined interference has a resistive or insulating joint in it. Such a case may occur in a pipeline feeding a protected tank. In one specific case a large circular tank bottom was being protected from a distributed groundbed as shown in Fig 263. A pipeline leading into the tank had an insulating joint in it near to the tank and was resistively bonded to avoid the effects of combined interference upon it. Induced corrosion was not suspected but occurred further along the pipe at another insulating coupling and was subsequently revealed at a later inspection of the cathodic protection installation.

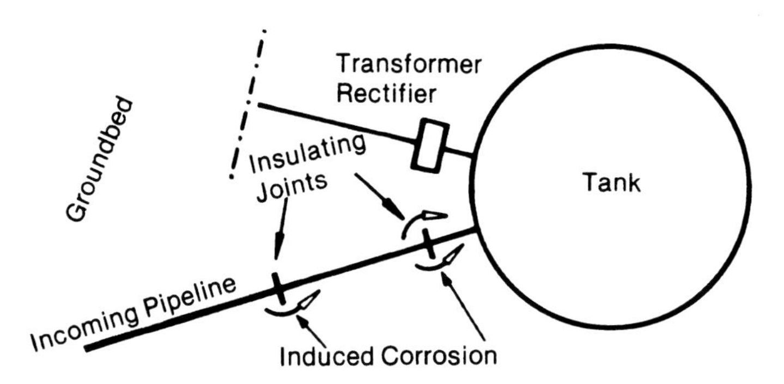

FIGURE 263 - Induced corrosion at insulating joint in pipeline.

Joint Interference

The last type of interference corrosion, for example that which occurs inside pipes carrying an electrolyte at insulating flanges, is difficult to control. If external interference is reduced by coating this will not mean that internal interference is equally reduced for the metal to metal potential will remain the same. Usually fresh water will be the electrolyte inside the pipe and the most successful method of reducing the interference corrosion will be to coat the cathodic side of the joint. If a resistive coupling is added to this then no joint corrosion should occur. In protecting joints, couplings and insulating valves it is necessary to bond not only the two pipes, but all the metal components of the joint.

An interesting case of corrosion interference occurred in the protection of some metal pipe work immersed in water contained in a concrete tank. The pipe work was electrically connected to a whole mass of other steel work lying in the ground so that the effect of the cathodic protection was to make the water potential positive rather than to depress the metal potential. The tank was made of lined concrete so that no current flowed through it, and was fitted with a steel drainage pipe which ran below ground to join a clay pipe. The positive potential of the water in the tank caused current to flow through this drainage pipe into the soil and the outer end of the pipe metal was rapidly corroded.

In studying stray current corrosion, it is often useful to be able to plot potential variations in the electrolyte. This can most easily be done by placing two half cells in the ground and measuring the voltage between them and to eliminate any voltages caused by differences in the cells these can be interchanged. By modifications to this technique potential gradients and lines of equipotential can be found.

Offshore

It was generally assumed that there would be little or no interference effect in sea water but the massive currents that are used in a number of installations can cause a considerable problem. The normal cathodic protection current, particularly flowing onto a structure that is not planar but has framing or corrugation, can cause a potential drop of 30 mV per ft. This was already described in jetty protection, where great care is taken to place the anodes within the jetty structure, the piles of the jetty acting as a Faraday cage. The problem with ships sitting in this area is considerable and great care must be taken not to use this form of design. Some jetties, were designed to provide cathodic protection to ships that were berthed alongside. The interference effect that such a system must create is enormous and there is a considerable hazard when large currents are flowing.

Offshore structures are sometimes protected, or protection is proposed, by the use of remote anodes. In a typical North Sea multitubular structure 8,000 amps are required for protection of a structure in 400 ft of water, that is a current of 20 amps per vertical foot of sea water. If the structure is considered to have a mean plan dimension of 100 ft × 100 ft, then there is 50 mA per sq ft of current flux at the cathode. Over a considerable distance from the structure this will lead to a voltage drop of 50 mV per linear ft, or 1 volt over a distance of 20 ft. This, obviously, will cause tremendous hazards, both to the pipes that come into the structure, Fig 264 and to anyone working in its vicinity.

A ship being unloaded by a crane from the platform Fig 265 could well be at a potential of 2 to 3 V different from the structure and if wire ropes are used to lift the cargo then a man steadying it may, on contact being broken, take the induced current through his body which could be fatal. Perhaps it is fortunate that most people working in these conditions are equipped with rubber boots and plastic clothing as the weather is usually inclement and these act as good insulators. Damage may occur to vessels that come alongside, to divers who work from bells or use any form of umbilical or to suspension ropes when the end shackle in particular may be subject to the worst attack if a non-metallic strop is used.

The use of remote anodes should be restricted to comparatively small currents. For example, if the current output had been 800, not 8,000 amps, then little or no ill effect would have been noted in any of these circumstances. This is particularly true if the anodes were used to augment the cathodic protection in general on the structure, thereby giving an even, reduced flux rather than to correct failure in one sector where a larger current flux could be expected. It is interesting that in none of the North Sea accidents has the author ever seen this noted as a major cause for concern.

Concrete Interference

The interference that is found on concrete structures can be particular-

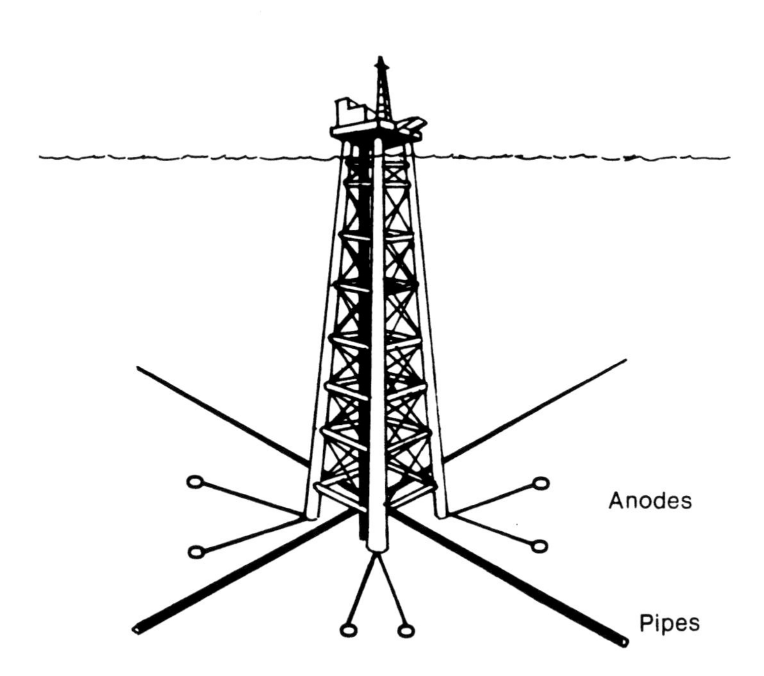

FIGURE 264 - Remote anode interference on offshore gravity platform.

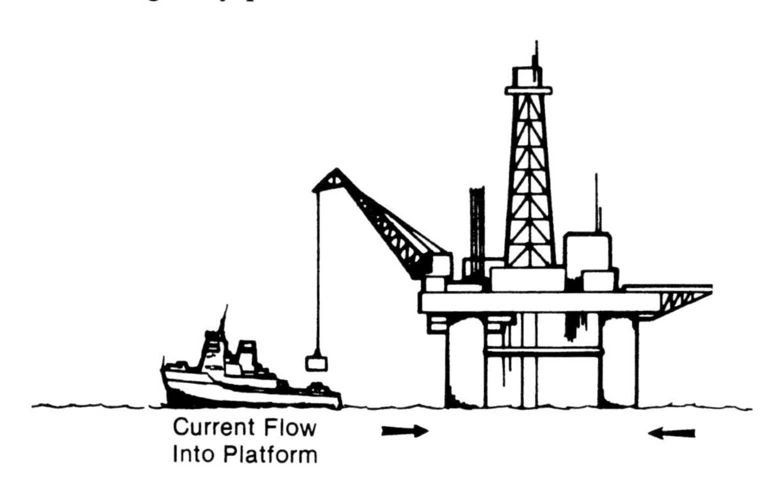

FIGURE 265 - Possible hazard on offloading cargo with large potential gradients in sea.

ly damaging if the reinforcing is continuous and of sensible cross section. While cathode current can be picked up over a comparatively large area the anodic discharge may well be concentrated at points where the reinforcing is fortuitously near the surface or in an area of abnormal micro-cracking.

This could lead to a very rapid breakdown at that point with failure of the reinforcing and spalling of the concrete.

One particular problem similar to the in-tank corrosion can be found where protection is applied to the risers and guide frames inside an offshore concrete tower, Fig. 266. The wells themselves will form part of the cathode and in general the potential of the sea water inside the tower will be raised so that current will try to flow from inside the tower to protect the outgoing lines but more particularly the wells. A potential of several hundred millivolts—indeed, upwards of 1 volt—can be generated across the concrete shell. This could lead to massive breakdown of parts of the concrete and can be prevented by the application of cathodic protection outside the structure which will protect the outgoing lines and wells. It would also, of course, protect the reinforcing that lies near the outer shell. By using a distributed system of anodes the potential of the water outside the concrete could be arranged to balance that inside.

Geomagnetic or Telluric Effects

Following periods of intense solar flare activity low frequency electrical

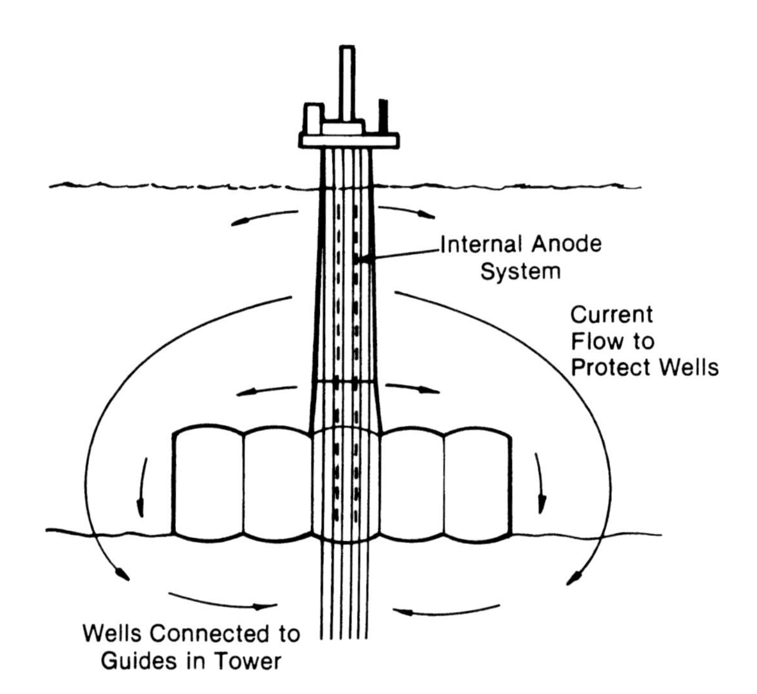

FIGURE 266 - Current flow through concrete tower platform from enclosed guide tube protection.

oscillations are found in the ground and can be detected as current in long distance pipelines. These have been known for many years and the variations in potential produced by these have been used by geologists to detect very large masses of ore. Long distance pipelines, particularly those that are well coated, are affected by these currents and though there is some evidence to suggest that in terms of net current flow the corrosion should be zero, this is not the case where wide changes in potential are imposed upon the pipeline as the anodic dissolution of the pipe metal is irreversible.

The potentials generated in the line upset cathodic protection potential measurements when these are sampled instantaneously or over a very short period of time. They make considerable differences to the values and comparisons between periodic readings are misleading. With slower reading techniques, for example using a moving coil meter, the currents and potential changes can be seen or are integrated. Many of these telluric effects are recorded and this information can be used to correct potential readings if these are time related. In many areas the telluric effect is more marked in one direction, for example in the United Kingdom in a north-south orientation, whereas in others it is affected by geological conditions; at an isthmus the principal fluctations will be across the land.

The potentials generated in the coated pipeline can be reduced by inserting insulating couplings in the line and are much less pronounced where sacrificial anodes provide the cathodic protection as these form an earth path for the currents. Potentials of several volts have been found and occasionally the voltages that are developed can cause damage to instruments, etc., and reach levels at which there is concern for safety.

The periods of major telluric activity can be predicted and this service is available from many geo-physical laboratories.

Electrolyte Corrosion

One effect of cathodic protection is that the electrolyte in the vicinity of the anode and at the cathode is altered. With an impressed current anolyte this is usually the production of a low pH or acidic environment with the volution of gas which may be chlorine. The acid of the anolyte can attack the supporting framework or dissolve a rock formation and the anode can be destroyed by this. The gases that are evolved can cause corrosion above the water line, particularly in enclosed spaces; impressed current systems are often not used because of this problem. Equally, sacrificial anodes can evolve hydrogen, (though they do not, of course, encourage an acidic environment) and this can be a hazard in enclosed spaces.

At the cathode there will be deaeration and the creation of an alkali environment with an increase in pH. Occasionally a coating will become disbonded or a recess in a structure may become filled with sediment and these areas will be de-aerated by the cathodic protection. Under these conditions it is possible for bacteria to grow and if the recess is inadequate for the flow of sufficient current, then corrosion can occur.

The alkali accumulation beneath disbonded coating on a warm pipeline can lead to stress corrosion cracking and this has been pinpointed as being caused by a bicarbonate/carbonate build up beneath the disbonded coating which, in the absence of cathodic protection, can cause cracking. The relationship between the cracking and the cathodic protection on a pipeline is most clearly demonstrated by the fact that the majority of the cracks start first at the bottom of the pipe as this is where the maximum cathodic protection current density will be arriving as described in Chapter 6. Even though this is a small effect because the conditions around the pipe are otherwise uniform, the majority of failures will occur where the environment is most favorable.

Safety

The majority of the effects have been from direct stray currents but there are a large number of alternating currents flowing in the ground. These may be caused by overhead transmission, by earth returns, and by faults. The current flowing in overhead lines will generate current in the ground and also in long conductors such as pipelines, both by the inductive effect and by the potentials from the aerial flow of current. In general a c corrosion is not a hazard and there is very little evidence except with some of the amphoteric metals that alternating current causes corrosion; certainly pipelines that are cathodically protected do not suffer in this way.

Some earth currents are caused by faults or by natural high energy phenomena such as lightning. These often have sufficient energy to cause direct damage to the structure, and more importantly, to be a hazard to personnel working on or near the pipeline. There are strict codes of practice for the laying of pipelines or any construction in the vicinity of high voltage transmission lines. The cathodic protection engineer is particularly susceptible to this hazard as he is dealing with the structure and with the very good earthing system of his anodes. A number of precautions are recommended to reduce this hazard: measurements taken on the line or to the cathodic protection system must be made very carefully and it is best practice to use earth mats whenever the engineer makes contact with any of the electrical equipment associated with the pipeline or the cathodic protection. Zinc ribbon earth plates can be used as measuring points. The cathodic protection techniques are described in Chapter 6.

In addition to these, lightning and fault current discharge points must be built into the pipeline. These are particularly essential at any insulating flange or where the pipe passes close to another structure, whether it is protected and, indeed, whether it is buried or not. The use of the earth electrode made of a pair of zinc anodes, of the spark gap or of the electrolytic cell is essential to reduce these hazards to life.

In addition to the dangers to life there are many places where an electric arc could constitute a serious fire hazard. The same techniques can be used to reduce this hazard, though the use of overhead transmission close to such areas or the location of such areas close to overhead transmission is in itself bad practice.

Cathodic Corrosion

Iron, steel and most common metals are corroded by an anodic reaction, that is by current flowing from the metal into the electrolyte. Lead and aluminum can be corroded under heavy cathodic conditions and interference on cable sheaths or structures made of these metals must be controlled both in the extent of the cathodic and anodic potential swings. Lead will be most susceptible to cathodic corrosion when it is laid in alkaline environments such as asbestos cement ducts, and in these circumstances only a very low cathodic swing, a 100 mV at the most, can be tolerated. Aluminum has been successfully protected at potentials of −1.4 V to copper sulfate, though it has corroded at −1.6 V in other sites. Preferably, aluminum should not be made more negative than −1.2 V to copper sulfate.

Explosions

There is a tendency to regard low voltage electricity as being harmless because, for example, a voltage of 1 or 2 V is difficult to detect and even to get some tingling sensation. However, at these low voltages massive currents can flow, particularly through good conductors, and aluminum gangway, perhaps connecting a work barge to a structure, could provide such a massive current path. Both structures are earthed and so at 1 V or 2 V the current flowing through such a connector could be many hundreds of amps. If there is a tidal flow then the gangway will move as the tide changes and this may mean the difference between it forming a good connection at the end on which it probably runs on rubber track wheels and it being in isolation. At the moment of isolation of the gangway a colossal inductive spark can occur and this would certainly have the power to ignite a gas/air mixture. Similarly, inductive surges can cause fatalities. It is only with the realization that although cathodic protection may be concerned with very small voltages, the currents can be large and the inertia of these currents can lead to large discharges of power.

Responsibility

No mention has so far been made of the ownership of the various structures that may be involved in the interference corrosion, as the problem of the degree of responsibility and the tolerance that one neighbor must have for another's activities are beyond the scope of this book. Certainly any discussion of such problems would be greatly helped if a relationship between the interference parameters and the corrosion suffered by a foreign structure could be ascertained. At present such work is being performed on the relation between potential changes and enhanced corrosion rates. This factor alone does not seem to be entirely adequate as potential changes will depend upon the structure shape and the amount of any depolarizer present. Homogeneous soils and water may generally, however, prove susceptible to such a criterion.

Instruments for Cathodic Protection

CHAPTER 10

Reference Potential Devices

The criterion for the cathodic protection of most metals is expressed as a potential of the metal relative to a reference device. Many other uses of the metal potential or change in potential are made in cathodic protection and corrosion studies. A constant potential device is required which would possess the following properties: its potential should be constant irrespective of the electrolyte in which it is used and it should not vary greatly with changes in electrolyte temperature or other parameters; any changes should be predictable and should have no hysteresis effect; the device should not polarize at small currents either when these are cathodic or anodic, equally it should have a low internal resistance. All these properties should be constant with time or the device should have a definable life. For use in most electrolytes the device should not be bulky so as to have an effect upon the current paths in the electrolyte and the point of measurement should be small so that precise positional determinations of potential can be made. The device should not alter nor contaminate the electrolyte in any way. Mechanically the field conditions in which it is likely to be used require a tough construction made of components readily available and it should be rechargable in the field if this becomes necessary.

The best devices for achieving the properties outlined above are the metal/electrolyte half cells. In these a piece of metal is in contact with a solution which contains some of its own ions at a fixed concentration. The potential of the device or half cell will depend upon the metal and metal ion chosen, varying several volts between the most noble and the most base. The potential of a particular metal/metal ion half cell will vary with the metal ion concentration, and in some cases these ions will be in a saturated solution when their concentration, or rather activity, will vary with the solubility of the salt and the solution temperature. The potential of the half cell will be influenced by the purity of the metal and the solution and some impurities will have a great effect upon it.

The metal electrode is immersed in a specific solution of its own ions and this solution must be connected to the corroding electrolyte through a salt bridge. This bridge may consist of some electrolyte-holding medium such as a gell, sintered or porous ceramic or wood saturated with either the half cell solution, the corroding electrolyte or a liquid compatible with both.

Half Cells

The absolute reference electrode is the hydrogen half cell, in which hydrogen at one atmosphere pressure is bubbled into a solution of unit activity of hydrogen ions; this electrode is not suitable for practical use either in the field or in the laboratory. It is often replaced in electro-chemical work by the calomel half cell which consists of a metal electrode of mercury surrounded by a solution of potassium chloride saturated with mercurous chloride. The potential of the electrode depends upon the concentration of the potassium chloride and two models are popular. A decinormal solution of potassium chloride gives the half cell a negligible temperature coefficient while the saturated solution of the salt is easier to make up and can be used with a saturated potassium chloride salt bridge. Care must be taken to exclude bromides and mercuric ions if the cell is to be accurate. This electrode is fragile and not easily repaired in the field, so finds only limited use. Where it is employed it is often held close to the voltmeter and connected to the electrolyte by a salt bridge of saturated potassium chloride contained in a clear plastic tube with a glass tube and sintered plug at the end. The potentials of the half cells are listed in Table XV.

The half cell generally used for cathodic protection in the field is the copper sulfate electrode. This consists of an electrode of electrolytic copper in a saturated solution of copper sulfate. The electrode can easily be made to have a large current capacity and will carry current better when it is acting as an anode than as a cathode. This capacity decreases with time by the formation of copper sulfate crystals and copper oxide on the metal so that after two to three months the cell needs to be cleaned to restore its former current carrying capacity. The cell is usually saturated by an excess of copper sulfate crystals and these may form a conglomerate of high resistance if very small crystals are used. The cell has a high temperature coefficient and displays some temperature and current hysteresis. The copper sulfate is easily contaminated, and though the potential of the cell might not alter, it becomes very susceptible to polarization, so that it is unsuitable for prolonged use in sea or estuary water. In the field, the cell is easily re-charged and commercially pure copper sulfate solutions give potentials consistent to

TABLE XV

Electrode	*Potential to Hydrogen Electrode at 25°C*	*Temperature Variation*
Calomel	Volts	mV/°C
0.1N KCl	-0.334	-0.07
1.0N KCl	-0.280	-0.24
Saturated KCl	-0.242	-0.76
Copper sulfate	-0.318	+0.9
Silver chloride, normal	-0.222	+0.6
Silver chloride, sea water . . .	-0.25	
Silver chloride saturated . . .	-0.225	+0.6

within 5 mV. Copper sulfate is an unpleasant liquid to handle as it will attack the clothes and stain the skin and, if it is allowed to leak into the electrolyte, copper may come out of the solution and form small copper cathodes on the structures whose potential is being measured. If this occurs, for example, in a cold water storage tank, then the effect is serious as the tank corrosion is greatly increased.

The most universal half cells are of the silver chloride type. These consist of a silver wire coated with silver chloride so that both silver and silver chloride are in contact with a chloride ion electrolyte. The potential of the half cell will depend upon this chloride ion concentration or more exactly upon the logarithm of the chloride ion activity. The half cell element can be constructed in a number of ways, two of which will be described. A silver wire spiral or welded gauze is cleaned and then dipped into molten silver chloride which has been heated some 50 °C above its melting point in a pure silver or quartz crucible. The silver is removed from the melt and the majority of the molten salt shaken off leaving only a thin silver chloride film. This is then made the cathode in a solution of sodium chloride and some of its surface is reduced to silver. The element is then placed in a chloride solution and after a few hours can be used to form a silver chloride half cell. The other method of making a half cell consists of plating a silver layer on to a platinum wire and then forming silver chloride on this by an anodic reaction in a chloride solution. The former method produces the more robust electrode and is recommended for practical work. The accuracy of silver chloride half cells has been greatly improved and various techniques are used that will produce a more massive electrode. Care is taken over the purity, both of the metal and of the silver chloride, and the use of sintered metal improves the technique. The electrolyte next to the cell is equally carefully prepared and is connected through a porous diaphragm to the sea water. This made flexible—or part of the cell is made flexible—so that it is not adversely affected by pressure. Skilled technicians are able to make cells which are within better than ± 10 microvolts and, by selection, accuracies of ± 2 microvolts are possible.

The accuracy of the electrode depends upon the amount of chloride ion that is present in the solution and the accuracy with which it is controlled. The normal cell, that is with a normal chloride ion concentration, has a potential about 95 mV different from the copper sulfate half cell at 25 °C and such that a cathodically protected piece of steel would display a potential of -0.755 V to it when displaying -0.85 V to a copper sulfate half cell. In practical applications two concentrations of chloride ion are popular. In the first the electrolyte surrounding the silver chloride element is sea water or a 3 per cent solution of common salt, and because some of these cells are made to be surrounded by the corroding sea water electrolyte, they are called equilibrium cells. The other type of cell uses a saturated sodium chloride solution. Both silver chloride cells have large temperature coefficients but there are no hysteresis or other effects so that these variations can be calculated. Most cells of higher resistance than the copper sulfate cells and this resistance is a function of the thickness of the silver chloride layer.

It is usual to ascribe a protective potential for steel of -0.80 V to a sea water equilibrium cell; normal sea temperatures are near 10 °C.

Because of the trouble that may be caused by spillage of copper sulfate in lead cable ducts, a lead/lead chloride series of half cells has been developed. The metal electrode is usually a lead alloy with 3 per cent of antimony to harden it, and this is surrounded by a saturated solution of lead chloride either in distilled water or in a potassium chloride solution. The potential of the cell will depend upon the concentration of lead ions in the solution. Most workers report that the saturated lead chloride cell is not as good as the lead chloride/potassium chloride mixed solution cell and a concentration of 400 gm per liter of potassium chloride is suggested. Cells made up to this specification will display potentials within 5 mV of each other. The potential of this cell is reported as 0.325 V to the hydrogen electrode and has little or no temperature coefficient. The potential relative to a lead cable is low and so small errors due to high circuit resistance can be tolerated. The potassium chloride solution reduces the cell resistance to about that of the copper sulfate half cell.

It would be an advantage to have a universal half cell which has a low potential to the structure. This led to the development of a cadmium sulfate half cell but it has not proved to be satisfactory and is not generally used.

An electrode that is more useful in this respect is the zinc half cell. The zinc metal used can be either one of the purer metal alloys used for anodes or 99.99 per cent pure zinc. The rod itself can be placed in a sodium chloride electrolyte; a number of other solutions are suggested including buffered sea water and sodium and calcium sulfate. The electrode possesses a potential of -1.10 V to a copper sulfate half cell and so will indicate a metal potential of $+0.25$ V for iron cathodically protected at -0.85 V to a copper sulfate cell.

A piece of zinc can be used as an electrode when immersed in an electrolyte that contains some quantity of chloride or sulfate ions or can be used in the ground if surrounded by a suitable backfill. The anode backfill for zinc is usually bentonite, sodium sulfate, gypsum and sodium chloride.

Zinc electrodes made from high purity anode material or ultra high pure zinc give a stable constant potential to better than 5 mV. The electrode may slowly polarize and may have marine growth on it in sea water but this is generally prevented if the zinc is allowed to act as an anode at a very low current density. With respect to steel, this will happen if the circuit resistance is of the order of tens of thousands of ohms as it would be, for example, with a moving coil voltmeter placed in the circuit. The electrode is particularly rugged. It is close to the steel protection potential so that percentage errors are low in magnitude and when used in the form of a portable half cell, the electrolyte can be replaced by tap water and table salt.

Lead can be used in a similar way and bare coupons of lead are sometimes used in duct experiments. The lead is usually brightened before use.

Practical Half Cells

The half cell must be so constructed that it is entirely separated from the corroding electrolyte except through the salt bridge. If this is not so then the half cell must be used with care so that only the salt bridge is in contact with the electrolyte, say the soil, and no uninsulated part is in contact with the soil or the operator's bare hand. A good method of testing the half cell to check if it is properly insulated is to measure a potential relative to the half cell when only the salt bridge is in the corroding solution and when the whole cell is completely immersed. The salt bridge connection is usually made as a porous plug and sintered glass, ceramic and wood are used. The first two are arranged to have a low resistance coupled with a slow percolation of liquid and care must be taken with thin plugs that the cell electrolyte does not crystallize within them as this may shatter the material. Wooden plugs are usually made from a straight grained wood which is non-resinous.

The half cell can be extended by a long salt bridge which consists of a flexible plastic tube filled with a high conductivity solution that is compatible with the half cell and the corroding electrolyte and has a porous plug at its end. The size of the plug will be decided by two considerations, the resistance of the plug through the electrolyte and the uniqueness of the potential that the cell measures. In some cases, for example in the field, a plug diameter of up to 2 in. can be treated as a point and so a low plug

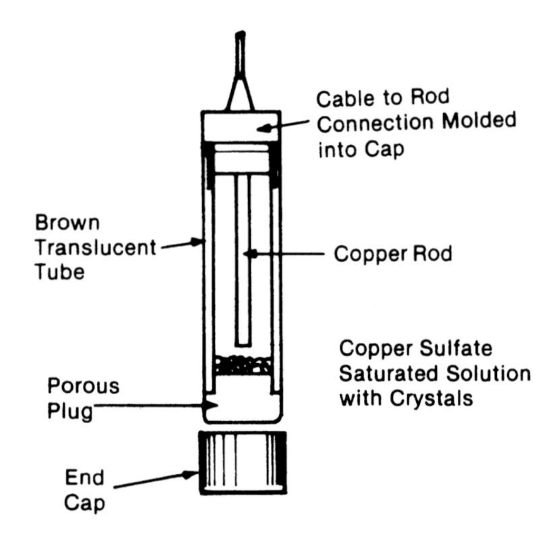

FIGURE 267 - Copper sulfate half cell.

resistance can be obtained, while in the laboratory the half cell must be small and this usually means that the plug resistance will be high.

Most electrodes will need inspection and re-charging of the electrolyte and some will need to have the metal cleaned or treated; preferably the half cell should be designed so that it can be taken apart for inspection and cleaning. The plastic, while being translucent, should be brown in color to reduce the effect of light and shadow on the cell and it is suggested that the addition of alcohol to the copper sulfate gives the cell a longer life.

The most popular shape for half cells is the long cylinder with electrical connection at one end and the porous plug at the other end. Modern plastics offer a variety of insulators which are tough and can be fabricated readily.

For use on high resistivity ground, and particularly on pavements, concrete etc., there is a need to enlarge the contact area. This can be done by the use of sponges. The author prefers the use of a shoe, and a typical one is shown, to which the cell can be clipped. One of its advantages is that it is not easily knocked over and the operator does not need to hold the electrode as is often the case with a sponge.

Figs 267 to 271 show typical electrodes and these are as described in their titles.

Zinc metal electrodes can be used, because of their ruggedness, in unusual applications, one example is the use of a pencil electrode attached to flexible cable that can be towed behind a ship or used where it will stream effectively. The electrode is shown in Fig 272.

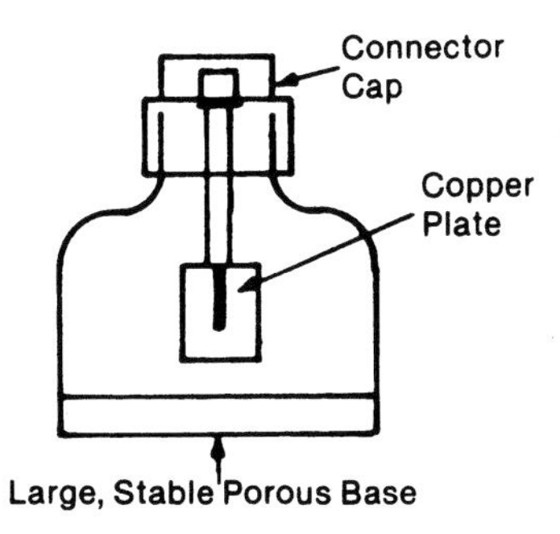

FIGURE 268 - Broad based or pavement half cell.

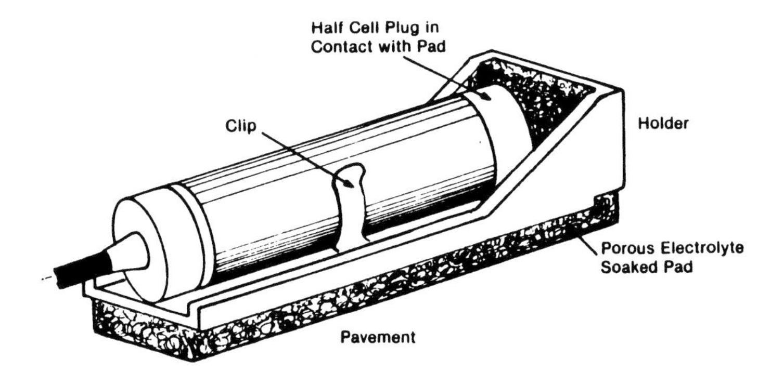

FIGURE 269 - Pavement shoe to hold half cell.

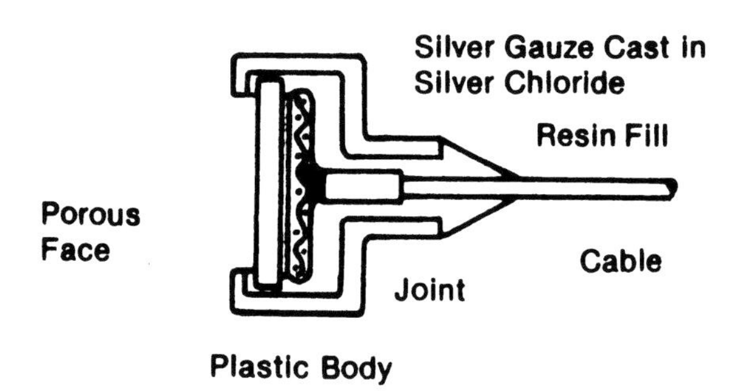

FIGURE 270 - Silver chloride permanent electrode.

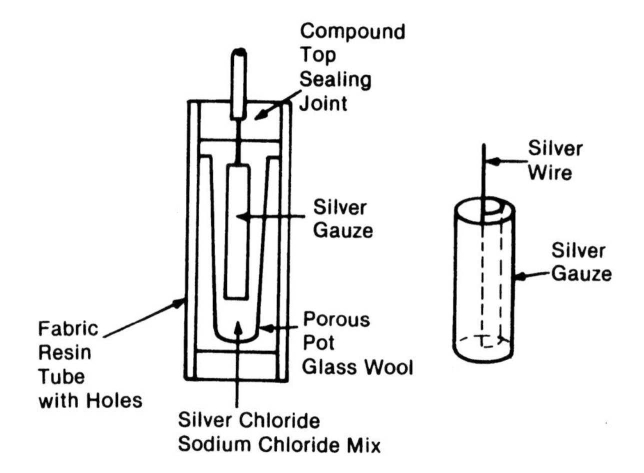

FIGURE 271 - Admiralty pattern silver chloride half cell.

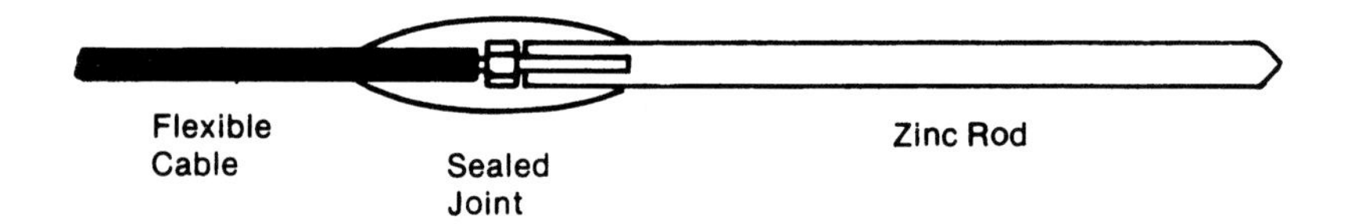

FIGURE 272 - Pencil zinc towable electrode.

In a number of environments the electrode will not be accessible for checking and in these circumstances the author has pioneered the use of a pair of electrodes, one silver chloride and the other zinc, which are placed effectively to be at the same point. The potential of each can then be read relative to the steelwork and also the potential relative to each other. It is difficult to envisage any fault in either electrode that would provide an increase in the potential between them. Therefore, if the potential is maintained between them, that is at -1.05 V, then the engineer is sure that both are working properly. This has proved invaluable on offshore platforms where electrodes are installed at considerable depths and in comparatively inaccessible locations, Fig 273.

It was found on one platform that there was a change of potential of the two electrodes with depth. This was not quite consistent and it was feared that some property of sea water pressure was affecting the electrodes. However, it was discovered that if the length of cable was plotted against the potential variation a straight line graph ensued and the effect was caused by the length of copper cable in the sea water. Presumably this

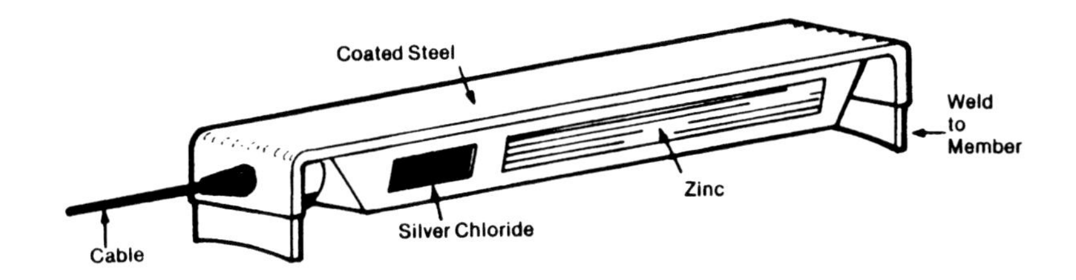

FIGURE 273 - Silver chloride and zinc electrode pair for mounting on offshore structure.

formed a copper sulfate half cell with a very high resistance though a large area of porous sheath. The cable which had been specified was found to have a water absorption which was too high and caused this change in potential. The use of polythene insulation on subsequent platforms removed this anomaly.

The use of zinc in salt water is not the only reference metal electrode. Bronze is used in sea water and metal in other media can be used either as an absolute reference electrode or as a swing electrode.

Special Salt Bridges

Often it is necessary to measure potentials where the electrolyte is not easily accessible as, for example, in the water box of a condenser or on the inside of a steel pipe carrying sea water. Permanent reference electrodes could be placed inside the box or pipe but this is expensive and no maintenance can be carried out on the electrode. The author developed a salt bridge consisting of a porous plug which can be inserted through the wall of the pipe in an insulated holder. This is shown in Fig 274. The measuring half cell is now placed in contact with a ceramic plug outside the vessel of pipe and the potential recorded is that relative to the inside end of the salt bridge plug. The point of measurement may be extended by the use of a short length of plastic tubing. The potential is then measured at the open end of the tube. This must be positioned so that air is not entrapped inside it, which would effectively isolate the plug.

Potential Measuring Instruments

The instruments used to measure the potentials discussed above will be chiefly employed for field determinations. The accuracy required in the measurements will call for a sensitive instrument while the use it receives will mean that it must be rugged. If a conventional moving coil instrument is used then it should be light in weight and easily carried, the movement

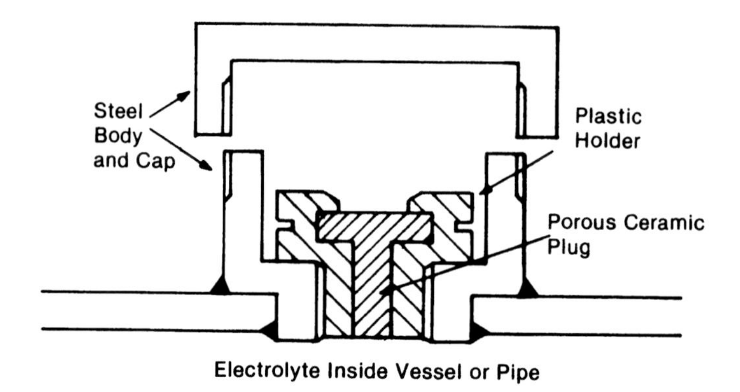

FIGURE 274 - Porous plug salt bridge to measure electrolyte potential through metal wall.

being of taut band pivotless suspension. To achieve sensitivity the actual movement will be light and this must be damped to give a rapid response time. The instrument should not require accurate leveling though it is no disadvantage if it has to be used 'flat'. For carrying, the movement should be shorted or held and this is best not incorporated in the lid movement of the meter case. Rain, frost and hot sun must not affect the movement and it should be surrounded by a moisture and dustproof case. If the instrument is to be used in urban areas it should be mounted on soft rubber feet to absorb traffic and other vibration.

The potential measuring instrument will be required to possess specific properties for each type of measurement undertaken and the range of potentials measured will extend from a few millivolts to several tens of volts. The accuracy of the potential reading will depend upon the meter resistance per volt or ohms per volt. The normal moving coil meter actually measures current and is used to measure voltage by measuring the current that flows through a large resistance. If the meter is sensitive to 1 μA then if one volt potential is applied to a 1 megohm resistor and the meter is connected in series with this, 1 μA will flow through it. If this combination is now used to measure potential then, as an instrument, it would have a resistance of 1 megohm per volt. Similarly, an instrument that needed 10 μA current to give a deflection which indicated 1 V would have a resistance of 100,000 ohms per volt and one with 100 μA deflection would have a resistance of 10,000 ohms per volt. For most cathodic protection measurements a high resistance voltmeter is required and this generally means a resistance of more than 100,000 ohms per volt. Moving coil meters with full scale deflection of 10 μA are available commercially as rugged instruments suitable for use in the field though they will require careful handling and

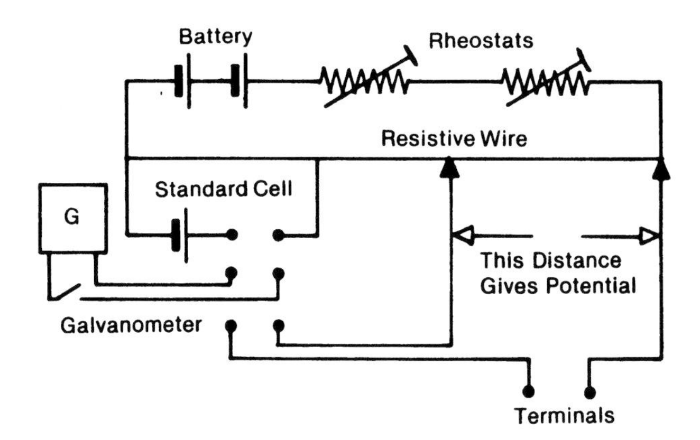

FIGURE 275 - Classical potentiometer circuit.

use. The degree of ruggedness of the instrument will not depend solely upon the current required for full scale deflection but will be a function of the power consumed in making the meter movement indicate and this will depend upon the current and voltage required.

In certain cathodic protection measurements a very high resistance voltmeter will be needed and, particularly, one that takes a negligible current from the measuring circuit. Moving coil instruments can be used to make such observations, when they are incorporated in potentiometers and similar devices. The classical potentiometer circuit is shown in Fig. 275 in which the measured potential is compared with that of a standard cell. The accuracy of this instrument and the amount of current drawn from the circuit depends upon the sensitivity of the galvanometer. High accuracy potential measurements can be made with the instrument in the field but it is not capable of rapid measurements, nor can it be used to follow small voltage fluctuations.

If the potentiometer is altered slightly by using a voltmeter instead of a galvanometer, then the degree of off-balance can be read and small fluctuations in the potential can be followed. Another variation is the backing-off voltmeter which measures potentiometrically, usually with a helical potentiometer and a digital readout. The instrument can be offset by a particular value and this is useful when there is a criterion of change in a swing potential. The accuracy of the instrument depends upon the sensitivity of the voltmeter and if this can be read to within 10 mV then the potential may be determined to this value. If only 0.1 μA is needed to cause a noticeable deflection then a 1 V reading can be made drawing only 0.1 μA from the circuit so that the input resistance is 10 megohms per volt.

While potentiometers can be balanced to take no appreciable current from the circuit when the reading is made, they may take from or put into the circuit quite large currents while these adjustments are being made. Modern solid state electronics can amplify the potential reading and increase the ohmic input resistance or, indeed, allow it to be selected to any particular value. This means that a more rugged meter can be used without loss of resistance and input impedances of up to 20 megohms are standard. The analogue output meter can be replaced by digital meters and these usually read from 0 to 1999. Positive and negative indication and decimal points can be inserted. With a digital volt meter small variations in potential will cause the numbers to change and this change may be rapid, making the display unreadable. Most instruments include within them a circuit that integrates over short periods of time or alternatively will give a specific random reading. This reading will change periodically, perhaps every second, so that a sample of readings will be displayed. The ability to latch the reading and retain it is built into many of the instruments.

The input circuit can be made to reject a c. The meter can automatically be zeroed and a standard cell can be used for calibration.

Instruments for Cathodic Protection Potential Measurements

Structure to Electrolyte Potentials

These are the most general measurements of potential in cathodic protection and the potential is usually measured relative to a half cell. The resistance of the structure to the electrolyte is usually low. It can be estimated from the current required to achieve protection and the potential change at the drainage lead; a protection current of 1 amp causing a 400 mV potential swing means that the structure has a resistance of 0.4 ohm.

The half cell resistance will comprise two parts. Firstly a resistance associated with the half cell itself, and varying with the different types from about 10 ohm to several hundred ohms, and secondly the resistance of the half cell through the electrolyte which may vary from a tenth to one times the electrolyte resistivity that is from 200 to 2,000 ohm in a soil whose resistivity is 2,000 ohm cm. This means that the total circuit resistance may vary from a few ohms in sea water to several tens of thousands of ohms in dry soils. Most half cells give a potential of about one volt relative to protected steel and an accuracy of 10 mV, that is one per cent, is generally required. To achieve this the meter must have a resistance of more than a hundred times that of the half cell circuit. In sea water this is easily obtained but in soils it may mean using a meter whose resistance is several megohms per volt. Such resistances can be obtained by using a potentiometer or solid state voltmeter, though the former instruments may initially draw a larger electronic current from the half cell circuit. This current is unlikely to polarize the half cell as it will have been chosen to have sufficient current capacity. The effect of this current on the structure may in exceptional cases be significant, for example, a buried aluminum cable sheathed with an extruded P.V.C. covering. The initial current will polarize the structure and the potential measured will be the post-polarization value. This effect can be simply demonstrated by attempting to measure the potential of a sheet, some two to three square feet in area, of heavily anodized aluminum.

The conventional instruments described, despite their limitations, may be the only ones available in the field. Often they can be made to indicate their limitations and sometimes give a reasonable estimate of the true potentials that they cannot directly measure. The low resistance moving coil meter which might show an error of 40 per cent on its one-volt-scale because of the high half cell circuit resistance will only show a 20 per cent error on its two-volt scale, that is a potential of one volt will be indicated as 0.6 V on the one-volt scale, and as 0.8 V on the two-volt scale. This change shows that there is a large error in the reading and that the instrument is of insufficient resistance to give a true reading. It could be estimated from the values given that the true potential would be very close to one volt. As most field instruments are of the multi-range type, this course of action is often available to the engineer. The effect of high initial currents from the potentiometric instruments can be eliminated by several techniques. Where the structure is suspected of having a high resistance the potentiometer can be set to what is the expected value of potential, so that the initial current is minimized. Alternatively, the instrument can be left to settle after the initial polarization and if constantly adjusted will return to its initial potential within several minutes. If there is some doubt about the completeness of any such change back to the initial potential, the potentiometer can be offset above and later below the structure potential so that a small polarizing current is caused to flow one way with the first reading and the other way with the second reading. Any reaction to this polarization current would be shown by a difference in the potential obtained by these two approaches.

Metal to Metal Potentials

The largest metal potentials that will occur in cathodic protection are the circuit driving voltages in impressed current installations. These vary from a few volts to about 50 V and as the source of potential is well regulated a low resistance meter can be used. Similar measurements can be made to determine the open circuit potential of sacrificial anodes, though in some cases the anode or structure resistance may be high and the current taken from the circuit in the potential measurement must be low and a high resistance meter used. The potential that exists between the various metal structures involved in interference of stray current corrosion testing may be large, up to several volts. Usually these structures are well earthed and a reasonably low resistance voltmeter can be used. Occasionally well wrapped cable sheaths or pipes are involved and these are susceptible to currents of less than 1 mA and so a medium resistance voltmeter should be used.

Two types of voltage determination require measurement of the potential between various parts of the same structure. These measurements can be made with a low resistance meter and in this case the lead and contact resistance of the meter must be known, and an allowance made for any voltage drop that might occur. The order of the voltages measured in this application are small, often no more than 1 mV. These measurements are used to determine the variations in the potential of a structure while investigating the spread of protection or in assessing the current flowing in a conductor or part of the structure whose resistance is known.

Voltmeters are used to measure the potential across shunts. These have low values particularly when inserted in sacrificial anode leads, and a maximum of 10 mV is permissible where the anode is to be compared with others that are not monitored. A high input impedance voltmeter of the solid state type is preferred for this work.

Electrolyte to Electrolyte Potentials

Stray current problems and the general appraisal of a cathodic protection system often require knowledge of the potential variations in the electrolyte. To do this either the potential between two half cells placed in the electrolyte is measured or the potential of the structure is measured relative to the same half cell severally placed in the electrolyte. In either method the circuit resistance will be high and little or no current should be taken from it. When measurements are made between half cells the meter must have a high resistance and a range of sensitivity giving a full scale deflection of less than 50 mV. Necessarily the half cells used in this work must be compared as their potentials may vary by five to six mV.

Alternatively, a bridge system can be used, particularly in liquid electrolytes, in which the half cell is placed in contact with two different areas of the structure. For example, the bridge may be moved or alternatively two concentric tubes rotated so that openings coincide at different points in the solution. This technique has been used to measure potential gradients in electrolytes close to anodes and cathodes.

Practical Potential Ranges

The uses described for the potential measuring voltmeter will give some useful ranges and facilities that the ideal cathodic protection meter might possess. For measuring metal to metal potentials the chief requirement will be a reverse polarity switch or a negative signal indicated on the display, and this will also be the chief requirement for electrolyte potential

TABLE XVI — Requirements of Cathodic Protection Voltmeter

Voltage Range	*Resistance*	*Measurements*
0-100 V		*Circuit Voltages*
0-10 V 0-1 V 0-100 mV	Very high or solid state	Structure to electrolyte Electrolyte to electrolyte using two half cells
0-100 mV 0-10 mV 0-1 mV	Low	Circuit voltages Volts drop in conductors and structures

gradient measurements. In the determination of structure to electrolyte potentials several refinements will be an advantage. The change in potential of a structure is often required and if the initial potential of the structure could be backed off either to zero or to some specific offset value, then changes in the structural potential could be read directly. This would be particularly useful if the change could be read on a more sensitive scale, say ± 50 mV about an initial 0.7 V potential. The ranges of potential measurements required are given in Table XVI which shows the appropriate use for each of these.

Circuit Measurements

Current Meters

The currents required for cathodic protection range from a few milliamps to several hundred amps. The measurement of these is simple and a conventional ammeter can be used as small voltage drops of the order of a few millivolts are easily tolerated. The other circuit measurements have already been mentioned and these involve the measurement of the current flowing in cables, bonds and structures. The first may be found by tong or clip-on testers. It is possible where a permanent measurement of current in a bond or in a structure such as an anode support is required that a hall effect circuit can be built-in. This can provide a more accurate reading and modern encapsulation techniques allow this to be used as a current measuring device in a variety of locations, Fig 276.

In structures and bonds where the introduction of a shunt or a meter would be difficult or might disturb the current distribution two methods are possible. Either the resistance of the section of the bond or structure can be measured or calculated and the voltage drop that occurs in it can be measured or a device called a Zero Current Meter can be used.

FIGURE 276 - Hall effect current transducer encapsulated for offshore use. (Photo reprinted with permission from HEME International, Ltd., England.)

In this instrument, current is caused to flow against that flowing in the structure or shunt until there is no voltage drop in the structure. The current that was flowing in the section of structure selected will now be flowing in the circuit which was used to drive the current in opposition to the original flow. This arrangement is shown in Fig 277. Besides the components of the current circuit the accuracy of the method depends upon the sensitivity of the voltmeter. This should be sensitive to better than 1 mV and preferably have a full scale deflection of 100 mV or less. At this order of potential, contact resistances and thermal e m f's become important and the zero current meter should be arranged so that the same potential difference occurs in the structure whether the cathodic protection is switched off or on.

The zero current ammeter can be arranged to operate automatically by using an amplifier and a controlled output from the power source. This

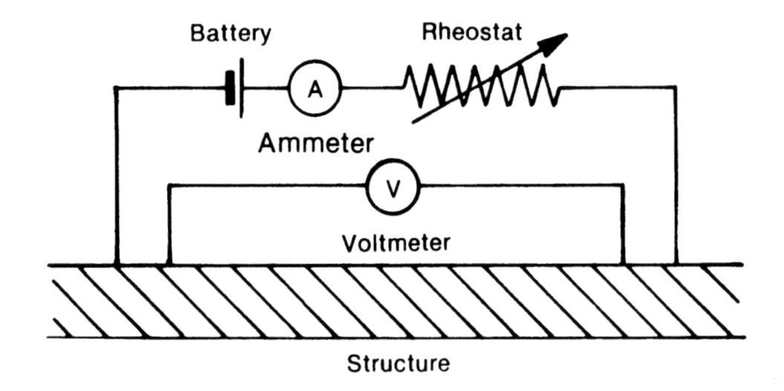

FIGURE 277 - Zero current ammeter.

technique is very useful in the measurement of the output of a selected number of anodes on a long life cathodic protection system as fitted to an offshore oil platform. The structure in Fig 277 is replaced by the anode support bracket, one of which is insulated and the other has a low resistance shunt in it. The zero current circuit improves the accuracy of the measurement over that found with the shunt. In addition to this the zero current ammeter can be run continuously so that there is never any potential developed across the shunt, hence the anode remains typical of all others, and the current flowing can be integrated so that the number of ampere hours that have been consumed can be determined. This not only indicates the continuity of the cathodic protection but is also useful in determining the anode efficiency.

The zero current ammeter electrical circuit can be extended so that no current flows from the anode as opposed to there being no voltage across the shunt. Under these conditions the voltage developed across the shunt will be the difference between the open circuit potential of the anode and the structure. This means that the anode will be acting as a reference electrode and will indicate the potential of the structure in its vicinity. A large current is required to do this and if the shunt resistance is 10 milliohms, then 25 amps would need to flow through it in order to achieve a 250 mV drop. This technique has been used successfully on some large structures and indicates both the anode output and the potential that is achieved on the structure. The anode can be 'switched-off' for some reasonable time and the potential in its absence, is that achieved by the other anodes on the structure. The system can be made to operate automatically where a power supply is available or an additional temporary circuit can be introduced to make the potential measurements which only need be taken occasionally. The circuits are shown in Fig 278 and 279 together with the arrangement of the anode and the cabling.

Direct current flowing in a cable can be measured by the use of a clip-on ammeter. These exploit the hall effect in a semi conductor and with a good operator can give accuracies of ±5%. The unit is portable and has the great advantage that there is no need to disconnect cables when making measurements, Fig 280.

Contact to Structures

At this point it might be appropriate to stress the importance of the electrical connections and contacts in these measurements. These must be well insulated from the electrolyte and any joints must be waterproof and dry. The effect of the latter can be seen if a copper or galvanized clip is touched to a wet steel surface without making metallic contact to it.

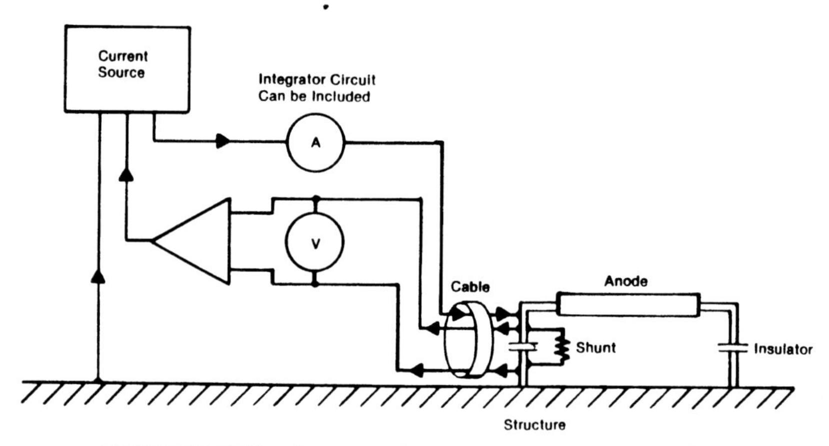

FIGURE 278 - Automatic zero current ammeter for offshore use.

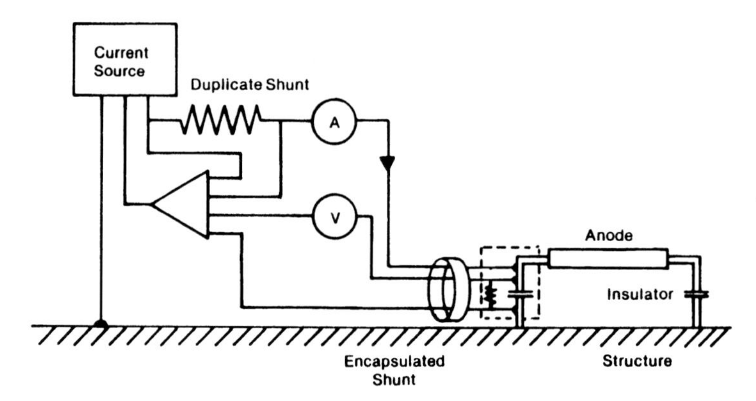

FIGURE 279 - Extension of zero current ammeter to use anode as reference electrode.

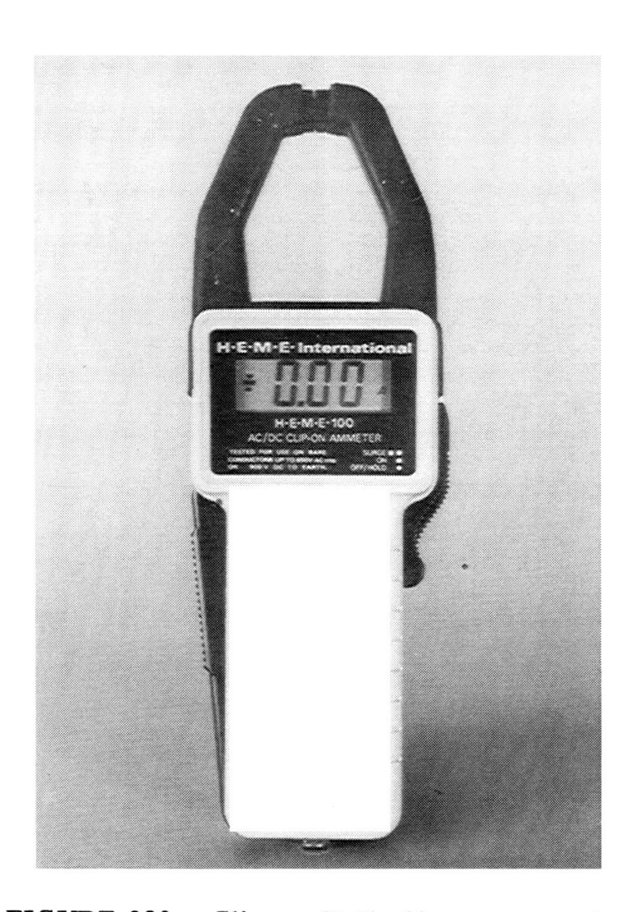

FIGURE 280 - Clip-on Hall effect ammeter for measuring conductor current. (Photo reprinted with permission from HEME International, Ltd., England.)

Good metal to metal contact must be achieved and high pressure contacts by toothed grips are preferable to contacts of low pressure and large area, as mud or clay may be trapped between these and create an electrolytic cell. In contacting a pipe it is field practice to drive a metal probe with a hardened steel point into the ground to hit the pipe. With large diameter bare pipe this is satisfactory providing the half cell is not placed close to the probe. Pipes that are reasonably well insulated should only be contacted with an insulated probe and its insulation should be of the order of that of the pipes otherwise false readings of potential and local current density may result.

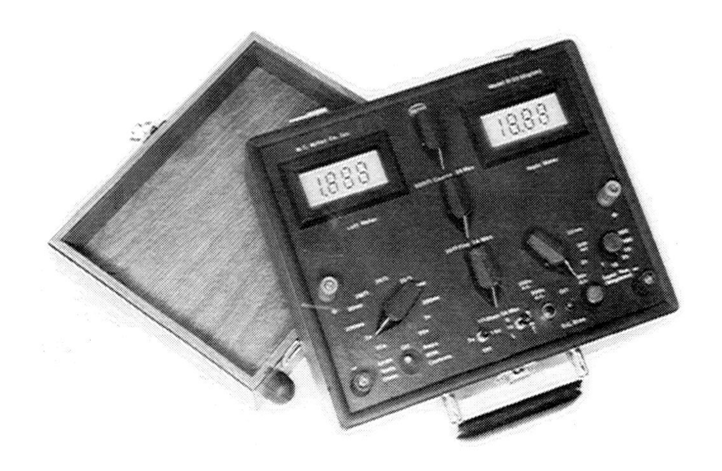

FIGURE 281 - Digital multi-combination meter. (Photo reprinted with permission from M.C. Miller Co., Inc. Upper Saddle River, NJ.)

Combination Meters

Many instruments have been developed which can measure both current and potential simultaneously. These are used to measure the current required to give protection and at the same time check its adequacy. Additionally, they can be used as potentiometer-voltmeters, zero-current meters and as ammeter and voltmeter singly. Best known of these instruments is the Miller multi-combination meter, Fig 281, which, as well as these refinements, can be used for swing tests at small currents from its own batteries and with greater currents from external batteries. In many of these instruments it would be an advantage to be able to select pairs from a series of half cells and structure contacts to measure the various potentials between them.

As a great number of transformer rectifiers are being used as the source of d c power, some instruments which measure the current and voltage of the a c supply will be useful and these are readily available as portable instruments. Where efficiency measurements on the transformer rectifier are to be made, then an a c watt-hour meter would be necessary and these are available.

Resistance and Resistivity Meters

Two Terminal Meter

The principal methods used to measure resistivity were described in Chapter 3. These could be divided into two types. Those that measure the resistance, or a current/voltage ratio, of some geometric configuration of the electrolyte and those that measure the property as such. The instruments that measure resistance can be sub-divided into those that measure resistance simply and those that use a four-terminal method measuring current and voltage by two separate pairs of leads. This latter class includes the measurement of resistance by the combined use of a voltmeter and an ammeter. The former instruments are simple in construction and either rely upon the measurement of current that is derived from a constant voltage source, employ bridges or use ratiometers, the latter being a meter that indicates the ratio of the current in two coils that are mounted to give opposing torque to the needle of an otherwise conventional current meter. Most of the methods that use d c cause the measuring electrodes to polarize and rapid readings must be made. Some of the resistance meters employ low frequency a c or reversed d c and avoid this difficulty.

Four Terminal Meter

The most useful resistance meters for use in corrosion studies employ the four terminal method of measurement. For some purposes the use of a voltmeter and ammeter with an appropriate driving current is sufficient and the combination meters described above can be used for this purpose. Again, polarization is a nuisance. Two types of instruments are available which employ a c methods and indicate the value of resistance, that is the voltage/current ratio, directly, Fig 282.

The first of these two types has a circuit as shown in Fig. 283, the current circuit is fed from a d c generator which is hand driven. The current from this is fed through a reversing switch and one coil of ratiometer. The potential circuit is connected to a similar reversing switch which is synchronized with the first and the current flowing in this circuit is fed to the other coil of the ratiometer. To ensure that the potential circuit gives a current which is proportional to the potential, the total circuit resistance is adjusted to a constant value. The most widely used model of this instrument is the Earth Megger and in this the resistance of the potential circuit is adjusted to 2,000 ohms. These instruments remain accurate in all soils although in some areas 2,000 ohm contacts are difficult to obtain without salting or watering the ground.

The other type of four terminal instrument employs an a c source, usually a battery driven inverter, to provide the current and this is fed through a transformer primary winding. The potential developed by this

FIGURE 282 - Megger digital 4-pin earth resistance tester. (Photo reprinted with permission from THORN EMI INstruments Ltd., Dover, Kent, England.)

transformer secondary is balanced against the potential developed across the potential terminals. The potential is balanced by various transformer tappings and by potential dividers. As the potential circuit draws no current it will tolerate quite large resistances in its circuit though adjustment of the balance will become difficult when this resistance is high, Fig 284.

The four terminal resistance meters are used for three purposes: to measure the resistivity of the electrolyte, to measure structure resistances relative to various distances and points, and to measure circuit resistance. The first is best achieved using the Wenner four pin method described earlier. The first is best achieved using the Wenner four pin method described earlier. It is often useful to measure the resistivity at a series of spacings between pins, and it is possible to use 10 pins to achieve five spacings

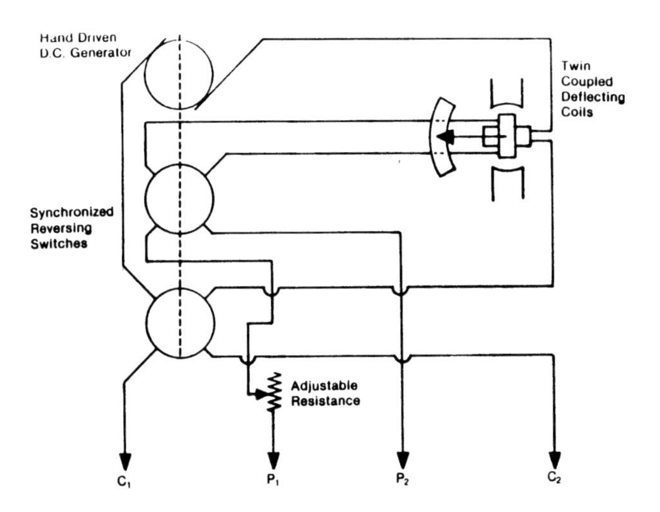

FIGURE 283 --"Earth megger" four terminal resistance meter.

of a, 2a, 3a, 4a and 6a. Since the meter reads the ratio of voltage to current as ohms and the resistivity of the electrolyte is required in ohms cm, the resistance must be multiplied by $2\pi a$ to give the resistivity. Now it would be useful to make $2\pi a$ some round figure, say 1,000, then 'a' will be $1{,}000/2\pi$ cm, i.e., 159.2 cm or 5 ft 2 1/2 in. The 10 pins can be arranged as in Table XVII and by simple switching, the four terminal meter can be successively connected to groups of four of the 10 pins to give the Wenner arrangement. Such switching has been incorporated into one of the models of the inverter type four terminal instrument. Other accessories that can be used with the four terminal meter include soil box or tube as described in Chapter 3. The four terminal instruments can be used in conjunction with the probes or cells commonly employed with the two terminal resistant meters. In this case the potential and current terminals P_1 and C_1 are shorted together as are the terminals P_2 and C_2. A portable four pin arrangement can be made using a sole plate with C_1, P_1 and P_2 pins built into it and a remote C_2 pin. A spacing of either 6 1/4 in. or 3 1/8 in. can be used to give resistivities in ohm cm which will be 200 and 100 times the resistance in ohms, this being because the Wenner formula is now modified to be $P = 4\pi aR$.

Structure and Circuit Resistance

Structure to earth resistances are measured by connecting the terminals C_1P_1 to the structure and connecting the terminals P_2C_2 to two pins placed some distance from the structure, as shown in Fig 285. The structure resistance measured is that relative to the distance P_1P_2 if C_2 is remote,

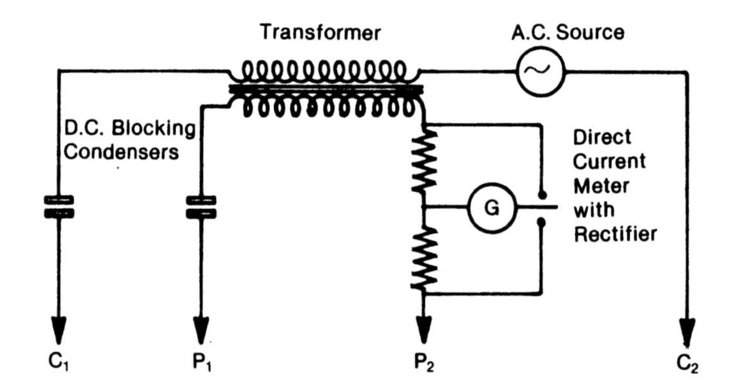

FIGURE 284 - Alternating current four terminal resistance meter.

TABLE XVII — Ten Probes Arranged to Give Five 4-Pin Arrangements of Various Spacings

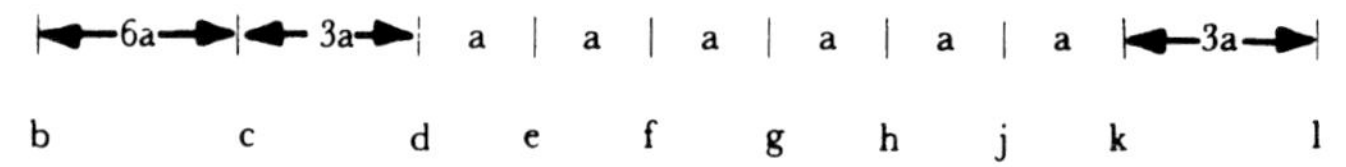

Probe

Probe Separations of "a", 2"a", 3"a", 4"a", and 6"a" are achieved by switching as below.

Meter Terminals / Spacing	C_1	P_1	P_2	C_3
"a"	e	f	g	h
2"a"	d	f	h	k
3"a"	c	d	g	k
4"a"	c	e	j	l
6"a"	b	c	g	l

Probe Connections to Meter Terminals

otherwise the resistance between a hemisphere of radius P_2C_2 whose center is at C_2 and the structure will be measured providing P_2 is remote from the structure. This means that to obtain a true structure to earth resistance then the distances P_1P_2 and P_2C_2 should be increased until the resistance given is constant or determinable.

A similar method can be given to estimate the current required to give cathodic protection to the structure; C_2 is placed at the anode position and

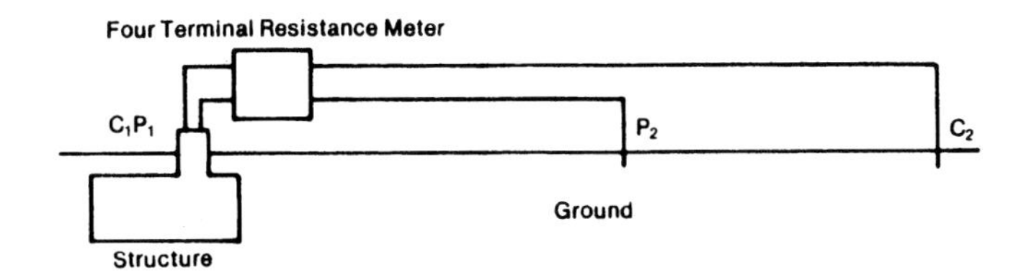

FIGURE 285 - Four terminal resistance meter measuring earth resistance of structure.

P_2 at the half cell position. The resistance measured will be the ratio of the cathodic protection potential change of the cathodic relative to the half cell and the current required to cause this. If the former is assumed to be 400 mV then a 0.1 ohm resistance would suggest that 4 amp will be required for protection. This point was discussed more fully in Chapter 2.

Circuit resistances can be measured by breaking the circuit and using an a c instrument to measure the resistance. This might be a four terminal meter with terminals C_1P_2 and P_2C_2 shorted or an a c two terminal meter. This technique can be used to determine the resistance between a structure and an anode or groundbed. In deducing either the current from a sacrificial anode or the voltage required to drive a permanent groundbed the effects of d c polarization must be considered and these will not be measured by the a c meters.

Four terminal meters can be used to check insulating couplings and cables brought to mimic board as in Fig 286.

Electro Magnetic or Inductive Resistivity Meters

The use of electro magnetic meters has been pioneered by one company who have developed instruments that operate on the principles described in Chapter 3 where resistivity is determined by the ratio of the primary magnetic field compared with the secondary magnetic field produced in a parallel coil. The transmitting coil is energized with audio frequency alternating current and the receiving coil is placed some distance away where it receives both a primary magnetic field and a secondary field from the currents induced in the ground. The resistivity is proportional to the ratio of the primary to secondary fields and to the square of the distance. The depth of exploration varies with the orientation of the coils' axes and if they are horizontal, that is the coils are placed as wheels, then the exploration depth is about three quarters of the inter-coil spacing while if they are vertical dipoles, that is held like haloes, then twice this distance or one and a half times the inter coil spacing is explored.

A portable instrument has been developed where the coils are 12 feet apart and are normally operated as vertical dipoles and give the mean

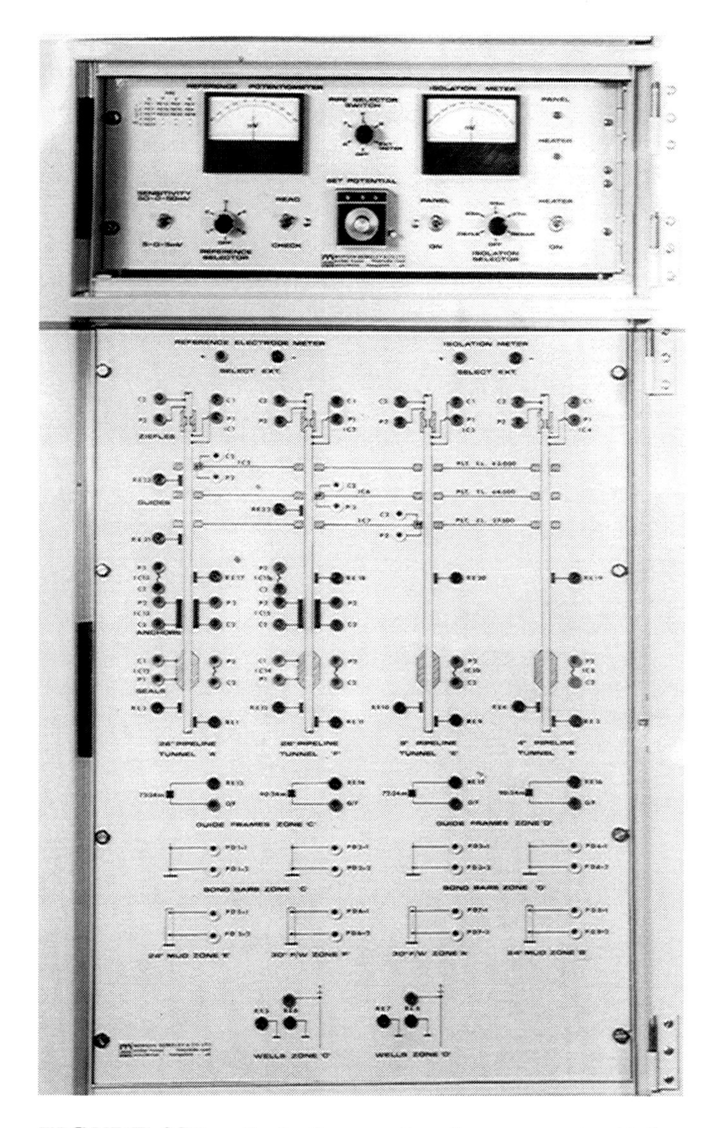

FIGURE 286 - Isolation and reference potential measurement panel with mimic diagram and terminals for external metering. (Photo reprinted with permission from CORRINTEC/UK, Ltd., Winchester, Hants, England.)

resistivity over a depth of 20 feet. By rotating the meter through 90° the mean to half this depth is given. The operation is simple and rapid surveys can be made.

Alternatively, two coils can be used separately and this requires two

TABLE XVIII — Depth to Which Resistivity is Averaged

	Exploration Depth—Feet Dipole Axes	
Coil Spacing Feet	Horizontal (Wheels)	Vertical (Haloes)
32	24	48
64	48	96
128	96	192

operators and an accurate measure of the distance between the coils. An instrument combining both these features is available and here very much larger spacings are possible, for example the makers recommend operating at 30 feet, 60 feet and 120 feet, and the exploration depth is shown in Table XVIII. Both instruments can be used to explore geological formations and the instruments are simple to use and much more rapid surveys can be carried out than with the Wenner method. The principal disadvantages of the system are that it loses accuracy below about 2000 ohm cms and secondly zero adjustment of the instrument has to be done with some care. At the moment there is limited vertical resolution though this should be sufficient for the vast majority of cathodic protection groundbeds. With the electrodes which are a fixed distance apart two layer resistivity configurations near the surface can be detected by taking measurements with the coils at various heights about the ground.

Transient techniques and radio transmission techniques can also be used, though these are not often employed in cathodic protection work.

Pipe locators that also measure resistivity directly usually employ high frequency radiation in the range 3 kHz to 50 kHz and indicate, usually audibly, the coupling that the conductor produces between two coils. The methods are generally comparative depending upon the geometry and use of the meter for the consistency of their accuracy and the resistivity is measured only to a comparatively shallow depth. The attenuation of the radio waves can be used to determine resistivities and suitable apparatus for doing this can be used from an aeroplane.

Coating Inspection

Cathodic protection is most easily and usually most economically employed as a complement to a good coating. The cost of a perfect coating, even if it could be obtained, would be prohibitive both because of the flaws and holidays that occur during the application of the coating and because of the damage that normal handling and installation involves. To ensure the best performance from a coating it is necessary to test for the presence of holidays and, with hot applied coatings, for any areas that may have been coked. The instruments used to perform this work are called holiday detectors and they can be classified into two types, those that use a high voltage spark discharge and those that measure the change in resistance between an electrode and the underlying metal when the coating is flooded with a low resistivity electrolyte.

High Voltage Holiday Detector

The high voltage spark models use either a c or d c sparks. The a c instruments usually employ a battery driven inverter with step-up transformer and operate up to 15 to 20 kV open circuit output. When in use the value of this voltage will be reduced depending upon the lead and any other stray capacitance. Some coatings, particularly those that are covered with asbestos fiber or similar outer wrap, have a high capacity when they are wet. This reduces the effective driving voltage considerably and makes the detector difficult to operate on large diameter pipes. The d c detectors usually use a high frequency current which is stepped up to a high voltage and then rectified. These models are available up to 30 kV and are less affected by lead or coating capacitance. They are usually more complicated and, as a result of this, require careful handling in the field. Both types of detector are fitted with short leads so that their output is not lethal.

The output voltage can be controlled in several ways. In some instruments this is achieved by using an adjustable spark gap in parallel with the coating while in others, usually the d c electronic type, simpler methods can be used to control the low voltage stages of the circuit. The detector will indicate a holiday when current flows through the coating at a weak point to form a break-down arc. Signals are operated from the increase in the output current or some similar sympathetic change, and include bells, lights, buzzers, etc. They are usually chosen for their contrast with the conditions in which the detector is used.

The detectors are designed either for use in the field or in a coating factory and the former are made to be light and operate off batteries. The greatest use of these detectors is made in pipe coating inspection and some field instruments are designed to run along the top of the pipe. The factory models are usually mounted on the final coating equipment or are trolley mounted for use in a yard or storage shed and may work either from batteries or directly from the a c mains. The earthing of the detector to the coated metal can either be achieved via a trailing earth lead, and rely upon the contact that is achieved through the coating rig or other earth connections, or can be made by a direct electrical contact. In either case the earth circuit must be tested for continuity as the detector depends upon this return path for its proper functioning, Figure 287.

FIGURE 287 - High voltage holiday detector using rolling spring electrode. (Photo reprinted with permission from Tinker & Rasor, San Gabriel, CA.)

The high voltage electrode that is drawn over the coating surface must be such that each part of the coating is swept by it. Two types of electrode are used: a brush which may be shaped to fit a particular curve, or a rolling spring which is used on pipelines. Most brushes are made with phosphor bronze bristles and these are made sufficiently close together to give a complete coverage of the surface swept. The brush may be curved either to fit a surface or to surround a pipe. Conducting synthetic rubber brushes are more durable as they have a high resistance which will limit the current output on short circuit whether this is by accidental contact or at a holiday.

The rolling springs are made to fit completely around the pipe and contact to them is made with a shoe which is machined to fit the spring and be just greater than semi-circular. The spring must be made so that neither is it too tight, which would make rolling difficult, nor too slack in which case the lower end might leave the pipe coating. The spring metal section must be rectangular so that the whole surface of the coating is covered otherwise with circular metal section springs small holes may pass between the coils and not be detected. The springs are not satisfactory for use on soft coatings as they tend to pinch and nip the material. Some brushes and a spring electrode are shown in Fig 288. The rate of traverse that can be

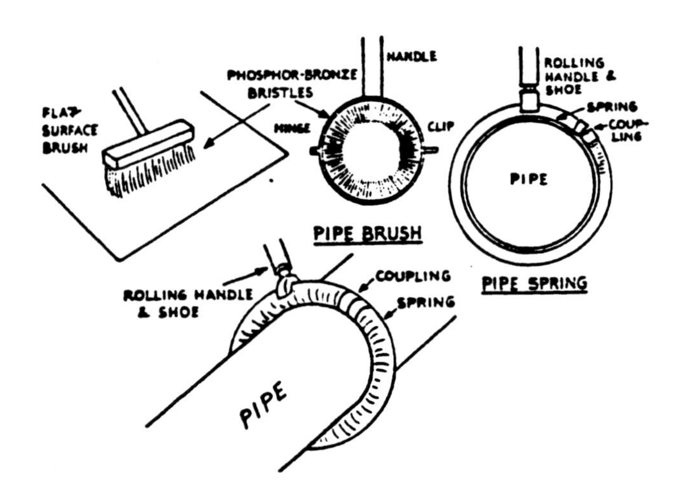

FIGURE 288 - High voltage holiday detector electrodes.

made with the electrode depends upon the response time of the detector. Usually the only time when a set speed has to be maintained is when the detector is following a mechanical wrapping machine. These travel at a slow walking pace and most commercial machines can maintain this speed.

The spark normally burns a hole in the coating which is often difficult to see after the detector has passed and so it should be marked by the detector operator. Each hole must be cut out and re-filled as the spark will probably coke the coating and the danger of this has already been described. When a section of pot-coked coating is discovered the whole area should be cleaned, re-coated and tested again.

Electrolytic Holiday Detectors

The electrolytic coating testers usually work at low voltages, less than 1 kV, and use an electrolyte-soaked pad of sponge or felt which is held on a plate electrode and the area under inspection is swept with it, Fig 289. It is useful to incorporate a dye in the electrolyte, as this will indicate the area of coverage. The detectors vary considerably but most are battery operated and employ an inverter-transformer circuit to give the medium voltage output. Their chief use is in the inspection of lightweight coatings and the method, unlike the high voltage technique, is non-destructive. This means the exact holiday is not apparent and to make detection easier a twin pad system can be used where a small electrode is mounted on the back of the large electrode, so that by turning the assembly over, the holiday can be pin-pointed more rapidly. Such an electrode scheme is shown in Fig 290.

FIGURE 289 - Non destructive low voltage holiday detector with electrolyte soaked sponge. (Photo reprinted with permission from Tinker & Rasor, San Gabriel, CA.)

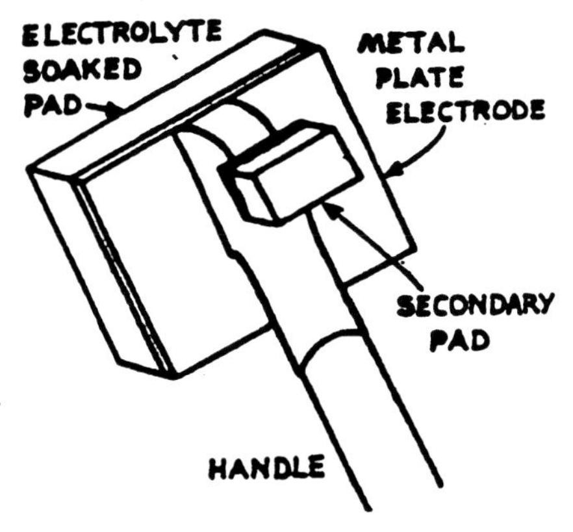

FIGURE 290 - Low voltage holiday detector electrolyte salt dual pad electrode.

The electrolytic holiday detector also gives an indication of the mean coating resistance and as such can be use to check the correct application of paints or coatings especially when these are filled.

Paint Thickness Gages

With modern fusion bonded coatings the thickness of the coating can be determined using standard paint thickness testers. These either work on eddy currents induced in the underlying metal through the coating substrate or they depend on the magnetic effects and, of course, they can only be used on magnetic materials. Both are available as portable, accurate devices and are invaluable for the rapid assessment of coating quality.

Pearson Detector

Pearson has devised a method of detecting small coating defects on a buried pipeline by a very similar method. In this an audio-frequency electric signal is fed between the pipeline and a distant earthed electrode. At holidays there will be an increased current flowing in the ground and this can be detected. If two point contacts are made to the ground and the signal that is developed between them amplified, then a change in signal strength will be found as the two contacts are moved, the maximum occurring when either point of contact is over the holiday. Pearson used two men with earthed contacts on their feet walking a fixed distance apart; the leading man traced the pipe route while the other operated the detector. Two methods of detection were used. Either both men walked over the pipe when two signals were received from each holiday, or one walked over the pipe and the other, who operated the detector, walked a parallel path at a fixed distance from the first, each holiday giving a single maximum. The system is illustrated in Fig 291. This method finds its chief use in detecting

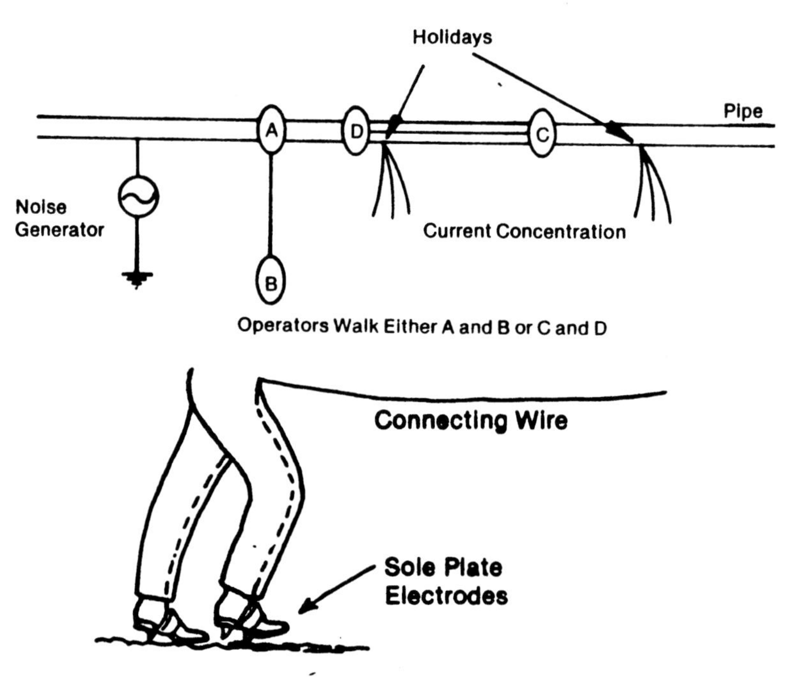

FIGURE 291 - Pearson buried pipe holiday detector.

FIGURE 292 - Current interrupter to interference or polarization testing - Quartz controlled. (Photo reprinted with permission from Tinker & Rasor, San Gabriel, CA.)

damage and deterioration of pipe coatings in the ground. It can detect small areas of damage due to soil stress or rocks and these can be repaired before or during cathodic protection.

Coating Resistance Measurements

The meters described above have been used to detect flaws in the coating. It is equally important to be able to make an estimate of the conductance of the coating. One method of doing this for a buried coating—by a current drainage survey—has already been described. A method has also been given for measuring the structure to earth resistance of a body and this could be modified to give an accurate value of the coating conductance. On many long structures the area considered will have to be limited either by using an insulated length of the structure or by limiting the effective area considered by careful anode positioning. When the coating is not buried or immersed two methods are available. The first uses an electrolyte soaked flannel to surround the pipe or to cover an area of the structure and this is held on by a metallic grid which acts as one electrode. Current is caused to flow between this electrode and the metal underlying the coating and from the current flowing and the driving voltage required the coating resistance can be found. The other method relies on laboratory determinations of the coating conductance at various thicknesses and in the field only the coating thickness is measured, there being several portable instruments that give an accurate measure of the coating thickness on metal surfaces.

One other coating property is of interest to the cathodic protection engineer: that is adhesion. This can be tested by several methods, most of which involve mechanically breaking away the coating to see if it comes away from the metal or if the coating material itself disintegrates.

On/Off Devices

With impressed current systems increasing use is made of the protection criterion based on the polarization immediately on switching off or on making a step-wise adjustment in the level of the cathodic protection current. A number of devices have been developed that will automatically switch the transformer rectifier on and off in a cyclic manner providing periods of on and off operation which can be set to different values, Fig 292. This criterion is being used increasingly with well coated pipelines and particularly at places where it is felt that there is some coating damage or deterioration. The off period is made very short so that polarization of the line is maintained and typically the system is left on for more than 80% of the time. With a modern coating the deterioration in the potential is extremely rapid and an off period of 2 seconds is normal. It is essential, therefore, that where there are a number of power supplies, as on a long pipeline, these are switched off simultaneously. Devices using radio signals and sometimes radio beacons have been developed and these produce a sharp, simultaneous cut off.

Accurate measurement and recording of the polarization potential of the pipe before, during switching, and immediately following is essential. Elegant electronic techniques have been used which can record and/or compute the results on the spot. These techniques have increased the popularity of this method and will continue to do so. The location of the half cell is an essential in this process and some systems employ a multiplicity of cells which provide the information to eliminate the volts drop in the soil by the techniques described by Pearson. Other methods have been developed in which a small coupon is separately connected, and can be separately disconnected from the pipe. In some of these the surface potential of the coupon can be measured and a technique using a porous plug salt bridge has been employed.

The other use for the On/Off devices is in interference testing where the effect of the cathodic protection current can most easily be interpreted when it is switched. In these surveys it is not essential to maintain the polarization of the protected structure and it is more usual to have an ar-

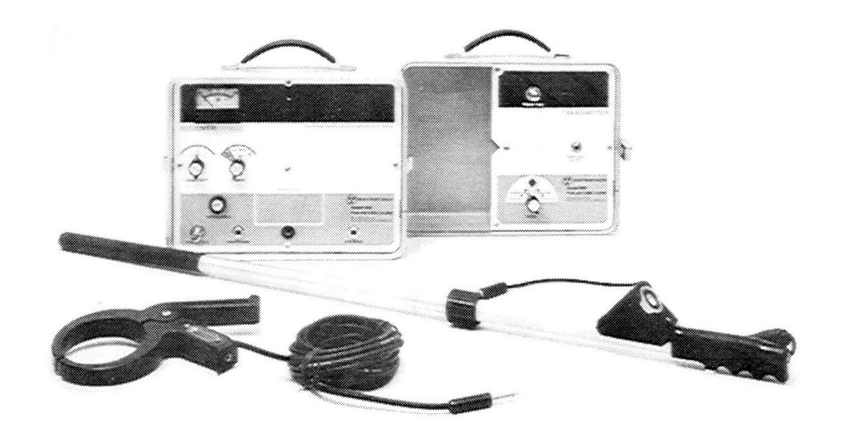

FIGURE 293 - Pipeline and cable locator with inductive probe and insulated inductive clamp for coupling signal to pipe, cable or metallic tape. (Photo reprinted with permission from Metrotech Corp., Mountain View, CA.)

rangement where the cathodic protection is switched on and off over periods of about 20 to 30 seconds duration. This is often standardized by a group of utilities in one area and is related to the immediate change in potential and/or current flowing in the affected structures. Often the utilities in an area will get together and decide on a pattern of switching and a related criterion of damage which is based on their experience.

Other Instruments

Coupling Locator

Structures that are to be cathodically protected must be electrically continuous. The resistance meters already described will measure the structure resistance, but it is equally necessary to ascertain the effects of very heavy currents flowing through the structure, and this is particularly true of pipelines that are made of lengths of tube joined by flexible couplings. Some joints will have metallic contact, while in others there will be a low resistance because of the close proximity of the two pipe ends in the electrolyte. To distinguish between these conditions and to reveal any but perfect

FIGURE 294 - Pipeline or cable automatic tracing instrument giving location and depth. (Photo reprinted with permission from Metrotech, Mountain View, CA.)

metal to metal joints, a large d c current must be made to flow through the structure when the resistive joints are detected by a potential survey. The apparatus used to do this testing is called a *coupling locator* as its chief use is detecting flexible couplings in pipelines. The only difference between this detector and the d c four terminal resistance meters is the large output of the current source which will give up to 100 amp.

Pipe Locators

The need for careful selection of the anode or groundbed site, particularly with regard to adjacent structures, was stressed in the last chapter.

It is not always possible to obtain accurate maps of the groundbed area which indicate other buried pipes and structures, and even if these are available they are not necessarily complete. As a pipe buried close to the groundbed may be subjected to serious corrosion, the area in the groundbed's immediate vicinity should be searched for buried metal. Pipe or metal detectors are available which will indicate the presence of pipe or metal at a considerable depth and which can be used to trace long lengths of pipeline. These usually operate on the increased linkage between two coils in a similar manner to the resistivity meter. Two coils are used, one as a transmitter and the other as a detector, and they are arranged to have their axes horizontal and vertical. When a pipe or other metal has been located its depth can be found by orienting the coil, whose axis was vertical, through 45° and again locating the metal. The distance between the surface location of the metal in both cases will give the depth of the metal beneath the former location, Fig 293.

A similar class of instruments can be used to trace pipelines and these usually employ a high frequency noise generator between the pipe and an earth electrode so that the pipe metal can be traced by a single detector coil and amplifier. Most instruments give an audio output which increases in magnitude to indicate the presence of the metal or pipe while others additionally give a metered output, Fig 294.

Monitoring

Test Points

Many structures that are cathodically protected are inaccessible or difficult of access for the measurement of protection potentials. Perhaps the most common structure to which it is difficult to make an electrical connection is a buried pipeline. The technique often adopted is to use a steel probe with a sharp hardened tip and to drive this through the ground to contact the pipe; where the pipe is laid near to the surface in soft ground and its position is accurately marked, this technique is successful. Under other conditions it is preferable to attach permanent cathode leads to the pipe and to bring these to the surface at marker posts. The leads are usually welded or brazed to a pad or stud on the pipeline and this junction and the electric cable are insulated. More than one lead may be brought to the surface and these may be connected to either end of a calibrated section of the line so that current flow in the pipeline may be determined. Similarly, the connection between a sacrificial anode and the structure can be brought to the surface so that the anode current can be measured. It is usual to use an 'elephant's foot,' concrete pillar or similar device both to hold these connections and to indicate the location of the structure.

The majority of other structures will be accessible so that electrical

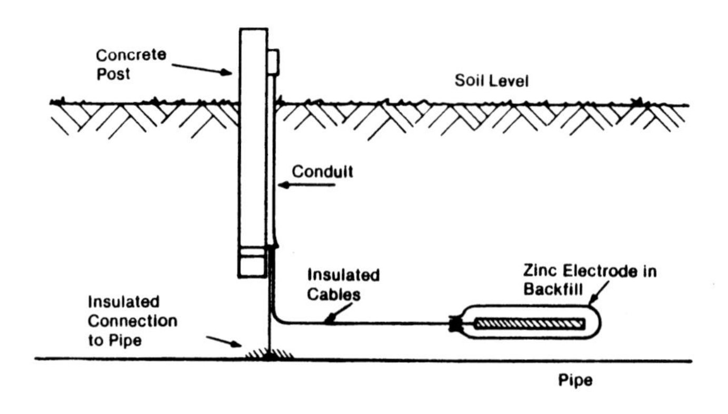

FIGURE 295 - Permanent buried zinc electrode close to pipeline.

connections can be made to them. One exception might be deep well casings where only the metal potential at the top end may be easily measured. The potential of the rest of the tube can be determined by lowering a contact 'pig' down the casing tube and measuring the contact potential. To ensure a good metal to metal contact some electro-magnetic or other force must be used to drive the point contacts home. The potentials measured by this method are small and an electronic voltmeter is essential.

With many structures there is not easy access to the metal electrolyte interface and a salt bridge device that overcomes this difficulty by piercing the wall of a pipe, tank or possibly a ship's hull has been described earlier. Similar salt bridges can be made in many other applications, for example, to determine the potential of the steel plates forming the bottom of an oil storage tank, a plastic tube could be laid beneath the tank to its center prior to the tank construction and then filled with an electrolyte; a half cell placed in the outer end of this tube will display the same potential as a half cell placed at the other end of the tube. The resistance of such a salt bridge would be high.

Where frequent potential measurements have to be made at a relatively inaccessible point then a half cell may be permanently located there. The best half cell for this purpose is the silver chloride type as these can be constructed to last several years; in the soil their chief limitation is the retention of the chloride environment. When this is difficult much advantage may be gained by using a zinc rod as a reference electrode. This is surrounded by a backfill mixture of gypsum, sodium sulfate, salt and clay as were the sacrificial anodes. The potential of the zinc will remain constant once it has settled down but it is advisable to check it absolutely against a half cell.

In sea water zinc has a constant and predictable potential so that pieces of zinc of anode purity can be used as reference standards. The metal must be massive to avoid polarization and should be used anodically if any current is drawn from it during the measurement. The potential between the cathode and the zinc or the silver chloride half cells can be used to drive a permanently installed meter, and if the reference electrode is large then a high resistance voltmeter could be used and read directly. Fig 295 shows a potential test point to which a cathode lead to a pipeline and a buried zinc reference electrode are connected.

Fig 273 showed a permanent reference electrode attached to a tubular member of an offshore platform. The silver chloride and zinc elements are shown, together with the anchoring arrangement for the cable. The particular shape of the mounting is designed to prevent damage and the electrodes are held so that the increase in potential by the voltage drop to the surface is matched by the shielding effect of the support. The unit is preassembled in the factory and can be welded to the structure either directly or onto small pads without damage.

Moving Electrodes

It has been assumed that the electrode is held stationary in the water but it is possible to move the electrode relative to the cathode to obtain a survey. One of the earliest uses of this is in studying the protection—and particularly the spread of protection—around a moving ship's hull. An electrode can be trailed in the water but it was found to be more satisfactory to use an electrode that would float at a specific distance below the water line and would run past the ship in the water. To do this the electrode was heaved, rather as a lead line is, forward from the bow of the ship and its potential recorded as the ship moved past it. The cable to the electrode was secured so that before it reached the propeller it was arrested and retrieved. By using an adjustable depth to the electrode below a small float a good potential survey of a moving ship can be made. This, unfortunately, is generally limited to speeds not in excess of 10 knots. An alternative technique which has been used is to survey from an electrode attached to the hull of a glassfiber or wooden hulled boat by moving this around the ship or by running alongside when the ship is underway. This can again only be used at slow speed and under these circumstances it is better that the cathode lead be used to interconnect the small boat and the vessel.

The other main use of traversing electrodes is in the offshore industry where they have been used to scan the potential of platforms. Taught wires are built into the platform and the electrode, or often a series of electrodes, are run up and down these wires. The use of plastic covered steel wire rope or of plastic rope itself controls the position of the electrode and its location can be accurately determined. The technique is very useful where there are doubts as to the efficacy of the protection and how this is affected by variations in the potential of the electrolyte and variations in the current demand of the structure.

In some processing and cooling water systems electrodes have been floated down pipelines and culverts in order to assess the potential profile inside the pipe.

Land Pipeline Surveys

In the analysis of the spread of protection along a pipeline it was suggested in Chapter 6 that measurement of the ground potential and of the pipe potential would give this information. In modern pipelines the parameters of the pipe itself are accurately known and so the electrical variations in the pipe metal can be accurately predicted. It would be useful to be able then to plot the potential of the surface of the ground over the pipe as this will confirm protection and will also indicate where there are changes in the coating conductance or in other factors that may affect the potential. The use of modern microprocessor equipment allows a survey to be carried out by integrating the potential found by walking one or more electrodes along the line. If a single electrode is used some technique of distance measurement is essential, as is some memory from the pipe. If two or more electrodes are used then the potential can be integrated along the route. A variety of methods is proposed, including ones in which the distance between the electrodes is accurately controlled, and others in which the integration itself is carried out against distance by a separate measuring technique. This type of survey is particularly useful where the line has been in service for some time and is in either inhomogeneous countryside or in areas that have been affected by development. The results can be used to indicate the potential and the local current density and local resistivity. 'Instant off' measurements can be included in the survey.

Remote Measurements

Many submarine pipelines that exist in the oil producing seas of the world have to be surveyed to test the efficacy of their cathodic protection. Techniques of measurement are difficult and some of the parameters that can be measured give a useful indication of the protection level of the anode output and of the quality of the coating. The author used a technique which was a development of Pearson's method by placing two zinc electrodes 2 meters apart on a stick with an amplifier connected to headphones so that a diver could travel along a sea bed pipeline and get indications of the potential gradient along the line and, incidentally, locate it by moving the stick so that it was radially oriented with the line. The movement through about 45° indicated the exact location of the line by the change in the signal in the headset.

The magnitude of the signal when the electrode was held radially showed the quality of the coating and also detected anodes if these were not visible.

A recent and elegant development of Pearson's method involves the use of two silver chloride electrodes mounted about a foot apart at either end of an arm which is rotated at high speed. The output from the electrodes in a potential field is sinusoidal and the direction of the field and its magnitude can be deduced from the phase and magnitude of the ac current. It is possible to use such a device to detect both changes in the current flux—and hence detect holidays and other breaks in the coating—and also to use it to locate the anode or the recipient of the cathodic protection, Fig 296.

This device has been developed to monitor subsea pipelines and the method works particularly well when there are only dc potentials. When traversing along a pipeline the speed is limited by the spiral path which the electrodes trace as the greater the pitch the less accurately and clearly will minor holidays in the pipe coating be detected. As a rule of thumb the spirals should not have a pitch of more than one half the pipe diameter.

There are small complications when the cathodic protection current is derived from automatically controlled single phase rectification of the ac mains. In these circumstances 100 Hz or 120 Hz ripple is imposed upon the measurements. This can be eliminated by the circuitry which computes the

FIGURE 296 - Rotating pair of electrodes for potential gradient mounted current on remotely controlled sub sea vehicle. (Photo reprinted with permission from Corrocean, Trondheim, Norway.)

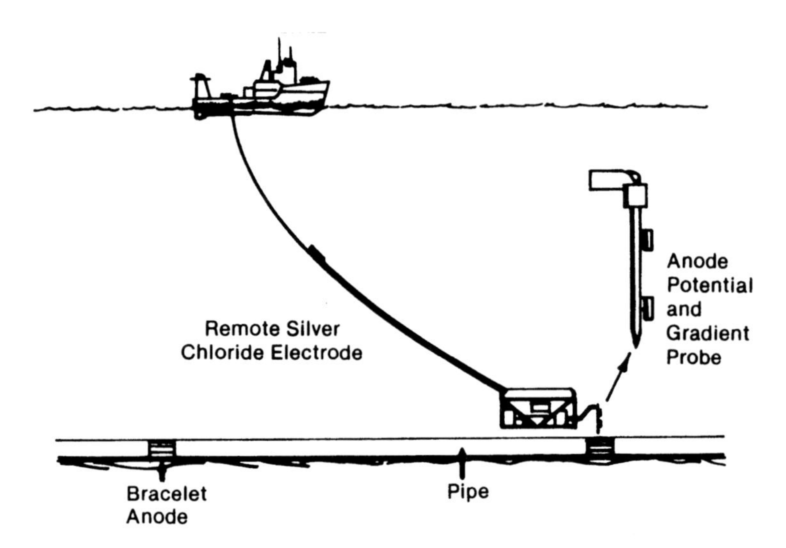

FIGURE 297 - Morgan technique of subsea pipeline potential survey.

pipeline data from the results providing it is of sensibly constant magnitude and frequency.

A sub-sea pipeline can be surveyed without contacting the pipe by a method proposed by the writer in which two electrodes are used. One is remote from the line and the other is carried close along the line's surface. The potential as the scanning electrode passes an anode is markedly different from the potential over the coating between anodes. As the anode potential is known, it would not be a very good sacrificial anode without it being so, then the potential at the least protected point on the coating can be determined. The bracelet anode on a pipeline is in electrical contact with the line and the potential of the anode and, indeed, of the pipe itself can be measured by making contact with the bracelet anode. The survey method can be used at high speeds with a submarine and by using two close electrodes or one electrode with a moving salt bridge, the potential gradient along the pipe can also be determined.

Fig 297 shows the system schematically with a pair of electrodes and a metal bip comprising the probe attached to the submersible and with a remote electrode fixed to the umbilical. The closest of the electrodes, coupled with the remote electrode, provides the potential scan along the pipe and occasionally at anodes the metal point can be used to check the pipe and anode potential absolutely. The current flux onto the pipe or off the anode is determined by the potential gradient between the two probe electrodes, Fig 298.

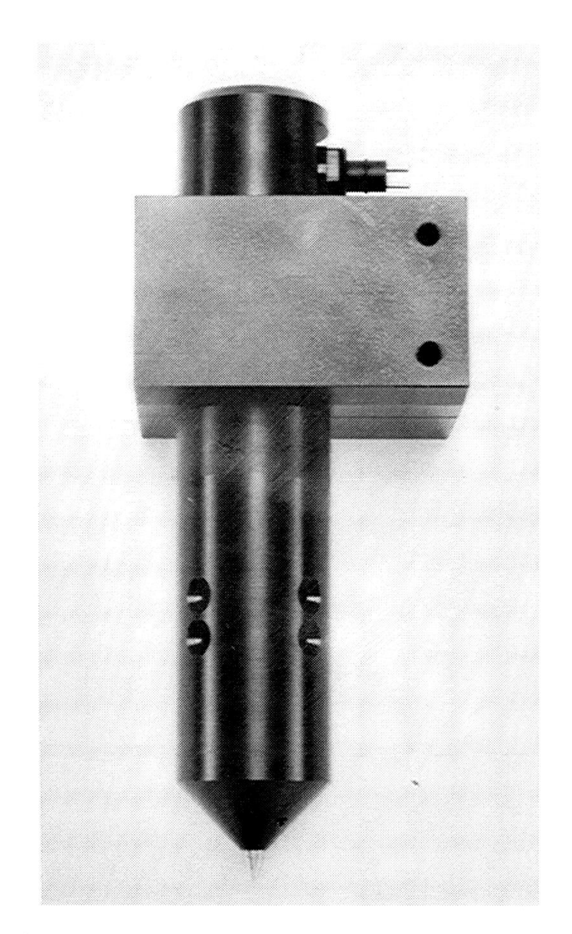

FIGURE 298 - Close positioned silver chloride electrode and probe for Morgan subsea survey (Photo reprinted with permision from Subspection, Ltd., UK).

The results that are obtained with the technique are shown in Fig. 299. These are transmitted to the surface from a submersible via the umbilical and from them the parameters relating to the pipeline can be computed; they are sufficiently good for minute holes and field wrapped joints to be detected.

The potential profile of a pipe approaching a structure or the shoreline usually indicates a voltage gradient in the pipe metal itself indicating that current is flowing either onto or away from the structure in the pipe metal. From the voltage gradient, the pipe size and metal resistivity this current can be calculated. In general the pipe anodes will be providing protection to the structure and from the potential gradient in the vicinity of the anodes,

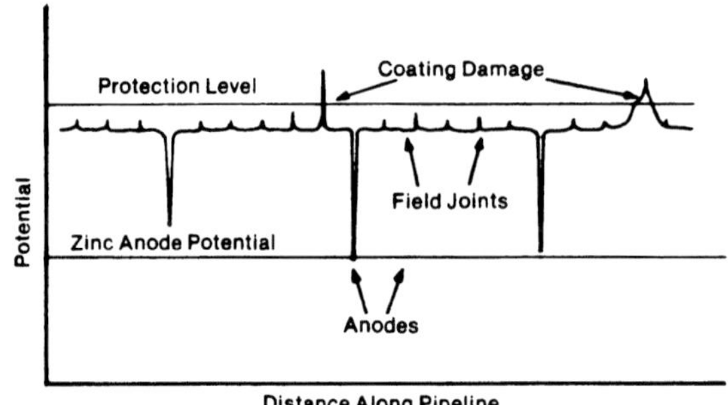

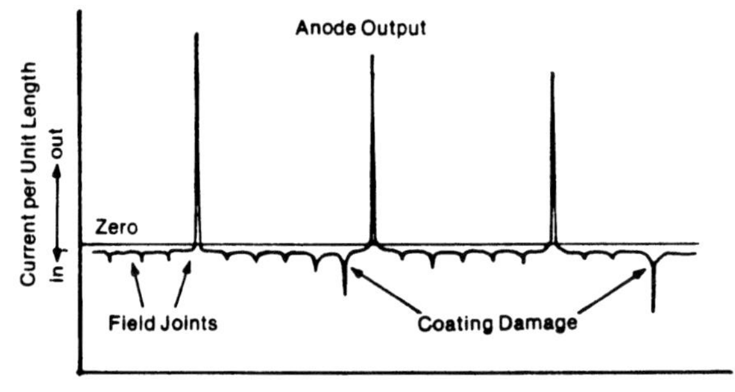

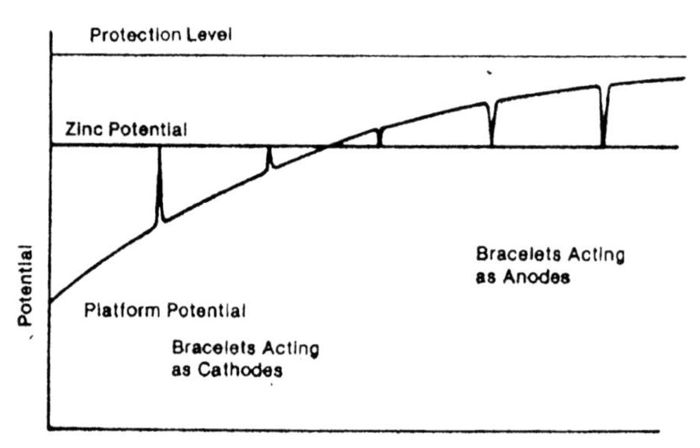

FIGURE 299 - Result of survey on zinc bracelet protected pipeline showing (a) anode and pipe potentials (b) current density and (c) potential profile close to platform.

their current output can be determined and the ratio of the current flowing to the structure and to the pipeline can be estimated. Where long lengths of pipe are protected by impressed current the same technique can be used to measure the pipeline potential. Zinc or aluminum anode bracelets are attached to the line and these will act both as sacrificial anodes and as reference electrodes. The same survey can now be carried out and the protection of the pipe determined many miles from the impressed current stations. Again, ac ripple on the system can be computed out or smoothing used to eliminate it at the groundbed.

Measurements Underwater

It would be useful if a diver could make potential measurements in a self-contained unit under water. The early attempts at this encased a moving coil meter and had an electrode and probe assembly. The visibility of the meter was poor and it suffered from parallax. The writer developed a much superior technique in which a silver chloride electrode and probe were mounted in what would be the muzzle of a gun and there was a digital bright display of the potential in the breach. The handle contained rechargeable batteries. The gun is shown in Fig. 300. The silver chloride electrode has a skirt around it so that the potential that is measured is that of the surface. A stainless steel tip which is replaceable is pressed onto the surface and the reading is given instantly.

The meter requires an integrating circuit in it so that the numerals are steady and change only slowly and it has an ON/OFF ring switch.

There were considerable problems in making the instrument watertight to 1,000 ft. The rechargeable batteries needed connections so one is made into the handle, but to reduce the connection problems, the other point for battery charging is the tip. The ON/OFF switch is magnetically controlled by moving the ring and the whole assembly is potted into a glass fiber resin body. The glass for the water is made from heavy armored glass. The unit has proved to be very reliable, extremely accurate, and has speeded up very considerably the surveying of platforms. The meter is accurate to 1 mV and the silver chloride electrode to about the same, Fig 301. To test the meter, a zinc electrode is used and this is a block of zinc which can be held in the diver's hand and by pressing the tip of the meter to it a reading is obtained. This calibrates the instrument though in general it is itself superior to the calibration block. Alternatively, a small nonconducting tank can be assembled containing a piece of metal whose potential is held constant by a dc device stabilized by an accurate silver chloride electrode. If smooth dc is required, then the anode-cathode arrangement must be such that the insertion of the instrument into the calibration vessel does not affect the potential. As an alternative to the use of a seawater container a similar device could be suspended over the side of the diver's boat, though this would only allow calibration on the instruments carried into the sea.

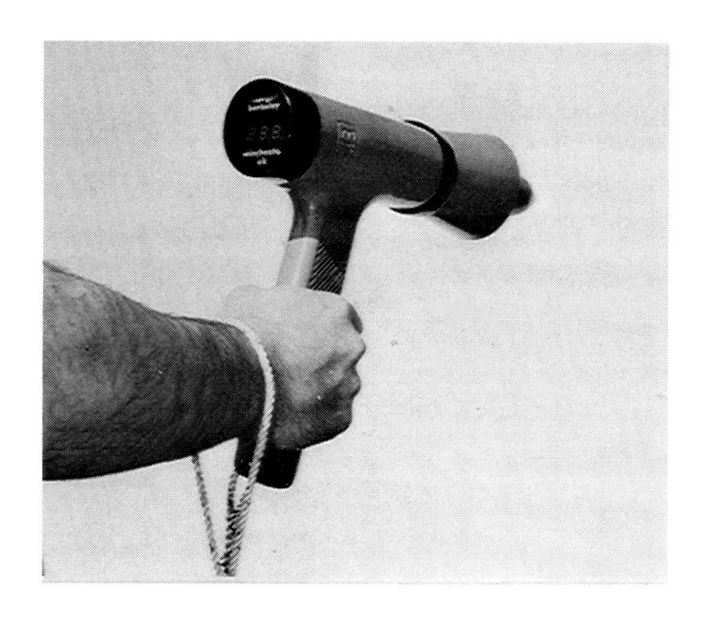

FIGURE 300 - Pistol-type integral/digital potential measuring instrument. (Photo reprinted with permission from CORRINTEC/UK, Ltd., Winchester, Hants, England.)

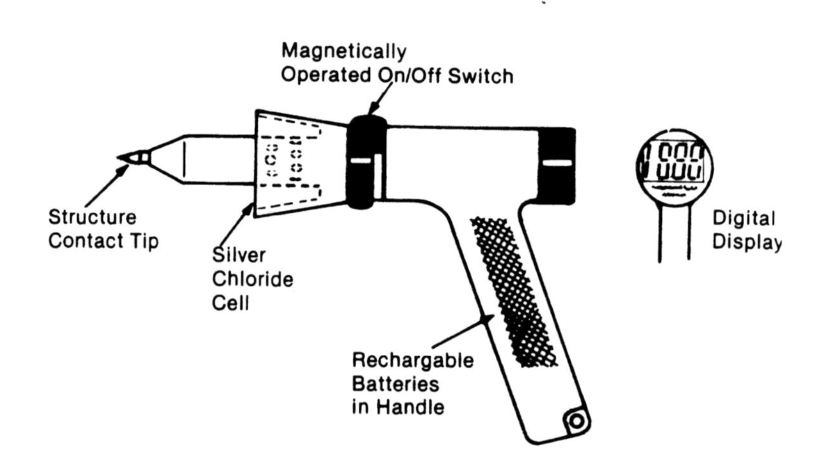

FIGURE 301 - Potential measuring "gun" for diver-held underwater survey.

The electrical and electronic components in the gun can be grouped and arranged so that it can be cut to replace a defective part. The connections are remade and the gun reassembled or encapsulated. Failure of most of the components is infrequent but loss of sectors in the digital display has proved troublesome. While encapsulation techniques are sound to about 1,000 feet at greater depths, the present trend is to use machined housings, often filled with oil, that are watertight at these pressures.

Variations and modifications to the gun-type technique can be used, one of which is to replace the tip with a second electrode assembly so that the displayed potential is that between the two silver chloride electrodes or the potential gradient in the sea. Any mis-match between the electrodes can be seen if the instrument is placed in an area where there is no potential field, which the diver can arrange by inserting the probe into a plastic bag in the sea or containing sea water.

Many measurements taken on offshore structures are used for purposes of comparison and it is essential that the instrument is sensitive and remains constant in its measurement of potential. In general the requirements are of an error of less than 2 mV, one of these being attributed to the digital readout and the other to the maximum tolerance of variations in the silver chloride cell. Where there are marked changes in sea water composition, such as in areas of stratification near an estuary or river mouth, then measurement of resistivity and temperature will allow a correlation between this and the silver chloride potentials.

Because many of the potentials that will be read will be comparisons over small distances, such as the potential gradient near an anode or the potential variation into a recess, the use of instruments that read to ± 10 mV will not be acceptable. Measurements that are more accurate than this or multiple range instruments may be useful if operator error does not increase.

As a result of such a detailed potential survey on an underwater structure the engineer will receive a mass of results which are difficult to interpret. One technique which the writer would propose for an analysis of these is simple and visual. The potentials are plotted in relation to the structure—either to its depth, to one of the faces or in relation to the plan—and each of the different types of measurement is given a different notation. Such a plot against depth of a typical North Sea platform is shown in Fig. 302. It is immediately apparent that there is a change in the cathodic protection results with depths and this change is reflected in the readings—both those taken in inaccessible areas, the general potentials, the potentials close to the anodes and the potential of the anodes themselves.

Analysis is now much easier and the results eliminate the discrepancy in the individual readings which may be caused by minor variations in the position of the electrode, the anodes, or in the condition of the steel surface.

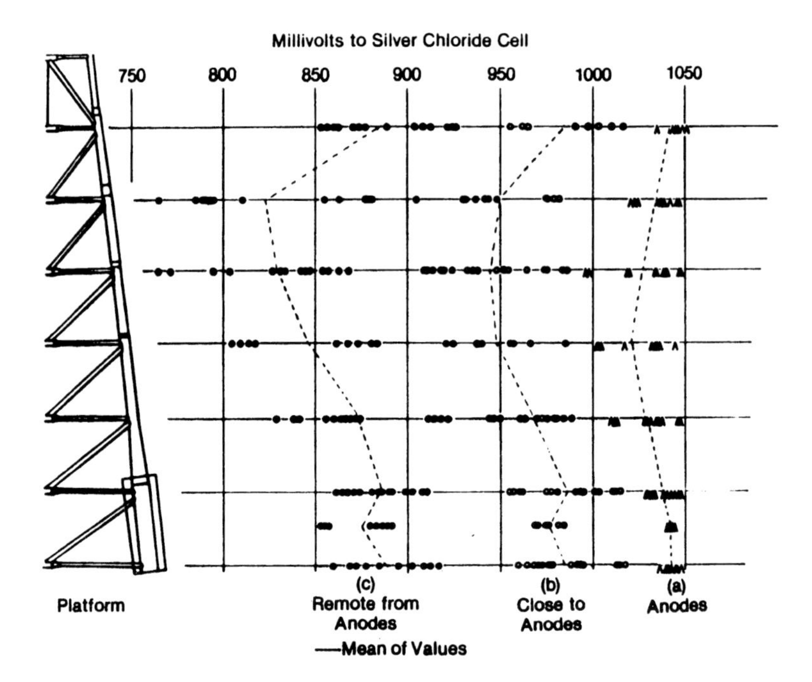

FIGURE 302 - Plot of potentials of (a) anodes, (b) close to anodes and (c) remote from anodes against depths on offshore structure.

If permanent reference electrodes and anode and currents monitoring are installed on the structure then the potentials at the electrode from these can similarly be plotted and immediately correlated to the wider information obtained from the survey.

The engineer requires accurate information because he has to base his decision on whether or not to augment the cathodic protection systems almost three years ahead of the event. This is because he needs to make an economic and technical decision in time to order the equipment and get it to site and then be able to miss one year through other essential work occupying the limited work window.

Remote Monitor Devices

In the pipeline industry increasing use is made of aerial surveys of the pipeline route both to assess the integrity of the line and to detect any actions that may cause damage or hazard to the line itself. The same survey can be used to monitor the cathodic protection. This may vary from a simple device which indicates adequacy or inadequacy at each cathodic protection station or it may send, by radio interrogation, information on the variety of parameters measured either at the transformer rectifier or at selected responder stations.

With modern microprocessors it is possible to use the simple indicator to show that the whole system is working properly and within the proper limits. Equally, the same technique can be used to provide an analysis for radio response and also to allow the over-flying aircraft or helicopter to make adjustments to the system. In remote locations it is probable that if such information is available, rather than wait until the aerial survey, the correction would be applied automatically to the equipment. The system is not limited to pipelines and similar monitoring can be used in other remote or semi remote locations such as, for example, a dolphin or trestle in an estuary.

For use in deep sea water the radio monitor can be replaced by an acoustic system in which an ultrasonic transmission triggers a response from the monitoring station. The pulse must be highly coded as ultrasonics are used for a wide variety of purposes around an offshore structure and it is advisable that the signal sent back contains not only the basic data but also some specific measurements as comparison. Because of the interference it is usual to send a train of pulses and to select from these those that lie within an acceptable band and to use the mean of these as the definitive signal. Such systems must be made to operate at great depths and will require their own battery supply. Interrogation, say at monthly intervals, should mean the system would have a life in excess of five years, and hopefully of ten, the condition of the batteries being one of the bits of information transmitted by the device.

These monitors are used principally to convey the potential of the structure to the surface. Some of them use a single electrode potential whereas others collect a series of potentials from a small area of the structure and this is transmitted to the surface. The monitors are expensive but in many areas where hard wiring has to be taken through the wind water line and is susceptible to damage they have been used with some success.

Anode Consumption

The consumption of sacrificial anodes on offshore structures or in any inaccessible location is a problem and often a diver is asked to interpret the rate of consumption of the anode. Because of the shape of the anodes and the change of shape as they are consumed this can be difficult both for the diver and for simple interpretation from photographs. One technique that gives a crude but immediate indication is the use of small threaded plastic inserts placed strategically in the anode at points of most rapid consumption, Fig 303. If these are made of contrasting color—say orange—then the diver can quickly count the number of exposed threads to indicate the amount of insert that is exposed.

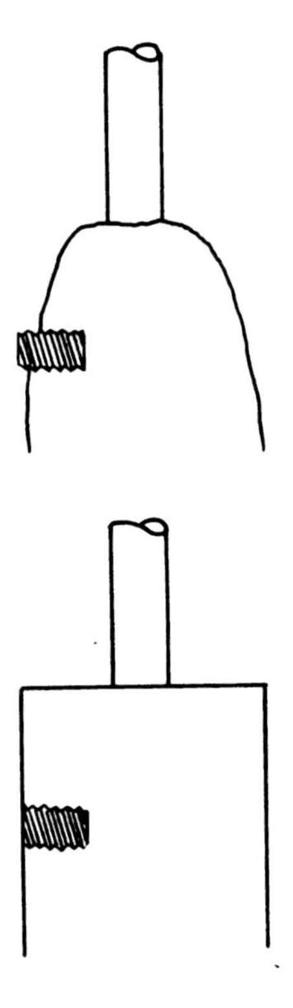

FIGURE 303 - Plastic insert to indicate extent of anode consumption.

Other techniques that can be used by the diver include graded straps which are placed around the anode and show the change in perimeter of the anode, Fig 304. These are useful where rectangular shapes rapidly lose their corners. By spacing these measurements on a rectangular anode the total consumption can be calculated. The use of devices which reproduce the anode shape such as those used for shape transfer in joinery can be used if adapted. Rolling measures can be used in place of straps and are useful for measuring the total length remaining of an anode.

Automatic Control

The conditions in the cathodic protection circuit will vary with time. It may be a change in demand by the cathode, a change in the electrolyte itself affecting either or both the anode and the cathode, or it can be by consumption of the anode. The various types of change have been discussed in the chapters on applications as have the users of automatic control.

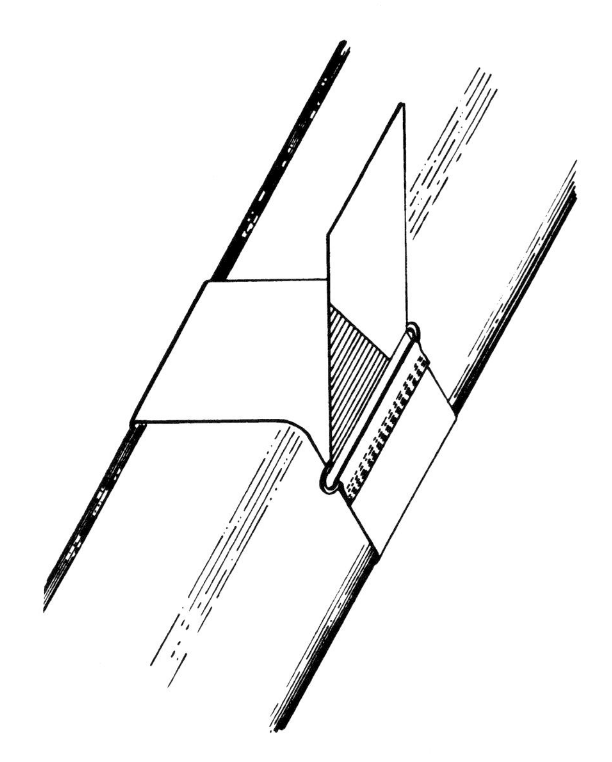

FIGURE 304 - Strap to measure change in circumference to compare anode consumption.

Sacrificial Anodes

It was shown that sacrificial anodes have a degree of self-regulation in their operations: as the cathode polarizes the driving voltage is reduced and the anode output is adjusted. This operates most markedly with low driving voltage anodes, particularly zinc and the aluminum alloys; here, with a driving voltage to protected steel of 250 mV, a change of 100 mV in the cathode potential can reduce the anode output by 40%.

To obtain the most value out of this system, the anode has to have a very low resistance and on average provides about 50 mV overprotection, the cathode potential varying from protected to 100 mV overprotected. This will have no adverse effect on the cathodic protection with most metals but it will increase the rate of consumption of the anode, the cathode current will double for every 100 mV increase in cathode potential depression. There will be a 40% increase in the amount of sacrificial anode required to achieve partial compensation.

Earlier a technique was suggested by which the anode driving voltage is further reduced, say to 50 mV, by inserting in the anode lead a device which will drop a constant 200 mV. This can be by means of a manually varied resistor across a meter or by the use of an automatic unit. The anode can now have a much lower resistance and to achieve the same degree of control will only add about 10% to the rate of anode consumption. A higher driving voltage anode can be used, allowing this type of control in environments other than sea water.

Impressed Current Protection

Impressed current protection is usually controlled by supplying a specific and near constant voltage through the anode or anode groundbed circuit. This will comprise the resistance and a back emf which will reduce the driving voltage available through the anode system. While the anode resistance stays the same the current through the circuit will be virtually constant. Changes in polarization of the cathode, a matter of a few hundred millivolts, will have very little effect on a net anode driving voltage of 20 volts.

One of the problems that will arise is with changes in the anode resistance. This may be by consumption, which would occur with an anode such as a steel rail, or with changes in the electrical resistivity of the anolyte by drying out of the ground or by a change of salinity in a tidal estuary. Under these conditions the cathode current demand may well not vary and automatic control can be used to provide a constant current in the circuit. This can be effected in applications that only require a small current by the use of a ballast resistor. If, for example, only 10 volts is required as the minimum driving voltage and 20 volts as the maximum driving voltage through the anode system, then a 50 volt source with a dropping resistor will stabilize the current. The power consumption will increase but this may be much cheaper than the cost of alternative automatic control equipment. Particularly this could be the case where the cost of bringing electricity to the cathodic protection system is large and the utility asks for a minimum payment for power.

More usually there will be a change in the current demand of the cathode. This may be a slow change such as the initial polarization of the structure, and as this may fall within the commissioning period, automatic control may not be necessary. With many structures, however, there will be a continuous variation in demand and this can be for a number of reasons. The amount of metal exposed to the electrolyte may change as in a tidal jetty or with different depths of loading on the hull of a ship. The coating on the cathode may deteriorate either through water absorption or by physical damage and the electrolyte in which the cathode is situated may change by variations caused by seasons or by location. The electrolyte may change physically, either by changing its velocity, changing its temperature, or by changing the aeration within it. All these are reflected in the electrical potential of the cathode and this can be used to control the current output thereby regulating the cathodic protection to a constant potential level. The most universal source of cathodic protection current is the ac mains and the majority of systems are designed to operate from them. In general the current will vary through an anode system which has a more or less constant resistance so that the power consumed will vary with the square of the current. The total circuit voltage will include the back emf and at low voltage there will be less power variation than in high voltage systems. In many systems there will be a change of anode resistance as well as a change in cathode current. However, the most common combination will be that there will be an increase in anode resistance and a decrease in cathode current demand.

In a large number of cases the power range will be les than 10:1, the main exception being the protection of ships' hulls. Here there will be a wide difference in the current required between the time when the ship is freshly painted and at rest to the demand when the ship is old, the paint is damaged and it is underway fully ladened in a hot saline sea. This may give a 10:1 change in current and probably a 40:1 change in power.

Most controllers operate by comparing the reference electrode signal with a generated stable potential, amplifying the error and using this to control the current output. Usually the current output will be achieved again by increasing the voltage with the anode system. This may mean at times that the automatic controller will maximize current output when the anode circuit resistance is at its lowest and either the system has to be considerably over-designed or it must incorporate current limiting so that the components are not burnt out.

The earliest type of systems used the saturable reactor which reduced the current through a transformer rectifier by saturating the magnetic field in the core of the reactor. The reactor core is made of a steel whose hysteresis curve gives maximum variation as the control winding is energized with direct current. A saturable reactor will not control power below 5% of maximum but this limitation can be overcome by the use of manual controls on the transformer rectifier. The principal problem with the saturable reactor is that it is slow and in many cases the cathode reaction time will match that of the reactor and the system can oscillate. An alternative technique of power control is to use a silicon controlled rectifier: this is a device which is gated open by an injected signal. It can be used in two ways; either the rectifier can be opened so that a number of cycles pass through it and then stopped so that there is a period of non-transmission. Alternatively, the rectifier can be gated open part way through a cycle so that only part of each cycle is transmitted. The former is called mark space control and produces a square wave which can vary from one or two cycles on with 20 to 30 cycles off to the reverse and gives a wide range of control.

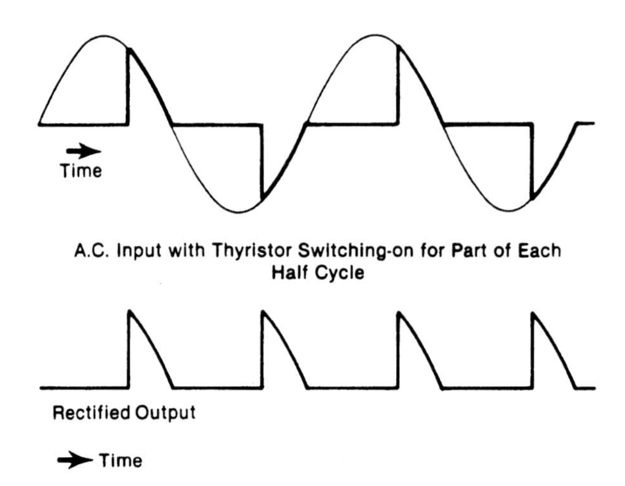

FIGURE 305 - Wave form of phase controlled thyristor alternating and rectified current.

With this type of output the circuit may ring, that is at the abrupt cut off of the dc there may be a brief period of current reversal. This will greatly increase the rate of corrosion of platinum surfaced anodes. It can also, through the change in demand, affect other equipment that is operating on the same power line where the cathodic protection is a sensible part of the total power load. Because of these problems this type of control has not been used in the last decade.

The second technique which is called phase control has its wave form distorted as in Fig. 305. It does confer a very wide degree of control, both with single phase and three phase equipment. The control can be exercised either in the incoming power line so that a series of pulses is fed through the transformer rectifier, or it can be operated in the rectification circuit when a similar wave form will result but the silicon controlled rectifier will be handling higher currents at lower voltage. In general the size and cost of the SCR depends on its current capacity so control of the ac power is employed rather than control of the low voltage output. Distortion of the wave form increases the heating effect in the transformer and when using SCR's the transformer has to be suitably de-rated and a low flux unit used. High frequency components are present and smoothing may be necessary. With three phase equipment at a low power outputs the phases must be well balanced or there will be a tendency to gate only one or two of the phases. In power control the SCR is used in a bridge and a device which incorporates two SCR's back to back called a triac is now generally substituted for it. Where large currents are involved it is usual to use single phase

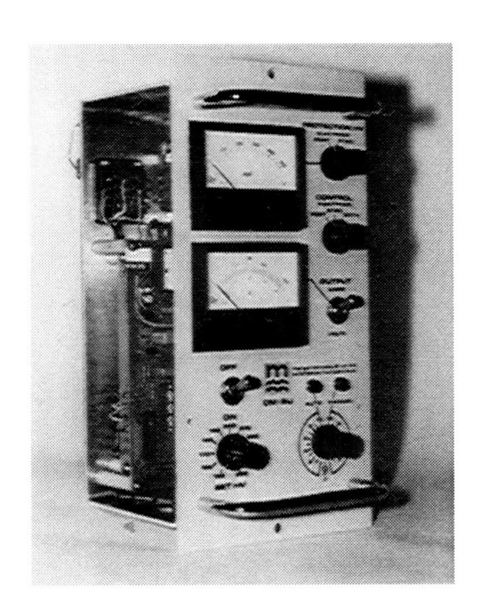

FIGURE 306 - Automatic control module with mv meter, amp/voltmeter and controls for level-of-protection, reference electrode selectio (both for control and monitoring) current limiting and manual control. (Photo reprinted with permission from CORRINTEC/UK, Ltd., Winchester, Hants, England.)

equipment and to distribute the power units complete with triac, transformer rectifier and current limit control. A dc signal is sent from the control unit and this allows the power units to be operated on different phases, reducing the out of balance in a three phase supply. The method also reduces the heavy current cabling, gives much better redundancy and better current limit control, especially where anode failure by short circuit may happen. It is usual to be able to adjust the potential at which the controller will operate and ideally full control is achieved over a 10 mV range. The actual potential achieved, both at the control electrode and at other electrodes, can be displayed and additional circuitry that will select the electrode where control is most difficult to achieve and yet rejecting false readings can be used. Average potentials can be used for control or any particular electrode selected. The level of current limit can be controlled, the current output either from each of the power units or from each of the anodes can be read. Fusing of the anode circuits can be a problem as on failure of one fuse the system will tend to increase the current through the other anodes and if close rated fusing is used then these will fail successively.

The same circuit that operates the current limit can be adapted to give manual current control, Fig 306.

Where direct current is available either the output from a transformer rectifier or from some other source, a power transistor can be used which will control the current through it. These devices are cost effective for small currents, usually not above 20 amps, and can be used to give individual control to anodes. This is useful in many circumstances such as the internal protection of sea water carrying pipelines, where local control of each anode or of each small section of plant is required. On a ship's hull or a long pipeline the cathodic protection will spread from the anodes almost irrespective of failure of one of them. In the internal protection of plant and machinery the failure of one anode will reduce the protection in its immediate vicinity. It is usual for the anodes to be placed so that they are able to spread to make up this loss and automatic control of small numbers of anodes, therefore, is required.

Where the cathodic protection current is obtained from other sources such as a motor generator, thermal generator, solar or wind power, then this can be controlled by other means. With a motor generator control can either be local by the use of transistors or it can be by control of the motor itself. Thermal, solar and wind power generators usually feed into a bank of batteries, the output from which can be controlled, for example by the use of a transistor controller. Devices such as wind power generators, where they may be subject to hurricanes, have their own control mechanisms built in, usually by automatically feathering of the vanes or blades.

Occasionally, where dc power is available as, for example, in a yacht, then a dc to dc converter can be used and this can be controlled automatically. In most cases the dc is converted to ac, the ac transformed and then rectified so that the input need not have the same polarity as the output and they are not interconnected. A number of commercial versions of this are available for small boat protection.

Data Processing

A vast amount of information is produced by an automatic cathodic protection system, particularly one that has several controllers and a multiplicity of anodes and reference electrodes. This information can be fed into data processing equipment which will indicate the source of malfunction, fault or operation beyond limits. The information as to a fault or malfunction is stored in the data processor or is displayed so that the operator has no need to scan the whole of the equipment to know that it is working properly. Similarly, the stored information can be analyzed and various parameters plotted against time or against each other. This adds significantly to the determination of the long life viability of the system. Alternatively, the data can be processed to improve the control. These systems can be incorporated in a small microprocessor which is part of the cathodic protection system or can be linked to a larger controlling computer facility.

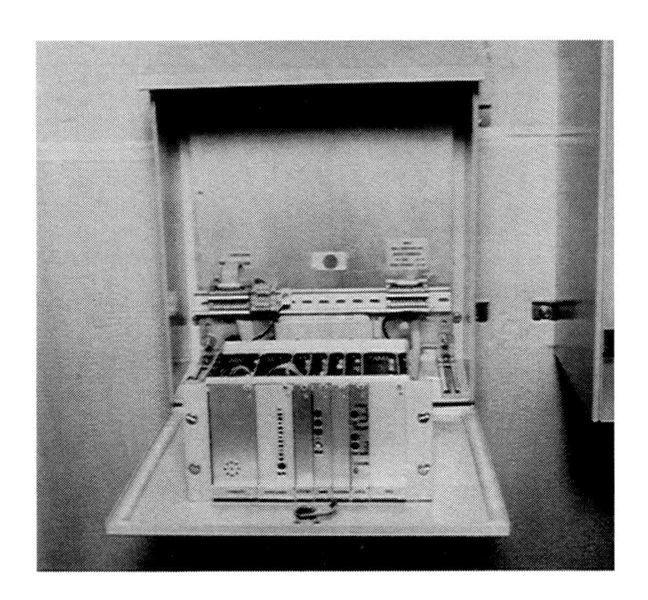

FIGURE 307 - Telephone interrogator fpr rural pipeline control and monitoring. (Photo reprinted with permission from Hockway, Ltd., Croyden, Surrey, England.)

Test Coupons and Corrosion Evaluation

The data can be converted into a synthesized voice and transmitted over the telephone network. This can be actuated by the engineer or in case of operation outside pre-determined limits for a specific period of time will itself telephone the fault information, Figure 307.

Often it is not possible to determine the sufficiency of a cathodic protection system by any of the accepted criteria. It may be that the electrolyte changes in resistivity, as it would do in an estuary, or that protection is being derived from some intermittent application of current. In these cases the actual rate of corrosion of the structure or the rate of corrosion of specially prepared plates called coupons or witnesses is measured.

The coupon technique has two chief uses. Firstly, they can be attached to the structure and their rate of corrosion measured and this will be easy because the coupon will start life as a clean, flat, smooth surface metal plate whose weight can be accurately determined. The efficacy of the protection can be judged rapidly and accurately. The coupon, when used in this way, must replace some part of the structure surface as closely as possible: if it is a piece of mild steel placed in an oil tanker compartment it must be flush with the tank wall, and be coated and sealed to the tank on its back face. Electric connections to the tank must be via a water-tight low resistance joint and the sealing compound must not extend over the tank wall. Secondly, coupons can be made to receive various degrees of protection as discussed in Chapter 2 and so determine a cathodic protection criterion.

The actual structure may itself be examined by several methods to determine its rate of corrosion. If the surface is new then visual observation might be sufficient to show that no corrosion is occurring, while if pitting occurs the pit depth can be measured and the depth of the 10 deepest pits per unit area recorded. If the corrosion attack is evenly distributed then the metal thickness could be measured and there are several non-destructive techniques available which give an accurate determination of the thickness. One of the easiest to use of these operates from one side of the metal only and measures the ultrasonic resonance of the metal between its two surfaces. The frequency at which this resonance occurs gives a measure of the plate thickness. Where the metal surface is already corroded the pit depth and plate thickness techniques can be used. A further technique is available which replaces the visual observation of the smooth surface and this consists of taking an impression or replica of the surface with some molding compound. A similar replica can be made some time later and the two compared to determine the extent of any further corrosion.

Two chemical methods are often used to determine the adequacy of the cathodic protection. The first uses a determination of the pH of the surface electrolyte and the second an analysis of the electrolyte to determine the amount of metal ion in it. Both techniques are improved by using them as a comparison with a series of earlier measurements. The pH value may be found by using indicating papers and wetting these with distilled water when they are in contact with the surface. Alkaline pH values of 8 or 9 indicate that protection is being achieved. The second method will only be applicable to circulating or enclosed water or possibly in some soils. The analysis must be sensitive, usually better than one part per million, as a loss of 1 lb. of steel in a storage tank of 10,000 cu. ft of water will only give 1.6 ppm of iron in the water. Protection might be considered successful if the rate of corrosion were reduced by 90 per cent and the analysis would not have to be sensitive enough to indicate the presence of less than 0.2 ppm of iron.

SUBJECT INDEX